SEED BIOLOGY
Advances and Applications

Seed Biology
Advances and Applications

Proceedings of the Sixth International Workshop on Seeds, Mérida, México, 1999

Edited by

Michael Black

Division of Life Sciences
King's College, London, UK

Kent J. Bradford

Department of Vegetable Crops
University of California
Davis, USA

and

Jorge Vázquez-Ramos

Departamento de Bioquímica
Universidad Nacional Autónoma de México
México DF

CABI *Publishing*

CABI *Publishing* is a division of CAB *International*

CABI Publishing
CAB International
Wallingford
Oxon OX10 8DE
UK

Tel: +44 (0)1491 832111
Fax: +44 (0)1491 833508
Email: cabi@cabi.org
Web site: http://www.cabi.org

CABI Publishing
10 E 40th Street
Suite 3203
New York, NY 10016
USA

Tel: +1 212 481 7018
Fax: +1 212 686 7993
Email: cabi-nao@cabi.org

A catalogue record for this book is available from the British Library, London, UK.

Library of Congress Cataloging-in-Publication Data
International Workshop on Seeds (6th : 1999 : Merida, Mexico)
Seed biology : advances and applications : proceedings of the sixth International Workshop on Seeds, Merida, Mexico, 1999 /editors Michael Black, Kent J. Bradford, Jorge Vazquez-Ramos.
p. cm.
Includes bibliographical references.
ISBN 0-85199-404-0 (alk. paper)
1. Seeds Congresses. I. Black, Michael. II. Bradford, K.J. (Kent J.). III. Vazquez-Ramos, Jorge. IV. Title.
SB113.3.I46 1999
575.6'8—dc21 99-42460
CIP

ISBN 0 85199 404 0

Typeset by AMA DataSet Ltd, UK
Printed and bound at the University Press, Cambridge

Contents

Contributors

S.W. Adkins, *School of Land and Food, The University of Queensland, St Lucia Campus, Brisbane, Qld 4072, Australia*

P.S. Allen, *Department of Agronomy and Horticulture, Brigham Young University, 275 Widstoe Building, Provo, UT 84602, USA*

C. Alonso-Blanco, *Department of Genetics, Wageningen University, Dreijenlaan 2, NL-6703 HA Wageningen, The Netherlands*

A.I. Batis, *Instituto de Ecología, Universidad Nacional Autónoma de México, Apartado Postal 70-275, Ciudad Universitaria, 04510 México D.F., México*

S.M. Bellairs, *School of Land and Food, The University of Queensland, St Lucia Campus, Brisbane, Qld 4072, Australia*

R.L. Benech-Arnold, *IFEVA, Facultad de Agronomia, Universidad de Buenos Aires, Av. San Martín 4453, (1417) Buenos Aires, Argentina*

M.A. Bennett, *Department of Horticulture and Crop Science, The Ohio State University, Columbus, OH 43210, USA*

L. Bentsink, *Department of Genetics, Wageningen University, Dreijenlaan 2, NL-6703 HA Wageningen, The Netherlands*

P. Berjak, *School of Life and Environmental Sciences, University of Natal, Durban 4041, South Africa*

I. Bernal-Lugo, *Departmento de Bioquímica, Facultad de Química, Universidad Nacional Autónoma de México, 04510 México D.F., México*

R.J. Bino, *Plant Research International, PO Box 16, NL-6703, Wageningen, The Netherlands*

M. Black, *Division of Life Sciences, King's College London, 150 Stamford Street, London SE1 8WA, UK*

A. Bochicchio, *Department of Agronomy and Land Management, University of Florence, Italy*

T.Y. Bogracheva, *John Innes Centre, Norwich Research Park, Colney, Norwich NR4 7UH, UK*

R. Bottini, *Laboratoires de Fisiologíca Vegetal, Facultad de Ciencias Exactas, Físicas, Químicas y Naturales, Universidad Nacional de Río Cuarto, Campus Universitaris, 5800 Río Cuarto, Provincia de Córdoba, Argentina*

I. Boubriak, *Oxford Research Unit, The Open University, Boars Hill, Oxford OX1 5HR, UK and Institute of Cell Biology and Genetic Engineering, Kiev 252143, Ukraine*

K.J. Bradford, *Department of Vegetable Crops, One Shields Avenue, University of California, Davis, CA 95616-8631, USA*

C.M. Bray, *School of Biological Sciences, University of Manchester, Oxford Road, Manchester M13 9PT, UK*

J. Buitink, *Laboratory of Plant Physiology, Department of Plant Sciences, and Laboratory of Molecular Physics, Department of Biomolecular Sciences, Wageningen University, Arboretumlaan 4, 6703 BD Wageningen, The Netherlands*

A. Camacho, *Departmento de Bioquímica, Facultad de Quimíca, Universidad Nacional Autónoma de México, 04510 México D.F., México*

D.J. Cantliffe, *University of Florida, IFAS, Horticultural Sciences Department, Gainesville, FL 32611-0690, USA*

I. Capron, *Laboratoire mixte CNRS/Rhône-Poulenc (UMR041), Rhône-Poulenc Agro, 14–20 rue Pierre Baizet, 69263 Lyon cédex 9, France*

A. Carballo, *Centro de Genética, Colegio de Posgraduados, Montecillo, Estado de México, 56230 México*

P. Carbonero, *Laboratorio de Bioquímica y Biología Molecular, Departamento de Biotecnología-UPM, E.T.S. Ingenieros Agrónomos, 28040 Madrid, Spain*

F. Carrari, *Laboratorio de Fisiología y Biología Molecular, FCEyN, Universidad de Buenos Aires, Ciudad Universitaria, Pabellón 11, (1428) Buenos Aires, Argentina*

M.P.M. Caspers, *Center for Phytotechnology RUL/TNO, TNO Plant Biotechnology Department, Wassenaarseweg 64, 2333 AL Leiden, The Netherlands*

P.B. Cavers, *Department of Plant Sciences, University of Western Ontario, London, Ontario, N6A 5B7, Canada*

F. Chen, *Department of Vegetable Crops, One Shields Avenue, University of California, Davis, CA 95616-8631, USA*

D. Côme, *Physiologie Végétale Appliquée, Université Pierre et Marie Curie, Tour 53, 1er étage, 4 place Jussieu, 75252 Paris cédex 05, France*

M.B. Cooley, *Department of Vegetable Crops, One Shields Avenue, University of California, Davis, CA 95616-8631, USA*

F. Corbineau, *Physiologie Végétale Appliquée, Université Pierre et Marie Curie, Tour 53, 1er étage, 4 place Jussieu, 75252 Paris cédex 05, France*

F. Dacher, *Physiologie Végétale Appliquée, Université Pierre et Marie Curie, Tour 53, 1er étage, 4 place Jussieu, 75252 Paris cédex 05, France*

P. Dahal, *Department of Vegetable Crops, One Shields Avenue, University of California, Davis, CA 95616-8631, USA*

P.J. Davidson, *School of Land and Food, The University of Queensland, St Lucia Campus, Brisbane, Qld 4072, Australia*

M.I. Daws, *School of Biological Sciences, University of Manchester, Oxford Road, Manchester M13 9PT, UK*

M. Dawson, *Royal Botanic Gardens, Kew, Wakehurst Place, Ardingly, West Sussex RH17 6TN, UK*

J. Dekker, *Weed Biology Laboratory, Agronomy Department, Iowa State University, Ames, IA 50010, USA*

I. Diaz, *Laboratorio de Bioquímica y Biología Molecular, Departmento de Biotecnología-UPM, E.T.S. Ingenieros Agrónomos, 28040 Madrid, Spain*

H. Donkers, *Plant Research International, PO Box 16, NL-6700 AA, Wageningen, The Netherlands*

B. Downie, *Department of Vegetable Crops, One Shields Avenue, University of California, Davis, CA 95616-8631, USA*

M.P. Downs, *Department of Plant Sciences, University of Western Ontario, London, Ontario, N6A 5B7, Canada*

N. Dudai, *Division of Aromatic Plants, ARO, Newe Yaar, PO Box 1021, Ramat Yishai 39005, Israel*

R.H. Ellis, *Department of Agriculture, University of Reading, Earley Gate, Reading RG6 6AT, UK*

S. Encisco, *IFEVA, Facultad de Agronomía, Universidad de Buenos Aires, Av. San Martín 4453, (1417) Buenos Aires, Argentina*

W.E. Finch-Savage, *Horticulture Research International, Wellesbourne, Warwick CV35 9EF, UK*

M.E. Foley, *United States Department of Agriculture, Agricultural Research Service, Biosciences Research Laboratory, PO Box 5674, State University Station, Fargo, ND 58105-5674, USA*

K.K. Fukunaga, *Department of Vegetable Crops, One Shields Avenue, University of California, Davis, CA 95616-8631, USA*

M. Gavilanes-Ruíz, *Departmento de Bioquímica, Facultad de Quimíca, Conjunto E., Universidad Nacional Autónoma de México, Ciudad Universitaria, 04510 México D.F., México*

O.H. Gee, *Department of Vegetable Crops, One Shields Avenue, University of California, Davis, CA 95616-8631, USA*

E.A. Golovina, *Laboratory of Plant Physiology, Wageningen University, Arboretumlaan 4, 6703 BD Wageningen, The Netherlands and Timiryazev Institute of Plant Physiology, Academy of Sciences, Botanicheskaya 35, Moscow, Russia*

R. J. Górecki, *Department of Plant Physiology and Biotechnology, Olsztyn University of Agriculture and Technology, Kortowo, 10-718 Olsztyn, Poland*

R.L. Grayson, *Department of Horticulture, Virginia Tech, Blacksburg, VA 24061-0327, USA*

S.P.C. Groot, *Plant Research International, PO Box 16, NL-6700 AA, Wageningen, The Netherlands*

M.K. Gunatilaka, *Department of Horticulture, Virginia Tech, Blacksburg, VA 24061-0327, USA*

S. Gurusinghe, *Department of Vegetable Crops, One Shields Avenue, University of California, Davis, CA 95616-8631, USA*

Y. Gutterman, *Wyler Department of Drylands Agriculture, Jacob Blaustein Institute for Desert Research and Department of Life Sciences, Ben-Gurion University of the Negev, Sede Boker Campus, 84990, Israel*

F.J.M. Harren, *Department of Molecular and Laser Physics, University of Nijmegen, Toernooiveld 1, 6525 ED Nijmegen, The Netherlands*

F. Hay, *Royal Botanic Gardens, Kew, Wakehurst Place, Ardingly, West Sussex RH17 6TN, UK*

C.L. Hedley, *Department of Applied Genetics, John Innes Centre, Norwich NR4 7UH, UK*

M.A. Hemminga, *Laboratory of Molecular Physics, Department of Biomolecular Sciences, Dreijenlaan 3, Wageningen University, 6703 HA Wageningen, The Netherlands*

D.R. Hill, *Department of Biochemistry, Virginia Tech, Blacksburg, VA 24061-0327, USA and The Procter and Gamble Co., Environmental Sciences Department, ITC, 5299 Spring Grove Avenue, Cincinnati, OH 45217-1087, USA*

G. Hoch, *Institute of Plant Physiology, University of Vienna, PO Box 285, A-1091 Vienna, Austria*

F.A. Hoekstra, *Laboratory of Plant Physiology, Department of Plant Sciences, Wageningen University, Arboretumlaan 4, 6703 BD Wageningen, The Netherlands*

D.J. Huber, *University of Florida, IFAS, Horticultural Sciences Department, Gainesville, FL 32611-0690, USA*

N. Iusem, *Laboratorio de Fisiología y Biología Molecular, FCEyN, Universidad de Buenos Aires, Ciudad Universitaria, Pabellón 11, (1428) Buenos Aires, Argentina*

C. Job *Laboratoire mixte CNRS/Rhône-Poulenc (UMR041), Rhône-Poulenc Agro, 14–20 rue Pierre Baizet, 69263 Lyon cédex 9, France*

D. Job, *Laboratoire mixte CNRS/Rhône-Poulenc (UMR041), Rhône-Poulenc Agro, 14–20 rue Pierre Baizet, 69263 Lyon cédex 9, France*

P. John, *School of Plant Sciences, University of Reading, Reading RG6 6AS, UK*

R. Johnson, *Department of Biology, Colby College, Waterville, ME 04901, USA*

A.D. Jones, *Department of Applied Genetics, John Innes Centre, Norwich NR4 7UH, UK*

H. Kato, *Department of Biological Sciences, Tokyo Metropolitan University, Minami-ohsawa 1-1, Hachioji, Tokyo 192-0397, Japan*

M.A. Khan, *Department of Botany, University of Karachi, Karachi, Pakistan*

J.I. Kioko, *School of Life and Environmental Sciences, University of Natal, Durban 4041, South Africa*

M. Koornneef, *Department of Genetics, Wageningen University, Dreijenlaan 2, NL-6703 HA Wageningen, The Netherlands*

L.B. Lahuta, *Department of Plant Physiology and Biotechnology, Olsztyn University of Agriculture and Technology, Kortowo, 10-718 Olsztyn, Poland*

P. Lara, *Laboratorio de Bioquímica y Biología Molecular, Departamento de Biotecnología-UPM, E.T.S. Ingenieros Agrónomos, 28040 Madrid, Spain*

A. Lara-Nuñez, *Departamento de Bioquímica, Facultad de Química, Universidad Nacional Autónoma de México, 04510 México D.F., México*

O. Larkov, *Division of Aromatic Plants, ARO, Newe Yaar, PO Box 1021, Ramat Yishai 39005, Israel*

O. Leprince, *Department of Plant Sciences, Wageningen University, Arboretumlaan 4, 6703 BD Wageningen, The Netherlands*

H.R. Lerner, *Department of Botany, The Hebrew University of Jerusalem, Jerusalem 91904, Israel*

D. Lijavetzky, *Instituto de Biotecnología, CICV, INTA Castelar, (1708) Morón, Provincia de Buenos Aires, Argentina*

O. Lorenzo, *Departmento de Fisiología Vegetal, Universidad de Salamanca, 37007 Salamanca, Spain*

L. Lyne, *Oxford Research Unit, The Open University, Boars Hill, Oxford OX1 5HR, UK*

R. Manku, *Department of Plant Sciences, University of Western Ontario, London, Ontario, N6A 5B7, Canada*

B. Mariotti, *Department of Agronomy and Land Management, University of Florence, Italy*

J. Marro, *Royal Botanic Gardens, Kew, Wakehurst Place, Ardingly, West Sussex RH17 6TN, UK*

L. Matthew, *School of Land and Food, The University of Queensland, St Lucia Campus, Brisbane, Qld 4072, Australia*

A.M. Mayer, *Department of Botany, The Hebrew University of Jerusalem, Jerusalem 91904, Israel*

U. Mayer, *Institute of Plant Physiology, University of Vienna, PO Box 285, A-1091 Vienna, Austria*

R.A. Mella, *Department of Vegetable Crops, One Shields Avenue, University of California, Davis, CA 95616-8631, USA*

M. Mena, *Laboratorio de Bioquímica y Biología Molecular, Departamento de Biotecnología-UPM, E.T.S. Ingenieros Agrónomos, 28040 Madrid, Spain*

S.E. Meyer, *USDA Forest Service Shrub Sciences Laboratory, Provo, Utah, USA*

T-G. Min, *Department of Agronomy, Taego University, Tinryang-myrn, Kyingsan-gin, Kyengbok 712-714, South Korea*

T. Minamikawa, *Department of Biological Sciences, Tokyo Metropolitan University, Minami-ohsawa 1-1, Hachioji, Tokyo 192-0397, Japan*

D.J. Murphy, *Department of Brassica and Oilseeds Research, John Innes Centre, Norwich Research Park, Norwich NR4 7UH, UK*

D.J. Mycock, *Department of Botany, University of Witwatersrand, Johannesburg 2050, South Africa*

W.M. Nascimento, *University of Florida, IFAS, Horticultural Sciences Department, Gainesville, FL 32611-0690, USA*

V. Naumenko, *Institute of Cell Biology and Genetic Engineering, Kiev 252143, Ukraine*

S.C. Navie, *School of Land and Food, The University of Queensland, St Lucia Campus, Brisbane, Qld 4072, Australia*

C. Nicolás, *Departmento de Fisiología Vegetal, Universidad de Salamanca, 37007 Salamanca, Spain*

G. Nicolás, *Departmento de Fisiología Vegetal, Universidad de Salamanca, 37007 Salamanca, Spain*

H. Nonogaki, *Department of Vegetable Crops, One Shields Avenue, University of California, Davis, CA 95616-8631, USA*

L. Oñate, *Laboratorio de Bioquímica y Biología Molecular, Departamento de Biotecnología-UPM, E.T.S. Ingenieros Agrónomos, 28040 Madrid, Spain*

A. Orozco-Segovia, *Instituto de Ecología, Universidad Nacional Autónoma de México, Apartado Postal 70-275, Ciudad Universitaria, 04510 México D.F., México*

D.J. Osborne, *Oxford Research Unit, The Open University, Boars Hill, Oxford OX1 5HR, UK*

N. Özbingöl, *Physiologie Végétale Appliquée, Université Pierre et Marie Curie, Tour 53, 1er étage, 4 place Jussieu, 75252 Paris cedex 05, France*

D.H. Paine, *Department of Horticultural Sciences, New York State Agricultural Experiment Station, Cornell University, Geneva, New York, NY 14456, USA*

N.W. Pammenter, *School of Life and Environmental Sciences, University of Natal, Durban 4041, South Africa*

E.W. Pavel, *Department of Horticulture, Virginia Tech, Blacksburg, VA 24061-0327, USA and Department of Plant Production and Soil Sciences, University of Pretoria, Pretoria 0002, South Africa*

L. Peach, *Horticulture Research International, Wellesbourne, Warwick CV35 9EF, UK*

L. Perez-Flores, *Departmento de Ciencias de la Salud, DCBS, Universidad Autónoma Metropolitana-Iztapalapa, Av. Michoacán y Pirísma, col. Vicentina, C.P. 09340, México D.F., México*

T. Peterbauer, *Institute of Plant Physiology, University of Vienna, PO Box 285, A-1091 Vienna, Austria*

K. Phelps, *Horticulture Research International, Wellesbourne, Warwick CV35 9EF, UK*

A. Poljakoff-Mayber, *Department of Botany, The Hebrew University of Jerusalem, Jerusalem 91904, Israel*

R. Probert, *Royal Botanic Gardens, Kew, Wakehurst Place, Ardingly, West Sussex RH17 6TN, UK*

M. Puschenreiter, *Institute of Plant Physiology, University of Vienna, PO Box 285, A-1091 Vienna, Austria*

E. Putievsky, *Division of Aromatic Plants, ARO, Newe Yaar, PO Box 1021, Ramat Yishai 39005, Israel*

M.M. Qaderi, *Department of Plant Sciences, University of Western Ontario, London, Ontario, N6A 5B7, Canada*

A. Richter, *Institute of Plant Physiology, University of Vienna, PO Box 285, A-1091 Vienna, Austria*

D. Rodríguez, *Departmento de Fisiología Vegetal, Universidad de Salamanca, 37007 Salamanca, Spain*

M. Rodríguez-Penagos, *Departmento de Bioquímica, Facultad de Química, Universidad Nacional Autónoma de México, 04510 México D.F., México*

R. Rodríguez-Sotres, *Departmento de Bioquímica, Facultad de Química, Universidad Nacional Autónoma de México, 04510 México D.F., México*

M. Rojas-Aréchiga, *Instituto de Ecología, Universidad Nacional Autónoma de México, Apartado Postal 70-275, Ciudad Universitaria, 04510 México D.F., México*

R.A. Sánchez, *IFEVA, Facultad de Agronomía, Universidad de Buenos Aires, Av. San Martín 4453, (1417) Buenos Aires, Argentina*

M. Sánchez-Coronado, *Instituto de Ecología, Universidad Nacional Autónoma de México, Apartado Postal 70-275, Ciudad Universitaria, 04510 México D.F., México*

E. Sánchez de Jiménez, *Departmento de Bioquímica, Facultad de Química, Universidad Nacional Autónoma de México, 04510 México D.F., México*

S. Sánchez-Nieto, *Departmento de Bioquímica, Facultad de Quimíca, Conjunto E., Universidad Nacional Autónoma de México, Ciudad Universitaria, 04510 México D.F., México*

U.R. Sinniah, *Department of Agriculture, University of Reading, Earley Gate, Reading RG6 6AT, UK*

E. Śliwińska, *Department of Genetics and Plant Breeding, University of Technology and Agriculture, Kaliskiego 7, 85-796 Bydgoszcz, Poland*

J.R.A. Steckel, *Horticulture Research International, Wellesbourne, Warwick CV35 9EF, UK*

H. Steinbach, *IFEVA, Facultad de Agronomía, Universidad de Buenos Aires, Av. San Martín 4453, (1417) Buenos Aires, Argentina*

Y. Sung, *University of Florida, IFAS, Horticultural Sciences Department, Gainesville, FL 32611-0690, USA*

A.G. Taylor, *Department of Horticultural Sciences, New York State Agricultural Experiment Station, Cornell University, Geneva, New York, NY 14456, USA*

I.N. Taylor, *School of Land and Food, The University of Queensland, St Lucia Campus, Brisbane, Qld 4072, Australia*

K. Tesnier, *Plant Research International, PO Box 16, NL-6700 AA, Wageningen, The Netherlands*

M. Tuena de Gómez-Puyou, *Departamento de Genética Molecular, Instituto de Fisiología Celular, Universidad Nacional Autónoma de México, Ciudad Universitaria, 04510 México D.F., México*

A.H.M. van der Geest, *Plant Research International, PO Box 16, NL-6700 AA, Wageningen, The Netherlands*

R.M. van der Meulen, *Center for Phytotechnology RUL/TNO, TNO Plant Biotechnology Department, Wassenaarseweg 64, 2333 AL Leiden, The Netherlands*

J.M. Vázquez-Ramos, *Departmento de Bioquímica, Facultad de Quimíca, Universidad Nacional Autónoma de México, Av. Universidad y Copilco, 04510 México D.F., México*

C. Vázquez-Yanes [deceased], *Instituto de Ecología, Universidad Nacional Autónoma de México, Apartado Postal 70-275, Ciudad Universitaria, 04510 México D.F., México*

J. Vicente-Carbajosa, *Laboratorio de Bioquímica y Biología Molecular, Departamento de Biotecnología-UPM, E.T.S. Ingenieros Agrónomos, 28040 Madrid, Spain*

D. Vinel, *Physiologie Végétale Appliquée, Université Pierre et Marie Curie, Tour 53, 1er étage, 4 place Jussieu, 75252 Paris cédex 05, France*

J. Vredenbregt-Heistek, *Center for Phytotechnology RUL/TNO, TNO Plant Biotechnology Department, Wassenaarseweg 64, 2333 AL Leiden, The Netherlands*

D. Vreugdenhil, *Department of Biomolecular Sciences, Wageningen University, Arboretumlaan 4, NL-6703 HA Wageningen, The Netherlands*

M. Walker, *School of Life and Environmental Sciences, University of Natal, Durban 4041, South Africa*

M.K. Walker-Simmons, *USDA Agriculture Research Service, Wheat Genetics, Quality, Physiology and Disease Research Unit, Washington State University, Pullman, WA 99164-6420, USA*

C. Walters, *USDA-ARS National Seed Storage Laboratory, 1111 South Mason Street, Fort Collins, CO 80521, USA*

M. Wang, *Center for Phytotechnology RUL/TNO, TNO Plant Biotechnology Department, Wassenaarseweg 64, 2333 AL Leiden, The Netherlands*

T.L. Wang, *John Innes Centre, Norwich Research Park, Colney, Norwich NR4 7UH, UK*

J.E. Warren, *Department of Horticulture and Crop Science, The Ohio State University, Columbus, OH 43210, USA; Present address: 374 Perry Highway, Harmony, PA 16037, USA*

W.M. Waterworth, *School of Biological Sciences, University of Manchester, Oxford Road, Manchester M13 9PT, UK*

G.E. Welbaum, *Department of Horticulture, Virginia Tech, Blacksburg, VA 24061-0327, USA*

J. Wesley-Smith, *School of Life and Environmental Sciences, University of Natal, Durban 4041, South Africa*

C.E. West, *School of Biological Sciences, University of Manchester, Oxford Road, Manchester M13 9PT, UK*

D.A. Wills, *School of Land and Food, The University of Queensland, St Lucia Campus, Brisbane, Qld 4072, Australia*

C-T. Wu, *Department of Vegetable Crops, One Shields Avenue, University of California, Davis, CA 95616-8631, USA*

H. Yang, *Department of Vegetable Crops, One Shields Avenue, University of California, Davis, CA 95616-8631, USA*

K-O. Yim, *Department of Vegetable Crops, One Shields Avenue, University of California, Davis, CA 95616-8631, USA*

Preface

The Sixth International Workshop on Seeds was held in Mérida, México in January 1999, following the tradition of previous workshops held in Jerusalem (1980), Wageningen (1985), Williamsburg (1989), Angers (1992) and Reading (1995), each one being recorded by the publication of the proceedings.

This book contains papers presented at the Sixth International Workshop on Seeds. Because of space limitations, the organizers and editors were forced to exercise some selection so the articles herein are not a record of all the papers that were presented. They do, however, represent all the areas in seed biology covered at the meeting and they reflect, as far as possible, major developments that have taken place in basic seed science over the past four years. Substantial progress has been made in numerous areas emphasizing the extremely important role that seed biology plays in advancing plant biotechnology, agriculture, and plant resource management and conservation.

The Workshop was attended by more than 200 participants who spent an enjoyable and scientifically profitable five days in the Yucatan peninsula of Mexico. It was unanimously agreed that the meeting was excellent and it was recognized that this was due in no small measure to the efficiency, planning and hard work of the local organizing committee from the Universidad Nacional Autónoma de México (UNAM), México, D.F: Jorge Vázquez-Ramos (Chairman), Rogelio Rodríguez-Sotres (Communications Technology), Irma Bernal-Lugo and Marina Gavilanes-Ruíz. Scientific planning of the meeting was carried out by the Local Organizing Committee in consultation with the International Organizing Committee: Michael Black (UK), Kent J. Bradford (USA), J. Derek Bewley (Canada), Daniel Côme (France), Yitzchak Gutterman (Israel) and Henk W.M. Hilhorst (The Netherlands).

The Workshop thanks the following for their considerable aid and sponsorship:

Facultad de Química, UNAM
Accessolab
Monsanto
Coordinacion de la Investigacion Cientifica, UNAM
CABI *Publishing*
Boehringer-Mannheim
Farmaceuticos Lakeside, S.A. de C.V., México
BQ-The Providers, México

It should also be noted that discussions at the meeting culminated in strong support for the establishment of an International Society for Seed Science (ISSS). Daniel Côme (France) was elected as the first President and Ralph Obendorf (Cornell, USA) as the President-elect and they were charged with the task of pursuing the establishment of this new scientific society. With the expectation that this will be successful, the ISSS will trace its foundation to the Sixth International Workshop on Seeds at Mérida, México, in January 1999. The Seventh International Workshop on Seeds will be held under the auspices of the ISSS in Salamanca, Spain, probably in the summer months of 2002 [local organizer: Professor Gregorio Nicolás-Rodrigo, Departamento de Fisiologia Vegetal, Universidad de Salamanca, 37007 Salamanca, Spain (gnr@gugu.usal.es)].

Michael Black
Kent J. Bradford
Jorge Vázquez-Ramos

I Opening Presentations

1 Travels with Seeds: a Cartography of Seed Science

M. BLACK

Division of Life Sciences, King's College London, 150 Stamford Street, London SE1 8WA, UK

In some ways, carrying out scientific research is rather like a journey of exploration. It involves making decisions about which roads to follow within the boundaries of the known 'world' and extending those boundaries by taking new directions into the unknown. Seed science is considered in such a context, tracing its development from the time when the map was restricted by the limits imposed by descriptive physiology, biochemistry and ecology through the period when the horizons were extended by more analytical approaches, to the present when progress in molecular and cell biology, genetics and biophysics offers opportunities to all branches of research in seed science for journeys into new worlds.

Introduction

Humans have always been intimately involved with seeds, from the time in our hunter-gatherer past when we learned about their nutritional properties, through the beginnings of agriculture when the manipulation of seeds formed the very basis of our social and cultural development, to the present time when we rely largely upon seeds to feed our vast population. But apart from their utilitarian value either as food or for growing crops, seeds have enormous biological interest. As the start of the next generation, they occupy a critical position in plant life history and in the survival of the species. They contain plants in miniature, possessing a remarkable capacity to ensure that the new individual makes a good start in life, in the right place and at the right time. Though many of the means for achieving this have been modified or removed from some of the popular research seeds by selection and breeding, this property is the essence of the seed's unique character and role. For example, by the use of light and temperature sensors the seed can detect how deep it is in the soil, where it is in relation to other plants, and what is the time of year; this

(eds M. Black, K.J. Bradford and J. Vázquez-Ramos)

information is integrated to enable it to decide if it is now suitable to send the new plant out into the world. If it is not, the seed postpones action and waits for the situation to improve. And to give the new plant the best possible start, it is bequeathed an abundance of food reserves to use until it becomes photosynthetically self-sufficient. All of its potential and benefits are established while the seed is developing on the parent plant. During the initiation, progress and maturation of the seed the patterns of embryo formation, differentiation and growth are tightly regulated, relatively massive amounts of reserves are laid down, dormancy is initiated and, in most cases, the seed is prepared to withstand the loss of water that terminates this phase of its life. Remarkably, after this severe dehydration the seed may survive for many years, subsequently to renew its metabolic activity and growth.

The Map of Seed Science

The unique biology of seeds has drawn researchers to study them for the intellectual and scientific challenge and also because of their profound importance in human affairs. Throughout my working life I have been involved with seeds and, like all biologists, I have seen my chosen area of science progress in ways that were unimaginable when I began. Research is an exploration which can be compared to making a journey. The metaphor is by no means original but it is nevertheless persuasive, and we can see that many of the actions and decisions required of the traveller – what the destination is, which direction to take, which map to follow or to make – have also to be made in research.

Our initial horizons might extend only for a short distance. Just as the known world (i.e. known to inhabitants of the Mediterranean region) in *c.* 250 AD was a bare fraction of what is known today, so the state of our knowledge about seeds was meagre when I started research (Fig. 1.1). Descriptive physiology, biochemistry, ecophysiology and ecology told us about water uptake and germination, the different kinds of dormancy and how they are broken, the chemical constituents of seeds, how some seeds were short- and some long-lived, and something about the behaviour of seeds in the natural environment, but precious little was known about the mechanisms involved, or about the processes occurring in seed development.

'La distance n'y fait rien; il n'y a que le premier pas qui conte' (the distance does not matter: it is only the first step that counts) – Marquise du Deffaud

Change began as a result of three discoveries, which opened up new roads for seed scientists to take – phytochrome, gibberellins and abscisic acid. The discovery of red, far-red reversibility led ecophysiologists to understand how seeds perceived the surrounding vegetation and the depth at which they were buried. For the first time, a naturally occurring chemical (a hormone) that could promote germination was found when the gibberellins became known, and great was the excitement when this substance was found to be involved in

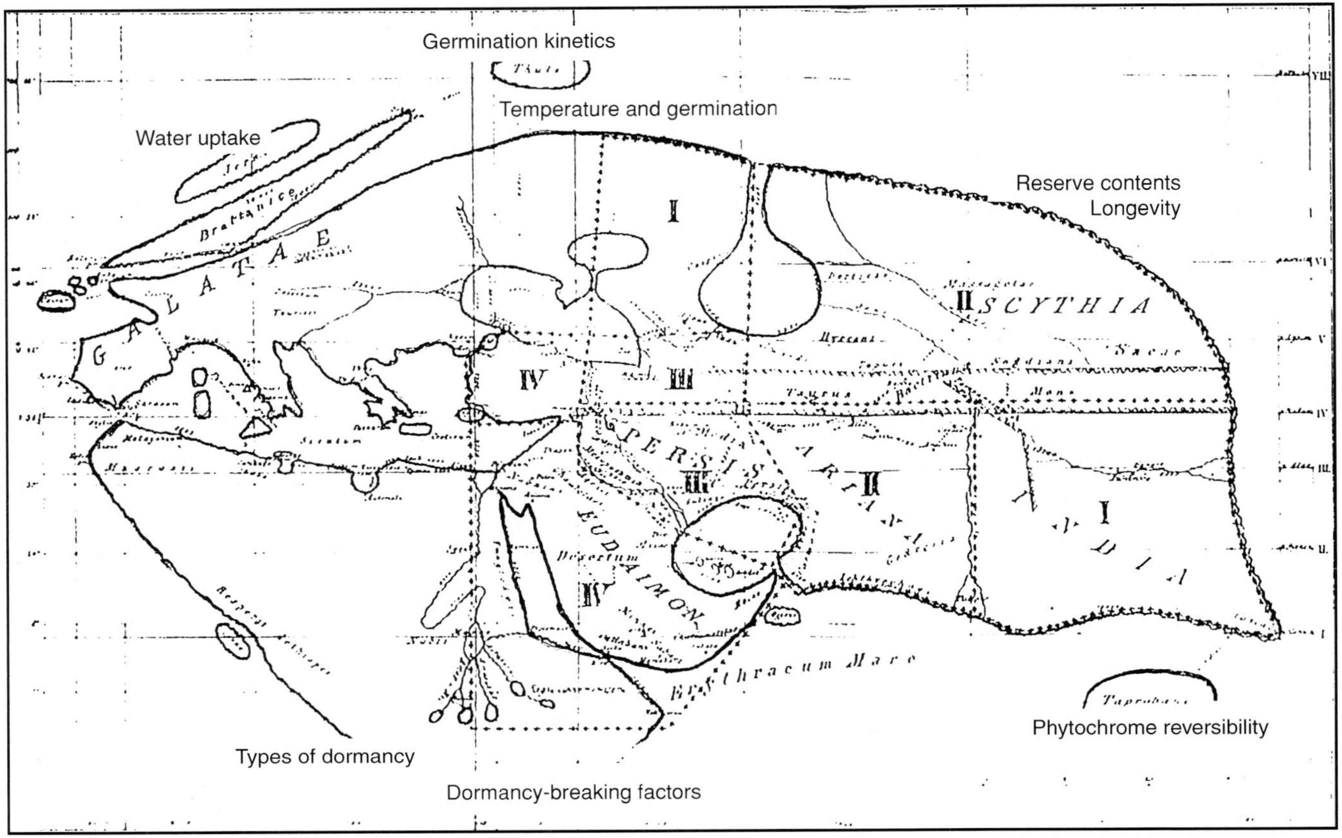

Fig. 1.1. An early map reflecting the world of seed science based on the known (old) world of *c.* 250 AD.

the mobilization of food reserves following germination of cereal seeds. Shortly afterwards, the newly-discovered abscisic acid was shown to be a potent inhibitor of seed germination and its possible role in seed dormancy was mooted. It is worth recalling that one of the routes leading to the identification of this hormone was being followed by investigators of dormancy in buds. So, though the journey ahead was long, the first steps were fortunate, and the routes to the destinations – germination, dormancy and food reserve mobilization were assured. The horizons had moved.

A major landmark in seed exploration arose when researchers caught sight of the *Arabidopsis* signpost (Karssen, 1993). The road was already crowded when seed scientists decided to follow it but it soon became clear that it was a profitable path to take. It led, most importantly, to knowledge about embryogenesis, the roles of gibberellin in germination, of abscisic acid in dormancy, desiccation tolerance and fatty acid and triacylglycerol deposition. The value of using mutants, first recognized in *Arabidopsis,* was extended to other species such as tomato, and it rapidly became clear that the route mapped by other plant scientists was worth following: thus, our knowledge about the roles of gibberellin and abscisic acid was consolidated.

'Nowadays a path is scarcely opened up when the crowd begins to pour in' – Jean Rostand, French biologist

Genetics was put on a molecular basis with the elucidation of the structure of DNA. The secrets of the gene, the genetic code, gene expression and protein synthesis shortly began to be revealed and plant scientists joined other biologists rushing to take this route to discovery: among them were those who started at the *Arabidopsis* signpost. The first on the road included those who used seeds especially to study storage protein synthesis, because the proteins themselves and the messenger RNAs were so abundant. Surprisingly, those whose interest centred particularly on seed biology (rather than on biochemistry) were noticeably slow to take the plant molecular biology road. It is only relatively recently that this approach has begun to contribute to our knowledge about seed maturation, desiccation tolerance, dormancy and endosperm degradation during germination, while other areas of great importance – germinative growth itself, viability and longevity, for example – still remain almost untouched.

'To boldly go where no man has gone before' – Star Trek (TV); 'Fools rush in where angels fear to tread' – Milton

Although seed scientists have begun to take the molecular biological route mapped out by plant scientists, albeit to a limited extent, there are other leads that we fail to follow. To illustrate this point I give three examples.

The sole event by which we can recognize germination is the elongation growth of the radicle/hypocotyl axis. Cell elongation is the culmination of germination: and one may argue that all the processes occurring in the seed after the start of imbibition are directed to this climactic consummation. Yet we know almost nothing about this defining event. As the pivotal phenomenon of plant growth, cell elongation has been intensively investigated. We know that the modification of cell-wall structure by, for example, the expansins and enzymes such as the xyloglucan endotransglycosylases is involved in wall extensibility (Cosgrove, 1997). Are these factors also implicated in embryo growth during seed germination? In one way, the germinating seed has unique value as a system in which to study cell elongation since this is 'switched' on in non-growing tissue, whereas in those organs which mostly have been used for studying elongation mechanisms (coleoptiles, hypocotyls, epicotyls) there is simply a rate change. It is possible, of course, that there are differences between the initiation of cell growth during germination of the embryo and the regulation of cell extension in growing tissues. In the latter, auxin plays an important part whereas there is no evidence that this hormone is involved in germinative growth: at least, embryo extension growth cannot be initiated by auxin, though it might be sensitive to low pHs (Footitt and Cohn, 1992) just as auxin-sensitive growth is. The fact that embryo axial elongation is more sensitive to another hormone, gibberellin, may indicate that somewhat different mechanisms for embryo cell-wall modifications are involved. Certainly, there is convincing evidence that native gibberellin is a requirement for germination, and while this is partly because degradation of the tissues enclosing the embryo is gibberellin sensitive (see, e.g. Welbaum *et al.*, 1998), direct effects of the hormone on the embryo also occur. In the same area, the effects of abscisic acid on cell elongation should not be overlooked. ABA reduces cell-wall extensibility (Schopfer and Plachy, 1985) and valuable clues as to the regulation of cell growth during germination might come from further investigation of this phenomenon.

Resolution of the regulation and mechanism of embryo cell-wall extensibility presents particular problems, not least because germinative growth initially involves only a relatively small population of cells, just behind the radicle tip. Large numbers of extending cells are not present unlike, for example, in a growing coleoptile. Nonetheless, available techniques for application at cell level should make it possible to extend to embryos the discoveries that have been made about cell growth in other plant parts.

This brings us to another lead that seed scientists seem slow to follow – gibberellin action itself. The role of this hormone in α-amylase production by aleurone cells is well known: gibberellin-regulated α-amylase genes have been sequenced and the promoters (GA-response elements) characterized. Similar molecular regulation most likely occurs in the case of GA-controlled germination but apart from preliminary studies on endosperm-degrading enzymes (e.g. endo-β-mannanases) (Bewley, 1997a,b) relatively little is known. This is a field of investigation from which seed science could profit substantially. For example, what are the GA-regulated genes in the embryo itself?

In this connection, there is no parallel in respect of the germination responses to GA and other factors to the progress that is being made in studies of signal perception and transduction in several plant systems. Knowledge about transduction mechanisms is expanding in all directions – for phytochrome, ethylene, abscisic acid, gibberellin, pathogenicity – except in seeds, which represent perhaps the most striking signal perception system in plants, involving light, low temperature, alternating temperatures and at least three hormones. It is now becoming clear that signal cascades involving protein kinases function in seeds (Walker-Simmons, 1998) and it will be interesting to see how far these are implicated in the different seed perceptive processes.

Seed explorers might profitably travel in these directions: hence the title of this section, chosen to attract the attention of intrepid adventurers.

'Long is the way and hard, that leads out of hell up to the light' – Milton

The seed science map has greatly extended since the time when I first tried to find a way, and the frontiers have been pushed back further and further. Seed scientists can take satisfaction from the inroads that have been made in many directions – embryogenesis, development, maturation, desiccation tolerance, recalcitrance, viability, longevity, storage, dormancy, aspects of germination physiology (especially endosperm or coat degradation), ecophysiology, seed enhancement and biotechnology. But, of course, there is still much that remains untouched and I comment on just a few cases. Linked with the point made above about signal transduction is the need for more knowledge about sensor mechanisms, particularly for temperature. This is a profoundly important determinant of germination and dormancy but there is a lack of information as to how it exerts its effect. Low temperature and alternating temperature are, together with light, the major environmental factors responsible for breaking dormancy. The light sensors are fairly well understood but not the temperature sensors: membrane phase changes may be involved in temperature sensing (Hilhorst, 1998) but our ideas about this remain speculative and there is room for extensive research.

A feature of seeds that has long puzzled me is the relatively fixed proportions of the various seed reserves in each species. Though during its development a seed may be able to accumulate triacylglycerols, protein and starch, one or two of these reserves generally predominate: how is this brought about? What determines, for example, whether metabolism in the plastids should be directed largely towards starch biosynthesis rather than to fatty acid production? Such questions about assimilate allocation are obviously extremely important in relation to the nutritional and industrial uses of seeds but they may also be relevant to the strategies adopted by seeds for their survival and success in plant establishment. It is to be hoped that future exploration will soon yield answers. The discovery in seeds of protein kinases that might participate in carbohydrate partitioning is a promising step (Walker-Simmons, 1998) and a good lead for further investigation.

Genetic transformation is already well established as a means of altering the reserve composition of seeds and for using them as production units for chemicals, medicinals and pharmaceuticals (e.g. Krebbers *et al.*, 1997). Transformation technology also offers enormous opportunities for the seed scientist to probe into the basic aspects of physiology, biochemistry and perhaps, even the ecology of seeds. For example, over- and under-expression of the appropriate genes might be used to clarify the relative roles of oligosaccharides and LEA proteins in desiccation tolerance and longevity and also to modify these aspects of seed behaviour. There are plenty of other targets and this kind of approach will bring valuable insights into many of the problems with which the seed researcher grapples. Biotechnologists have already begun to interfere with the basic, biological role of the seed – to start the next generation – by the so-called terminator technology in which seeds are engineered to 'commit suicide' during their maturation (Black, 1998). In theory, there is no physiological or biochemical property of seed which is not amenable to such kind of molecular manipulation.

'Many shall run to and fro' and knowledge shall be increased' – (The Bible, Book of Daniel)

There are exciting times ahead for the cartographer of seed science. As the directions taken by researchers increase and change new maps will be drawn and the frontiers of the seed world will be extended (Fig. 1.2). Sometimes the chosen path will lead the wrong way and the route will have to be altered. But, '*no matter what happens, travel will give you a good story to tell' (Jewish saying).*

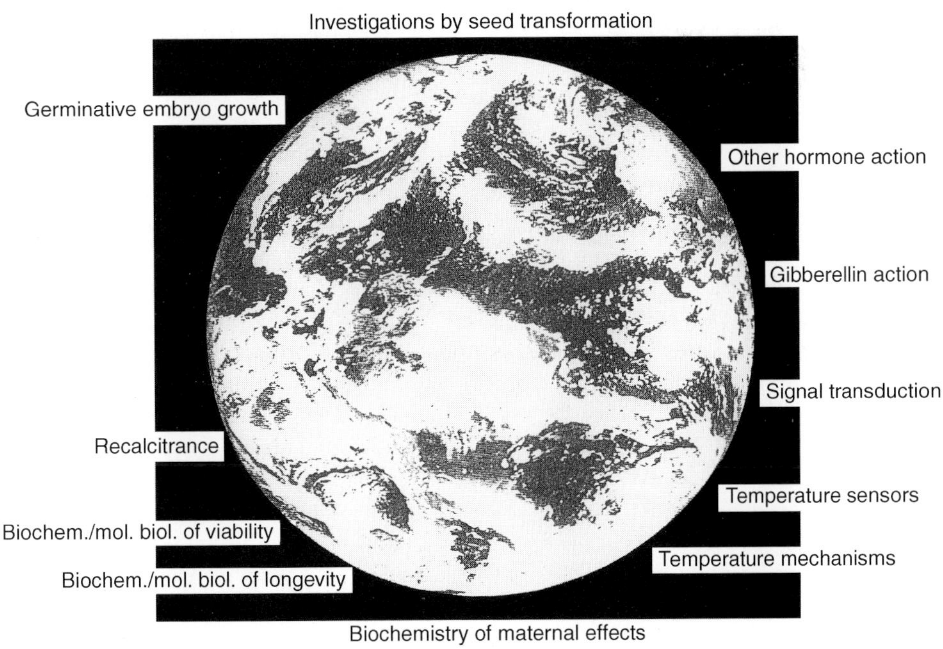

Fig. 1.2. Some possible destinations of future seed explorers.

References

Bewley, J.D. (1997a) Breaking down the walls – a role for endo β-mannanase in release from seed dormancy. *Trends in Plant Science* 2, 464–469.

Bewley, J.D. (1997b) Seed germination and dormancy. *Plant Cell* 9, 1055–1066.

Black, M. (1998) Knocking 'em dead. *Biologist* 45, 99.

Cosgrove, D.J. (1997) Relaxation in a high-stress environment: the molecular bases of extensible cell walls and cell enlargement. *Plant Cell* 9, 1031–1041.

Footitt, S. and Cohn, M.A. (1992) Seed dormancy in red rice VIII. Embryo acidification during dormancy-breaking and subsequent germination. *Plant Physiology* 100, 1196–1202.

Hilhorst, H.W.M. (1998) The regulation of secondary dormancy. The membrane hypothesis revisited. *Seed Science Research* 8, 77–90.

Karssen, C.M. (1993) Seed science: characteristics and objectives. In: Come, D. and Corbineau, F. (eds) *Fourth International Workshop on Seeds.* ASFIS, Paris, pp. 3–9.

Krebbers, E., Broglie, R., Hitz, B., Jones, T. and Hubbard, N. (1997) Biotechnological approaches to altering seed composition. In: Larkins, B.A. and Vasil, I.K. (eds) *Cellular and Molecular Biology of Plant Seed Development.* Kluwer Academic Publishers, Dordrecht, Boston, London, pp. 595–634.

Schopfer, P. and Plachy, C. (1985) Control of seed germination by abscisic acid III. Effect on embryo growth potential (minimum) turgor pressure and growth coefficient (cell wall extensibility) in *Brassica napus. Plant Physiology* 77, 676–686.

Walker-Simmons, M.K. (1998) Protein kinases in seeds. *Seed Science Research* 8, 193–200.

Welbaum, G.E., Bradford, K.J., Kyu-Ock Yim, Booth, D.T. and Oluoch, M.O. (1998) Biophysical, physiological and biochemical processes regulating seed germination. *Seed Science Research* 8, 161–172.

2 Protein Synthesis in Seed Germination: Connecting External Cell Signals with the Ribosome

E. SÁNCHEZ DE JIMÉNEZ

Departamento de Bioquímica, Facultad de Química, Universidad Nacional Autónoma de México 04510, México D.F., México

Specific mechanisms of translational regulation are known to be of relevance in eukaryotic organisms, particularly at specific stages of development and differentiation. During seed germination, translation of stored and newly synthesized mRNAs is tightly regulated in order for this process to be successfully accomplished. Data are presented that reveal a novel mechanism of translational control in germinating maize axes, based on the activation of a signal transduction pathway that targets proteins related to the translational apparatus. In the experimental approach, insulin was used as the putative effector that stimulates this pathway, and also the specific inhibitors, wortmannin and rapamycin. Phosphorylation of the S6 ribosomal protein (rp) in the 40S subunit was assessed in maize embryonic axes by measuring the amount of [^{32}P] incorporated into this protein. Results indicated a large increase of S6 ribosomal protein phosphorylation after insulin stimulation. Insulin-stimulated axes showed a selective increase in the synthesis of ribosomal proteins and iso4E initiation translation factor as determined by the amount of [^{35}S]-methionine incorporated into isolated rps and purified iso 4E. Recruitment of S6rp and of iso4E mRNAs into polysomes was determined by Northern blot analysis. Selective mobilization of these transcripts into polysomes was observed as a consequence of insulin application to the axes. These data suggest that the translation of these mRNAs is regulated by the translational mechanism described for the 5′TOP mRNAs, reported to be selectively recruited into S6rp phosphorylated polysomes, after insulin stimulation of target cells in animals. Current research oriented to further characterize this signal transduction pathway in maize axes is reported.

Introduction

Seeds represent a unique plant physiological stage characterized by a quiescent metabolic status. Early in seed germination, internal cell structures are reorganized, particularly the membranes, and many metabolic activities reinitiate followed by developmentally regulated biochemical and molecular processes, aimed at a successful germination.

Several studies have been performed to understand the biochemical and molecular events taking place during seed germination. Among them, it has been established that embryonic axes of cereal seeds reinitiate protein synthesis early during seed germination, based on translation of the stored mRNAs present in the embryonic axes (Sánchez de Jiménez and Aguilar, 1984; Pramanik and Bewley, 1996). In this system, the flux of genetic information is primarily based on translation and at later times on transcription-translation. These events are of utmost importance and are precisely regulated in order to lead axis cells to the acquisition of G1/S cell competence. In our laboratory, working with maize embryonic axes, it has been found that, after protein synthesis has been clearly established, transcription of new mRNAs occurs followed by DNA synthesis. Towards the end of germination, cell elongation takes place almost simultaneously with mitosis. This period lasts approximately 24 h.

The large amount of stored mRNAs present in the quiescent axes assures fast and sustained progress of protein synthesis early in germination. However, the central and relevant goal to pursue during this process consists of the programmed selectivity of translation for the stored and later on the new mRNAs. Indeed, the type and amount of the synthesized proteins in the cells, and thus the progress in successive appearance of different cell processes, would depend on cell fidelity following this programme. This picture underlies the relevant role that mechanisms of translational control might have in regulating gene expression, particularly during the first hours of germination.

Specific mechanisms for translational regulation are increasingly being recognized in eukaryotes (Cheatham *et al.*, 1994; Mathews *et al.*, 1996), including higher plants (Gallie, 1993; Marcotte, 1998), as a means of regulation of gene expression. These mechanisms are of particular relevance at specific physiological stages and/or developmental periods of the organisms, as in the particular case of the seed germination period.

The Translational Apparatus

To describe precisely the type of regulatory mechanism we are concerned with, let us first briefly review the organization of the translational apparatus in eukaryotic cells.

Protein synthesis is a very complex process comprising the participation of many different molecules and reactions. It basically takes place on the ribosomes, that constitute the main organelle for translation. This particle is formed by two subunits, 40S and 60S, the former containing the 18S rRNA plus approximately 33 proteins, and the latter, the 28S, 5.8S and 5S rRNAs plus about 53 proteins, depending on the organism.

The translation process can be divided into three main stages: (i) the initiation step that comprises the formation of the active initiation complex made up of the mRNA on the ribosome with the AUG position on the correct site, together with its counterpart, the methionyl tRNA molecule; (ii) the elongation phase that involves the formation of the peptide bond between one amino acid and the next, as dictated by the oligonucleotide sequences present in the mRNA; and (iii) the third step corresponding to the recognition of the termination triplet of the mRNA by the corresponding terminator factors.

Within this complicated scheme, it is clear that there are many possible sites for exerting translational control. The main issue behind this type of control is: 'How do cells know when, which ones, and how much of each mRNA shall be translated?' There are mechanisms designed to turn on and off the whole process, and thus they are responsible for regulating the general rate of cell protein synthesis, such as the phosphorylation–dephosphorylation process of the initiation factor-2B (eIF-2B) (Welsh *et al.*, 1997). Others, however, are designed to introduce selectivity in the process. Up to now, most of these mechanisms have been found to occur at different levels of the initiation step (Jefferies and Thomas, 1996; Sonenberg, 1996).

Indeed, the first and main point to be decided for protein synthesis initiation is to select, from the many mRNAs present in a cell pool, which ones will be engaged in the ribosome for translation. The recognition process seems to depend on the presence of the precise initiation factors, and on specific signals for translation, mainly on proteins called 'trans-acting' factors, which bind specifically to mRNA 'cis' sequences present at the 3′ and/or 5′UTR regions of the mRNA structure (Meyuhas *et al.*, 1996). It has to be borne in mind that protein factors and both mRNA and ribosome structures would have a preeminent role in the speed of specific mRNA recruitment into polysomes for translation at specific physiological or developmental stages of the organism. This selection process enables cells to define specific functional protein patterns that finally might help to define cell growth and differentiation.

Insulin Signal Transduction Pathway

Among the many mechanisms of translational control described in eukaryotes, the role that transduction of external signals induced by growth factors might have on translational regulation has been recently recognized (Jefferies and Thomas, 1996). In this system, effector–receptor interaction induces a cascade of biochemical reactions that transmit the signal to a precise target point of the translational apparatus. This signal transduction pathway regulates translation in a selective manner. The pathway might be stimulated by different effectors: insulin, growth factors, IGF I and II family members, mitogens. The signal originated by this interaction is internalized into the cell to a target site, through a cascade of reactions involving phosphorylation of different protein kinases and producing specific metabolites that function as second messengers (Cheatham *et al.*, 1994). The main target site stimulated by this pathway is the phosphorylation of the S6 ribosomal protein (rp) on the 40S ribosomal subunit (Fig. 2.1). By this means, a specific group of transcripts that contain a pyrimidine-rich sequence at their 5′UTR region, 5′TOP (track oligopyrimidines) mRNAs (Meyuhas *et al.*, 1996), are recognized by the phosphorylated ribosome and selectively mobilized into polysomes for translation. Within this group of transcripts, the mRNAs coding for many of the ribosomal proteins and some translational factors have been identified (Jefferies *et al.*, 1994; Pierandrei-Amaldi and Amaldi, 1994).

Within this picture, several questions arise. Is this signal transduction pathway a universal mechanism for regulating ribosomal protein synthesis in eukaryotes? Do plants regulate translation by a similar signal transduction pathway? In an attempt to approach these questions, our research has been centered on the search for a similar mechanism for regulating ribosomal protein synthesis within a specific and unique stage of plant development: seed germination.

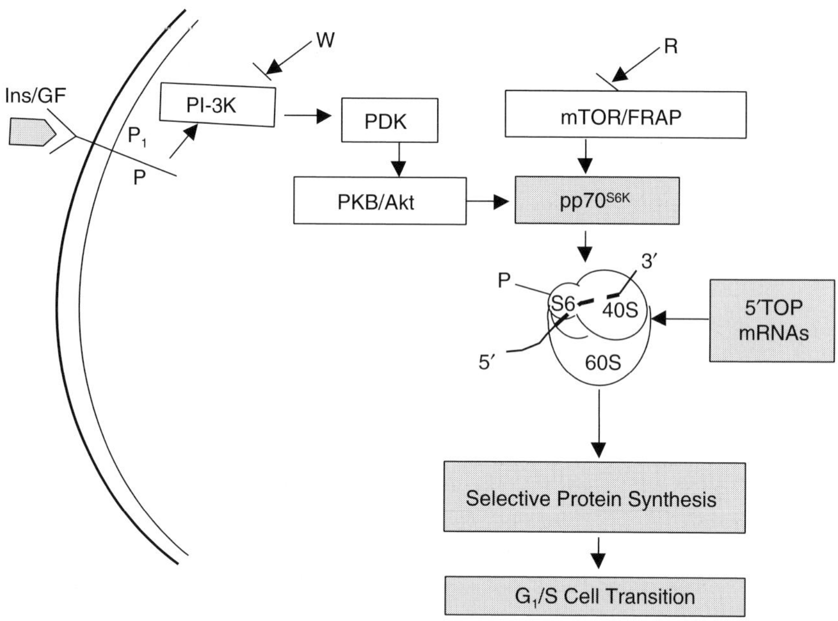

Fig. 2.1. Postulated scheme for insulin/GF signal transduction pathway. Insulin/Growth Factor (Ins/GF)-receptor recognition at the cell membrane initiates a cascade of phosphorylation/dephosphorylation reactions inside the cells. Phosphorylation of the S6 ribosomal protein constitutes the target of this pathway. Selective mobilization of 5′TOP mRNAs into polysomes for translation has been found after S6rp phosphorylation. Many of the ribosomal protein mRNAs contain the 5′TOP signal and are therefore selectively translated after Ins/GF stimuli. PI-3K, phosphatidylinositol 3-kinase; PDK, PKB/Akt and mTOR/FRAP, intermediate protein kinases of the transduction process; pp70^{S6k}, protein kinase that specifically phosphorylates S6rp; W, wortmannin, and R, rapamycin, are inhibitors of the insulin signal transduction pathway.

Antecedents supporting this hypothetical pathway in maize

1. Previous work in our laboratory has demonstrated that maize axes contain ribosomal protein mRNAs among their stored set of mRNAs. Further, active synthesis of ribosomal proteins has also been observed in axes during germination (Beltrán-Peña *et al.*, 1995).
2. The discovery of a group of cDNA sequences in plants coding for proteins homologous to cell surface receptors with intrinsic protein kinase activity provides some grounds for this possibility (Walker, 1996). One such receptor-like cDNA sequence codes for a protein with a large extracytoplasmic domain, a single membrane-spanning segment and a cytoplasmic domain with protein kinase activity, similar to the receptor described for insulin/GF/mitogen effectors. The diversity found among this kind of plant receptors implies a broad new area of possible regulatory processes in cellular signalling in plants, and brings about new implications for the mechanisms by which plant cells perceive and respond to extracellular signals (Walker, 1996).
3. The structure and function of the ribosomal proteins are universally conserved among eukaryotic organisms. The need for equimolecular assembly of these proteins into ribosomes supports the possibility for universally conserved regulatory mechanisms of ribosomal protein synthesis that ensure similar levels of these proteins for new ribosomal production.

Based on these antecedents, it was considered that the main objective of this research was to test whether maize axes would respond to external growth factor stimuli by inducing S6rp phosphorylation and specific translation of ribosomal protein mRNAs.

Experimental approach

Lacking, at the initial period of this research, the putative endogenous effector of the potential signal transduction pathway, it was decided instead to test insulin in all the experiments. Several reasons for testing insulin in maize germination were considered. Insulin is the most widely spread regulatory growth factor known in eukaryotes. In many cellular systems insulin is known to induce resting cells to move into the G1/S stage. This effect is accompanied by protein synthesis stimulation, particularly of ribosomal proteins that form the translational apparatus. Moreover, two growth factor families of peptides recognized by insulin antibodies (insulin-like peptides) have also been identified, namely IGF-I and IGF-II. They promote growth and differentiation by stimulating a similar signal transduction pathway as insulin in a wide variety of eukaryotic organisms (Parrizas *et al.*, 1997).

As a first approach, insulin effects on two physiological processes were assayed: maize seed germination and seedling growth. Fast germination and seedling development were observed in the presence of 300 to 500 μU insulin. Seedling length among plant populations was clearly larger for the insulin-stimulated seeds than for controls (Fig. 2.2).

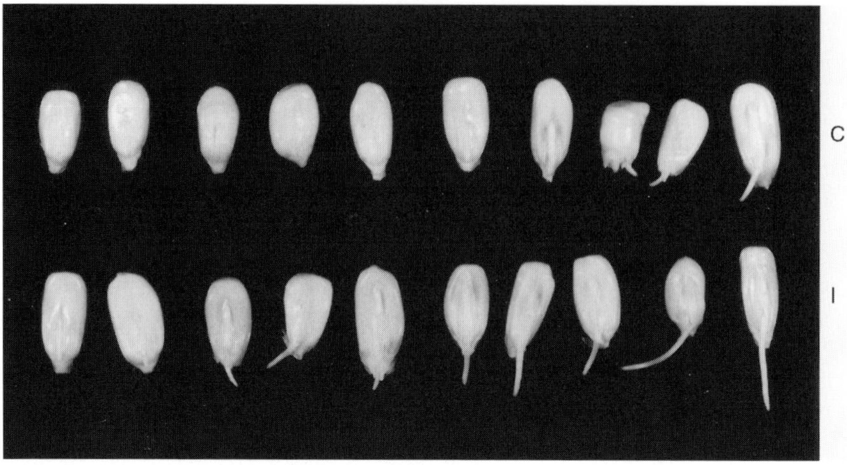

Fig. 2.2. Insulin effect on seed germination. Representative seeds from two 50-seed sets germinated under normal conditions for 24 h at 25°C under darkness. C, control seeds germinated with water; I, seeds germinated with 300 μU of insulin per ml of water.

The insulin effect on ribosomal protein synthesis was tested. For all the following experiments, excised axes from maize seeds at different germination periods were used. [^{35}S]-methionine pulse-labelling of germinating insulin-stimulated maize axes demonstrated more rapid incorporation of radioactive methionine into ribosomal proteins as compared with non-stimulated axes. This effect was blocked by application to the axes of either insulin mixed with its antibody, heat-denatured insulin or insulin treated with DTT. Specific inhibitors of protein kinases acting within the insulin signal transduction pathway, such as wortmannin or rapamycin, proved to inhibit insulin stimulation of ribosomal protein synthesis (Sánchez de Jiménez *et al.*, 1999).

To recognize the target proteins stimulated by insulin on the translational apparatus in maize axes, the effect of insulin on the phosphorylation of the S6rp in the 40S ribosomal subunit was analysed. The amount of [^{32}P]-orthophosphate incorporated on S6rp (32 kDa) at the 40S subunit increased after insulin stimulation (Fig. 2.3) (Sánchez de Jiménez *et al.*, 1997). Associated with this phenomenon, specific recruitment of 5′TOP mRNAs was expected. Using maize S6rp cDNA as a probe, the insulin effect on the S6rp mRNA (5′TOP-mRNA) recruitment into polysomes was tested. Northern blot analysis of polysomal RNA demonstrated selective mobilization of S6rp mRNA into polysomes by the insulin stimulus. Specific inhibitors of the insulin transduction pathway, wortmannin and rapamycin, blocked both the S6rp phosphorylation and the S6rp mRNA recruitment into polysomes (Fig. 2.3). This correlation also held when S6rp phosphorylation was induced by okadaic acid, an inhibitor of PP1 and PP2 A phosphatases, or treatment by heat shock, known to cause a decrease of S6rp phosphorylation. In both cases, the corresponding increase and decrease of S6rp mRNA recruitment into ribosomes was produced (Sánchez de Jiménez *et al.*, 1997).

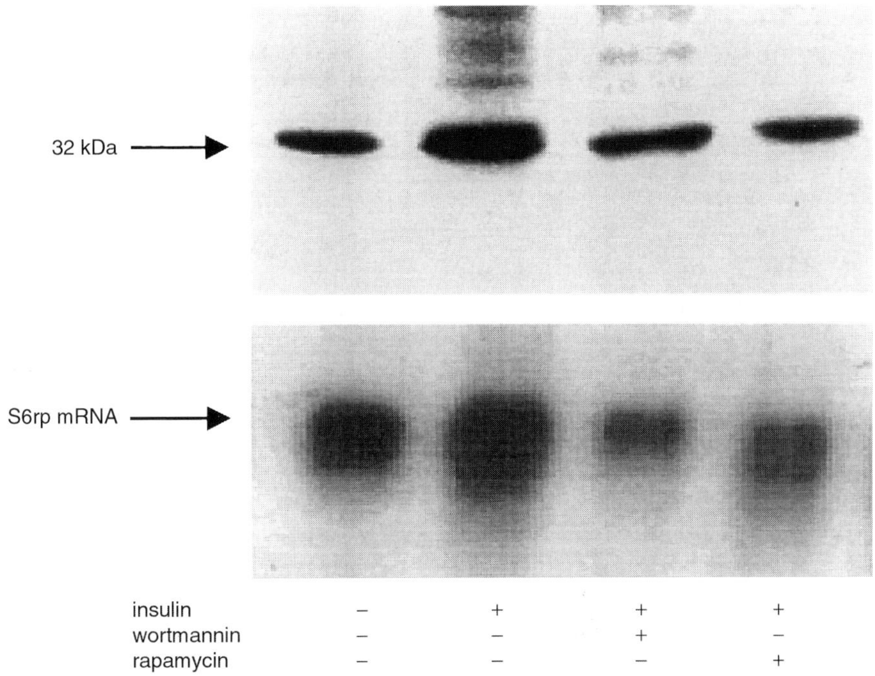

Fig. 2.3. Insulin effect on S6rp phosphorylation and S6rp mRNA recruitment into polysomes. Maize axes already germinated for 22 h were excised and incubated under sterile conditions with [^{32}P]-orthophosphate (500 μCi) for 2 h. Ribosomal proteins were extracted, electrophoresed by SDS-PAGE and analysed by autoradiography. The 32 kDa band corresponds to the S6rp (top). A similar set of axes obtained as above were incubated for 2 h with insulin (300 μU), or with wortmannin (0.1 μM) or rapamycin (0.1 μM) plus insulin (300 μU), homogenized and centrifuged on a 1.5 M sucrose cushion. From the pellet the polysomal RNA was extracted. This RNA was hybridized under stringent conditions with a labelled cDNA probe encoding maize S6rp. The autoradiography is presented in the lower panel. When present (+), wortmannin or rapamycin were added 10 min previous to insulin.

From the above data, it might be concluded that an insulin-responsive transduction pathway is present in maize axes, which seems to follow similar steps to that reported for other eukaryotes. It is proposed that this transduction pathway connects external signals induced by an insulin-like growth factor (mIGF) to a target factor on the translational apparatus, precisely the S6 protein of the ribosome, and selectively regulates by this means the synthesis of ribosomal proteins (5′TOP-mRNAs) in maize axes during germination.

Insulin-induced regulation of eIF-4E (Cap-binding protein)

Another important branch of the insulin transduction pathway has been recently demonstrated (Gingras *et al.*, 1998). Translation initiation factor 4E is the protein that recognizes the Cap structure in mRNAs. In mammals, it has been demonstrated that the amount of free eIF-4E is tightly regulated by trapping 4E in an inactive complex with a 4E-binding protein (PHAS) (Pause *et al.*, 1994). To release free 4E from the complex, 4E-BP phosphorylation must occur. A protein kinase present in a branch of the main insulin/GF transduction pathway is responsible for this phosphorylation reaction. Thus, activation of this kinase by insulin/GF stimuli would determine the level of free 4E in the cell, and so the rate of translation of Cap-dependent mRNAs. Overexpression of 4E in cells causes malignant transformation (Lazaris-Karatzas *et al.*, 1990).

eIF-4E protein is part of a large initiation complex known as eIF-4F. Interaction of this protein with the Cap structure would drive the mRNA for translation initiation. Several structures for Cap have been described, containing different numbers and/or positions of the methyl group in the mRNA molecule (Fig. 2.4); these variations have some effect on the affinity for 4E-binding protein, and thus might represent ways of regulating translation of specific mRNAs.

In plants, some interesting differences exist with regard to this initiation factor. First, two isoforms of the 4E factor have been found in wheat, rice (Browning *et al.*, 1996) and maize (Dinkova and Sánchez de Jiménez, 1996) of larger molecular masses than their animal counterpart (26 to 32 kDa vs. 24 kDa, respectively). Secondly, and most important, no 4E-binding protein seems to be present in plant tissues. Therefore, within the scope of our research, the question arises regarding the regulatory mechanism of expression for these translation initiation factors in maize during germination and its relation to the insulin/GF signal transduction pathway.

In maize axes, it was found that both 4E isoforms are present in axes from quiescent seeds, the iso4E form being more abundant than the 4E. Later during

Different structural features at the 5′UTR-mRNA
Cap-dependent translation • ^{7}Methyl GpppN-oligonucleotide • ^{2,7}di-Methyl GpppN-oligonucleotide • ^{7}Methyl Gppp-m-N-oligonucleotide • ^{2,7}di-Methyl GpppN-highly structured oligonucleotide Cap-independent translation • NNN-oligonucleotide. Internal AUG entree

Fig. 2.4. mRNA recognition sites for translation initiation. mRNAs with different requirements for translation have been recognized depending on the structural characteristics of their 5′UTRs (Un-Translated Regions). N, any nucleotide; m, internal methylation.

germination, both remain at approximately similar levels. By feeding axes for one hour with [^{35}S]-methionine and measuring its incorporation into each one of these factors, it was found that [^{35}S]-methionine incorporation into the iso4E protein during germination was faster and more steady than that into the 4E protein, which remained almost without label until 12 h of germination. To confirm the kinetics of *de novo* synthesis of the two 4E isoforms, *in vitro* translation of polysomes from axes at 6 and 12 h of germination was performed. Results indicated a very similar pattern as the one observed in the above experiment; that is, early translation of the iso4E and practically no labelled precursor incorporation into the 4E peptide up to 12 h of germination.

Looking for an explanation of this effect, analysis of the initiation factor transcripts was performed. Northern analysis of total RNA of quiescent axes showed two transcripts of 1.1 and 1.2 kb for the iso4E and the 4E proteins, respectively, indicating that both mRNAs were among the stored set of transcripts. Then, recruitment of these transcripts into polysomes was followed during the germination period by Northern blot of polysomal RNA. The iso4E mRNA was rapidly and increasingly mobilized into polysomes during germination. 4E transcript, on the other hand, was not mobilized until after 24 h of germination. These data indicate a differential regulation of 4E and iso4E expression in maize axes during germination.

The effect of insulin on both the rate of synthesis of these factors and recruitment of their corresponding transcripts was also tested. Following the same experimental design described above, the labelled methionine experiments were performed again in the presence of insulin. A large stimulation of [^{35}S]-methionine incorporation was found selectively for the iso4E protein (more than 100% at 24 h of germination) that was not observed for the 4E peptide. The effect of insulin on iso4E mRNA recruitment into polysomes was then analysed at 24 h. Results indicated that insulin induced an increase of iso4E transcript recruitment into polysomes as compared with the control (Fig. 2.5). This insulin-stimulated mobilization of iso4E into polysomes was specific and similar to the one observed for the S6 mRNA (5′TOP mRNAs), as indicated by the strong inhibition observed when either wortmannin or rapamycin were added to the system. The 4E protein did not show this behaviour under the same experimental conditions. The preferential recruitment of the iso4E transcript into polysomes by insulin stimulation suggests that iso4E mRNA might be under regulation of the 5′TOP signal.

As a conclusion from these experiments, it can be stated that iso4E expression in maize axes is regulated at the translational level during germination. This mechanism might in turn be controlled by activation of the insulin signal transduction pathway that regulates 5′TOP mRNA translation.

Current Research

At present, our research group is working on analysing a possible point of 'cross-talk' between the transduction pathway described above and an auxin-induced signal transduction pathway.

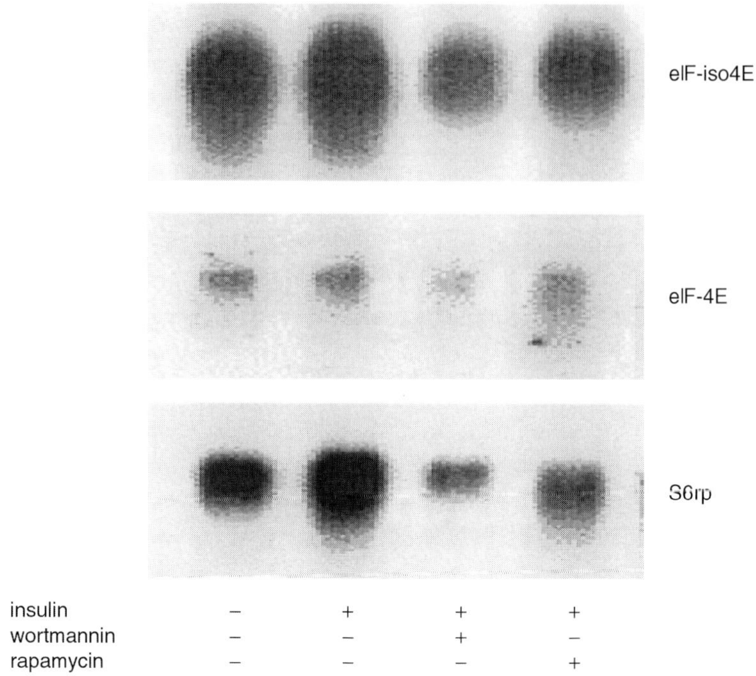

Fig. 2.5. Effect of insulin stimulation on eIF-4E and eIF-iso4E mRNA recruitment into the polysomal fraction and inhibition by wortmannin and rapamycin. Polysomal RNA was purified from 24 h germinated axes, stimulated or not by insulin for the last 2 h (as in Fig. 2.3). Two sets of axes were used to test either of the inhibitors, which were added (0.1 μM) to the axes 10 min before the insulin (300 μU). The RNA samples were resolved in 1.3% agarose gels and analysed by Northern blot. The RNAs were hybridized under stringent conditions with rice cDNA probes encoding eIF-4E or eIF-iso4E. The autoradiography shows the hybridized bands. Presence and absence indicated by + and –, respectively. S6rp hybridization is shown at the bottom for comparison.

Some considerations have to be made for the presentation of this proposal. Four main reasons can be stressed for suggesting a 'cross-talk' point between these two transduction pathways. They are the following:

1. Auxins play a central role in plant growth and development, regulating many physiological and biochemical responses (Abel and Theologis, 1996). Among these, stimulation of ribosomal protein synthesis has been demonstrated (Gantt and Key, 1985). However, the precise mechanism(s) of auxin action is (are) at present not fully understood (Venis and Napier, 1997).

2. The notion of an auxin-stimulated signal transduction pathway leading to different target points in plant cells is currently becoming a plausible explanation for the multiplicity of physiological effects of auxin (Zbell and Walter-Back, 1988; van der Zall *et al.*, 1996; Macdonald, 1997; Venis and Napier, 1997; Guilfoyle *et al.*, 1998). For some post-transcriptional auxin responses, a transduction pathway initiated by an auxin-receptor interaction at the cytoplasmic membrane, followed by G-protein activation has been suggested.

This pathway stimulates a phosphatidyl lipase PLC that produces inositol-3-phosphate (IP_3), and this in turn releases Ca^{2+} from the endoplasmic reticulum (Ettlinger and Lehle, 1988; Bush, 1993; Verhey and Lomax, 1993; Lommel and Felle, 1997; Venis and Napier, 1997).

3. Two very recent reports demonstrated that $pp70^{S6K}$, the kinase responsible for S6rp phosphorylation, is activated *in vivo* by Ca^{2+} in animal cells (Graves *et al.*, 1997; Conus *et al.*, 1998).

4. Previous results from our laboratory showed that auxin stimulates S6rp phosphorylation in maize tissues (Pérez *et al.*, 1990; Sánchez de Jiménez *et al.*, 1997).

Putting this information together, we hypothesize that a 'cross talk' between the two pathways might occur at the level of $pp70^{S6K}$ activation, induced in one case by the mIGF and, in the other, by the increment of free Ca^{2+} in the cytoplasm after auxin stimulation. In both cases, this activation would induce S6rp phosphorylation and in consequence, the two regulators would converge for the rest of the transduction pathway.

To test this hypothesis, the effect of auxin, IP_3 and Ca^{2+} on ribosomal protein synthesis is being analysed in maize axes. Ribosomal protein synthesis, measured by [^{35}S]-methionine incorporation after stimulation either with auxin, IP_3, Ca^{2+} or Nifedipine plus Ca^{2+} was performed. Preliminary data have shown an increased [^{35}S]-incorporation into ribosomal proteins from stimulated axes with either of the effectors tested, particularly for IP_3, as compared with the control (Fig. 2.6). Application of Nifedipine, an inhibitor of Ca^{2+} uptake, to the axes before Ca^{2+} addition blocked this effect. At present we are characterizing $pp70^{S6K}$ activity in maize axes to analyse whether IP_3 or Ca^{2+} stimulate this activity. Full identification of this point of interaction between the two pathways might provide grounds for postulating the mechanism of auxin action in regulating ribosomal protein synthesis.

Summary

1. A main pathway for selective translational regulation has been demonstrated in maize axes. This corresponds to a signal transduction pathway that connects external growth factor signals with the translational apparatus on the ribosome. This pathway introduces selectivity to the translation process since it stimulates preferential translation of target mRNAs containing the 5′TOP UTR region.

2. Specific regulation of iso4E mRNA translation by this pathway was demonstrated, suggesting that this transcript is a 5′TOP mRNA. This point of regulation of iso4E introduces another level of translational regulation, since this initiation translation factor might preferentially recognize some specific mRNAs required at certain plant developmental stages.

3. A point of cross interaction between two different signal transduction pathways has been postulated. The activation of these pathways might result in a synergistic effect on the process of protein synthesis. On the other hand, it has to be considered that second messengers that transduce the IGF signal are

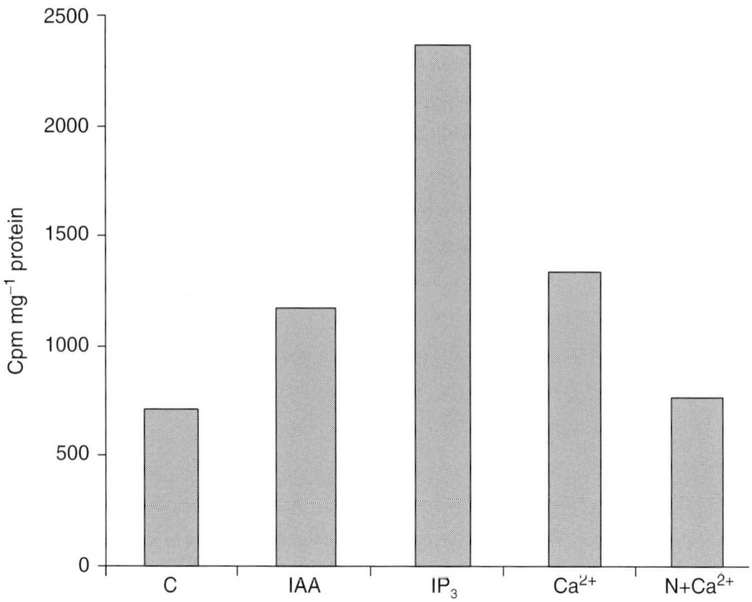

Fig. 2.6. Effect of auxin, IP_3, and Ca^{2+} on ribosomal protein synthesis. Maize seeds were germinated for 22 h and the axes were then excised and incubated under sterile conditions in the presence of [^{35}S]-methionine (300 μCi) and either: C, water; IAA, 20 μM IAA; IP_3, 20 μM IP_3; Ca^{2+}, 2 mM Ca^{2+}; N+Ca^{2+}, 10 μM Nifedipine + 2 mM Ca^{2+}. The ribosomes were then isolated, the ribosomal proteins purified, an aliquot counted in a scintillation counter and another measured in a spectrophotometer for protein concentration. Bars represent average of two values of cpm incorporated per mg of ribosomal protein.

phosphatidyl derivatives, products of PI-3K activity. Some of these metabolites are in turn substrates for the PLC enzyme from the auxin pathway. Therefore, competition between the two pathways could also be established at this level.

Further research has to be done to answer the questions raised by this postulated mechanism for protein synthesis regulation.

Acknowledgement

This work has been supported by Dirección General de Asuntos de Personal Académico (DGAPA), Universidad Nacional Autónoma de México, Grant IN217496.

References

Abel, S. and Theologis, A. (1996) Early genes and auxin action. *Plant Physiology* 111, 9–17.

Beltrán-Peña, E., Ortíz-López, A. and Sánchez de Jiménez, E. (1995) Synthesis of ribosomal proteins from stored mRNAs early in seed germination. *Plant Molecular Biology* 28, 327–336.

Browning, K.S. (1996) The plant translational apparatus. *Plant Molecular Biology* 32, 107–144.

Bush, D.S. (1993) Regulation of cytosolic calcium in plants. *Plant Physiology* 103, 7–13.

Cheatham, B., Vlahos, C.J., Cheatham, L., Wang, L., Blenis, J. and Kahn, C.R. (1994) Phosphatidylinositol 3-kinase activation is required for insulin stimulation of pp70S$_6$ kinase, DNA synthesis and glucose transporter translocation. *Molecular and Cellular Biology* 14, 4902–4911.

Conus, M.N., Hemmings, B.A. and Pearson, R.B. (1998) Differential regulation by calcium reveals distinct signaling requirements for the activation of Akt and s6kp70. *Journal of Biological Chemistry* 273, 4776–4782.

Dinkova, T. and Sánchez de Jiménez, E. (1996) Maize translation initiation factor eIF-4F: activity, synthesis and phosphorylation during germination. *Plant Physiology* 114, 245.

Ettlinger, C. and Lehle, L. (1988) Auxin induces rapid changes in phosphatidylinositol metabolites. *Nature* 331, 176–178.

Gallie, D.R. (1993) Posttranscriptional regulation of gene expression in plants. *Annual Review of Plant Physiology and Plant Molecular Biology* 44, 77–105.

Gantt, J.S. and Key, J.L. (1985) Coordinate expression of ribosomal protein mRNAs following auxin treatment of soybean hypocotyls. *Journal of Biological Chemistry* 260, 6175–6181.

Gingras, A.C., Kennedy, S.G., O'Leary, M.A., Sonenberg, N. and Hay, N. (1998) 4E-BP1, a repressor of mRNA translation is phosphorylated and inactivated by the Akt (PKB) signaling pathway. *Genes and Development* 12, 502–513.

Graves, L.M., He, Y., Lambert, J., Hunter, D., Li, X. and Earp, H.S. (1997) An intracellular calcium signal activates p70 but not p90 ribosomal S6 kinase in liver epithelial cells. *Journal of Biological Chemistry* 272, 1920–1928.

Guilfoyle, T.J., Hagen, G., Ulmasov, T. and Murfett, J. (1998) How does auxin turn on genes? *Plant Physiology* 118, 341–347.

Jefferies, H.B.J. and Thomas, G. (1996) Ribosomal protein S6 phosphorylation and signal transduction. In: Hershey, J.W.B., Mathew, M.B. and Sonenberg, N. (eds) *Translational Control.* Cold Spring Harbor Laboratory Press, Cold Spring Harbor, New York, pp. 389–409.

Jefferies, H.B.J., Reinhard, C., Kozma, S.C. and Thomas, G. (1994) Rapamycin selectively represses translation of the 'polypyrimidine tract' mRNA family. *Proceedings of the National Academy of Sciences USA* 91, 4441–4445.

Lazaris-Karatzas, A., Montine, K.S. and Sonenberg, N. (1990) Malignant transformation by eukaryotic initiation factor subunit that binds to mRNA. *Nature* 345, 544–547.

Lommel, C. and Felle, H.H. (1997) Transport of Ca^{2+} across the tonoplast of intact vacuoles from *Chenopodium album* L. suspension cells: ATP-dependent import and inositol-1,4,5-trisphosphate-induced release. *Planta* 201, 447–486.

Macdonald, H. (1997) Auxin perception and signal transduction. *Physiologia Plantarum* 100, 423–430.

Marcotte, W.R. Jr. (1998) Developmental regulation of translation and mRNA stability. In: Bailey-Serrés, J. and Gallie, O.R. (eds) *A Look Beyond Transcription.* American Society of Plant Physiology, Rockville, Maryland, pp. 64–67.

Mathews, M.B., Sonenberg, N. and Hershey, J.W.B. (1996) Origins and targets of translational control. In: Hershey, J.W.B., Mathews, M.B. and Sonenberg, N. (eds)

Translational Control. Cold Spring Harbor Laboratory Press, Cold Spring Harbor, New York, pp. 1–30.

Meyuhas, O., Avni, D. and Shama, S. (1996) Translational control of ribosomal protein mRNAs. In: Hershey, J.W.B., Mathews, M.B. and Sonenberg, N. (eds) *Translational Control.* Cold Spring Harbor Laboratory Press, Cold Spring Harbor, New York, pp. 363–368.

Parrizas, M., Saltiel, A.R. and Leroith, D. (1997) Insulin-like growth factor 1 inhibits apoptosis using the phosphatidylinositol 3′-kinase and mitogen-activated protein kinase pathways. *Journal of Biological Chemistry* 272, 154–161.

Pause, A., Belsham, G.J., Greigas, A.C., Ponze, O., Liu, T.A., Laurence, H.C.J. and Sonenberg, N. (1994) Insulin-dependent stimulation of protein synthesis by phosphorylation of a regulator of 5′ cap function. *Nature* 371, 762–767.

Pérez, L., Aguilar, R., Pérez-Méndez, A. and Sánchez de Jiménez, E. (1990) Phosphorylation of ribosomal proteins induced by auxins in maize embryonic tissues. *Plant Physiology* 94, 1270–125.

Pierandrei-Amaldi, P. and Amaldi, F. (1994) Aspects of regulation of ribosomal protein synthesis in *Xenopus laevis. Genetica* 94, 181–193.

Pramanik, S.K. and Bewley, J.D. (1996) Post-transcriptional regulation of protein synthesis during alfalfa embryogenesis: protein associated with the cytoplasmic polysomal and non-polysomal mRNAs (messenger ribonucleoproteins complex). *Journal of Experimental Botany* 44, 1871–1879.

Sánchez de Jiménez, E. and Aguilar, R. (1984) Protein synthesis patterns: relevance of old and new mRNA in germinating maize embryos. *Plant Physiology* 75, 231–245.

Sánchez de Jiménez, E., Beltrán-Peña, E. and Ortíz-López, A. (1997) Translation of ribosomal protein mRNAs in maize axes. In: Ellis, R.H., Black, M., Murdoch, A.J. and Hong, T.D. (eds) *Basic and Applied Aspects of Seed Biology.* Kluwer Academic Publishers, Dordrecht, pp. 385–394.

Sánchez de Jiménez, E., Beltrán-Peña, E. and Ortíz-López, A. (1999) Insulin-stimulated ribosomal protein synthesis in maize embryonic axes during germination. *Physiologia Plantarum,* 105, 148–155.

Sonenberg, N. (1996) mRNA 5′ Cap-binding protein eIF4E and control of cell growth. In: Hershey, J.W.B., Mathews, M.B. and Sonenberg, N. (eds) *Translational Control.* Cold Spring Harbor Laboratory Press, Cold Spring Harbor, New York, pp. 245–271.

van der Zall, B.J., Droog, F.N.J., Pierterse, F.J. and Hooykaas, P.J.J. (1996) Auxin-sensitive elements from promoters of tobacco *GST* genes and a consensus *as*-1-like element differ only in relative strength. *Plant Physiology* 110, 79–88.

Venis, M.A. and Napier, R.M. (1997) Auxin perception and signal transduction. In: Aducci, P. (ed.) *Signal Transduction in Plants. Molecular and Cell Biology Updates.* Birkhauser Verlag, Basel/Switzerland, pp. 45–63.

Verhey, S.D. and Lomax, T.L. (1993) Signal transduction in vascular plants. *Journal of Plant Growth Regulation* 12, 179–195.

Walker, J.C. (1996) Structure and function of receptor-like protein kinases of higher plants. *Plant Molecular Biology* 26, 1599–1609.

Welsh, G.I., Stokes, C.M., Wang, X., Sakane, H., Ogawa, W., Kasuga, M. and Proud, C.G. (1997) Activation of translation initiation factor eIF2B by insulin requires phosphatidyl inositol 3-kinase. *FEBS Letters* 410, 418–422.

Zbell, B. and Walter-Back, C. (1988) Signal transduction of auxin on isolated plant cell membranes: indications for a rapid poly-phosphoinositide response stimulated by indolacetic acid. *Journal of Plant Physiology* 133, 353–357.

II Development and Quality

3 bZIP and DOF Transcription Factors in the Regulation of Gene Expression in Barley Endosperm

P. CARBONERO, J. VICENTE-CARBAJOSA, M. MENA, L. OÑATE, P. LARA AND I. DIAZ

Laboratorio de Bioquímica y Biología Molecular, Departamento de Biotecnología-UPM, E.T.S. Ingenieros Agrónomos, 28040 Madrid, Spain

The protein quality of cereal grain is determined by the quantity and composition of the prolamin storage proteins (hordeins in barley) that are specifically synthesized in the endosperm and that are regulated at the transcriptional level. Most prolamin gene promoters contain a bipartite sequence motif in the –300 region that becomes occupied early in endosperm development by nuclear *trans*-acting factors. cDNAs have been characterized from barley endosperm encoding DNA-binding proteins with affinity towards the two halves of the –300 motif: 5′TGTAAAG-3′ prolamin box; PB) and 5′-GTGAGTCAT-3′(GCN4-like; GLM), respectively. The barley PB binding factor is a zinc finger of the DOF class while the GLM binding factors belong to the bZIP class of transcription factors. Experiments *in vivo* and *in vitro* are described that strongly implicate these factors as important activators of hordein gene expression and suggest their evolutionary conservation among small grain cereals.

Introduction

Cereal grains (wheat, rice, maize, barley, etc.) constitute the world's primary crops and the cereal endosperm provides the major source of carbohydrates and proteins for human food and livestock feed. Although embryo and endosperm in these grains have a similar genetic origin, the embryo is diploid whereas the endosperm that arises as a result of the fusion of a generative nucleus from the pollen tube with the 2n central cell of the embryo sac is normally triploid. After a series of coenocytic mitoses followed by a cellularization stage that starts with periclinal cell wall formation between nuclei at the periphery of the endosperm cavity, there is a commitment to particular cell types, such as aleurone, starchy endosperm and basal transfer layer cells. Within a given cell type, development is position dependent. For example, in the internal endosperm cells more starch is accumulated than in the cells at the periphery, but the mRNAs encoding reserve proteins and enzymes, such as the hordeins and the sucrose synthase isozymes Ss1 and Ss2 (Vicente-Carbajosa *et al.*, 1992; Guerin and Carbonero, 1997), are more abundantly expressed in the cells at the periphery.

Nitrogen and sulphur are stored in the starchy endosperm cells mainly in the form of a complex group of proteins, the prolamins, characterized by their alcohol-soluble properties. Synthesis of the cereal prolamins occurs only in these cells where they are under tissue-specific and temporal transcription control. Although the prolamin genes were among the first plant genes to be cloned and characterized, the molecular mechanisms underlying the precise regulation of their expression are still poorly understood and only partial information is available concerning the transcription factors involved (Shewry *et al.*, 1995).

The barley prolamins (hordeins) are classified, according to their mobility in SDS-electrophoretic gels, into three major classes: B-, C-, and D-hordeins. The B fraction represents approximately 75% of the total hordein content in most barley cultivars (cvs). All hordeins are structurally related and their genes presumably derive from a common ancestor by gene duplication and subsequent divergent evolution (Kreis *et al.*, 1985). The coordinate expression of all hordein genes in the starchy endosperm from 8–12 days after pollination (dap) suggests common regulatory mechanisms of transcription that should involve both *cis*-acting motifs and *trans*-acting factors.

A conserved *cis*-acting motif (Fig. 3.1) found in most storage protein gene promoters of seeds is the endosperm box (EB; Sorensen *et al.*, 1989; Kreis and Shewry, 1992). The EB is a bipartite motif located around 300 bp upstream of the translation initiation codon that contains two distinct nuclear protein binding sites: the prolamin box (PB = 5′TGTAAAG3′), also called the endosperm motif (EM) that resembles the viral SV40 enhancers sequence, and a GCN4-like motif (GLM = 5′(G/A)TGA(G/C)TCA(T/C)3′) that resembles the binding site of the yeast bZIP transcription factor GCN4 (Hill *et al.*, 1986; Müller *et al.*, 1995). Both PB and GLM are present in B- and C-hordein gene promoters whereas only the PB is present in the D-hordein promoters. Interestingly, B- and C-hordein synthesis is severely depressed in the high lysine-mutant Riso 1508 while the content of D-hordeins is similar to that found in the wild-type Bomi from which it derives. Although this matter is far from being settled, it is tempting to postulate that a transcription factor of the bZIP class (or some post-translationally modified form of it) recognizing the GLM must be involved (Rodriguez-Palenzuela *et al.*, 1989).

Functional analysis of a native C-hordein promoter by particle bombardment of developing barley endosperms (Müller and Knudsen, 1993) demonstrated that the GLM is the dominant *cis*-acting element, and that the PB exerted a silencing effect on the activity of that particular promoter. However, both GLM and PB from the promoter of a *Hor-2* gene that encodes a B1-hordein were essential positive elements conferring a high level of transcriptional activity to a minimal (–90 bp) 35S cauliflower mosaic virus (Δ35S CaMV) promoter in microparticle-bombarded developing endosperms from barley (Vicente-Carbajosa *et al.*, 1998).

PLANT /GENE	ENDOSPERM BOX	POSITION
	PB **GLM**	
Barley		
B-Hordein	TGACA**TGTAAAG**TGAATAAGG**TGAGTCA**TGCAT	-300
C-Hordein	TGTAG**TGTAAAG**TGAA-AAAA**TGAGTCA**-TCAT	-317
γ-Hordein	TGAGA**TGTAAAG**TGAATAAGA**TGAGTCA**-GCAC	-297
D-Hordein	TGTTT**TGCAAAG**CTCCAATTCCTCCTTGCTTAT	-242
Wheat		
α-Gliadin	TGAGC**TGTAAAG**TGAATAAGA**TGAGTCA**TGCAT	-315
LMW-Glutenin	TGACA**TGTAAAG**TTAATAAGG**TGAGTCA**TATGT	-303
Rye		
ω-Secalin	TGTAG**TGTAAAG**TGAAAAAAA**TGAGTCA**TCAGT	-319
Maize		
22 kDa Zein	ACATG**TGTAAAG**GT.(N)20.TCC**ACGT**AGATGA	-330
19 kDa Zein	ACATG**TGTAAAG**GT.(N)20.CCCATGTATTTGG	-326
Rice		
T-II Glutelin	ATAT**TGCAAAAAG**AG.(N)19.A**TGACTCA**CAAA	-306
	TGTAAAG TGACTCA	

Fig. 3.1. Alignment of sequences in the endosperm box region of prolamin gene promoters from cereals. The prolamin box (PB) and the GCN4-like (GLM) motifs are in bold. Position is indicated with respect to the ATG translation initiation codon.

Much of the pioneering work concerning the genetic and molecular mechanisms regulating cereal seed storage genes came from research done in maize, where a bZIP protein, Opaque 2 (O2; Hartings *et al.*, 1989; Schmidt *et al.*, 1990), was shown to bind to and activate transcription from an imperfect palindromic ACGT core (5′TCCACGTAGA3′) in the promoter of the 22-kDa class of zein genes. This motif is adjacent to and downstream of the PB-motif (5′TGTAAAG 3′) present in all zein genes (Fig. 3.1). The 22-kDa class of zein genes encodes one of the more abundant classes of zeins in maize endosperm. Here, as in other storage protein genes, this bipartite *cis*-acting motif is located about 300 bp upstream of the ATG translation initiation codon. The ACGT core is similar to the sequence recognized by many bZIP proteins in plants and other systems (Katagiri and Chua, 1992). The observation that not only the 22 kDa zeins but also their transcripts were severely depressed in the Opaque 2 (*O2*) mutant, while other classes of zein genes, not having this ACGT core in their promoters, were not affected (Motto *et al.*, 1989), provided circumstantial evidence that O2 was a regulatory protein of the bZIP class of transcription factors. Cloning and functional characterization of the *O2* gene fully demonstrated that indeed this was the case (for reviews see Schmidt *et al.*, 1994; Müller *et al.*, 1995).

Cloning and Characterization of bZIP Transcription Factors from Barley Endosperm

Functional analysis of the promoters of several genes expressed in barley endosperm, such as those encoding sucrose synthase (*Ss1, Ss2*) and trypsin inhibitor BTI-CMe (*Itr1*), had shown the presence of putative bZIP binding motifs in their promoters (Diaz *et al.*, 1995; Royo *et al.*, 1996). The crucial role of such motifs (Fig. 3.1) had also been recognized in the hordein genes by us (Vicente-Carbajosa *et al.*, 1992) and others. This led us to investigate the possible involvement of bZIP factors in their regulation.

As a first step in the isolation of barley bZIP-related genes, a barley bZIP probe was generated by PCR (Fig. 3.2). Primers were derived from conserved regions of plant bZIPs (SNRESA and KVMA/GE for sense and antisense primers, respectively) and used to amplify homologous sequences from barley endosperm cDNA. Sequencing of PCR products identified a 209 bp fragment with homology to bZIP factors. After screening endosperm cDNA and genomic libraries from barley using this fragment as a probe, several cDNA and genomic clones were identified. One such cDNA and two overlapping genomic clones corresponded to the same bZIP gene and were further characterized. A 4726 bp fragment of the genomic sequence, containing the complete coding region of the gene (hereafter designated *Blz1* gene; Vicente-Carbajosa *et al.*, 1998), contained six exons and five introns, as determined by comparison with the cDNA clone, and encoded a protein of 391 amino acid residues with a calculated molecular mass of 46,920 Da. Introns, although different in size, appeared at equivalent positions with respect to those in the maize *O2* and

---

CLONING OF bZIP GENES FROM BARLEY

I. cDNA synthesis from total RNA from developing endosperm of barley cv. Bomi.

II. PCR amplification of bZIP domain with appropriate primers.

```
                BASIC REGION                                LEUCINE ZIPPER

O2      MPTEERVRKRKESNRESARRSAYRKAAHLKELEDQVAULKAENSCLLRRIAALNQKYNTANVDNRVLRADMRTLRAKVKMGE

OHP1    N.VQQ.LQR..Q..........S......N...A.....RV...S....L.DV...F.E.A......K...VE.......A.

CPRF2   D.SDAKRVR.ML..........R..Q..MT...T..S..RV...S..K.LTDIS.R....A......K...IE.M.....A.
```

III. Selection of a truncated clone with sequence homology to bZIP transcription factors.

IV. Use of this clone as a probe for the screening of cDNA and genomic libraries.

Fig. 3.2. General strategy used for the isolation of endosperm cDNA and genomic clones from barley belonging to the bZIP class of transcription factors.

OHP1 genes (Pysh and Schmidt, 1996), and were flanked by typical gt/ag boundaries. Two putative nuclear localization signals (Varagona *et al.*, 1992) and a serine-rich phosphorylation site (Hunter and Karin, 1992) were also found. It is worth mentioning the presence of four ATG codons in the mRNA leader sequence, which determine four short upstream open reading frames (uORF), a feature shared by O2 and other transcription factors, which have been shown to be involved in the regulation of translation (Lohmer *et al.*, 1993).

Northern blot analysis of *Blz1* showed that this gene was ubiquitously expressed, since its mRNA of ~1.8 kb appeared not only in endosperm but also in roots and leaves. The temporal pattern of *Blz1* expression in developing endosperm was such that it preceded (10–15 DAP) those of *Itr1* and *Hor2* (encoding trypsin inhibitor CMe and a B-hordein, respectively), which showed maximum mRNA steady-state levels at later stages (15–20 DAP).

More recently, a second cDNA encoding an endosperm-specific bZIP protein (hereafter *Blz2*) has been characterized (Oñate *et al.*, 1999). The BLZ2 protein is probably the homologue of wheat SPA (Albani *et al.*, 1997) because they share 77.5% identical residues along the whole protein and 94.8% in their bZIP domains. BLZ2 is also related to the barley BLZ1 protein (34.3% identity over the whole protein; 70.1% in the bZIP domain), to O2 from maize, coix and sorghum, to REB from rice and to OHP1 from maize, and has limited but significant homology with CPRF2 from parsley and RITA1 from rice. A phylogenetic dendrogram based on the comparison of the whole proteins and the sequence alignments clearly indicates that these proteins form a well-defined subfamily of plant bZIPs (Fig. 3.3).

In order to investigate if the BLZ1 protein could function as a transcriptional activator, a yeast reporter system was used (Vicente-Carbajosa *et al.*, 1998). A series of constructs, involving the entire BLZ1 and various regions derived from it, were prepared in a yeast expression plasmid as fusions to the Gal4 DNA binding domain (DBD). These were tested as effectors for their ability to transactivate reporter genes under the control of promoters containing Gal4 binding sites. These constructs were designated to detect putative domains with activating capacity along the tested protein, specially the three stretches with a high content of acidic residues, as activation has been previously associated with this type of domain. Both a qualitative reporter gene (histidine auxotrophy; *His3*) and a quantitative one (β-galactosidase activity; *LacZ*) were used to detect the transactivating capacity. The BLZ1 protein was a potent activator in these yeast systems. Quantitatively, the N-terminal portion spanning residues 1–142 (AD1) retained 63% of the LacZ activation produced by the entire protein. Addition of the AD2 acidic domain (residues 143–203) increased the activity to 85%, although this domain was unable to activate the system by itself. The C-terminal AD3 domain (residues 292–391), either alone or in combination with the basic leucine zipper region, allowed growth in the absence of histidine and was responsible for about 10% of the total lacZ activity. As expected, the basic leucine zipper region (residues 204–291) by itself did not promote transcription. The activity mediated by the complete BLZ1 in the yeast system was ~50% of that obtained with the maize O2 factor,

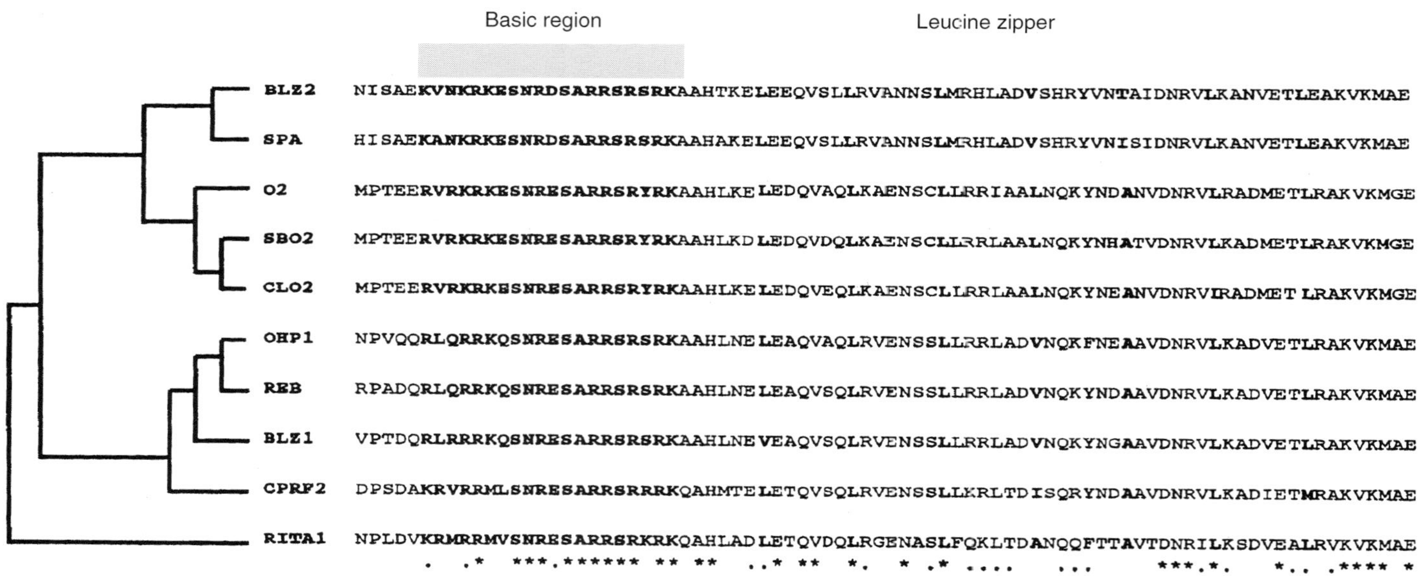

Fig. 3.3. Dendrogram and alignment of the deduced amino acid sequences of the bZIP domains of BLZ1 and BLZ2 with those of related plant bZIP proteins. The basic region is indicated with a shaded bar, and both the basic region and the leucine repeats are in bold. Asterisks correspond to identical residues in the ten sequences compared, and dots represent conserved substitutions: SPA from wheat, O2 from maize, sorghum (SBO2) and coix (CLO2), OHP1 from maize, REB from rice, CPRF2 from parsley, and RITA1 from rice (all these sequences are in the EMBL GenBank).

whereas the O2 equivalent of the AD3 domain of BLZ1 was completely inactive.

As we have stated previously, we could anticipate the kind of sequences able to be bound by BLZ1 and BLZ2 within promoters of genes expressed in barley endosperm. One such sequence shown in Fig. 3.1 is the GLM (5′GTGAGTCAT3′) present in the promoters of B- and C-hordein genes (Müller and Knudsen, 1993). The potential involvement of these bZIP proteins in the regulation of hordein gene expression was investigated both by electrophoretic mobility shift assays (EMSA) and by transient co-bombardment of barley developing endosperm.

For the electrophoretic mobility shift assays (EMSA), BLZ1 and BLZ2 were produced in *Escherichia coli*. The ^{32}P-labelled DNA fragments used were a 43 bp-long endosperm box oligonucleotide (HOR) of the *Hor2* promoter (Vicente-Carbajosa *et al.*, 1992) and two mutated versions of it containing alterations within the GLM (hor1) or within the PB (hor2) motifs. Retarded bands were produced when the HOR or the hor2 probes were incubated with protein extracts derived from bacterial cells expressing either the BLZ1 or BLZ2 recombinant proteins, while no retarded bands were detected when the labelled oligonucleotide used was the hor1, or when the protein extract was from bacteria transformed with the plasmid without insert. Binding specificity was demonstrated by competition titrations up to 100× concentration with the same unlabelled oligonucleotides used for binding; the mutated hor1 did not compete with the binding of HOR or of hor2 to the BLZ1 or BLZ2 proteins, whereas the hor2 oligonucleotide not only interacted with them but was as good a competitor as the original HOR probe (Vicente-Carbajosa *et al.*, 1998; Oñate *et al.*, 1999).

To investigate the roles of BLZ1 and BLZ2 *in planta*, we used transient expression assays of appropriate constructs in microbombarded developing endosperms from barley, using a biolistic helium gun device (Dupont PSD-1000). A model experiment appears in Fig. 3.4. When a reporter gene controlled by the bipartite endosperm box, containing intact PB and GLM motifs from the *Hor-2* promoter, fused to the 5′-end of the minimal Δ35S promoter (HOR-Δ35S in Fig. 3.4) was co-transfected with *Blz1*, *Blz2* or an equimolar mixture of the two, as effectors, a 3-fold increase in GUS activity was observed (Fig. 3.4C). As expected, mutations in the GLM (hor1-Δ35S) that prevented *in vitro* binding by these bZIP proteins, abolished GUS activation (data not shown). Mutations in the contiguous PB (hor2-Δ35S) sequence that did not interfere with the binding *in vitro* supported lower levels of BLZ1/BLZ2 activation *in planta* (~50%). These results indicate that BLZ1 and/or BLZ2 mediates transcriptional activation in barley endosperm through specific interaction with the GLM sequence. An intact PB is also essential for full transactivation. Moreover, in the absence of the exogenous effector, the reporters with mutations, either in the GLM or in the PB promoter sequences, display a much lower basal GUS activity compared with that sustained by the HOR-Δ35S promoter (Vicente-Carbajosa *et al.*, 1998; Oñate *et al.*, 1999). Thus, our results also suggest that a positive relationship between bZIP and the transcription factors recognizing the PB is necessary for high expression levels to

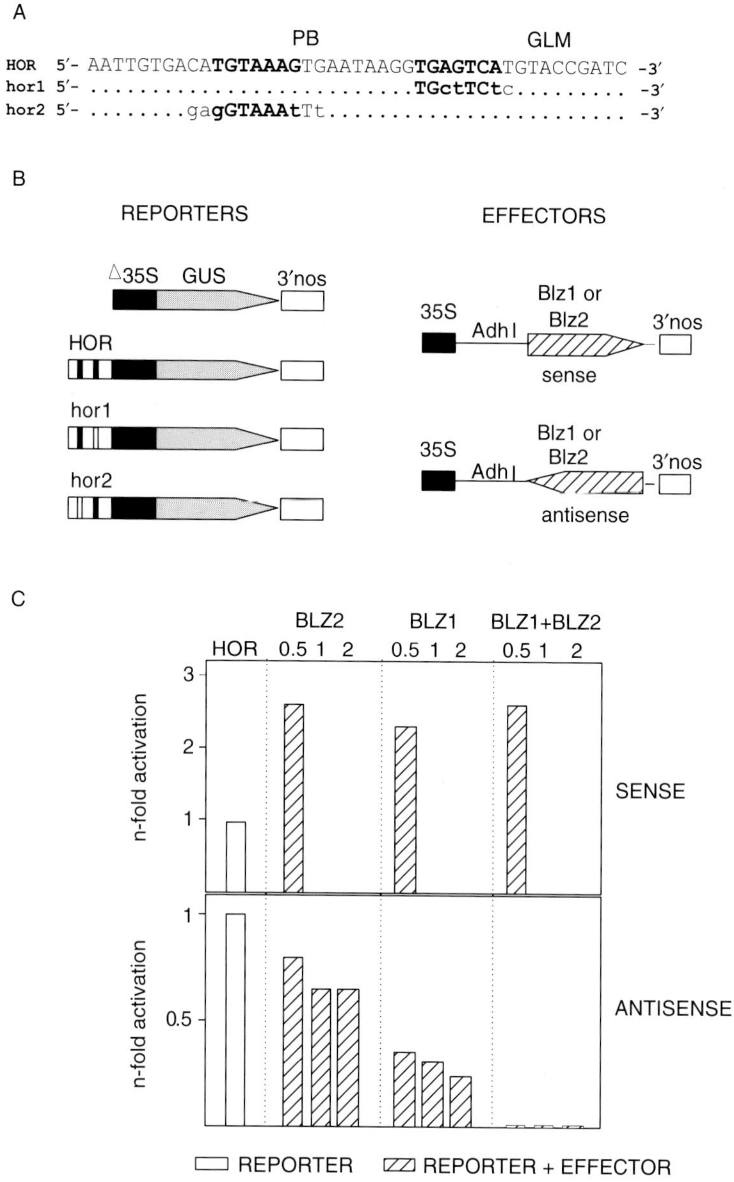

Fig. 3.4. Transient expression assays by co-bombardment of developing barley endosperm (15 DAP) using as effectors both the sense and antisense *Blz1* and *Blz2* as effectors. A. Sequences of the 43-mer oligonucleotides, derived from the endosperm box region of the *Hor-2* gene promoter (HOR) and derived mutants in the GLM (hor1) and in the P-box (hor2) that were linked to minimal (–90) Δ35S promoters fused to the β-glucuronidase (GUS) reporter gene. B. Schematic representation of the reporter and effector constructs used. AdhI, first intron of the maize *AdhI* gene. C. Transient expression assays of co-transfection of developing barley endosperms (15 DAP) with 150 ng of the reporter construct and 1:0.5 ratio of effector in the sense orientation. For the antisense experiments 250ng of the reporter and the indicated reporter: effector ratios (1:0.5, 1:1, and 1:2) were used. GUS activity was detected by histochemical staining and subsequent counting of blue dots per endosperm and expressed as n-fold activation relative to the HOR-Δ35S control construct without effector. Standard error of the mean for triplicate independent bombardments, with the same particle to plasmid suspension ratio, was < 15%.

be obtained from the EB of hordein promoters in barley endosperm. It should be noted that transient expression data indicate that BLZ1 and BLZ2 must not be saturating in barley endosperm or we would not have seen stimulus upon adding the BLZ1 and BLZ2 effector plasmid together with the HOR-Δ35S-GUS reporter. This suggests that overexpression of BLZ2 in transgenic barley might lead to increased levels of storage protein gene expression, an important agronomic goal.

Transient expression assays were also conducted with effector plasmids carrying *Blz2* and/or *Blz1* constructs in antisense orientation in order to investigate the effects of their depletion (Fig. 3.4C). The mRNA expression of *Blz2* in developing endosperm was only partially counteracted by the antisense approach, and a more effective GUS reduction was obtained with the *Blz1* antisense construct. However, co-transforming with both *Blz2* and *Blz1* antisense effectors resulted in a dramatic reduction of the basal GUS activity, even at the lowest effector/reporter ratio tested. This suggests a synergistic effect by a possible BLZ2/BLZ1 heterodimer in barley endosperm. It is worth noting that the GLM sequences present in most hordein promoters are of the AP-1 type (ACGT cores), this being a constraint for the binding by other plant bZIP factors belonging to the ATF/CREB group that recognize the ACGT core. This observation, together with the data concerning the transcriptional properties *in planta* of BLZ2 and BLZ1, strongly support that both proteins are significant, if not the unique, components of the machinery that mediates transcriptional activation through the GLM. However, we cannot rule out the contribution of other factors to the endosperm box complex. In this context, interactions with the barley PB-binding factor, as we will discuss later, could account for major differences in their mode of action as compared to homo- or heterodimer bZIP formation.

Cloning and Characterization of a DOF Transcription Factor from Barley Endosperm

The strong circumstantial evidence supporting an important role of the P-box (5′TGTAAAG3′) in barley hordein gene expression and the recent cloning of a maize endosperm-specific cDNA, encoding a protein with the expected characteristics for such a P-box binding factor that interacted *in vitro* with the bZIP Opaque 2 protein (Vicente-Carbajosa *et al.*, 1997), prompted us to try the cloning of the P-box binding factor from barley endosperm through homologous screening.

The sequence of such a barley PBF cDNA (hereafter BPBF; Mena *et al.*, 1998) encodes a protein of 333 amino acid residues with a predicted molecular mass of 34,113 Da. Near its N-terminus, it contains a conserved stretch of 50 amino acids (DOF domain), containing a putative CX_2-$CX_{21}CX_2C_{21}$ motif (Fig. 3.5), that may form a single zinc finger that is essential for DNA recognition. Previous experiments had shown that mutations in the cysteine residues, or the presence of metal chelators disrupting Zn coordination, abolished

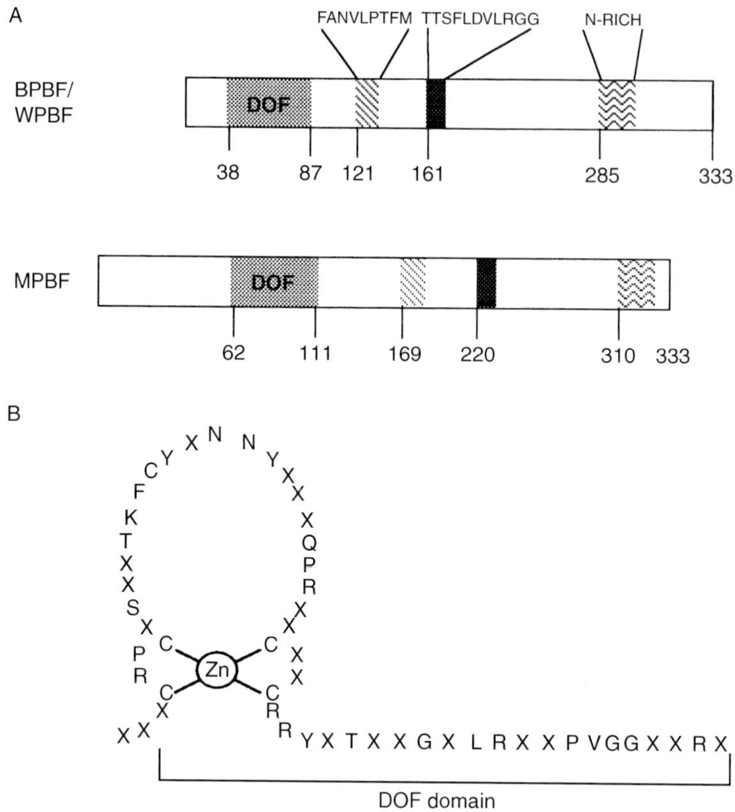

Fig. 3.5. A. Structural domains conserved in the prolamin-box binding factors from barley (BPBF), wheat (WPBF) and maize (MPBF). B. Schematic representation of the $CX_2CX_{21}CX_2CX_{21}$ DOF domain with its presumptive zinc finger.

DNA interaction with DOF proteins (Chen *et al.*, 1996; de Paolis *et al.*, 1996; Yanagisawa, 1996, 1997; Shimofurutani *et al.*, 1998).

Preliminary Southern blot studies indicated the presence in wheat of a gene closely related in sequence to barley *Pbf*. To search for this potential wheat *Pbf* homologue, an RT-PCR based strategy was applied. Total RNA from immature wheat endosperm was used as template and oligo-dT as primer for first-strand cDNA synthesis. An approximately 1 kb PCR product was obtained with primers derived from the N-terminal (sense) and C-terminal (antisense) coding sequence of the barley PBF cDNA. Sequencing confirmed the isolation from wheat of a partial cDNA spanning the whole coding region of a DOF protein with a clear homology to BPBF, hereafter designated WPBF (Wheat Prolamin-box Binding Factor). Through the DOF domain, BPBF and WPBF are identical, sharing 94% of sequence identity with the maize PBF (MPBF). This percentage never falls bellow 72% when comparisons are made with the other DOF proteins described thus far. When considering their whole sequence, BPBF and WPBF share 85% identical amino acid residues, and although they have only around 30% of conserved positions with respect to MPBF, these three proteins share three other distinctive regions of extensive homology, besides the DOF domain (Fig. 3.5A), such as an asparagine-rich stretch at the C-terminus that may constitute the activation domain. Outside the DOF domain, neither BPBF, WPBF nor MPBF show any significant homology with the other DOF proteins in the SwissProt databank.

Northern and EMSA experiments, similar to those previously described with the bZIP factors, demonstrated that the *Pbf* gene from barley encodes an endosperm-specific protein that binds in a sequence-specific manner to the P-box motif of a hordein gene promoter (Mena *et al.*, 1998). By transient expression assays in microparticle-bombarded barley endosperm, we have also shown that this interaction occurs in the homologous tissue and in the context of the native *Hor2-184* promoter. Direct binding of BPBF to the P-box element was necessary for transactivating a GUS reporter from the promoter of a B-hordein gene. These results, together with the conservation in sequence and position of the P-box element among prolamin promoters, suggest that BPBF plays a pivotal role in coordinating the activation of hordein genes in the endosperm. The fact that the barley *Pbf* transcripts accumulate in the endosperm before and during hordein gene expression strongly supports this idea. Two additional observations suggest that BPBF is likely to be the *in vivo* P-box regulatory factor from barley endosperm nuclei. *In vitro* experiments showed that the core sequence AAAG within the P-box motif was critical for specific recognition by the recombinant BPBF protein as well as by the nuclear P-box binding activity present in barley and wheat endosperm nuclei. *In vivo*, the mutated P-box form (AgAc) of the *Hor2-184* promoter (pBhor*) was not transactivated in microparticle-bombarded endosperm, either by the transiently overexpressed BPBF protein or by the endogenous P-box *trans*-acting factor, further supporting their functional similarity.

In addition to the P-box element, most prolamin promoters of the *Pooideae* grasses contain a second cis-acting sequence, the GLM. Both P-box and GLM are conserved in barley B- and C-hordeins (Brandt *et al.*, 1985; Forde *et al.*, 1985; Entwistle *et al.*, 1991), in the wheat α/β gliadins and LMW-glutenins (Sumner-Smith *et al.*, 1985; Colot *et al.*, 1987), and in the rye ω-secalins (Hull *et al.*, 1991). In barley endosperm, a positive interaction between these two elements appears necessary for high expression levels driven by C-hordein promoters (Müller and Knudsen, 1993). In a previous study from our laboratory, we have shown that an intact P-box site is essential for transactivation through the GLM site, by the barley bZIP proteins BLZ1 and BLZ2 (Vicente-Carbajosa *et al.*, 1998; Oñate *et al.*, 1999). Interestingly, the maize PBF interacts *in vitro* with the O2 bZIP protein (Vicente-Carbajosa *et al.*, 1997).

BPBF is one of the few DOF proteins as yet shown to mediate transcriptional activation, and this was found to occur in the context of a native promoter of a likely target gene in the homologous tissue where it is naturally expressed. Other maize DOF proteins, Dof1 and Dof2 (Yanagisawa and Izui, 1993; Yanagisawa, 1995), were reported to have transcriptional activity in transfected leaf protoplasts: Dof1 acting as a transcriptional activator, while

Dof2 repressed the activation mediated by Dof1 in the light response of the C4PEPC gene promoter (Yanagisawa and Sheen, 1998). Putative target promoters have been suggested for other previously characterized DOF proteins. The OBP1 *Arabidopsis* protein recognizes the 5′CTTT3′ core (5′AAAG3′ in the antiparallel DNA strand) and is known to interact with the bZIP factors OBFs involved in the expression control of the glutathione-S-transferase-6 gene (Zhang *et al.*, 1995; Chen *et al.*, 1996). The tobacco NtBBF1 protein also recognizes the 5′TAAAGT3′ sequence in the promoter of the *rolB* gene of *Agrobacterium rhizogenes* within a region that regulates its expression in meristems (De Paolis *et al.*, 1996). So it appears that the DOF proteins characterized so far regulate plant genes through recognition of the same core sequence in promoters (5′(A/T)AAAG3′) and several of them do so through interactions with bZIP transcription factors.

The strong conservation of PBF-like factors isolated from wheat, barley and maize supports the idea of a general mechanism in which DOF proteins play an important role as common regulators in the concerted interaction of multiple factors with the *cis*-elements responsible for seed-specific gene expression.

Acknowledgements

This work was supported by grant DGES (PB97-0561) to P.C. from the Ministerio de Educación y Cultura (Spain). We thank Lola Lamoneda for technical assistance. Luis Oñate and Pilar Lara were the recipients of PhD fellowships (FPI) from Ministerio de Educación y Cultura.

References

Albani, D., Hammond-Kosack, M.C.U., Smith, C., Conlan, S., Colot, V., Holdsworth, M. and Bevan, M. (1997) The wheat transcriptional activator SPA: a seed-specific bZIP protein that recognizes the GCN4-like motif in the bifactorial endosperm box of prolamin genes. *Plant Cell* 9, 171–184.

Brandt, A., Montembault, A., Cameron-Mills, V. and Rasmusen, S.K. (1985) Primary structure of a B1 hordein gene from barley. *Carlsberg Research Communications* 50, 333–345.

Chen, W., Chao, G. and Singh, K.S. (1996) The promoter of a H_2O_2-inducible Arabidopsis glutathione S-transferase gene contains closely linked OBF- and OBP1-binding sites. *Plant Journal* 10, 955–966.

Colot, V., Robert, L.S., Kavanagh, T.A., Bevan, M.W. and Thompson, R.D. (1987) Localization of sequences in wheat endosperm protein genes which confer tissue-specific expression in tobacco. *EMBO Journal* 6, 3559–3564.

De Paolis, A., Sabatini, S., De Pascalis, L., Costantino, P. and Capone, I. (1996) A *rolB* regulatory factor belongs to a new class of single zinc finger plant proteins. *Plant Journal* 10, 215–223.

Diaz, I., Royo, J., Sanchez de la Hoz, P. and Carbonero, P. (1995) Gene specificity is maintained in transient expression assays with protoplasts derived from different tissues of barley. *Euphytica* 85, 203–207.

Entwistle, J., Knudsen, S., Müller, M. and Cameron-Mills, V. (1991) Amber codon suppression: the *in vivo* and *in vitro* analysis of two C-hordein genes from barley. *Plant Molecular Biology* 17, 1217–1231.

Forde, B.G., Heyworth, A., Pywell, J. and Kreis, M. (1985) Nucleotide sequence of a B1-hordein gene and the identification of possible upstream regulatory elements in endosperm storage protein genes from barley, wheat and maize. *Nucleic Acids Research* 13, 7327–7339.

Guerin, J. and Carbonero, P. (1997) The spatial distribution of sucrose synthase isozymes in barley. *Plant Physiology* 114, 55–62.

Hartings, H., Maddaloni, M., Lazzaroni, N., Di Fonzo, N., Motto, M., Salamini, F. and Thompson, R.D. (1989) The *O2* gene which regulates zein deposition in maize endosperm encodes a protein with structural homologies to transcriptional activators. *EMBO Journal* 8, 2795–2801.

Hill, D.E., Hope, I.A., Macke, J.P. and Struhl, K. (1986) Saturation mutagenesis of the yeast *his3* regulatory site: requirements for transcriptional induction and for binding by GCN4 activator protein. *Science* 234, 451–457.

Hull, G.A., Halford, N.G., Kreis, M. and Shewry, P.R. (1991) Isolation and characterization of genes encoding rye prolamins containing a highly repetitive sequence motif. *Plant Molecular Biology* 17, 1111–1115.

Hunter, T. and Karin, M. (1992) The regulation of transcription by phosphorylation. *Cell* 70, 375–387.

Katagiri, F. and Chua, N. (1992) Plant transcriptional factors: present knowledge and future challenges. *Trends in Genetics* 8, 22–26.

Kreis, M. and Shewry, P.R. (1992) The control of protein synthesis in developing barley seeds. In: Shewry, P.R. (ed.) *Barley: Genetics, Biochemistry, Molecular Biology and Biotechnology.* CAB International, Wallingford, UK, pp. 319–333.

Kreis, M., Forde, B.G., Rahman, S., Miflin, B.J. and Shewry, P.R. (1985) Molecular evolution of the seed storage proteins of barley, rye and wheat. *Journal of Molecular Biology* 183, 499–502.

Lohmer, S., Maddaloni, M., Motto, M., Salamini, F. and Thompson, R.D. (1993) Translation of the mRNA of the maize transcriptional activator Opaque–2 is inhibited by upstream open reading frames present in the leader sequence. *Plant Cell* 5, 65–73.

Mena, M., Vicente-Carbajosa, J., Schmidt, R.J. and Carbonero, P. (1998) An endosperm-specific DOF protein from barley, highly conserved in wheat, binds to and activates transcription from the prolamin-box of a native B-hordein promoter in barley endosperm. *Plant Journal* 16, 53–62.

Motto, M., Di Fonzo, N., Hartings, H., Maddaloni, M., Salamini, F., Soave, C. and Thompson, R.D. (1989) Regulatory genes affecting maize storage protein synthesis. In: Miflin, B. (ed.) *Oxford Surveys of Plant Molecular and Cell Biology,* Vol. 6, pp. 87–114.

Müller, M. and Knudsen, S. (1993). The nitrogen response of a barley C-hordein promoter is controlled by positive and negative regulation of the GCN4 and endosperm box. *Plant Journal* 6, 343–355.

Müller, M., Muth, J.R., Gallusci, P., Knudsen, S., Maddalomi, M., Mottom, M., Schmitz, D., Sørensen, M.B., Salamini, F., Wettstein, D. and Thompson, R.D. (1995) Regulation of storage protein synthesis in cereal seeds: developmental and nutritional aspects. *Journal of Plant Physiology* 145, 606–613.

Oñate, L., Vicente-Carbajosa, J., Lara, P., Diaz, I. and Carbonero, P. (1999) Barley BLZ2: a seed-specific bZIP protein that interacts with BLZ1 *in vivo* and activates transcription from the GCN4-like motif of B-hordein promoters in barley endosperm. *Journal of Biological Chemistry* 274, 9175–9182.

Pysh, L.O. and Schmidt, R.J. (1996) Characterization of the maize OHP1 gene: evidence of gene variability among inbreds. *Gene* 177, 203–208.

Rodriguez-Palenzuela, P., Royo, J., Gomez, L., Sanchez-Monge, R., Salcedo, G., Molina-Cano, J.L., García-Olmedo, F. and Carbonero, P. (1989) The gene for trypsin inhibitor CMe is regulatory *in trans* of the *lys3a* locus in the endosperm of barley (*Hordeum vulgare* L.). *Molecular and General Genetics* 219, 474–479.

Royo, J., Diaz, I., Rodriguez–Palenzuela, P. and Carbonero, P. (1996) Isolation and promoter characterization of barley gene *Itr1* encoding trypsin inhibitor BTI–CMe: differential activity in wild type and mutant *lys3a* endosperm. *Plant Molecular Biology* 31, 1051–1059.

Schmidt, R.J., Burr, F.A., Aukerman, M.J. and Burr, B. (1990) Maize regulatory gene opaque–2 encodes a protein with a 'leucine zipper' motif that binds to zein DNA. *Proceedings of the National Academy of Sciences, USA* 87, 46–50.

Schmidt, R.J., Pysh, L.D., Ketudat, M., Parsons, R.L. and Hoshek, G. (1994) BZIP proteins regulating gene expression in maize endosperm. In: Coruzzi, G. and Puigdomenech, P. (eds) *Plant Molecular Biology*. NATO ASI Series, Vol. H 81. Springer-Verlag, Berlin, pp. 310–322.

Shewry, P.R., Napier, J.A. and Tatham, A.S. (1995) Seed storage protein: structures and biosynthesis. *Plant Cell* 7, 945–956.

Shimofurutani, N., Kisu, Y., Suzuki, M. and Esaka, M. (1998) Functional analysis of the Dof domain, a zinc finger DNA-binding domain, in a pumpkin DNA-binding protein AOBP. *FEBS Letters* 430, 251–256.

Sørensen, M.B., Cameron-Mills, V. and Brandt, A. (1989) Transcriptional and post-transcriptional regulation of gene expression in developing barley endosperm. *Molecular and General Genetics* 217, 195–201.

Sumner-Smith, M., Rafalski, J.A., Sugiyama, T., Stoll, M. and Söll, D. (1985) Conservation and variability of wheat α/β-gliadin genes. *Nucleic Acids Research* 13, 2905–3916.

Varagona, M.G., Schmidt, R.J. and Raikhel, N.V. (1992) Nuclear localization signal (s) required for nuclear targeting of the maize regulatory protein OPAQUE–2. *Plant Cell* 4, 1213–1227.

Vicente-Carbajosa, J., Beritashvili, D.R., Kraev, A.S. and Skryabin, K.G. (1992) Conserved structure and organization of the B hordein genes in the Hor2 locus of barley. *Plant Molecular Biology* 18, 453–458.

Vicente-Carbajosa, J., Moose, S.P., Parsons, R.L. and Schmidt, R.J. (1997) A maize zinc-finger protein binds the prolamin box in zein gene promoters and interacts with the basic leucine zipper transcriptional activator Opaque2. *Proceedings of the National Academy of Sciences, USA* 94, 7685–7690.

Vicente-Carbajosa, J., Oñate, L., Lara, P., Diaz, I. and Carbonero, P. (1998) Barley BLZ1: a bZIP transcriptional activator that interacts with endosperm-specific gene promoters. *Plant Journal* 13, 629–640.

Yanagisawa, S. (1995) A novel DNA-binding domain that may from a single zinc finger motif. *Nucleic Acids Research* 23, 3403–3410.

Yanagisawa, S. (1996) Dof DNA-binding proteins contain a novel zinc finger motif. *Trends in Plant Sciences* 1, 213–214.

Yanagisawa, S. (1997) Dof DNA-binding domains of plant transcription factors contribute to multiple protein–protein interactions. *European Journal of Biochemistry* 250, 403–410.

Yanagisawa, S. and Izui, K. (1993) Molecular cloning of two DNA-binding proteins of maize that are structurally different but interact with the same sequence motif. *Journal of Biological Chemistry* 268, 16028–16036.

Yanagisawa, S. and Sheen, J. (1998) Involvement of maize Dof zinc finger proteins in tissue-specific and light-regulated gene expression. *Plant Cell* 10, 75–89.
Zhang, B., Chen, W., Foley, R.C., Buttner, M. and Singh, K.B. (1995) Interactions between distinct types of DNA binding proteins enhance binding to ocs element promoter sequences. *Plant Cell* 7, 2241–2252.

4 Impact of Amphiphile Partitioning on Desiccation Tolerance

F.A. HOEKSTRA[1] AND E.A. GOLOVINA[1,2]

[1]*Laboratory of Plant Physiology, Wageningen University, Arboretumlaan 4, 6703 BD Wageningen, The Netherlands;* [2]*Timiryazev Institute of Plant Physiology, Academy of Sciences, Botanicheskaya 35, Moscow, Russia*

Sugars are reputed to protect membranes in dehydrated desiccation-tolerant organisms, such as seeds and pollens. They interact with the polar headgroups of the membrane phospholipids and control the gel-to-liquid crystalline transition temperature (T_m). Because the amount of sugar may be insufficient for full interaction in some organisms, another mechanism of membrane protection was sought. A mechanism is proposed that is based on the partitioning of amphiphilic compounds into membranes depending on the water available. This mechanism was tested with an amphiphilic nitroxide spin probe, using EPR spectroscopy. It was found that, apart from the spin probe, endogenous amphiphiles may also partition into membranes during dehydration. The amphiphiles reduce the dehydration-induced increase of T_m and cause fluidization. The advantages and disadvantages of such a mechanism are discussed. The proposed mechanism is extremely effective at automatically inserting amphiphilic antioxidants into membranes with dehydration, which could promote desiccation tolerance and extend storage longevity.

Introduction

Dehydrated, desiccation-sensitive organisms leak almost all of their cytoplasmic solutes when they are rehydrated. This points to the inability of their plasma membranes to cope with dehydration. Apparently, some irreversible damage has occurred in these membranes.

Under normal conditions of sufficient hydration, the hydrophobic effect of water forces phospholipid molecules into the bilayer structure, with the headgroups directed outward to the aqueous phase and the acyl chains inward (Tanford, 1978). On dehydration, the bilayer structure may be lost as a result of the disappearance of this hydrophobic effect. In the case of liposomes made of phospholipids, all the entrapped solutes are released to the surrounding water upon rehydration. In desiccation-tolerant organisms (anhydrobiotes), membranes remain intact on dehydration and subsequent rehydration. Apparently, there is a mechanism in such organisms which protects membrane structure.

Role of Sugars in the Protection of Dehydrated Membranes

The ample occurrence of non-reducing sugars in anhydrobiotes has been linked with the ability of membranes to retain the bilayer structure (Crowe *et al.*, 1984). It has been demonstrated that in the presence of trehalose or sucrose, liposomes are able to maintain their structure and size, and to retain trapped solutes inside upon rehydration (Crowe *et al.*, 1986). Hydrogen bonding of the sugar OH-groups with the phosphate of the polar headgroups has been proposed as a likely mechanism for this protection (Crowe *et al.*, 1987), which has been confirmed by different methods (reviewed in Crowe *et al.*, 1997). Due to this hydrogen bonding with the sugars, the phospholipid molecules in the membranes remain spaced. As a result, there is less opportunity for Van der Waals interactions between the acyl chains, and the dehydration-induced increase of T_m of as much as 70°C does not occur. Membranes thus remain in the liquid-crystalline phase (Crowe *et al.*, 1987). Sugars can therefore be considered to replace water with regard to hydrogen bonding capability. T_m of dry membranes in the presence of sugars may be even lower than that of hydrated membranes, because sugars may space the headgroups more than water does.

This mechanism was hypothesized also to work *in vivo* to retain membrane structure and to prevent dehydration-induced phase transitions (see Crowe *et al.*, 1992 for a review). Supportive data for this hypothesis have been presented for *Typha latifolia* pollen using *in situ* Fourier transform IR spectroscopy (FTIR) (Hoekstra *et al.*, 1991). Whereas T_m of dry isolated membranes was close to 60°C, the membranes inside ultra-dry pollen had T_m values of approximately 30°C. This reduction of the *in situ* T_m has been attributed to the ample sucrose present in this pollen (23% of the dry weight). However, T_m in dry pollen was not reduced to the level of the hydrated control (–6°C). In dry *E. coli* bacteria, T_m could be reduced by loading the bacteria with trehalose prior to dehydration (Leslie *et al.*, 1995).

Inspection of a number of dehydrated anhydrobiotes with FTIR has revealed that the increase in T_m with drying is generally modest (Hoekstra *et al.*, 1997; Linders *et al.*, 1997). Therefore, the occurrence in the cytoplasm of substances protecting the dry membranes must be common. The question is whether exclusively sugars play such a role.

Air-drying of liposomes with full retention of the entrapped solutes requires about five times more sugar than phospholipids present (Hoekstra *et al.*, 1997). Although this may be the case for the above-mentioned pollen, some other anhydrobiotes may have insufficient amounts of sugars to fulfil this requirement. Moreover, the possibility for hydrogen bonding with the polar headgroups depends on the sugar's coincidental proximity to the membrane. Compounds other than sugars therefore have been suggested also to interact with membranes. It was hypothesized that certain endogenous amphiphilic compounds could accomplish this by migrating from the aqueous cytoplasm into the lipid phase as dehydration progresses (Golovina *et al.*, 1998; Hoekstra *et al.*, 1999). This seems an attractive mechanism, because it proceeds spontaneously with the membranes being the target of the amphiphiles. In addition, the amounts of amphiphilic substances required to perturb the membranes can be relatively small.

Partitioning of Amphiphiles with Dehydration

To explore the possibility of amphiphilic compounds partitioning into the lipid phase with dehydration, we have infiltrated pollen with the amphiphilic nitroxide spin probe, TEMPONE (4-oxo-2,2,6,6-tetramethyl-1-piperidinyloxy), and analysed the properties of the direct environment of the spin probe with electron paramagnetic resonance (EPR) spectroscopy (Golovina *et al.*, 1998). The contributions of the TEMPONE signals from the aqueous phase and the lipid phase to the spectrum are well resolved in the high field region (Fig. 4.1, right side of spectrum a). The aqueous contribution of the spin probe to the spectrum disappeared upon dehydration of the pollen and the lipid contribution increased (compare spectra a and b). One can argue that the probe could have been trapped in the dried glassy cytoplasm. However, a similarly shaped spectrum as shown in Fig. 4.1 (spectrum c), representing TEMPONE trapped in a sucrose glass, would then have been expected, which is not the case in spectrum b. The peaks seen in the spectrum of TEMPONE in dried pollen are those of superimposed spectra, one from the mobile lipid environment and another from an immobile environment (outer extrema in the spectrum), the probable origin of which is discussed later.

A similar disappearance of the aqueous signal and increase of the lipid signal can be observed in dehydrating seeds (Fig. 4.2). On rehydration, the lipid signal is reduced and the aqueous signal re-appeared, which indicates that the partitioning is reversible.

Quantification of TEMPONE in the Different Phases during Dehydration and Rehydration

Using special subtraction techniques to decompose signals of complex spectra to their respective contributions, the relative integral intensity of the different contributions were calculated. Or, in other words, the relative amount of TEMPONE in the aqueous, lipid and immobile environments during dehydration and rehydration were calculated.

Figure 4.3 shows the relative distribution of TEMPONE over these environments during dehydration of *T. latifolia* pollen. The changes in relative distribution of the spin probe with drying reflect the dynamics of partitioning. Rehydration in humid air caused the spin probe to repartition according to a pattern that was essentially similar to that seen in Fig. 4.3 for the partitioning

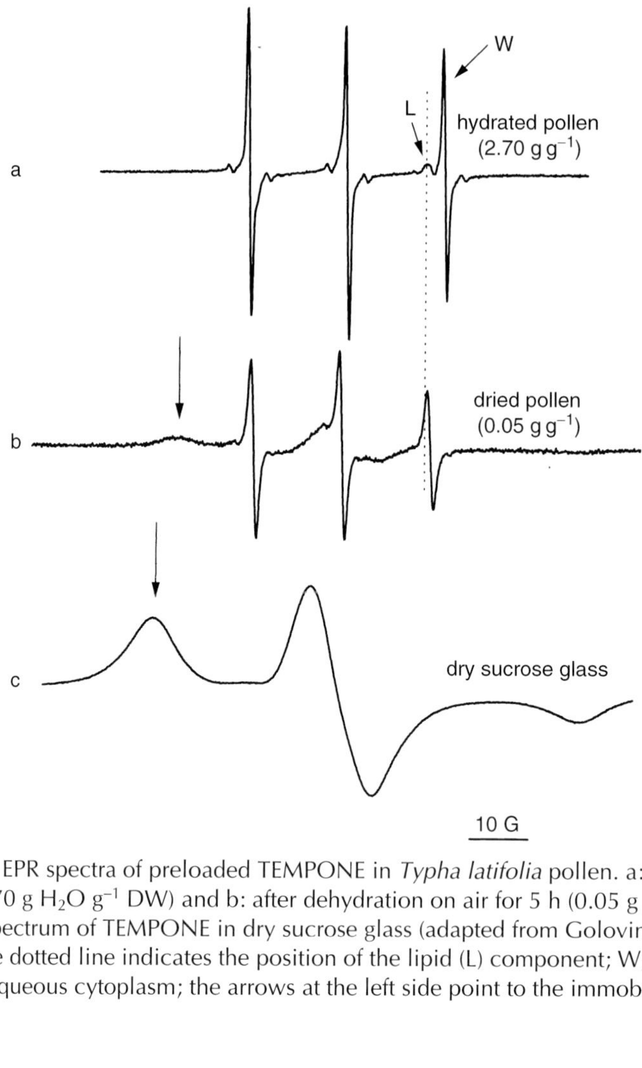

Fig. 4.1. EPR spectra of preloaded TEMPONE in *Typha latifolia* pollen. a: Hydrated pollen (2.70 g H_2O g^{-1} DW) and b: after dehydration on air for 5 h (0.05 g H_2O g^{-1} DW); c: spectrum of TEMPONE in dry sucrose glass (adapted from Golovina *et al.*, 1998). The dotted line indicates the position of the lipid (L) component; W, the signal from the aqueous cytoplasm; the arrows at the left side point to the immobile spectrum.

(data not shown). It has been proposed that the changing size of the water pool versus the constant size of the lipid pool during dehydration and rehydration is the driving force of the partitioning and repartitioning (Golovina *et al.*, 1998).

Three stages during the partition process can be distinguished in Fig. 4.3. First, the loss of water from 3 to 1 g H_2O g^{-1} DW is characterized by slow and equal increases in both the amounts of spin probe in the lipid and immobile environments, together with a gradual decrease of the amount in the aqueous environment. Between 1 and 0.3 g H_2O g^{-1} DW, the redistribution of the spin probe accelerated. Below 0.3 g H_2O g^{-1} DW, there was no water component left at all. A steep further increase of the immobile component occurred at the expense of the lipid component, which rapidly decreased. The question is whether the phenomena below 0.3 g H_2O g^{-1} DW are attributable to

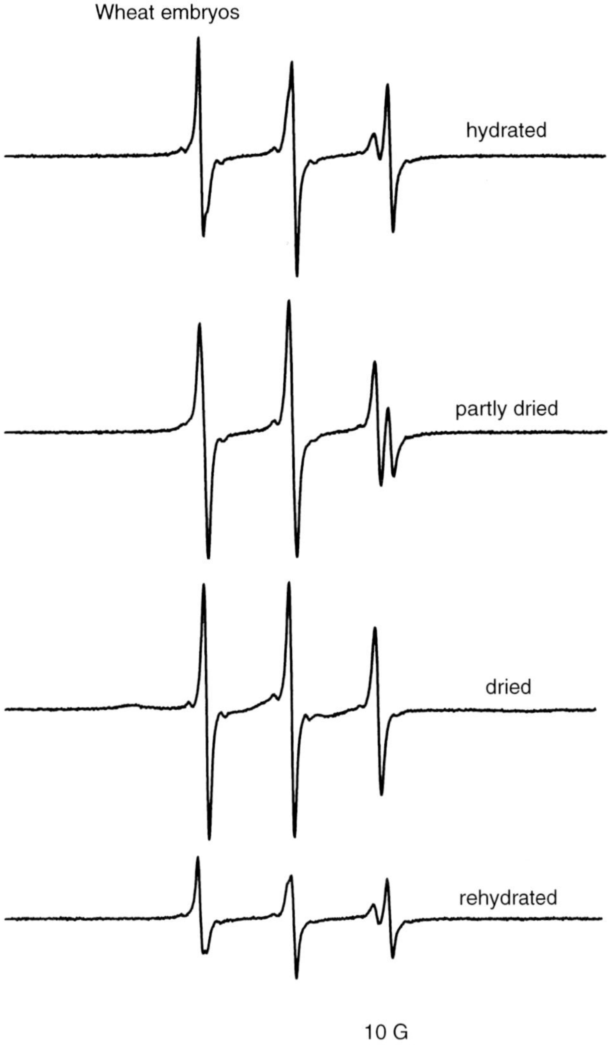

Fig. 4.2. EPR spectra of preloaded TEMPONE in dehydrating wheat embryos. a: After 4 h of hydration; b: after 45 min of air-drying of the 4 h hydrated specimen; c: after 5 h of air-drying of the 4 h hydrated specimen; d: after rehydration of sample c in minute amounts of water (containing 120 mM ferricyanide).

partitioning or immobilization. It is proposed that the proportional increase of the immobile component and decrease of the lipid component are the result of immobilization of TEMPONE in membranes, for the following reasons. The polarity of the surroundings of the spin probe determines the hyperfine splitting constant (distance between peaks). In the case of the spin probe in the dry sucrose glass, the hyperfine splitting was 71 G (distance between outer extrema), whereas it was 66–67 G in the case of the peaks in the immobile spectrum of dried pollen. This suggests the immobilized TEMPONE molecules

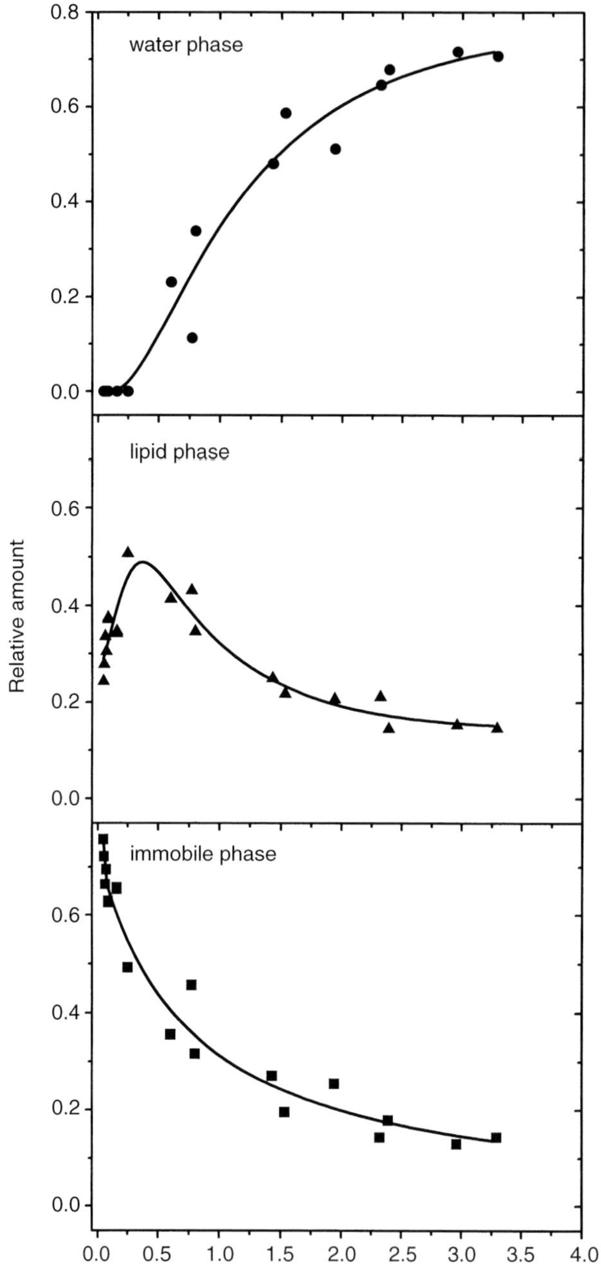

Fig. 4.3. Relative integral intensities (relative amounts) of the preloaded spin label, TEMPONE, in the aqueous, lipid and immobile environments during dehydration of *Typha latifolia* pollen. The integral intensity was calculated after decomposition (by subtraction techniques) of complex spectra to their respective contributions.

in the pollen to sense an environment that was considerably more apolar than those in the sucrose glass, most likely lipid environment. If the probe were to be located in the oil bodies, one would not expect immobilization on further drying at room temperature, because T_m of oil is insensitive to the presence or absence of water. The immobile spectra perfectly fitted those of doxylstearates in the pollen (data not shown), which are specific membrane probes. Also, it is not to be expected that TEMPONE molecules would repartition from the lipid phase back into the cytoplasm, because the viscosity of the cytoplasm is extremely high at moisture contents of ≤ 0.3 g H_2O g^{-1} DW (Leprince and Hoekstra, 1998). The above arguments contribute to the concept that TEMPONE molecules partitioned mainly into the membranes upon dehydration.

Occurrence of Endogenous Amphiphilic Compounds in Pollen

According to the principle described above, any amphiphile would partition into the lipid phase when the bulk water is removed. Depending on its polarity, it will end up in membranes or oil, or both. When amphiphiles are inserted in membranes, they may increase disorder and permeability (Casal *et al.*, 1987).

It has been shown that *T. latifolia* pollen extracts contain substances that particularly interact with dehydrated membranes, but fail to do so with hydrated membranes (Golovina *et al.*, 1998). This has been studied by assaying the permeability of egg phosphatidylcholine liposomes that were air-dried in the presence of these extracts. Table 4.1 shows that the entrapped dye, carboxyfluorescein (CF), is lost from the rehydrated liposomes to a considerable extent in the presence, but not in the absence of the extracts. This also held true for added TEMPONE. When the liposomes were kept hydrated over the entire time during which the other samples were dehydrated (3 h), the losses of CF were minimal. Sucrose was always present in the liposome samples to prevent fusion and dehydration-induced increases of T_m. It is clear that compounds in the extracts cause disturbance of bilayers to such an extent

Table 4.1. Loss of carboxyfluorescein from large unilamellar egg phosphatidylcholine (egg PC) vesicles as influenced by TEMPONE and amphiphilic compounds extracted from *Typha latifolia* pollen. The vesicles (90 nm) were always in a sucrose solution (mass ratio sucrose/egg PC = 5:1) and were air-dried + rehydrated, or kept hydrated.

	Loss of carboxyfluorescein (%)	
Addition	Vesicles kept hydrated	Vesicles dried + rehydrated
No addition	1	14
TEMPONE	1	41
Extract (apolar solvent extraction)	9	61
Extract (aqueous extraction)	20	65

during dehydration, that the CF could escape from the interior of the vesicles upon rehydration. The high permeability of the rehydrating vesicles was restored to the original low value within 5 min of rehydration, which indicates that the vesicle structure remained intact during partitioning (Golovina *et al.*, 1998).

Can the Transient Leakage During Imbibition of Anhydrobiotes be Explained by Partitioning?

If insertion of endogenous amphiphiles in the plasma membranes causes increased permeability, then this may explain the transient leakage from imbibing anhydrobiotes. During repartitioning from the plasma membrane into the aqueous phase, the leakage rate rapidly decreases (Hoekstra *et al.*, 1999). It was observed that this process takes less than 1 min during rehydration of prehumidified *T. latifolia* pollen (0.3 g H_2O g^{-1} DW) in germination medium. Actually, the plasma membranes were more or less closed already after the first 10 s. This again emphasizes the reversible nature of the phenomenon.

When the pollen was rehydrated in humid air to water contents as high as 1.1 g H_2O g^{-1} DW, the permeability of the plasma membranes at imbibition was almost as low as that for pollen incubated in germination medium for 5 min. This can be derived from Fig. 4.4, showing the negative correlation between plasma membrane permeability at imbibition and initial moisture content of the pollen. The permeability after 5 min in germination medium was also measured. These data show that the high permeability at the onset of imbibition was reversible within 5 min, except for initial moisture contents of less than 0.08 g H_2O g^{-1} DW, where irreversible damage had occurred. This damage may be associated with plasma membranes that were in the gel phase before imbibition (Crowe *et al.*, 1989; Hoekstra *et al.*, 1992) and may be explained by an insufficient flexibility of such membranes upon exposure to bulk water (Hoekstra *et al.*, 1999).

In an attempt to determine whether the permeability level at imbibition is associated with the amount of endogenous amphiphiles in the plasma membranes, the permeability data of Fig. 4.4 were plotted against the amount of TEMPONE that had repartitioned into the aqueous fraction (Fig. 4.5). The curve clearly shows that the more TEMPONE reappeared in the aqueous environment, the less was the imbibitional leakage. This correlation supports the hypothesis that imbibitional leakage depends on the amount of amphiphiles that resides in the plasma membrane at imbibition.

Effect of Flavonoids on Fluidity and T_m of Membranes

It has been shown previously that a flavonoid such as quercetin interacts with dry liposomes in that it depresses T_m and increases fluidity (Hoekstra *et al.*, 1997). The focus has been on quercetin, because this strong antioxidant may

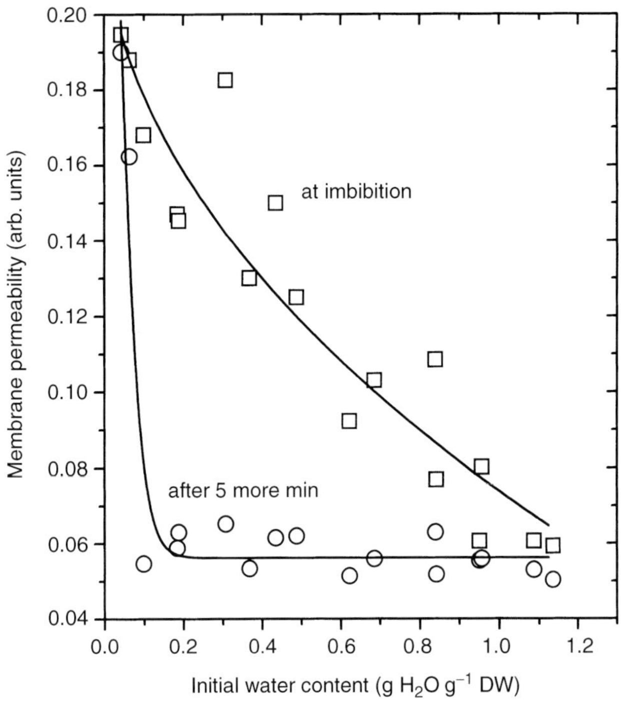

Fig. 4.4. Permeability of plasma membranes of *Typha latifolia* pollen of different initial moisture contents at imbibition and after 5 min of incubation in germination medium. Plasma membrane permeability was assayed by a nitroxide spin probe technique according to Golovina *et al.* (1997). Data from Golovina *et al.* (1998).

be present in pollen up to few percent of the DW (see Golovina *et al.*, 1998, for references). Quercetin is considerably apolar, which causes it to interact with both dry membranes (Hoekstra *et al.*, 1997) and hydrated membranes (Saija *et al.*, 1995). Later, it became clear that the reported occurrence of quercetin in pollen actually concerned rutin, a glycosylated form of quercetin (quercetin-3-rutinoside). During the analytical procedures, rutin apparently had been hydrolysed to give the aglycon quercetin. The amphiphilic rutin does not seem to interact with hydrated membranes, in contrast to quercetin, which does (Saija *et al.*, 1995). Also rutin has strong antioxidant properties.

The effect of rutin on palmitoyl-oleoyl phosphatidylcholine (POPC) vesicles during drying was studied using FTIR. The position of the band around 2852 cm^{-1}, which reflects the vibrational freedom of the CH_2 symmetric stretch, can give information about the packing density of the acyl chains of POPC. Figure 4.6 shows the shifts in band position with temperature for hydrated and dehydrated POPC vesicles, either in the presence or in the absence of rutin. A phase transition from gel to liquid crystalline is characterized by a discrete upward shift of approximately 2–4 wavenumbers. The middle of the melting transition is considered as T_m. Hydrated POPC vesicles had a T_m around 0°C. Adding rutin did not lead to a change of this value, and the band positions

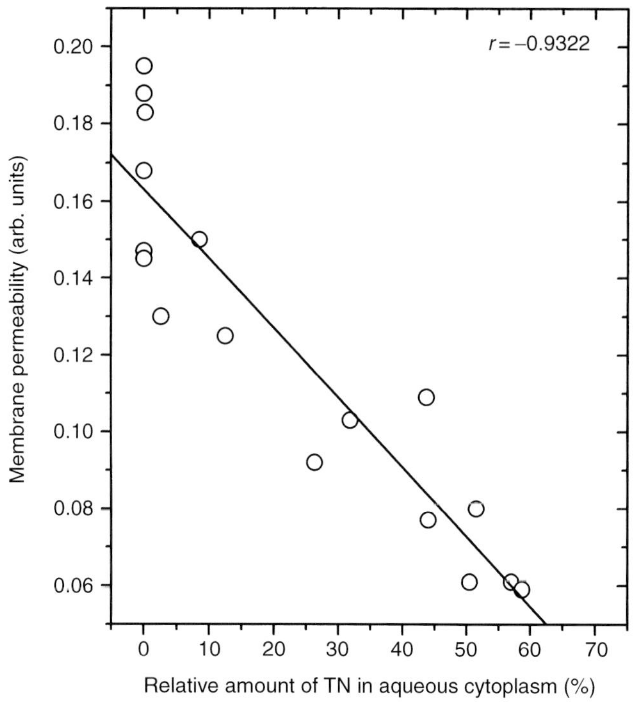

Fig. 4.5. Plot of the plasma membrane permeability at imbibition (data from Fig. 4.4) *versus* the relative amount of TEMPONE in the aqueous phase in *Typha latifolia* pollen just before imbibition. The relative amount of the spin probe in the different environments was calculated similarly as for Fig. 4.3.

were very similar, confirming that rutin does not interact with hydrated membranes. Dehydrated vesicles had an increased T_m of 48°C and low wavenumber positions in the gel phase, which is indicative of a generally higher packing density (less fluid) of the acyl chains below T_m. However, after dehydration in the presence of rutin, T_m was depressed. The vesicles had the main transition around 16°C and a minor transition around −10°C. This inhomogeneous melting behaviour might be explained by different types of interaction of rutin with the membrane. The sugar moiety of rutin may interact with the polar headgroups of the dry POPC to give the lowest T_m (comparable with the depression evoked by sugars). The major transition at approximately 16°C may be due to insertion of the apolar part of rutin at the surface of the membrane. The mole ratio of rutin to POPC (1:3) was of the same order of magnitude as that expected for intact pollen. This ratio would be by far insufficient to completely depress T_m to −30°C as usually observed with sugars at a mole ratio of 10:1 (sugar:POPC). The elevated wavenumber positions below T_m found in the dry, rutin-treated vesicles as compared to those of the dry control without any addition (difference of 0.8 cm^{-1}), may point to disturbance of the acyl chain packing evoked by rutin. This is interpreted as fluidization of the dry bilayer caused by partitioning.

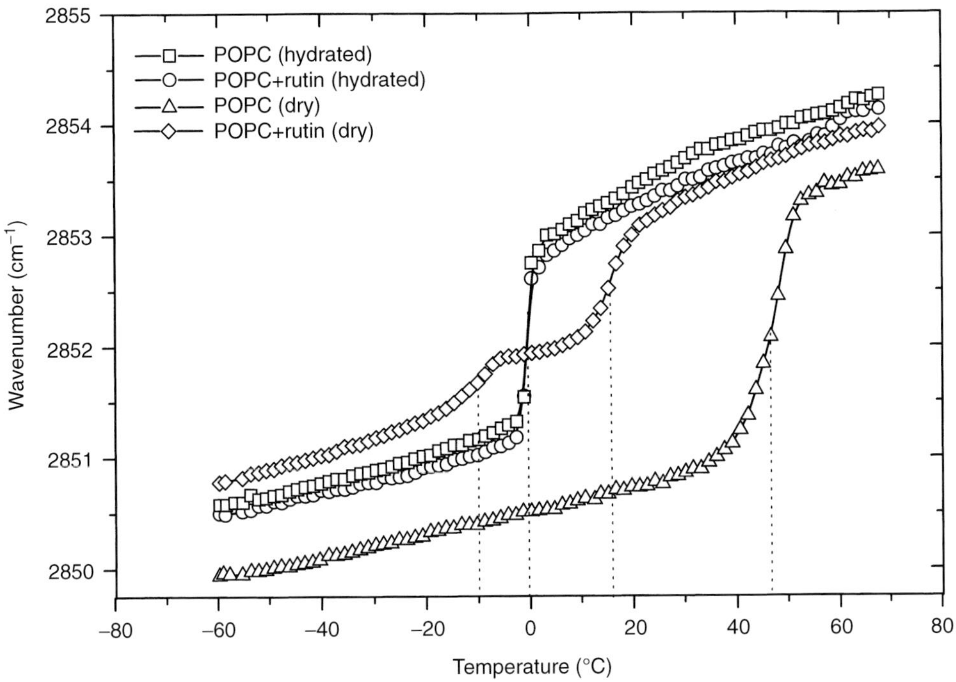

Fig. 4.6. Effect of rutin on the temperature dependence of the CH_2 symmetric stretching vibration band of palmitoyl-oleoylphosphatidylcholine (POPC) vesicles. Wavenumber versus temperature plots of hydrated and dehydrated POPC vesicles either in the presence or in the absence of rutin. FTIR spectra were recorded every min at temperature increments of 1.5°C min^{-1}.

Importance of Amphiphile Partitioning into Membranes for Desiccation Tolerance

The beneficial effect of endogenous amphiphiles in anhydrobiotes may include the ability to depress T_m and to increase fluidity of dry membranes and, thus, to prevent the formation of a gel phase. A number of amphiphilic flavonoids have strong antioxidant activity. A partitioning behaviour as described above would be most effective to give automatic antioxidant protection to membranes in dehydrating organisms. This may contribute to desiccation tolerance and extend storage longevity. That these membrane-fluidizing compounds also increase permeability may be an inevitable disadvantage. Too early an insertion into membranes during dehydration may have harmful effects, such as uncoupling of respiration due to collapse of the mitochondrial proton gradient, and may be the cause of desiccation sensitivity. Because of these dual effects, we suggest that dehydration-induced amphiphile partitioning is tightly controlled in desiccation-tolerant organisms.

Acknowledgements

We thank Mark Alberda for his help with the FTIR work and acknowledge the financial support of the Netherlands Organization for Scientific Research (NWO) to E.A. Golovina.

References

Casal, H.L., Martin, A. and Mantsch, H.H. (1987) Infrared spectroscopic characterization of the interaction of lipid bilayers with phenol, salicylic acid and o-acetylsalicylic acid. *Chemistry and Physics of Lipids* 43, 47–53.

Crowe, J.H., Crowe, L.M. and Chapman, D. (1984) Preservation of membranes in anhydrobiotic organisms: the role of trehalose. *Science* 223, 701–703.

Crowe, J.H., Crowe, L.M., Carpenter, J.F. and Aurell Wistrom, C. (1987) Stabilization of dry phospholipid bilayers and proteins by sugars. *Biochemical Journal* 242, 1–10.

Crowe, J.H., Hoekstra, F.A. and Crowe, L.M. (1989) Membrane phase transitions are responsible for imbibitional damage in dry pollen. *Proceedings of the National Academy of Sciences of the United States of America* 86, 520–523.

Crowe, J.H., Hoekstra, F.A. and Crowe, L.M. (1992) Anhydrobiosis. *Annual Review of Physiology* 54, 570–599.

Crowe, J.H., Crowe, L.M., Carpenter, J.F., Prestrelski, S.J., Hoekstra, F.A., de Araujo, P.S. and Panek, A.D. (1997) Anhydrobiosis: cellular adaptations to extreme dehydration. In: Dantzler, W.H. (ed.) *Handbook of Physiology, Section 13, Comparative Physiology, Vol. II.* Oxford University Press, Oxford, pp. 1445–1477.

Crowe, L.M., Womersley, C., Crowe, J.H., Reid, D., Appel, L. and Rudolph, A. (1986) Prevention of fusion and leakage in freeze-dried liposomes by carbohydrates. *Biochimica et Biophysica Acta* 861, 131–140.

Golovina, E.A., Tikhonov, A.N. and Hoekstra, F.A. (1997) An electron paramagnetic resonance spin probe study of membrane permeability changes with seed aging. *Plant Physiology* 114, 383–389.

Golovina, E.A., Hoekstra, F.A. and Hemminga, M.A. (1998) Drying increases intracellular partitioning of amphiphilic substances into the lipid phase: impact on membrane permeability and significance for desiccation tolerance. *Plant Physiology* 118, 975–986.

Hoekstra, F.A., Crowe, J.H. and Crowe, L.M. (1991) Effect of sucrose on phase behaviour of membranes in intact pollen of *Typha latifolia* L., as measured with Fourier transform infrared spectroscopy. *Plant Physiology* 97, 1073–1079.

Hoekstra, F.A., Crowe, J.H. and Crowe, L.M. (1992) Germination and ion leakage are linked with phase transitions of membrane lipids during imbibition of *Typha latifolia* pollen. *Physiologia Plantarum* 84, 29–34.

Hoekstra, F.A., Wolkers, W.F., Buitink, J., Golovina, E.A., Crowe, J.H. and Crowe, L.M. (1997) Membrane stabilization in the dry state. *Comparative Biochemistry and Physiology* 117A, 335–341.

Hoekstra, F.A., Golovina, E.A., van Aelst, A.C. and Hemminga, M.A. (1999) Imbibitional leakage from anhydrobiotes revisited. *Plant Cell and Environment* 22, 1121–1131.

Leprince, O. and Hoekstra, F.A. (1998) The responses of cytochrome redox state and energy metabolism to dehydration support a role for cytoplasmic viscosity in desiccation tolerance. *Plant Physiology* 118, 1253–1264.

Leslie, S.B., Israeli, E., Lighthart, B., Crowe, J.H. and Crowe, L.M. (1995) Trehalose and sucrose protect both membranes and proteins in intact bacteria during drying. *Applied and Environmental Microbiology* 61, 3592–3597.

Linders, L.J.M., Wolkers, W.F., Hoekstra, F.A. and Van't Riet, K. (1997) Effect of added carbohydrates on membrane phase behavior and survival of dried *Lactobacillus plantarum*. *Cryobiology* 35, 31–40.

Saija, A., Scalese, M., Lanza, M., Marzullo, D., Bonina, F. and Castelli, F. (1995) Flavonoids as antioxidant agents: importance of their interaction with biomembranes. *Free Radicals in Biology and Medicine* 19, 481–486.

Tanford, C. (1978) The hydrophobic effect and the organization of living matter. *Science* 200, 1012–1018.

5 Unravelling the Responses of Metabolism to Dehydration Points to a Role for Cytoplasmic Viscosity in Desiccation Tolerance

O. LEPRINCE[1], F.A. HOEKSTRA[1] AND F.J.M. HARREN[2]

[1]Department of Plant Sciences, Wageningen University, Arboretumlaan 4, 6703 BD Wageningen, The Netherlands; [2]Department of Molecular and Laser Physics, University of Nijmegen, Toernooiveld 1, 6525 ED Nijmegen, The Netherlands

Tolerance of desiccation in seeds is a multifactorial trait in which the synthesis of protective substances and the repression of degradative processes that are induced during dehydration are equally critical. The most documented degradative reaction linked with desiccation sensitivity in seeds is the accumulation of peroxidative damage following drying. Such oxidative stress is likely to originate from reactive O_2 species (ROS) that are generated as a result of uncontrolled respiration and impaired electron flow in mitochondria. Under physiological conditions, the generation of ROS depends on the redox states of the electron-transfer components and the availability of ADP and O_2. However, it is not known whether these metabolic conditions prevail in drying seed tissues. Therefore this study focuses on the significance for energy metabolism of the changes in physical properties of water during drying. The effects of the loss of water on the redox states of cytochromes and energy metabolism during drying of imbibed cowpea cotyledons and pollens of *Typha latifolia* and *Impatiens glandulifera* were assessed non-invasively. Using electron spin resonance spectroscopy, it was found that viscosity is a valuable parameter to characterize the relation between desiccation and decrease in respiration and energy metabolism. During drying, mitochondrial cytochromes are progressively and synchronously reduced in conjunction with a decline in energy charge. The reduction of mitochondrial cytochromes probably originates from a decreased O_2 availability as a result of the increased viscosity and impeded diffusion. The metabolism in desiccation-intolerant tissues shifted partially to ethanolic fermentation during drying, indicating that they experience anoxia during drying. Such metabolic perturbations were absent in desiccation-tolerant tissues. Thus a regulated viscosity increase and controlled O_2 diffusion may be regarded as mechanisms of desiccation tolerance.

Introduction

To be desiccation tolerant, developing embryos rely on two components: the synthesis of protective mechanisms and the capability of evading free radical damage during drying (Leprince *et al.*, 1993; Vertucci and Farrant, 1995). This damage probably results from the formation of reactive O_2 species (ROS) during dehydration. Previous studies on desiccation-intolerant embryos (Leprince *et al.*, 1994, 1995) showed that respiration is involved in free radical processes leading to lipid peroxidation and membrane damage, suggesting that the tight control of ROS production is lost during drying. Several studies on both orthodox and recalcitrant seeds have suggested that a controlled down-regulation of metabolism must occur during drying to avoid overproduction of ROS and free-radical damage (Leprince *et al.*, 1994; 1995; Vertucci and Farrant, 1995). Although this hypothesis is mentioned regularly in the literature (e.g. Vertucci and Farrant, 1995; Berjak and Pammenter, 1997), the mechanisms that control the generation of ROS during dehydration have not yet been investigated. Studies on isolated mitochondria and animals have identified several metabolic conditions that dramatically increase the probability of ROS formation (Cadenas, 1989; Skulachev, 1996): (i) the nature and concentration of mitochondrial electron carriers (ubiquinones, Cyt *b*, flavins and non-haem iron proteins are prone to produce ROS when they are highly reduced); (ii) depletion of the ADP pool by phosphorylation to ATP (state 3–state 4 transition), as in ADP-limiting conditions, Cyt *b* and ubiquinone are maintained reduced for longer periods of time than when ADP is not limiting phosphorylation; (iii) the rate of O_2 consumption by cytochrome oxidase (COX); and (iv) the O_2 availability. This study examines whether dehydration may lead to potential metabolic conditions promoting ROS formation. The relations between desiccation, the redox states of cytochromes, energy charge and respiration were characterized in cowpea cotyledons that retain cellular integrity during drying. Furthermore, the response of alcoholic fermentation to drying was compared in desiccation-tolerant and -intolerant germinating axes of pea.

The significance of the changes in the physical properties of water during drying for metabolism has received little attention. One can envisage that the removal of water will induce an increase in the cytoplasmic viscosity. Viscosity is known to influence both the diffusion of metabolites within the mitochondria and cytoplasmic matrix and the mobility of the electron carriers within the mitochondrial lipid bilayer (Fato *et al.*, 1993). Therefore, the relation between loss of water and rise in viscosity during drying was investigated using electron spin resonance (ESR) spectroscopy.

In situ Analysis of the Redox States of Cytochromes Reveal that Mitochondrial Cytochromes are Reduced in Dried Cowpea Cotyledons

The redox states of cytochromes were studied non-invasively using low temperature UV-visible spectrophotometry as described by Leprince and Hoekstra (1998). The effects of decreasing water content (WC) on the difference spectra of powdered cotyledons are shown in Fig. 5.1A. Difference spectra were obtained by subtracting the absorbance spectrum of a hydrated sample from the spectrum of samples having different WC. Previous experiments

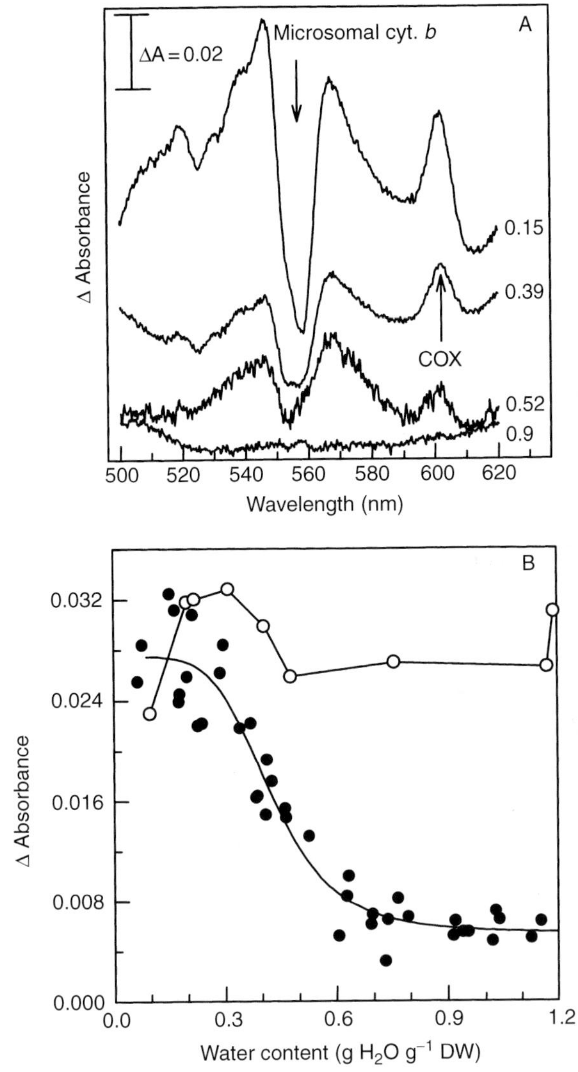

Fig. 5.1. The effect of drying on the redox states of cytochromes in cowpea cotyledons. Before drying, seeds of cowpea were imbibed at 15°C overnight. Before scanning, material was powdered in liquid N_2 and loaded in a pre-cooled spectroscope cuvette equipped with a low temperature attachment (Leprince and Hoekstra, 1998). A. Dried minus hydrated difference spectra from powdered samples at various WC (indicated on the right and expressed as g H_2O g^{-1}DW). The peak and trough is indicated by an upward and downward arrow, respectively. B. The relation between WC and changes in the relative absorbance of the α-band of COX in the absence (●) and presence (○) of 1 mM KCN. Adapted from Leprince and Hoekstra (1998).

demonstrated that the peak at 599–603 nm is attributable to haem *a*–a_3 of the COX and its amplitude is proportional to the reduction level of COX (Leprince and Hoekstra, 1998). The trough between 558 and 560 nm indicates the presence of *b*-type cytochromes, the origin of which is microsomal (Hendry *et al.,* 1981). To study the response of the Cyt redox states to water loss, the absorbance of mitochondrial Cyt in cotyledons was plotted as a function of WC obtained at intervals during drying (Fig. 5.1B). In imbibed cotyledons, the absorbance at 599 nm was low. The *a*–a_3 complex is mostly oxidized in steady-state physiological conditions (Leštan *et al.*, 1993). Therefore, the absorbance values in the hydrated cotyledons correspond to near 100% oxidized levels. During drying, the absorbance at 599–603 nm did not increase substantially until a WC of 0.6 g H_2O g^{-1} DW (g g^{-1}) was reached. Thereafter, it increased markedly until 0.2–0.3 g g^{-1}. Below these WC, the absorbance reached a constant value corresponding to 100% reduction. KCN, an inhibitor blocking electron transfer within the COX, kept COX reduced regardless of WC (Fig. 5.1B), indicating that the increase in absorbance was not due to synthesis of cytochromes during drying. In intact organisms, the only conditions known to induce total reduction in the terminal portion of the cytochromes are anoxia and treatment with inhibitors that block the transfer of electrons to COX.

Effects of O_2 Availability on COX Reduction during Drying

To test whether the COX reduction stemmed from a lack of O_2 availability during drying, hydrated cotyledons of cowpea were dried in a sealed chamber with 100% N_2 or 100% O_2. Under anoxia, only 60% reduction in COX was observed in the hydrated state (Fig. 5.2). At a WC of 0.7 g g^{-1}, the extent of reduction increased to reach 100% at 0.3 g g^{-1}. Drying cotyledons in the presence of 100% O_2 did not affect greatly the pattern of COX reduction (Fig. 5.2). A reduction of approximately 70% was achieved in the dry state, despite the large gradient of the O_2 concentration from outside to inside the tissues. This observation suggests that there is a barrier limiting O_2 availability to COX. This barrier could be due to a decrease in the O_2 diffusion rates inside the cotyledons during drying. Since the rate of molecular diffusion is directly related to viscosity, we investigated the relation between viscosity and the COX redox state during drying.

Changes in Cytoplasmic Viscosity and O_2 Diffusion Coefficient during Drying

To estimate cytoplasmic viscosity, we studied the effects of drying on the rotational motion of 3-carboxyl-proxyl (CP) that was introduced into the cytoplasm of cotyledon cells according to the methods described in Buitink *et al.* (1998) and Leprince and Hoekstra (1998). From the rotational motion of the spin probe, the cytoplasmic viscosity η can be determined using a modified

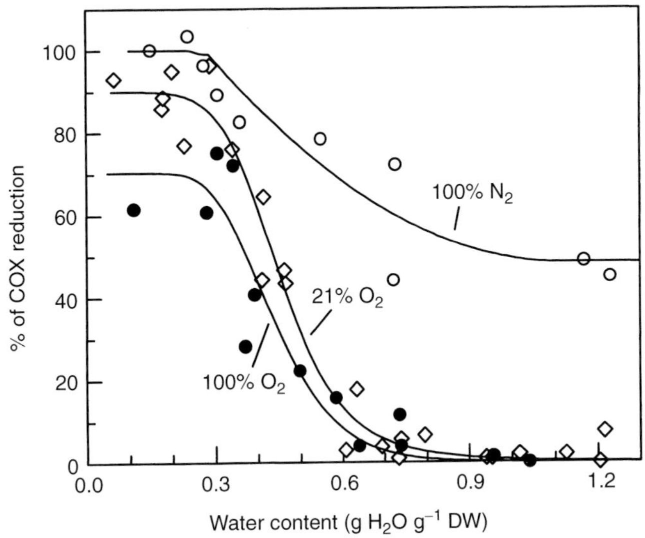

Fig. 5.2. The effect of O_2 concentration on the reduction of COX during drying of cowpea cotyledons. Treatments were 100% N_2 (○), air (◊) and 100% O_2 (●). Data were fitted with 3rd order polynomial regressions or asymmetric functions as an aid to the eye.

Stokes-Einstein equation (Hemminga *et al.*, 1993; Leprince and Hoekstra, 1998). Before drying, the cytoplasmic viscosity of cowpea cotyledons was around 0.15 Poise (Fig. 5.3). No change in viscosity was observed until the cotyledons reached a WC of 0.7 g g^{-1}; thereafter viscosity increased exponentially with further drying. A viscosity of 10 Poise in cotyledons was reached around 0.3 g g^{-1} (Fig. 5.3). At this WC, glass transition temperature (Tg) is approximately at −70°C as determined by differential scanning calorimetry (data not shown). As Tg progressively approaches a value close to 22°C with decreasing WC, viscosity is expected to rise further.

The O_2 diffusion coefficient is inversely proportional to the viscosity of the medium and can be estimated using the following empirical relation between viscosity of aqueous solutions and temperature for a binary mixture (Reid *et al.*, 1987):

$$D_{O_2} = 7.410^{-8}\left[T(2.26\,M_{H_2O})^{1/2}/\eta\,V_{O_2}^{0.6}\right] \tag{5.1}$$

where T is the absolute temperature, M_{H_2O} the MW of water and V_{O_2} the molar volume of O_2. It must be noted that the above relation is only valid for binary systems, and becomes less accurate with increasing viscosities. Therefore, the relation between O_2 diffusion coefficients versus WC is regarded as semi-quantitative. The O_2 diffusion coefficients were around 9.5×10^{-9} m^2 s^{-1} prior to drying and declined exponentially during dehydration below 0.8 g g^{-1} (Fig. 5.3). These observations suggest that O_2 diffusion through the tissues is progressively impeded upon the loss of water and increase in viscosity. This suggestion is confirmed by the plot of O_2 diffusion coefficients versus the

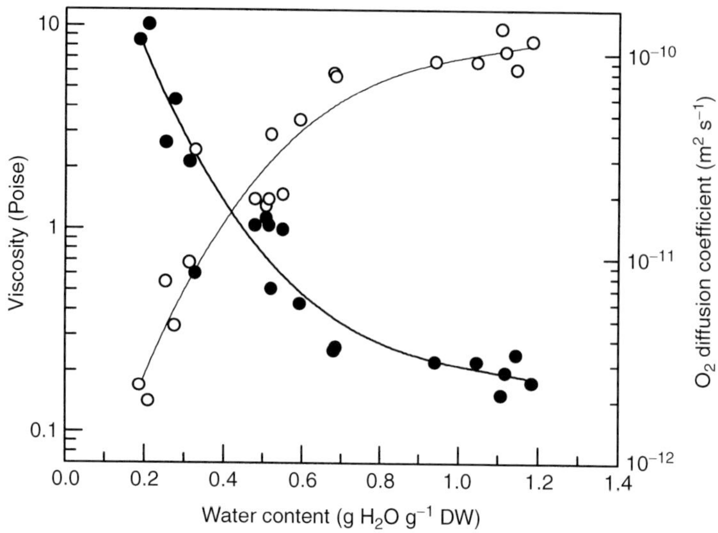

Fig. 5.3. The effects of dehydration on cytoplasmic viscosity (●) and O_2 diffusion coefficient (○) in cowpea cotyledons. Data were fitted with a 3rd order polynomial regression (r^2 = 0.908).

levels of COX reduction which show a significant linear relation (Fig. 5.4). Therefore, we suggest that the O_2 availability plays a critical role in the COX reduction upon drying.

The Response of Fermentation, Respiration and Energy Metabolism to Desiccation

The above results suggest that dehydrating cotyledons experience anoxic conditions during drying. Therefore, it was interesting to know whether the loss of water also induces alcoholic fermentation. For this purpose, we resorted to laser photoacoustics which allowed us to measure, on-line and non-invasively, acetaldehyde and ethanol at a much higher sensitivity than by gas chromatography. A review of photoacoustic techniques and their applications can be found in Harren and Reuss (1997). In desiccation-tolerant pea axes, the amounts of acetaldehyde and ethanol released by the drying tissues declined following a single exponential decay (Fig. 5.5). This was attributed to a passive release of acetaldehyde that was present in the tissues before drying. Desiccation-intolerant pea axes exhibited a different pattern: a peak of both acetaldehyde and ethanol appeared during drying between 2 and 1 g g^{-1} (Fig. 5.5). This production preceded the onset of the desiccation-induced membrane damage as shown by the arrow on Fig. 5.5. These observations indicate that desiccation-intolerant tissues experience anoxia during drying. Such metabolic perturbations are absent in desiccation-tolerant tissues.

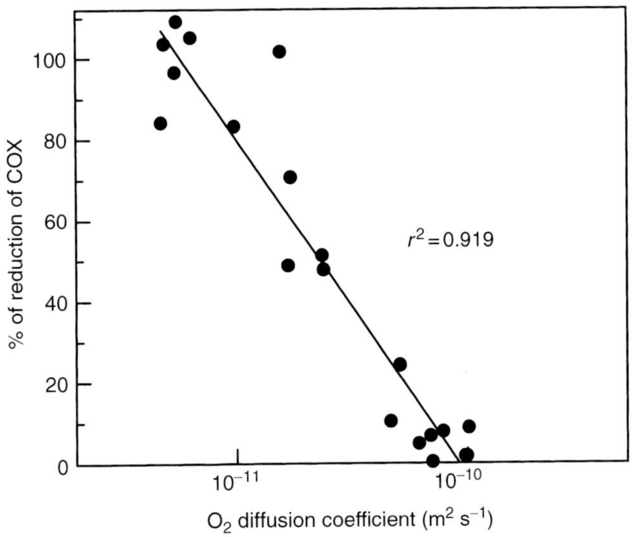

Fig. 5.4. The relation between the reduction levels of COX and O_2 diffusion coefficients that were estimated from the regression in Fig. 5.3.

To examine the relationship between changes in viscosity and energy metabolism, O_2 uptake and CO_2 release rates were measured in the gas phase using gas chromatography (Leprince and Hoekstra, 1998). Furthermore, the adenylate energy charge (AEC) values were also determined as an estimate of the total metabolic activity (Pradet and Raymond, 1983). Respiration rates declined linearly as a function of WC (Fig. 5.6A). Interestingly, the decline of respiration rates was biphasic when they were plotted as a function of viscosity (Fig. 5.6B). The onset of drying was accompanied by a 50% decrease in respiration rates. Above 0.3 Poise, the respiration rates decreased linearly with the logarithmic increase in viscosity. During the onset of dehydration, the AEC values remained constant. They declined at 0.5 g g^{-1}, approximately concomitantly with the increase in reduced levels of COX (Fig. 5.6C). When plotted as a logarithmic function of viscosity, AEC declined linearly during drying (Fig. 5.6D). These observations suggest that rates of metabolism during dehydration are related to the cytoplasmic viscosity as previously demonstrated in isolated mitochondria (Fato *et al.*, 1993). However, the 50% fall in respiration during the initial loss of water was not accompanied by changes in the Cyt redox states nor by significant changes in AEC values. This indicates that there appears to be an active down-regulation of metabolism which is not dependent on cytoplasmic viscosity.

Conclusions and Future Research

Our results indicate that viscosity can be regarded as a valuable parameter to characterize the relations between desiccation and decreased metabolism during drying. Furthermore, we suggest that a regulated increase in viscosity

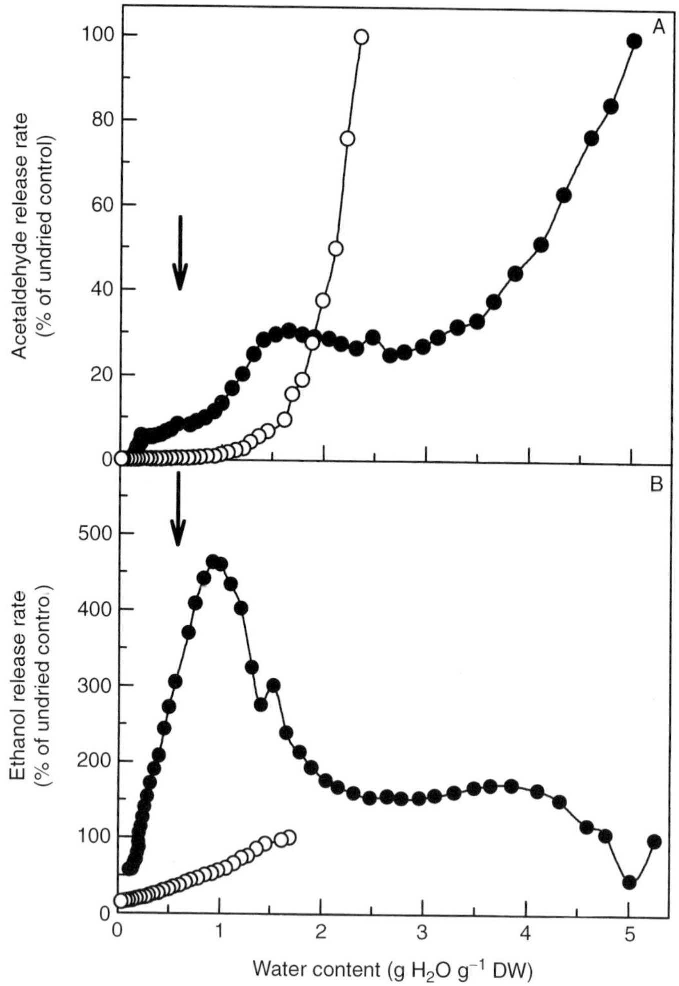

Fig. 5.5. The effects of drying on the release rates of acetaldehyde (A) and ethanol (B) during drying of germinating pea desiccation-tolerant (○) and intolerant axes (●). Acetaldehyde and ethanol were measured using laser photoacoustic as described in Harren and Reuss (1997). Data are expressed as % of undried material. The arrows indicate the WC corresponding to the onset of plasma membrane damage in desiccation-intolerant tissues.

and O_2 diffusion may be regarded as a mechanism of desiccation tolerance. We are currently comparing the desiccation-induced rise in viscosity and changes in O_2 solubility between desiccation-tolerant and -intolerant tissues.

Acknowledgements

The authors wish to thank J. Buitink for her assistance with the ESR and stimulating discussions during the experiments, and Drs G. Cotti, S. te Linkel Lekkert

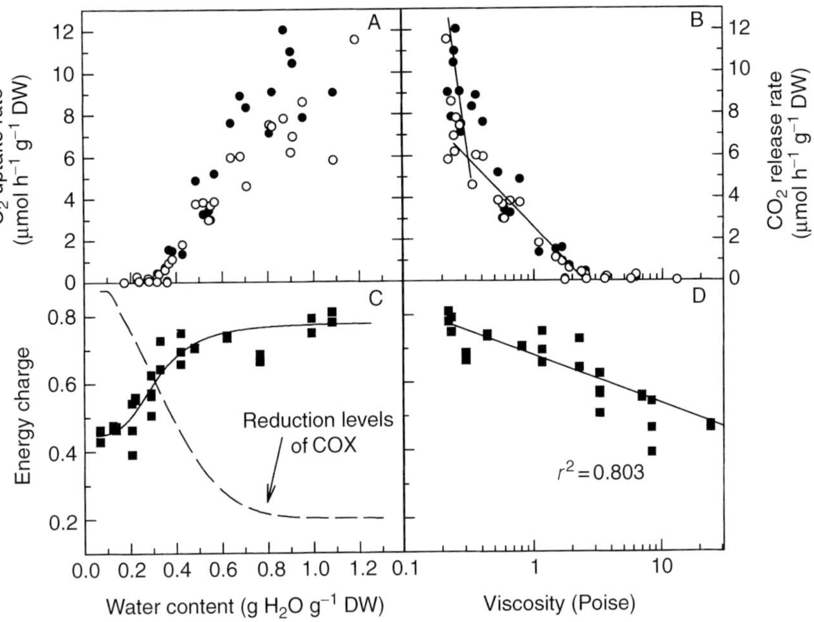

Fig. 5.6. The effects of drying on rates of O_2 uptake (●) and CO_2 release (○) (A–B) and AEC (■) (C–D) in cowpea cotyledons. Data were plotted as a function of WC (A,C) and viscosity (B,D) during dehydration. Viscosity values were obtained from the equation fitting the viscosity versus WC plot in Fig. 5.3. In panel B, the lines are aids to the eye to show the biphasic response of respiration rates to the increase in viscosity. In panel C, the broken line indicates the levels of COX reduction and corresponds to the fit shown in Fig. 5.2 (adapted from Leprince and Hoekstra, 1998).

and S. Persijn (University of Nijmegen) for their assistance with the photoacoustic measurements. This project was supported by a grant from the Wageningen Agricultural University and by the Dutch Technology Foundation which is subsidized by the Netherlands Organization for Scientific Research.

References

Berjak, P. and Pammenter, N.W. (1997) Progress in the understanding and manipulation of desiccation-sensitive (recalcitrant) seeds. In: Ellis, R.H., Black, M., Murdoch, A.J. and Hong, T.D. (eds) *Basic and Applied Aspects of Seed Biology.* Kluwer Academic Press, Dordrecht, pp. 689–703.

Buitink, J., Claessens, M.M.A.E., Hemminga, M.A. and Hoekstra, F.A. (1998) Influence of water content and temperature on molecular mobility and intracellular glasses in seeds and pollen. *Plant Physiology* 118, 531–541.

Cadenas, E. (1989) Biochemistry of oxygen toxicity. *Annual Review of Biochemistry* 58, 79–110.

Fato, R., Cavazzoni, M., Castlluccio, C., Parenti Castelli, G., Palmer, G., Degli Eposti, M. and Lenaz, G. (1993) Steady-state kinetics of ubiquinol-cytochrome c reductase in bovine heart submitochondrial particles. *Biochemical Journal* 290, 225–236.

Harren, F. and Reuss, J. (1997) Spectroscopy, photoacoustics. *Encyclopedia of Applied Physics* 19, 413–435.

Hemminga, M.A., Roozen, M.J.G.W. and Walstra, P. (1993) Molecular motions and the glassy state. In: Blanshard, J.M.V. and Lillford, P.J. (eds) *The Glassy State in Foods*. Nottingham University Press, Nottingham, pp. 157–171.

Hendry, G.A.F., Houghton, J.D. and Jones, O.T. (1981) The cytochromes in microsomal fractions of germinating mung beans. *Biochemical Journal* 194, 743–751.

Leprince, O. and Hoekstra, F.A. (1998) The responses of cytochrome redox state and energy metabolism to dehydration support a role for cytoplasmic viscosity in desiccation tolerance. *Plant Physiology* 118, 1253–1264.

Leprince, O., Hendry, G.A.F. and McKersie, B.D. (1993) The mechanisms of desiccation tolerance in developing seeds. *Seed Science Research* 3, 231–246.

Leprince, O., Atherton, N.M., Deltour, R. and Hendry, G.A.F. (1994) The involvement of respiration in free radical processes during loss of desiccation tolerance in germinating *Zea mays* L. An electron paramagnetic resonance study. *Plant Physiology* 104, 1333–1339.

Leprince, O., Vertucci, C.W., Hendry, G.A.F. and Atherton, N.M. (1995) The expression of desiccation-induced damage in orthodox seeds is a function of oxygen and temperature. *Physiologia Plantarum* 94, 233–240.

Lestan, D., Podgornik, H. and Perdih, A. (1993) Analysis of fungal pellets by UV-visible spectrum diffuse reflectance spectroscopy. *Applied and Environmental Microbiology* 59, 4253–4260.

Pradet, A. and Raymond, P. (1983) Adenine nucleotide ratios and adenylate energy charge in energy metabolism. *Annual Review of Plant Physiology* 34, 199–224.

Reid, R.C., Prausnitz, J.M. and Poling, B.E. (1987) *The Properties of Gases and Liquids*, 4th edn. McGraw-Hill, New York.

Skulachev, V.P. (1996) Role of uncoupled and non-coupled oxidation in maintenance of safely low levels of oxygen and its one-electron reductants. *Quarterly Reviews of Biophysics* 29, 169–202.

Vertucci, C.W. and Farrant, J.M. (1995) Acquisition and loss of desiccation tolerance. In: Kigel, J. and Galili, G. (eds) *Seed Development and Germination*. Marcel Dekker, New York, pp. 237–271.

6 Soluble Sugars in Maturing Pea Seeds of Different Lines in Relation to Desiccation Tolerance

R.J. GÓRECKI[1], L.B. LAHUTA[1], A.D. JONES[2] AND C.L. HEDLEY[2]

[1]Department of Plant Physiology and Biotechnology, Olsztyn University of Agriculture and Technology, Kortowo, 10-718 Olsztyn, Poland; [2]Department of Applied Genetics, John Innes Centre, Norwich NR4 7UH, UK

The relationship between the accumulation of soluble non-reducing sugars and the acquisition of desiccation tolerance during maturation of four pea lines differing in the content and composition of soluble carbohydrates was analysed. Maturing seeds were harvested at 4-day intervals, tested for desiccation tolerance (measured as the ability to germinate after slow and fast drying) and the soluble carbohydrates in axis and cotyledons were assayed. Immature seeds of RR RbRb and SD9 lines containing the lowest levels of raffinose were more sensitive to rapid drying as compared with seeds of other lines, but finally reached maximum desiccation tolerance at full maturity. Acquisition of desiccation tolerance was preceded by accumulation of sucrose and its development was accompanied by the accumulation of raffinose family oligosaccharides. Seeds of SD1 and SD6 lines did not accumulate verbascose, but this had no effect on their desiccation tolerance.

Introduction

One of the major reasons for limited consumption of pea seeds is the presence of relatively high quantities of the raffinose series of oligosaccharides (RFO) including raffinose, stachyose and verbascose that result in flatulence (Saini and Gladstones, 1986). Therefore, pea plant breeders search for genotypes with low or even no RFO content in the seeds (Jones *et al.*, 1999). However, there is evidence that α-galactosides are beneficial to plants. It is believed that in addition to LEA proteins and ABA, oligosaccharides play a key role in the acquisition of seed desiccation tolerance (Obendorf, 1997). Legume seeds accumulate sucrose during development, and during maturation drying they also accumulate raffinose, stachyose and verbascose (Dornbos and McDonald, 1986; Kuo *et al.*, 1988; Horbowicz and Obendorf, 1994; Lahuta *et al.*, 1995; Obendorf *et al.*, 1998). Soybean seeds naturally develop desiccation tolerance in correlation with the accumulation of raffinose and stachyose in the axis

 Seed Biology: Advances and Applications (eds M. Black, K.J. Bradford and J. Vázquez-Ramos)

(Blackman *et al.*, 1992). Acquisition of desiccation tolerance and germinability of lupin seeds is accompanied by accumulation of sucrose and RFO with stachyose as a predominant sugar (Górecki *et al.*, 1997). Other studies indicate that RFO function in the same way in pea, field bean (Lahuta *et al.*, 1995) and lentil (Piotrowicz-Cieślak *et al.*, 1995). Also, recent evidence suggests that sucrose and RFO are required for long-term storability of orthodox seeds (Steadman *et al.*, 1996; Obendorf, 1997).

The objective of the current study was to examine the desiccation tolerance of maturing pea seeds differing in the content and composition of soluble carbohydrates.

Materials and Methods

Four genotypes of pea seeds, that is RR RbRb, SD1, SD6, and SD9 were used. The RR RbRb refers to the round seeded wild type pea which makes up most of the commercially grown pea crop (Wang and Hedley, 1993). The SD lines are of exotic genetic stock from the John Innes Pisum Germplasm Collection identified recently by Jones *et al.* (1999). Plants were grown in a greenhouse at 20–22°C day (16 h) and 18°C night (8 h). Seeds were harvested at 4-day intervals beginning from 10 days after flowering (DAF) until full maturity. Freshly harvested seeds were analysed for fresh and dry weights, germination, desiccation tolerance and soluble sugar content and composition.

Seed dry mass was determined after drying at 100°C to constant weight. For the germination test, three to four replications of 20 seeds each were placed on wet paper towels (Anchor Paper Company) for 21 days at 22°C in the dark. To test desiccation tolerance, immature seeds were exposed to slow and fast drying treatments. For fast drying, seeds were dried at 12% RH and 22°C over a saturated solution of LiCl. For slow drying, seeds were dried for 7 days by transferring them daily to progressively lower RHs (from 92.5 to 12% maintained over saturated salt solutions as described by Górecki and Obendorf, 1997). After drying, seeds were examined for germinability as above.

At harvest times, seeds were separated into cotyledons and axes and stored at −80°C until analysis. Five axes or 30 to 50 mg of cotyledons were homogenized in ethanol:water (1:1, v/v) containing phenyl α-D-glucoside as internal standard, heated and centrifuged. The pellet was re-extracted and supernatants pooled. Aliquots of clear supernatants were passed through 10,000 MW cut-off filters and evaporated to dryness under a stream of nitrogen. Residues were derivatized with trimethylsilylimidazole:pyridine (1:1, v/v) for analysis of soluble carbohydrates by high resolution gas chromatography (Górecki *et al.*, 1997).

Results

There was an evident variation in the composition and the contents of soluble carbohydrates in mature seeds among the analysed lines (Table 6.1). Total

Table 6.1. Soluble carbohydrates in axes, and cotyledons of pea seeds. Values (mg g^{-1} dry weight) are means ± SE for four replicate analyses.

	Line			
Soluble carbohydrates	RR RbRb	SD1	SD6	SD9
Axis				
Fructose	1.79 ± 0.31	0.96 ± 0.01	0.00	1.05 ± 0.06
Maltose	0.22 ± 0.02	0.25 ± 0.03	0.00	0.25 ± 0.03
Sucrose	69.68 ± 5.78	43.01 ± 3.23	40.79 ± 1.80	22.81 ± 3.78
Raffinose	19.64 ± 0.94	44.66 ± 3.79	39.19 ± 1.28	16.67 ± 1.94
Stachyose	65.58 ± 3.69	83.12 ± 12.00	56.63 ± 3.73	59.53 ± 5.11
Verbascose	45.61 ± 3.22	0.00	2.30 ± 0.02	52.28 ± 1.29
myo-Inositol	4.58 ± 0.44	5.29 ± 0.17	5.43 ± 0.09	3.27 ± 0.19
Galactinol	1.09 ± 0.06	7.08 ± 0.01	6.89 ± 1.24	3.07 ± 0.91
Total	208.19	184.46	151.23	159.11
Sucrose:RFO ratio	0.53	0.34	0.42	0.18
Cotyledons				
Fructose	0.64 ± 0.17	0.73 ± 0.01	0.00	0.71 ± 0.08
Maltose	0.05 ± 0.01	0.08 ± 0.02	0.12 ± 0.02	0.27 ± 0.03
Sucrose	26.63 ± 0.30	21.52 ± 2.05	18.20 ± 0.51	22.19 ± 0.67
Raffinose	4.89 ± 0.04	6.31 ± 1.00	5.69 ± 0.51	2.69 ± 0.34
Stachyose	13.74 ± 0.12	36.09 ± 2.68	27.05 ± 1.91	14.94 ± 1.67
Verbascose	34.84 ± 3.25	0.00	0.00	43.44 ± 6.79
myo-Inositol	1.55 ± 0.07	0.87 ± 0.10	0.85 ± 0.02	0.94 ± 0.06
Galactinol	0.87 ± 0.04	1.64 ± 0.25	1.29 ± 0.02	0.99 ± 0.06
Total	83.21	67.24	52.35	86.17
Sucrose:RFO ratio	0.49	0.51	0.56	0.36

contents of soluble carbohydrates in axis tissues ranged from 20.8% in RR RbRb to 15.1% in SD6, including fructose, sucrose, maltose, raffinose, stachyose, verbascose, *myo*-inositol, and galactinol. Cotyledon tissues contained about one-third of the content of soluble sugars as did the axes. The amounts of individual sugars varied in axes and cotyledons. Axes of the SD9 line contained the lowest amount of sucrose, which resulted in the smallest sucrose:RFO ratio. Seeds of the SD1 line were deficient in verbascose. Similarly, seeds of the SD6 line had only trace amounts of verbascose. The mass ratios of sucrose to RFO in axis tissues ranged from 0.53 in RR RbRb to 0.18 in SD9. In cotyledons the ratios were more uniform and varied between 0.56 (SD6) and 0.36 (SD9).

During seed development maximum seed fresh weight occurred between 26 and 30 DAF (Fig. 6.1). Seeds reached maximum dry weight (mass maturity or physiological maturity) at 30–34 and 38 DAF for SD lines and RR RbRb line, respectively.

Freshly harvested seeds started to germinate at about 14 DAF (Fig. 6.2). The time needed for germination of fresh immature seeds of RR RbRb and SD9 lines was evidently longer as compared with seeds of SD1 and SD6 lines.

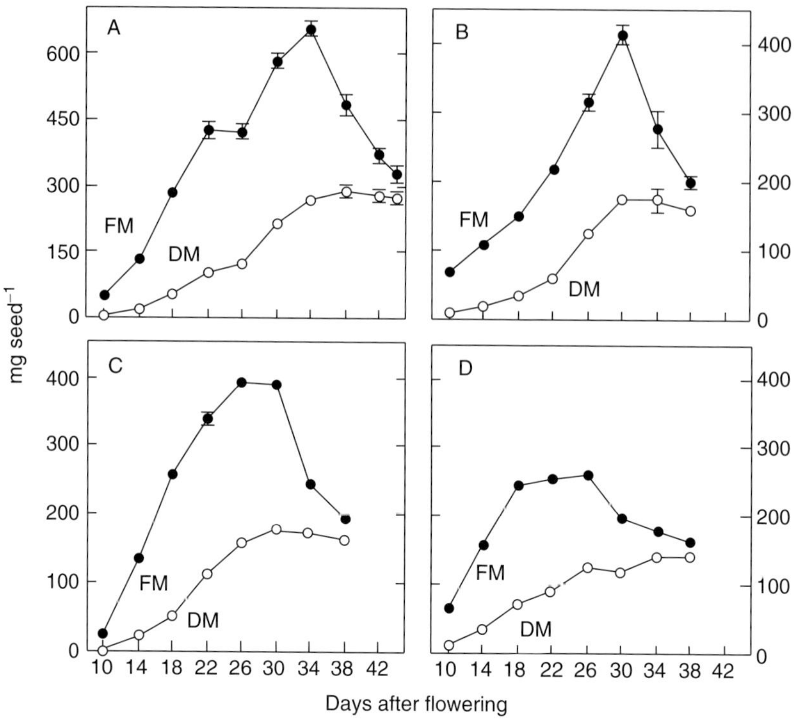

Fig. 6.1. Fresh (●) and dry mass (○) of maturing pea seeds of RR RbRb (A), SD1 (B), SD6 (C) and SD9 (D) lines. Values are means ± SE of five replications.

Drying treatments resulted in the reduction of germinability of immature (14–22 DAF) seeds of RR RbRb and SD9 genotypes, indicating their sensitivity to desiccation, but this had little or no effect on germination of SD1 and SD6 seeds.

Accumulation of sucrose (Fig. 6.3) preceded accumulation of RFO and the acquisition of desiccation tolerance. Raffinose accumulation started at 14–18 DAF in seeds of SD6 and SD9 and 22–26 DAF in RR RbRb and broadly accompanied acquisition of desiccation tolerance. Stachyose accumulation started from 2 to 4 days later as compared with raffinose. Finally, the start of verbascose accumulation was after that of stachyose and in seeds of RR RbRb and SD9 continued until full maturity. In contrast, in axes of SD6 seeds verbascose accumulation reached a maximum at 30 DAF, and afterwards decreased to small amounts. There was no evidence of verbascose accumulation in developing seeds of SD1 genotype.

The mass ratio of sucrose to RFO in axis tissues dropped rapidly before acquisition of desiccation tolerance and reached values less than 1.0 in all genotypes when seeds reached maximum germinability (Fig. 6.4). A drop of mass ratio of sucrose to RFO was evident when immature seeds were exposed to fast- and slow-drying treatments. In addition, a lower value of the ratio of sucrose to RFO corresponded with increasing vigour of immature seeds, both before and after drying (data not shown).

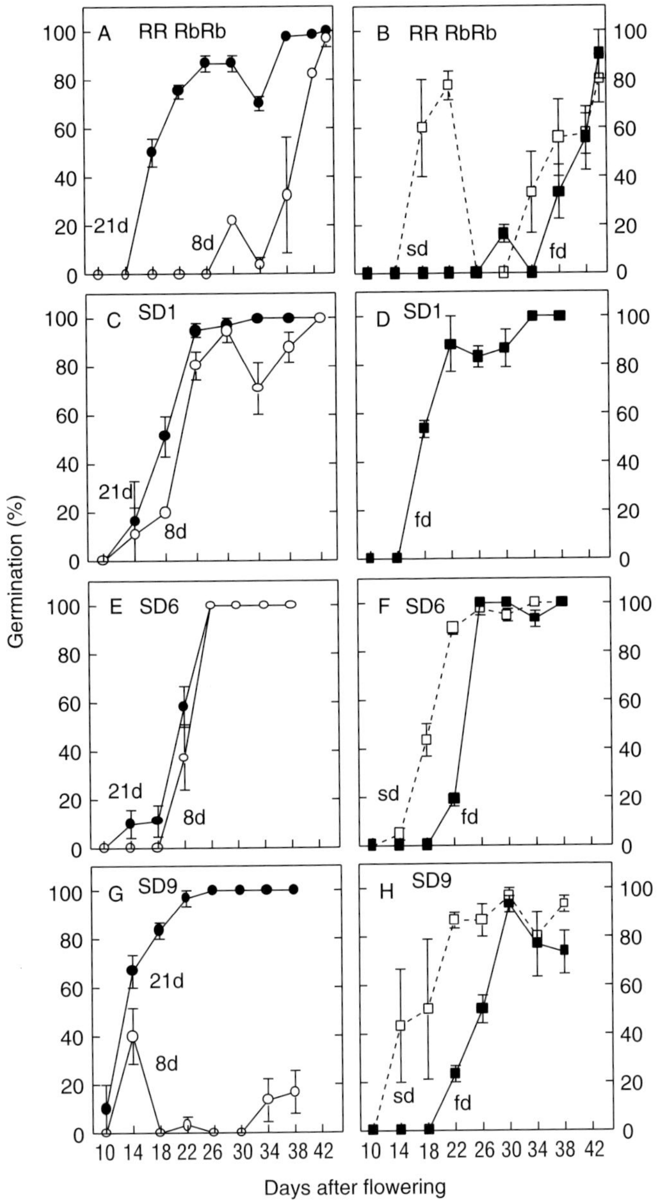

Fig. 6.22. Germination of maturing pea seeds. Germination of freshly harvested seeds (left column, A, C, E, G) scored after 8 days (○), after 21 days (●); desiccated rapidly at 12% RH and 22°C for 6 days (■, fd), or slowly over saturated salt solutions (□, sd) prior to testing for germinability (8 days, right column, B, D, F, H). Values are means ± SE of three replications.

Discussion

The data presented here provide evidence that pea seeds of all analysed genotypes gradually developed desiccation tolerance during maturation. Except in RR RbRb seeds, the maximum desiccation tolerance in seeds of SD1, SD6 and SD9 was achieved 8 to 12 days before harvest maturity. The increase in desiccation tolerance was broadly accompanied by the accumulation of raffinose family oligosaccharides. Among these, raffinose seems to be the most

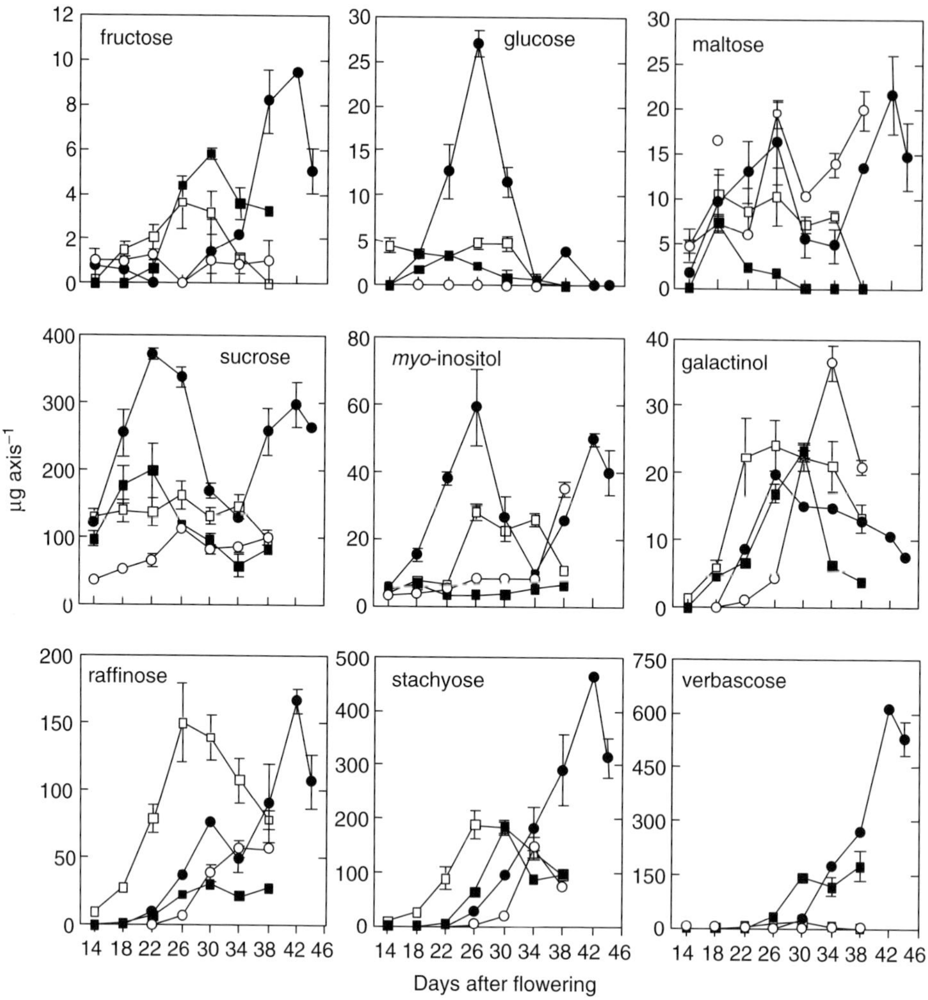

Fig. 6.3. Contents of soluble carbohydrates in axis of maturing pea seeds of RR RbRb (●), SD1 (○), SD6 (□) and SD9 (■) lines. Values are means ± SE of three replications.

important sugar. It is interesting to note that the lack or very low level of verbascose (the most abundant α-galactoside in cultivated peas) in seeds of SD1 and SD6 had no effect on desiccation tolerance. This suggests that verbascose is not a prerequisite for pea seed desiccation tolerance. On the other hand, breeding pea plants for the elimination of verbascose creates the possibility of improving the nutritional value of seeds without decreasing the physiological quality of the seeds. Other studies (Blackman *et al.*, 1992) indicate that in soybean stachyose and sucrose are required for the acquisition of seed desiccation tolerance. In some other seeds, for example in buckwheat, the role of the raffinose series of oligosaccharides in seed desiccation tolerance is replaced by galactosyl cyclitols – fagopyritols (Horbowicz *et al.*, 1998).

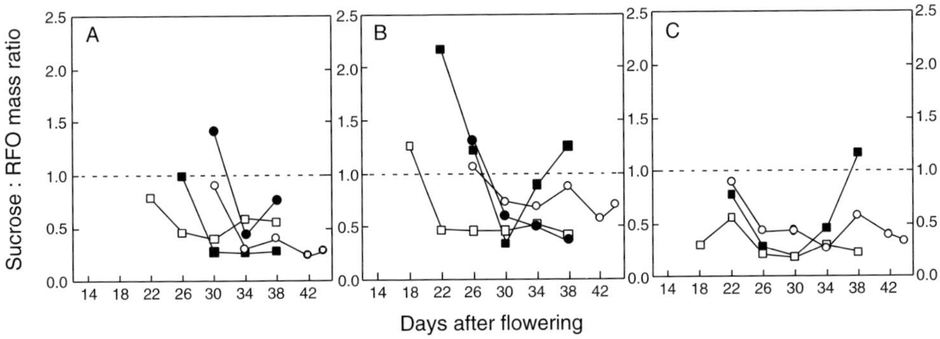

Fig. 6.4. Sucrose to RFO mass ratios in axis of maturing pea seeds of RR RbRb (○), SD1(●), SD6 (□) and SD9 (■) lines. A, freshly harvested seeds; B, after rapid drying; C, after slow drying.

The prominent role of sugars in desiccation tolerance of seeds is based on the stabilization of membranes and protein macromolecules (Vertucci and Farrant, 1995). Accumulation of sucrose correlates with the early stages of desiccation tolerance, but sucrose alone is not sufficient for desiccation tolerance. The presence of raffinose, stachyose and/or galactosyl cyclitols is required to prevent crystallization of sucrose and to promote formation of the glassy state (Koster and Leopold, 1988; Obendorf, 1997). Therefore, acquisition of seed desiccation tolerance is closely related to the sucrose:RFO mass ratio (Obendorf, 1997). In many orthodox seeds, the lower the mass ratio of sucrose to RFO, the longer is the seed storability (Horbowicz and Obendorf, 1994; Steadman *et al.*, 1996). During maturation of pea seeds, a rapid decrease of sucrose:RFO mass ratio was observed. Full desiccation tolerance of pea seeds was acquired when the mass ratio of sucrose to RFO was less than 1. This relationship applies to freshly harvested seeds and to seeds exposed to artificial drying. An elevated mass ratio of sucrose to RFO (Fig. 6.4) or sucrose to raffinose (data not shown) corresponded to an increase in seed sensitivity to fast drying.

Acknowledgements

This work was financially supported by the State Committee for Scientific Research grant No. 5 P0 6B 03012. We thank Dr M. Horbowicz for supplying the galactinol standard and Mrs T. Jagielska for technical assistance.

References

Blackman, S.A., Obendorf, R.L. and Leopold, A.C. (1992) Maturation proteins and sugars in desiccation tolerance of developing soybean seeds. *Plant Physiology* 100, 225–230.

Dornbos, D.L. and McDonald, M.B., Jr (1986) Mass and composition of developing soybean seeds at five reproductive growth stages. *Crop Science* 26, 624–630.

Górecki, R.J., and Obendorf, R.L. (1997) Galactosyl cyclitols and raffinose family oligosaccharides in relation to desiccation tolerance of pea and soybean seedlings. In: Ellis, R.H., Black, M., Murdoch, A.J. and Hong, T.D. (eds) *Basic and Applied Aspects of Seed Biology: Proceedings of the Fifth International Workshop on Seeds*, Reading, 1995. Kluwer Academic Publishers, Dordrecht, pp. 119–128.

Górecki, R.J., Piotrowicz-Cieślak, A.I., Lahuta, L.B. and Obendorf, R.L. (1997) Soluble carbohydrates in desiccation tolerance of yellow lupin seeds during maturation and germination. *Seed Science Research* 7, 107–115.

Horbowicz, M. and Obendorf, R.L. (1994) Seed desiccation tolerance and storability: dependence on flatulence-producing oligosaccharides and cyclitols – review and survey. *Seed Science Research* 4, 385–405.

Horbowicz, M., Brenac, P. and Obendorf, R.L. (1998) Fagopyritol B1, O-α-galactopyranosyl-(1→2)-D-*chiro*-inositol, a galactosyl cyclitol in maturing buckwheat seeds associated with desiccation tolerance. *Planta* 205, 1–11.

Jones, D.A., Dupont, M.S., Frias, J., Rhodes, M.J.C., Fenwick, G.R. and Hedley, C.L. (1999) The discovery of variation for α-galactoside composition in pea seeds. *Seed Science Research* 9, 305–310.

Koster, K.L. and Leopold, A.C. (1988) Sugars and desiccation tolerance in seeds. *Plant Physiology* 88, 829–832.

Kuo, T.M., Van Middlesworth, J.F. and Wolf, W.J. (1988) Content of raffinose oligosaccharides and sucrose in various plant seeds. *Journal of Agricultural and Food Chemistry* 36, 32–36.

Lahuta, L.B., Rejowski, A. and Górecki, R.J. (1995) Accumulation of sugars in maturing field bean (*Vicia faba minor* L.) and pea seeds (*Pisum sativum* L.) in relation to seed viability. In: AEP (ed.) *2nd European Conference on Grain Legumes*, Copenhagen, Denmark, p. 44.

Obendorf, R.L. (1997) Oligosaccharides and galactosyl cyclitols in seed desiccation tolerance. *Seed Science Research* 7, 63–74.

Obendorf, R.L., Horbowicz, M., Dickerman, A.M., Brenac, P. and Smith, M.E. (1998) Soluble oligosaccharides and galactosyl cyclitols in maturing soybean seeds *in planta* and *in vitro*. *Crop Science* 38, 78–84.

Piotrowicz-Cieślak, A.I., Górecki, R.J. and Frias, J. (1995) Changes in galactosides content during maturation of yellow lupin (*L. luteus* L.) and lentil (*Lens culinaris* L.) seeds. In: AEP (ed.) *2nd European Conference on Grain Legumes*, Copenhagen, Denmark, p. 378.

Saini, H.S. and Gladstones, J.S. (1986) Variability in the total and component galactosyl sucrose oligosaccharides of *Lupinus* species. *Australian Journal of Agricultural Research* 37, 157–166.

Steadman, K.J., Pritchard, H.W. and Dey, P.M. (1996) Tissue-specific soluble sugars in seeds as indicators of storage category. *Annals of Botany* 77, 667–674.

Vertucci, C.W. and Farrant, J.M. (1995) Acquisition and loss of desiccation tolerance. In: Kigel, J. and Galili, G. (eds) *Seed Development and Germination.* Marcel Dekker, Inc., New York, pp. 237–271.

Wang, T.L. and Hedley, C.L. (1993) Seed development in peas: knowing your three 'r's' (or four, or five). *Seed Science Research* 1, 3–14.

7 The Role of Stachyose Synthase in the Oligosaccharide Metabolism of Legume Seeds

A. RICHTER, G. HOCH, M. PUSCHENREITER, U. MAYER AND T. PETERBAUER

Institute of Plant Physiology, University of Vienna, PO Box 285, A-1091 Vienna, Austria

Legume seeds contain two major classes of soluble α-galactosides, namely the raffinose family of oligosaccharides and galactosyl cyclitols. They are biosynthetically linked by the enzyme stachyose synthase, which not only forms stachyose from raffinose and galactinol, but also catalyses the galactinol-dependent synthesis of glactosyl cyclitols. Stachyose synthase enzymes purified from seeds of adzuki bean and lentil, respectively, were similar with respect to their physicochemical properties but differed substantially in their substrate specificity. The enzyme from lentil (a pinitol producing plant) was able to catalyse the formation of galactopinitol A from D-pinitol, while stachyose synthase from adzuki bean was able to synthesize galactosyl ononitol. The lentil enzyme was additionally able to form ciceritol (a digalactosyl pinitol) from galactopinitol A and galactinol. Moreover, both galactosyl cyclitols could substitute for galactinol as galactosyl donors in the synthesis of stachyose. Thus, stachyose synthase accounts for the synthesis of most galactosyl cyclitols and holds a central position in the interconversion of oligosaccharides in legume seeds. Analysis of mRNA and protein abundance, enzyme activity and metabolite levels during seed development suggested that the oligosaccharide composition of mature seeds is not only determined by total stachyose synthase activity, but also by the concentration of precursors (cyclitols, galactinol, and raffinose).

Introduction

Mature legume seeds contain substantial amounts of soluble α-galactosides. Among these substances the raffinose family of oligosaccharides (RFO), i.e. raffinose, stachyose, verbascose and higher homologues, are common constituents of the carbohydrate fraction of seeds (Avigad and Dey, 1997). The biosynthesis of RFO proceeds by the sequential action of a set of galactosyl transferases, which utilize galactinol [*O*-α-D-galactopyranosyl-(1→1)-L-*myo*-inositol] as galactosyl donor. This rather uncommon donor is formed by galactinol synthase (EC 2.4.1.123) from UDP-galactose and *myo*-inositol. Galactinol itself belongs to a second group of soluble α-galactosides frequently found in legume seeds, termed galactosyl cyclitols. Besides *myo*-inositol, three cyclitols, namely D-ononitol (1D-4-*O*-methyl-*myo*-inositol), D-pinitol (1D-3-*O*-methyl-*chiro*-inositol) and D-*chiro*-inositol have been found as aglycons in galactosyl cyclitols (Obendorf, 1997). The cyclitol moiety is attached to up to three galactosyl residues, yielding several homologue series of galactosyl derivatives. Although galactosyl cyclitols are not carbohydrates *sensu strictu*, they are often classed as oligosaccharides due to their structural similarity to RFO.

Galactosyl cyclitols accumulate alongside RFO during seed development. While galactosyl cyclitols constitute up to 3% of dry matter, stachyose is the predominant oligosaccharide in mature seeds of most legume species (Quemener and Brillouet, 1983). The first step in the synthesis of stachyose is catalysed by raffinose synthase (EC 2.4.1.82), which forms raffinose from galactinol and sucrose. The enzyme stachyose synthase (STS, EC 2.4.1.67) then catalyses the reversible transfer of a galactosyl moiety of galactinol to raffinose, yielding stachyose and the co-product *myo*-inositol (Tanner and Kandler, 1968). We have recently demonstrated that STS is also involved in the metabolism of galactosyl cyclitols (Peterbauer and Richter, 1998; Hoch *et al.*, 1999). In this paper we discuss the importance of STS-catalysed reactions in the oligosaccharide composition of legume seeds.

Purification and Characterization of STS from Seed Sources

We have purfied STS from mature seeds of adzuki bean (Peterbauer and Richter, 1998) and lentil (Hoch *et al.*, 1999), two legume species that differ in their oligosaccharide pattern. STS was purified to apparent homogeneity from both species by conventional techniques (protamine and ammonium sulphate precipitation, hydrophobic interaction and anion-exchange chromatography) combined with native preparative gel electrophoresis. Polyclonal antibodies have been prepared and the gene has been cloned from adzuki bean (Peterbauer *et al.*, 2000).

Adzuki bean STS had an apparent molecular mass of about 90 kDa, while the lentil enzyme was slightly smaller (Peterbauer and Richter, 1998; Hoch *et al.*, 1999). The pH optima (around pH 7.0) and pI values (pI 4.7 to 5.1) of STS from lentil and adzuki bean fit well with reports on purified and partially purified STS preparations from seeds and leaves (Tanner and Kandler, 1968; Gaudreault and Webb, 1981; Holthaus and Schmitz, 1991).

Antibodies raised in rabbits against the adzuki bean enzyme exhibited cross-reactivity with STS from lentil and other legumes, such as soybean and chick pea (Fig. 7.1A). This indicates that the enzymes from different legume seeds are closely related. Nevertheless, a variation of molecular masses (between 85 and 95 kDa) was observed, even within the genus *Vigna* (Fig. 7.1B). It is interesting to note that the antibodies raised against STS from adzuki bean seeds also cross-reacted with a polypeptide of about 87 kDa from

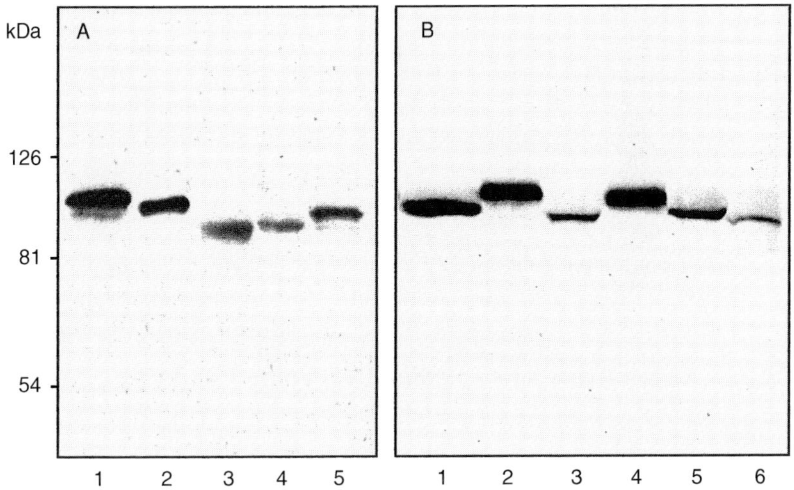

Fig. 7.1. Immunoblot analysis of total soluble protein from various plant species. Samples were subjected to SDS-PAGE on 7.5% gels, blotted onto PVDF membranes and probed with rabbit polyclonal antibodies against STS from adzuki bean. The total IgG fraction was purified by affinity chromatography and used at a dilution of 1:10,000. Bound antibodies were visualized with goat anti-rabbit IgG conjugated with alkaline phosphatase and a colour reaction using NBT/BCIP as substrate. A, samples from seeds and leaves of various genera. Lane 1, adzuki bean (seeds); 2, lentil (seeds); 3, soybean (seeds); 4, chickpea (seeds); 5, zucchini (leaves). B, seed samples from species of the genus *Vigna*. Lane 1, *V. radiata*; 2, *V. umbellata*; 3, *V. mungo*; 4, *V. sinensis*; 5, *V. unguiculata*; 6, *V. marina*. Lanes were loaded with 20 μl crude protein extract (equivalent to 2.5 mg fresh mass).

leaves of zucchini (Fig. 7.1A). STS purified from leaves of squash, a closely related member of the Cucurbitaceae, was reported to consist of two subunits of 45 and 50 kDa, respectively, present in unequal amounts (Holthaus and Schmitz, 1991). Therefore, a reinvestigation of leaf STS seems necessary.

Reactions and Enzyme Mechanism of STS

Kinetic data and the demonstration of partial reactions indicated that stachyose synthesis by STS proceeds by an enzyme substitution mechanism (Tanner and Kandler, 1968; Peterbauer and Richter, 1998; Hoch *et al.*, 1999). First, galactinol binds to the enzyme to form a galactose-enzyme complex, while *myo*-inositol is liberated. In a second step, raffinose reacts with the enzyme-bound galactose and stachyose is released. It has been convincingly demonstrated that the partial reactions are fully reversible. For example, STS catalysed an exchange reaction between galactinol and labelled *myo*-inositol (Peterbauer and Richter, 1998):

$$\text{Galactinol} + \textit{myo}\text{-}[^3\text{H}]\text{inositol} \rightleftharpoons \textit{myo}\text{-inositol} + [^3\text{H}]\text{galactinol} \qquad (7.1)$$

This exchange reaction is consistent with the fact that *myo*-inositol acted as an competitive inhibitor with respect to raffinose (Gaudreault and Webb, 1981; Holthaus and Schmitz, 1991; Peterbauer and Richter, 1998). In the exchange reaction (7.1), no net synthesis of a product occurs. However, substitution of *myo*-inositol by other cyclitols results in the formation of corresponding galactosyl cyclitols:

$$\text{Galactinol + cyclitol} \rightleftharpoons \textit{myo}\text{-inositol + galactosyl cyclitol} \quad (7.2)$$

Additionally, we have demonstrated that galactosyl cyclitols produced by reaction (7.2) were also able to substitute for galactinol as galactosyl donor in the formation of stachyose (Peterbauer and Richter, 1998; Hoch *et al.*, 1999):

$$\text{Galactosyl cyclitol + raffinose} \rightleftharpoons \text{cyclitol + stachyose} \quad (7.3)$$

Substrate Specificity of STS and Galactosyl Cyclitol Pattern

D-Ononitol, derived from the ubiquitous *myo*-inositol (Wanek and Richter, 1995), is the most widespread cyclitol in legumes, since it not only accumulates in certain legume species, but also serves as a precursor for D-pinitol biosynthesis. D-Pinitol is the major methylated inositol in many important crop species, such as soybean, alfalfa, chickpea and lentil, while many species of the genus *Vigna* contain exclusively ononitol (Quemener and Brillouet, 1983; Yasui *et al.*, 1985). The cyclitol composition of a certain species is reflected in the pattern of galactosyl cyclitols in its seeds, suggesting that all naturally occurring cyclitols serve as substrates for the formation of galactosyl cyclitols. Seeds of adzuki bean contain galactosyl ononitol [*O*-α-D-galactopyranosyl-(1→3)-4-*O*-methyl-D-*myo*-inositol] (Richter *et al.*, 1997) and angularitol [*O*-α-D-galactopyranosyl-(1→6)-*O*-α-D-galactopyranosyl-(1→3)-4-*O*-methyl-D-*myo*-inositol] (Peterbauer *et al.*, 1997) in addition to galactinol, while lentil seeds exhibit a highly complex oligosaccharide composition with galactopinitol A [*O*-α-D-galactopyranosyl-(1→2)-4-*O*-methyl-D-*chiro*-inositol], ciceritol [*O*-α-D-galactopyranosyl-(1→6)-*O*-α-D-galactopyranosyl-(1→2)-4-*O*-methyl-D-*chiro*-inositol], galactopinitol B [*O*-α-D-galactopyranosyl-(1→2)-3-*O*-methyl-D-*chiro*-inositol] and fagopyritol B1 [*O*-α-D-galactopyranosyl-(1→2)-D-*chiro*-inositol] (Quemener and Brillouet, 1983; Obendorf, 1997).

Stachyose synthase from both sources converted D-ononitol to galactosyl ononitol at rates comparable to stachyose synthesis, but only the lentil enzyme utilized D-*chiro*-inositol and its methylated derivative D-pinitol (Table 7.1). This remarkable difference in substrate specificities perfectly matches the endogenous cyclitol composition of the respective species. Stachyose synthase from lentil galactosylated D-pinitol yielded two products, galactopinitol A and low amounts of a positional isomer, galactopinitol B (Table 7.1). Furthermore, galactopinitol A itself was utilized as acceptor resulting in the di-galactopinitol ciceritol, and D-*chiro*-inositol was converted to fagopyritol B1. Although lentil STS was highly active on D-ononitol, only traces of galactosyl ononitol could be found in lentil seeds, most probably due to the presence of only minute amounts of free D-ononitol. In summary, these results demonstrate that the

Table 7.1. Substrate specificities of STS purified from adzuki bean and lentil.

Galactosyl acceptor	Activity (%) Adzuki bean	Lentil
Carbohydrates		
Raffinose	100.0[a]	100.0[b]
Raffinose[c]	37.8	69.9
Raffinose[d]	n.a.[e]	13.8
Sucrose	n.d.[f]	n.d.
Stachyose	n.d.	n.d.
Cyclitols		
myo-Inositol	224.8[g]	n.a.
D-Ononitol	103.5	109.8
D-*chiro*-Inositol	n.d.	26.3
D-Pinitol	0.9[h]	40.4[h](4.3)[i]
Galactosyl cyclitols		
Galactosyl ononitol	n.d.	n.a.
Galactopinitol A	n.a.	26.4

[a]Corresponding to an activity of 11.2 nkat mg^{-1} protein. [b]Corresponding to an activity of 9.1 nkat mg^{-1} protein. [c]Assayed with 10 mM galactosyl ononitol. [d]Assayed with 10 mM galactopinitol A. [e]n.a., not analysed. [f]n.d., not detected. [g]Assayed with *myo*-[^{3}H]inositol. [h]Product galactopinitol A. [i]Product galactopinitol B.
Data compiled from Peterbauer and Richter (1998) and Hoch *et al.* (1999). Reactions were assayed with 10 mM galactinol as galactosyl donor unless otherwise stated. Only substrates naturally occurring in legumes are listed.

entire galactosyl cyclitol spectrum of seeds may be derived solely by the action of STS.

Galactosyl ononitol and galactopinitol A could substitute for galactinol in the biosynthesis of stachyose from raffinose (Table 7.1). Considering the kinetic data and the similar galactosyl group transfer potential of the investigated galactosyl cyclitols (Peterbauer *et al.*, 1998; Hoch *et al.*, 1999), we have suggested that galactosyl ononitol and galactopinitol A may contribute considerably to stachyose synthesis during certain stages of seed development (Peterbauer and Richter, 1998; Hoch *et al.*, 1999). Using a partially purified enzyme preparation of vetch seeds, Tanner *et al.* (1967) have suggested that STS is responsible for the galactinol-dependent synthesis of verbascose from stachyose. However, STS from adzuki bean and lentil did not utilize stachyose as a substrate at measurable rates (Table 7.1). Our results are not necessarily in conflict with the results of Tanner *et al.* (1967), since there is a high variation in the verbascose content of legume seeds. For example, seed verbascose contents ranged from undetectable levels to 1% of dry matter in a survey of 16 lentil genotypes (Frias *et al.*, 1994). In this context, it would be interesting to see whether the properties of STS are responsible for differences in the quantitative oligosaccharide composition of seeds of various species and cultivars.

Stachyose Synthase and Oligosaccharide Accumulation during Seed Development

Galactosyl cyclitols and RFO are generally absent during the early development of legume seeds, while sucrose, monosaccharides and cyclitols are dominant during these growth stages (Lowell and Kuo, 1989; Obendorf *et al.*, 1998). During seed development of adzuki bean, galactinol and galactosyl ononitol appeared simultaneously at about 14 days after flowering (DAF) (Fig. 7.2). Raffinose was initially detected 16 DAF, followed by stachyose, which started to accumulate 20 DAF. It is interesting to note that galactosyl ononitol was detected several days prior to stachyose, although both are synthesized by STS. Similar observations have also been reported for rice bean (Peterbauer *et al.*, 1998).

It has been proposed that galactinol synthesis plays the key role in the formation of stachyose (Saravitz *et al.*, 1987). However, our results suggest that

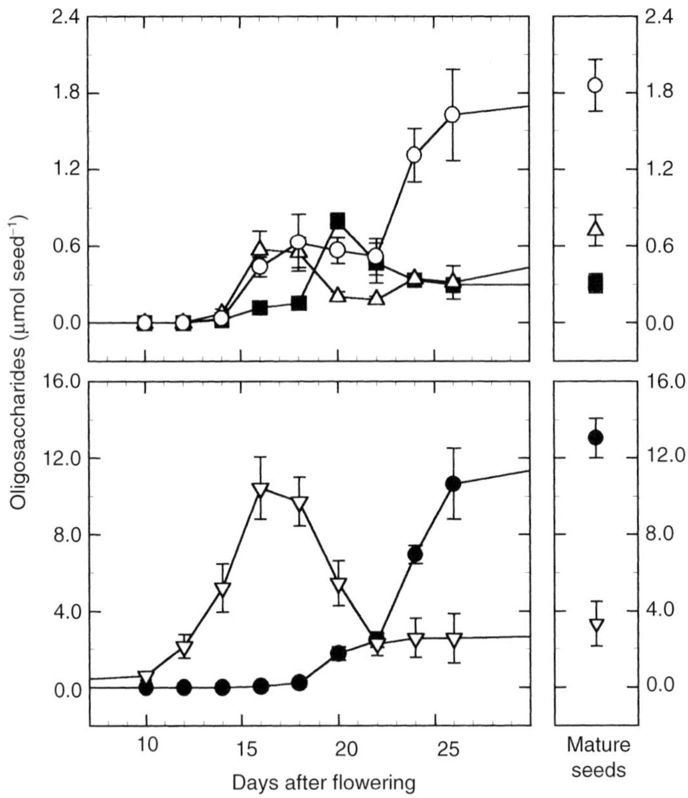

Fig. 7.2. Changes in galactosyl cyclitols and RFO during seed development of adzuki bean. Seeds were harvested and extracted with 50% (v/v) ethanol. Aliquots of the extracts were deionized and analysed by capillary gas chromatography as previously described (Peterbauer *et al.*, 1998). △, galactinol; ○, galactosyl ononitol; ▽, sucrose; ■, raffinose; ●, stachyose. Symbols are means ± SE of four replicates.

the supply of the galactosyl acceptor raffinose rather than that of the galactosyl donors (i.e. galactinol and galactosyl ononitol) were limiting the biosynthesis of stachyose, at least during early seed development. In maturing seeds (22–26 DAF) the content of both stachyose and galactosyl ononitol increased dramatically (Fig. 7.2). This is in good agreement with changes in STS protein levels, as shown by Western analysis of representative growth stages (Fig. 7.3A). RT-PCR was used to assess the abundance of transcripts encoding STS from adzuki bean. Intense amplification of transcripts was detected at growth stages IV and V (Fig. 7.3B), parallel to STS protein abundance. However, no amplification was observed at growth stage III, although low amounts of STS protein were already present (Fig. 7.3A) and traces of transcripts were detectable by Northern analysis (unpublished). The reason for this discrepancy is currently unclear.

Stachyose synthase protein was found to be present in substantial amounts in mature seeds (Fig. 7.3A). Furthermore, STS activity and protein levels were fully preserved during the first 72 h of germination (Peterbauer *et al.*, 2000). The physiological significance of this observation is not fully understood at present. The equilibrium constant of the galactosyl transfer from galactinol or galactosyl ononitol to raffinose is close to four (Tanner and Kandler, 1968; Peterbauer *et al.*, 1998) favouring stachyose synthesis. However, the reaction is readily reversible and may proceed in the reverse direction (i.e. stachyose degradation), when stachyose concentrations are high. Thus, STS may also participate in the mobilization of oligosaccharides during germination.

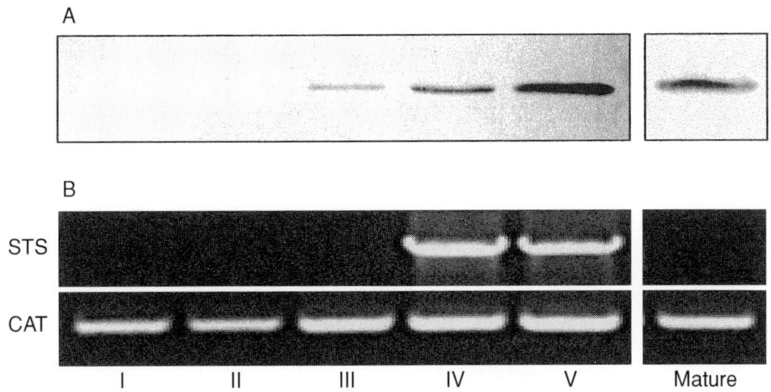

Fig. 7.3. Changes in STS protein levels and gene expression during seed development of adzuki bean. A, Western analysis. Total protein was extracted from seeds at selected growth stages, subjected to SDS-PAGE, blotted and probed with polyclonal antibodies against adzuki bean STS. For details, see legend to Fig. 7.1. B, RT-PCR analysis. Oligo-(dT)-primed cDNA synthesis was performed on total RNA and equal aliquots of each reaction were used in PCR with specific primers that will amplify a 2950-bp target sequence of the cDNA encoding adzuki bean STS. The primer pair used encompasses an intron and, hence, allows us to distinguish the amplified cDNA from any potential genomic DNA contamination. For a positive RT-PCR control, a primer pair directed at a constitutively expressed catalase (CAT) was used. The growth stages were: I, 8–10 DAF; II, 12–14 DAF; III, 16–18 DAF; IV, 20–22 DAF; V, 24–26 DAF.

Stachyose Synthase, a Central Enzyme in Seed Oligosaccharide Metabolism

We have demonstrated that STS is not only responsible for the synthesis of stachyose, but holds a central position in a complex network of galactosyl transfer reactions in developing seeds. With the exception of galactinol synthesis, STS is responsible for the synthesis of many, if not all galactosyl cyclitols found in legumes. Therefore, STS represents the link of the biosynthetic pathways of RFO and galactosyl cyclitols (Fig. 7.4). Another unique feature of STS is its ability to redistribute galactosyl residues stored in galactosyl cyclitols to form further products. It utilizes a range of galactosyl cyclitols as galactosyl donors in the synthesis of stachyose or as galactosyl acceptors to form higher galactosyl cyclitol homologues (Fig. 7.4). A possible involvement of STS in verbascose synthesis needs further investigation.

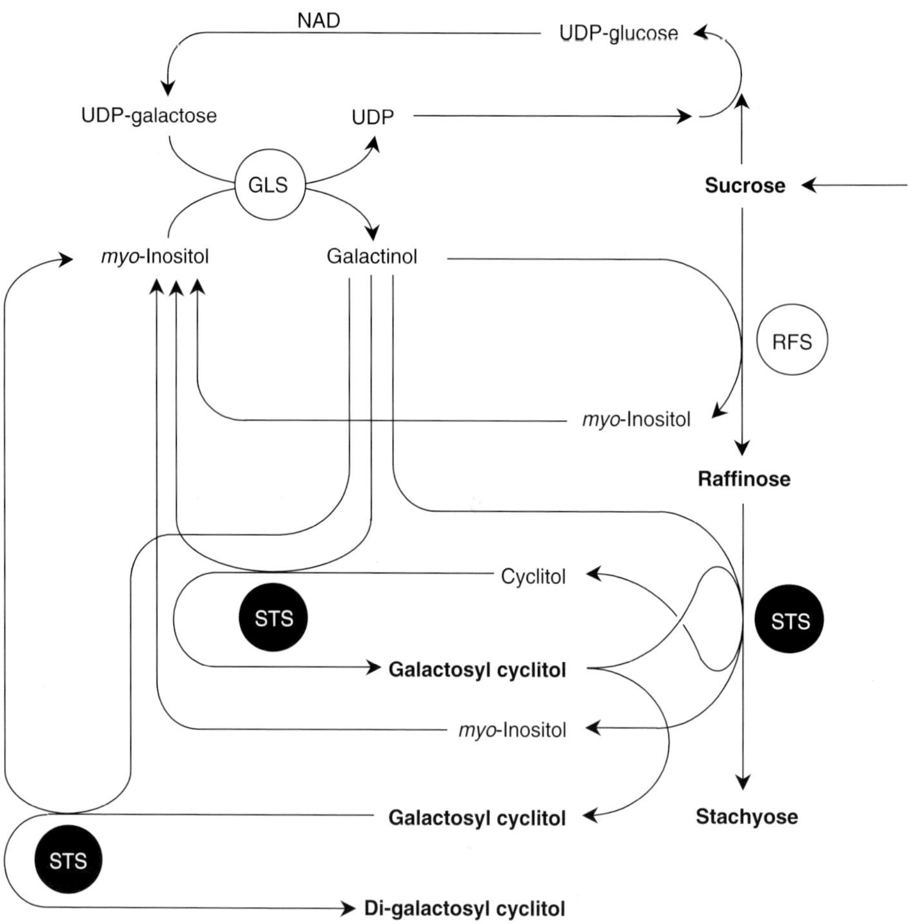

Fig. 7.4. Pathways of oligosaccharide synthesis in developing legume seeds. Participating enzymes: GLS, galactinol synthase; RFS, raffinose synthase; STS, stachyose synthase.

Stachyose synthase appears to be transcriptionally controlled, although much more information on its molecular regulation is required. However, the accumulation of RFO and galactosyl cyclitols follows complex kinetics *in situ*, due to the presence of multiple substrates and intermediates. Thus, the oligosaccharide composition of mature seeds is determined not only by total STS activity, but also by the concentration of precursors (i.e. cyclitols, galactinol, and raffinose) during seed development.

Acknowledgements

We wish to thank Dr J. Mucha for help with RT-PCR and Dr W. Wanek for valuable comments on the manuscript. This work was supported by the Austrian Science Foundation (project number 10917-BIO).

References

Avigad, G. and Dey, P.M. (1997) Carbohydrate metabolism: storage carbohydrates. In: Dey, P.M. and Harborne, J.B. (eds) *Plant Biochemistry*. Academic Press, San Diego, pp. 143–204.

Frias, J., Vidal-Valverde, C., Bakhsh, A., Arthur, A.E. and Hedley, C. (1994) An assessment of variation for nutritional and non-nutritional carbohydrates in lentil seeds (*Lens culinaris*). *Plant Breeding* 113, 170–173.

Gaudreault, P.-R. and Webb, J.A. (1981) Stachyose synthesis in leaves of *Cucurbita pepo*. *Phytochemistry* 20, 2629–2633.

Hoch, G., Peterbauer, T. and Richter, A. (1999) Purification and characterization of stachyose synthase from lentil (*Lens culinaris*) seeds. Galactopinitol and stachyose synthesis. *Archives of Biochemistry and Biophysics* 366, 75–81.

Holthaus, U. and Schmitz, K. (1991) Stachyose synthesis in mature leaves of *Cucumis melo*. Purification and characterization of stachyose synthase (EC 2.4.1.67). *Planta* 184, 525–531.

Lowell, C.A. and Kuo, T.M. (1989) Oligosaccharide metabolism and accumulation in developing soybean seeds. *Crop Science* 29, 459–465.

Obendorf, R.L. (1997) Oligosaccharides and galactosyl cyclitols in seed desiccation tolerance. *Seed Science Research* 7, 63–74.

Obendorf, R.L., Horbowicz, M., Dickerman, A.M., Brenac, P. and Smith, M.E. (1998) Soluble oligosaccharides and galactosyl cyclitols in maturing soybean seeds *in planta* and *in vitro*. *Crop Science* 38, 78–84.

Peterbauer, T. and Richter, A. (1998) Galactosylononitol and stachyose synthesis in seeds of *Vigna angularis*. Purification and characterization of stachyose synthase. *Plant Physiology* 117, 165–172.

Peterbauer, T., Puschenreiter, M., Richter, A. and Brereton, I. (1997) A novel series of galactosides of D-ononitol (1D-4-*O*-methyl-*myo*-inositol) in seeds of *Vigna angularis*. In: Vliegenthart, H., Kamerling, H. and Veldink, G. (eds) *Proceedings of the 9th European Carbohydrate Symposium*, Utrecht, p. 284.

Peterbauer, T., Puschenreiter, M. and Richter, A. (1998) Metabolism of galactosylononitol in seeds of *Vigna umbellata*. *Plant and Cell Physiology* 39, 334–341.

Peterbauer, T., Mucha, J., Mayer, U., Popp, M,. Glössl, J. and Richter, A. (2000) Stachyose synthesis in seeds of Adzuki bean (*Vigna angularis*): molecular cloning and functional expression of stachyose synthase. *The Plant Journal* (in press).

Quemener, B. and Brillouet, J.-M. (1983) Ciceritol, a pinitol digalactoside from seeds of chickpea, lentil and white lupin. *Phytochemistry* 22, 1745–1751.

Richter, A., Peterbauer, T. and Brereton, I. (1997) Structure of galactosylononitol. *Journal of Natural Products* 60, 749–751.

Saravitz, D.M., Pharr, D.M. and Carter, T.E. (1987) Galactinol synthase activity and soluble sugars in developing seeds of four soybean genotypes. *Plant Physiology* 83, 185–189.

Tanner, W., Lehle, L. and Kandler, O. (1967) *myo*-Inositol, a cofactor in the biosynthesis of verbascose. *Biochemical and Biophysical Research Communications* 29, 166–171.

Tanner, W. and Kandler, O. (1968) *myo*-Inositol, a cofactor in the biosynthesis of stachyose. *European Journal of Biochemistry* 4, 233–239.

Wanek, W. and Richter, A. (1995) Purification and characterization of *myo*-inositol 6-*O*-methyltransferase from *Vigna umbellata* Ohwi et Ohashi. *Planta* 197, 427–434.

Yasui, T., Tateishi, Y. and Ohashi, H. (1985) Distribution of low molecular weight carbohydrates in the subgenus *Ceratotropis* of the genus *Vigna* (Leguminosae). *Botanical Magazine Tokyo* 98, 75–87.

8 Compartmentation of Abscisic Acid in Developing Muskmelon (*Cucumis melo* L.) Seeds

G.E. Welbaum[1], E.W. Pavel[1,3], D.R. Hill[2,4], M.K. Gunatilaka[1] and R.L. Grayson[1]

Departments of Horticulture[1] and Biochemistry[2], Virginia Tech, Blacksburg, VA 24061-0327, USA; [3]Department of Plant Production and Soil Sciences, University of Pretoria, Pretoria 0002, South Africa; [4]The Procter and Gamble Co., Environmental Sciences Department, ITC, 5299 Spring Grove Avenue, Cincinnati, OH 45217-1087, USA

Abscisic acid (ABA) levels vary dramatically during seed development, but little is known about changes in ABA compartmentation at the cellular and subcellular level. Soluble ABA was measured in developing muskmelon (*Cucumis melo* L.) endocarp, testa, endosperm, cotyledon, and embryonic axis tissue in fresh and dried seeds using an indirect enzyme-linked immunosorbant assay. Abscisic acid was visualized in immature 25 day after anthesis (DAA) and mature 55 DAA seed tissues at the subcellular level using transmission electron microscopy to detect gold particles conjugated to a polyclonal antibody raised against conjugated BSA-ABA. Soluble ABA concentrations were highest (250 ng g^{-1} DW) in embryonic axis tissue at 25 DAA and corresponded with rapid dry weight accumulation and development of desiccation tolerance. Soluble ABA concentrations in the embryonic axis dropped rapidly to 25 ng g^{-1} DW at 55 DAA as seeds reached maximum DW and developed full germinability. At 25 DAA, the highest density of gold particles was found in the nuclei, cytoplasm, and cell walls of the embryonic axis. In concert with the decline in ELISA measurements of ABA, the amount of gold labelling at 55 DAA was also reduced in embryonic axis tissue. Protein bodies were heavily labelled at 55 DAA, but little labelling was evident in the nucleus. Also at 55 DAA, labelling in the cytoplasm and cell wall was still evident. These results, along with recent studies with pea and lavender, show that ABA accumulates in the nucleus during periods of active gene expression.

Introduction

In dry-seeded crops such as wheat, rape and beans, the fruit tissues and the seed inside desiccate together, preventing precocious germination during the later stages of development (Walbot *et al.*, 1972; Barlow *et al.*, 1980; Schopfer

 Seed Biology: Advances and Applications (eds M. Black, K.J. Bradford and J. Vázquez-Ramos)

and Plachy, 1985). However, in fleshy fruits such as tomatoes or melons, germinable seeds are held at relatively high water contents for extended periods after the completion of seed development (Berry and Bewley, 1992; Welbaum, 1993). Viviparous germination of seeds in overripe ABA-deficient mutant tomato fruits, but not in the wild-type, suggests that ABA may prevent precocious germination (Groot *et al.*, 1991; Groot and Karssen, 1992). Immature seeds from ABA-deficient mutant lines of *Arabidopsis* will germinate precociously when isolated from the silique and incubated under moist conditions (Karssen *et al.*, 1983).

A number of studies have indicated that seed water relations may be as critical as ABA in regulating seed development and preventing precocious germination (Xu *et al.*, 1990; Galau *et al.*, 1991; Berry and Bewley, 1992, 1993). For example, seed water relations may also inhibit precocious germination in alfalfa, tomato, and muskmelon (Welbaum *et al.*, 1990; Xu and Bewley, 1991; Berry and Bewley, 1992). In a previous study, the water potential inside muskmelon fruits was sufficiently low to prevent precocious seed germination, although ABA levels during development were not measured (Welbaum *et al.*, 1990). By incubating seeds in combinations of ABA and polyethylene glycol (PEG), the same study concluded that ABA and osmoticum influenced germination rates and percentages by different mechanisms. Abscisic acid increases the minimum water potential threshold, while low water potential reduces turgor by reducing seed water content. However, endogenous seed ABA concentrations were not quantified in this study.

In addition to inhibiting precocious germination, ABA is also involved in the regulation of sink strength, primary dormancy, desiccation tolerance, and seed storage-protein accumulation as well as other developmental functions (Kermode, 1990; Hilhorst, 1995). Little is known about the compartmentation of ABA in seeds and how it changes during development. In this study, endogenous levels of ABA in embryonic axes, cotyledons, testae, endosperm, endocarp, and mesocarp were compared in immature 25 and fully mature 55 DAA seeds using ELISA and localized using immunocytochemistry.

Materials and Methods

Plant material

Commercial muskmelon (*Cucumis melo* L. cv. Top Mark, Asgrow Inc., Vineland, New Jersey) seeds were field grown as previously described, flowers were tagged at anthesis and fruits were harvested at 10 day intervals from 20 to 50 or 60 DAA (Welbaum and Bradford, 1988). Sprinkler irrigation was supplied at the rate of at least 2.5 cm a week to prevent drought stress. Fruits were harvested and immediately transported to the laboratory where seeds were removed within 2 h of harvest. Seeds were forced-air dried for 3 h and then dried in a desiccator to a water content of approximately 6% prior to either germination testing or storage.

Germination tests

For germination testing, two replications of 25 seeds or embryos each were placed in 9-cm wide, square Petri dishes on two thicknesses of germination blotter paper (Anchor Paper Co., St Paul, Minnesota) saturated with 10 ml of water or 50, 100, and 200 μM solutions of cis-trans (±) ABA (Sigma Chemical Co., St Louis, Missouri). Each group of seeds or embryos were incubated in the dark at 25 ± 1°C. Boxes were placed in self-sealing plastic bags to reduce evaporation. To determine the effects of water stress on ABA content of muskmelon seed tissues, intact dried 50 DAA seeds were imbibed for 12 h in 10 ml solutions of polyethylene glycol with initial water potentials of 0, −0.4, −0.6, −0.8 or −1.2 MPa prepared according to Michel (1983). The actual water potentials of the germination blotters were obtained from the average of three separate measurements made at the end of the experiment by osmometry (model 5100C, Wescor Inc., Logan, Utah).

Abscisic acid measurement

Abscisic acid was measured in three separate years. Two replications of 25 seeds or 5 g of mesocarp tissue were randomly selected from each fruit at 20, 30, 40, 50 and 60 DAA. At least two fruits were sampled from each stage of development. Endocarp, testa, endosperm envelope, cotyledon and embryonic axis were rapidly dissected from intact seeds in a cold room at 3°C and > 90% relative humidity. It took approximately 30 s to dissect each seed and all samples were covered to prevent moisture loss during preparation. Each replication of 25 seeds was stored on ice during collection, frozen in liquid nitrogen, and lyophylized. Duplicate samples of 25 seed tissues each were forced-air dried for 30 min, stored in a desiccator for 24 h, frozen in liquid nitrogen and lyophylized. Liquid endosperm was collected by the centrifugation of 100 decoated seeds in filtration tubes (Centrex 0.8 μM, Schleicher and Schuell, Inc., Keene, New Hampshire) at 5000 × *g* harvested either 10 or 15 DAA. In addition to the seed tissue samples, a 2 × 4 cm piece of mesocarp tissue was cut from around the locule of newly harvested fruits and immediately frozen in liquid nitrogen and lyophilized. All lyophilized samples were extracted in 80% methanol containing 100 mg l^{-1} of butylated hydroxytoluene for 2 h at 4°C in the dark and centrifuged for 30 min at 10,000 × *g*. The supernatant was dried and stored at −80°C until further processing.

Indirect ELISA and calculation of ABA concentrations in samples were as described by Walker-Simmons (1987). Monoclonal antibodies (MAb) were purchased (Idetek, San Bruno, California), and ABA-C4-BSA conjugate was made according to Quarrie and Galfre (1985) except that the dialysis volume was increased to 5 l (D. Still, personal communication). Standard solutions of (S)-(+)-ABA were prepared in 1 ml of methanol and diluted with TBS. A series of eight ABA standards over the concentration range of 50 to 400 pg 100 μl^{-1} was assayed on each microtitre plate.

Lyophilized tissue samples (1 g) were extracted in 80% methanol containing 10% butylated hydroxytoluene for 2 h at 4°C in the dark and centrifuged for 30 min at 10,000 × *g*. The supernatant was dried, redissolved in 1 ml of 10% methanol in water, and serial dilutions assayed to ensure the sample would fall within the range of the standard curve. Three separate aliquots were measured for each sample. Flat bottom, 96-well, microtitre plates (Immulon 4, Dynatech Laboratories, Alexandria, Virginia) were incubated in the dark at 25°C, until the absorbance of wells containing no ABA was approximately 1.0 at 405 nm (E-max, Molecular Devices, Menlo Park, California). Triplicate ABA standards for each concentration were assayed per plate, and the outer rows and columns were not used, to improve uniformity.

Electron microscopy

Tissues were excised from hydrated seeds and placed in ice-cold 2% EDC (1-(3-dimethylaminopropyl)-3 ethylcarbodiimide hydrochloride, Aldrich Chemical Co., Milwaukee, Wisconsin) in 10 mM PBS (phosphate-buffered saline) pH 7.4 to reduce the diffusion of ABA in the sample (Bertrand *et al.*, 1992). After 2 h incubation in EDC at 4°C, the tissues were placed in 1% glutaraldehyde, 8% paraformaldehyde in 10 mM PBS, pH 7.4, for 4 h at 4°C. Following fixation, samples were rinsed in 10 mM PBS, pH 7.4 for 5 min and dehydrated in a graded ethanol series. Tissues were infiltrated and embedded in Spurr's resin. Ultra-thin sections were placed on nickel grids and immunolabelled with polyclonal anti-ABA. Grids were stained with 2% uranyl acetate/lead citrate prior to viewing on a Zeiss 10CR transmission electron microscope. In some cases, grids were viewed without staining in order to distinguish the gold particles from the electron dense components of the cells.

Immunolabelling

Ultra-thin sections on nickel grids were blocked with 5% BSA (bovine serum albumin) in 10 mM PBS for 10 min. All incubations were performed at room temperature using 10 mM PBS, pH 7.4. A polyclonal anti-ABA antibody (PAb) was generated in rabbit against a (+)-2 *cis*, 4 *trans*-ABA-C4-BSA conjugate (Cocalico Biologicals, Reamstown, Pennsylvania). Grids were immediately incubated for 1 h with either anti-ABA PAb or MAb, 1:3000 dilution with 1% BSA in PBS. Following primary antibody incubation, grids were jet washed in PBS with 0.5 M NaCl and 0.05% Tween 20 followed by 15 min incubation in 0.05% Tween 20. Grids were blocked for 10 min with BSA prior to secondary antibody incubation with protein A gold (10 nm; Sigma, St Louis, Missouri) at 1:200 dilution for 20 min. Following secondary antibody incubation, grids were jet washed and incubated in Tween 20. Grids were jet washed with PBS, distilled water, and blotted dry.

Endo-β-mannanase assay

A solution of 0.05% locust bean gum (Megazyme, North Rocks, Sydney, Australia) was dissolved in 0.1 M citrate–0.2 m Na_2HPO_4, pH 5.5, by heating at 60°C. The suspension was cooled, centrifuged at 1700 × *g*, and the supernatant collected. Agarose (Phytogel, Sigma) (0.8%, w/v) was dissolved in the supernatant by heating in a microwave oven, and 20 ml was pipetted into 23 × 142 mm Petri plates and allowed to harden for 1 h. Wells were cut into the gel using a 3 mm diameter cork-borer. Fifteen microlitres of purified protein or commercially prepared endo-β-D-mannanase from *Aspergillus niger* (Megazyme) was placed in each well. Each plate was covered, sealed with Parafilm, and incubated for 22 h at 25°C. Plates were stained in 10 ml of a 0.5% (w/v) Congo Red dye in 0.2 M K_2HPO_4 and 0.05% (w/v) NaN_3, and the plates were shaken at 80 rpm for 20 min. Plates were destained in 0.1 M citrate–0.2 M Na_2HPO_4 pH 7.0 until the mannanase controls were transparent. The diameters of the transparent areas were measured with digital calipers in two directions and averaged. The activity in nKatals was calculated from a standard curve with four concentrations on each plate.

Results

Composition of muskmelon seeds

The testa and cotyledons comprise over 90% of the dry mass of muskmelon seeds (Table 8.1). The endocarp, or mucilaginous material that surrounds the outside of the seed, contributes little to the total dry mass but has the highest water content (Table 8.1). The cotyledons are the main source of storage reserves and had the lowest water content of any tissue (Table 8.1).

Endogenous ABA in undried, developing seed tissues

Abscisic acid concentrations in fresh, undried testae were the lowest of any seed tissue and varied little during development (Fig. 8.1A). Concentrations of ABA were much higher in endosperm tissue, peaking at 25 DAA then declining to a minimum at 50 DAA (Fig. 8.1B). The cotyledons contained comparatively less ABA with concentrations peaking at 25 DAA before steadily declining to low values during the later stages of development (Fig. 8.1C). The embryonic axis tissue had the highest ABA concentrations. ABA peaked in the embryonic axis at 25 DAA then declined sharply until 50 DAA (Fig. 8.1D). Liquid endosperm, isolated from 15 and 20 DAA immature seeds, contained 34 and 39 ng g^{-1} FW of ABA, respectively (data not shown).

A gelatinous layer of endocarp tissue that slowly decomposes during development surrounds immature muskmelon seeds. Abscisic acid concentrations in the endocarp were higher than in any other seed tissues at 20 DAA. Endocarp ABA declined sharply at 25 DAA and remained constant until

Table 8.1. Tissue composition of mature 50 DAA seeds.

Seed tissue	Percentage of total	Seed water content (% of DW)
Intact	100.0	83
Testa	42.0	104
Cotyledon	51.1	64
Embryonic axis	4.4	89
Endosperm	2.1	189
Endocarp	0.4	348

Values are expressed as a percentage of total intact seed weight after full hydration (cv. Top Mark). Water content of fully imbibed samples is expressed as a percentage of the oven DW (130°C, 1 h). Values represent the average of 50 seeds.

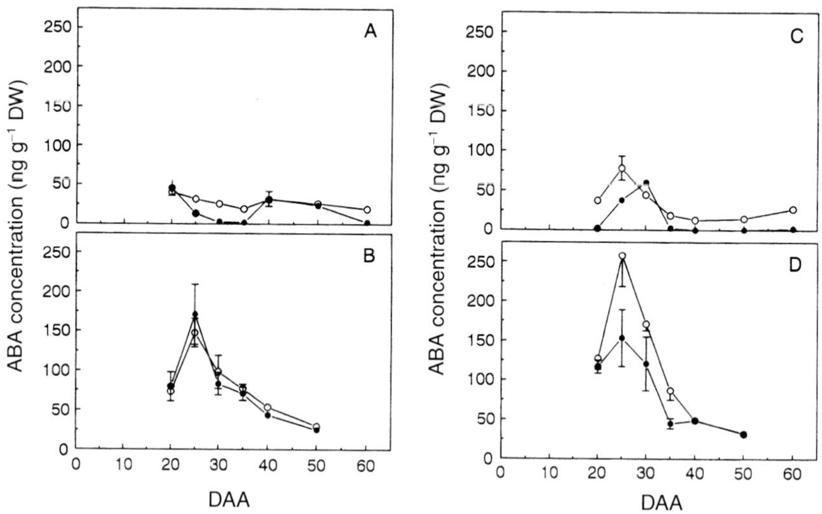

Fig. 8.1. Abscisic acid concentrations of fresh (○) and dry (●) muskmelon testae (A), endosperms (B), cotyledons (C), and embryonic axes (D) at various stages of development. Data for at least two fruits each from each stage of development were measured in three separate years and averaged. Error bars represent ± SE when larger than the symbols.

40 DAA then dropped to essentially nil at 60 DAA (Fig. 8.2). Abscisic acid concentration in fruit mesocarp tissue peaked at 30 DAA. Mesocarp ABA was less than 15% of that detected in adjacent endocarp tissue and was lower than in most seed tissues (Fig. 8.2).

Endogenous ABA after drying and rehydration

Rapid drying reduced ABA concentrations in testae at 25 to 35 and 60 DAA (Fig. 8.1A). Drying had little effect on ABA concentrations in the endosperm envelope (Fig. 8.1B). During the early stages of development, drying reduced ABA concentrations in the cotyledons and the embryonic axis (Fig. 8.1C, D).

The ABA content in cotyledons dissected from 50 DAA seeds imbibed in water for 20 h was slightly less than in cotyledons of newly harvested seeds from the same stage of development (cf. Table 8.2, Fig. 8.1C). Abscisic acid concentrations in 50 DAA embryonic axes dissected from seeds incubated in water for 20 h were higher than in 50 DAA fresh seeds (cf. Table 8.2, Fig 8.1B, D). Incubation of dried seeds at reduced water potential increased ABA in the cotyledons over the first 36 h of imbibition, but at 120 h concentrations had declined to the same level as cotyledons imbibed for 20 h in water (Table 8.2).

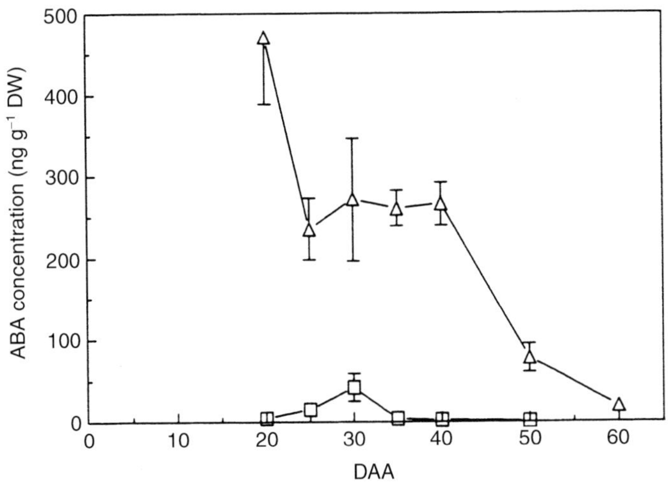

Fig. 8.2. Abscisic acid concentrations in muskmelon endocarp (△) and mesocarp (□) at various stages of development. Error bars represent ± SE when larger than the symbols.

Table 8.2. Effect of reduced water potential on ABA content in 50 DAA muskmelon seed tissues.

Incubation period (h)	Initial water potential (MPa)	Final water potential (MPa)	ABA concentration (ng g^{-1} DW)	
			Cotyledons	Embryonic axis
20	0.0	−0.2	45.9	101.2
20	−0.4	−0.6	67.5	129.8
20	−0.8	−1.2	85.3	156.7
36	−1.2	−1.7	87.6	165.9
120	−1.2	−2.0	40.8	155.9

$LSD_{0.05}$ = 37.2.
Intact dried 50 DAA seeds were imbibed for 12 h in 10 ml solutions of PEG with initial water potentials of 0, −0.4, −0.6, −0.8, or −1.2 MPa prepared according to Michel (1983). The actual water potentials of the germination blotters were obtained from the average of three separate measurements made by osmometry at the end of the experiment.

The ABA concentration of the embryonic axis increased during incubation at reduced water potential, but unlike the cotyledons, concentrations remained elevated after 120 h (Table 8.2).

Immunolocalization of ABA in immature and mature seed tissues

Embryonic axis tissue from immature, fresh 25 DAA seeds was randomly labelled throughout the cytoplasm (Fig. 8.3A). There was no labelling inside organelles except for the nucleus, where the region of heterochromatin appeared to be especially heavily labelled. The majority of label in the endosperm appeared in the nucleus and cytoplasm (Fig. 8.3B). In anticlinal section, the endosperm is primarily composed of thick cell walls. Gold label accumulated in the cell wall at the boundary between the endosperm and cotyledon tissues (Fig. 8.3B). Labelling in the endosperm was only visible in the pockets of cytoplasm scattered throughout the tissue. Abscisic acid was fixed in the tissue using EDC (Fig. 8.3A, B). However, 25 DAA embryonic axis tissue (Fig. 8.3C) and endosperm tissue (Fig. 8.3D) processed without EDC were labelled in the cytoplasm and nucleus indicating that ABA was tightly bound.

Tissue from undried, fully mature, 55 DAA muskmelon seeds was densely cytoplasmic with prominent liposomes, protein bodies, and nuclei (Fig. 8.4A, B). In agreement with ELISA measurements of soluble ABA, the amount of label in embryonic axis tissue at 55 DAA was greatly reduced compared to 25 DAA seeds (Fig. 8.4B). Labelling was most evident in protein bodies in the embryonic axis. The amount of labelling in the nucleus at 55 DAA was greatly reduced compared to 25 DAA when the nucleus was heavily labelled. Labelling in the cytoplasm was also greatly reduced compared to the earlier stage of development. In 40 DAA seeds imbibed in 100 μM ABA to full hydration, labelling increased in the cytoplasm and protein bodies of the embryonic axis but not in other organelles (Fig. 8.4C). The endosperm tissue at 55 DAA was primarily cell wall material with little cytoplasm present (Fig. 8.4A, D). Labelling in the endosperm was primarily confined to the cell walls with little labelling in the symplast (Fig. 8.4D). The endosperm tissue in 55 DAA seeds was labelled in the cytoplasm and nucleus (Fig. 8.4E). The degree of labelling in the endosperm of seeds incubated in 100 μM ABA was similar to that in mature fresh seeds, indicating that exogenous ABA did not accumulate in the endosperm (Fig. 8.4F).

Effect of ABA and reduced water potential on endo-β-mannanase activity

The endo-β-mannanase activity measured in 50 or 55 DAA endosperm tissue adjacent to the radicle was highest in seeds incubated in water for 20 h (Table 8.3). Incubating seeds for 20 h at either −0.4 or −0.6 MPa reduced endo-β-mannanase activity by more than half as did incubation for 36 h at −1.2 MPa. Incubating seeds for 20 h in 100 or 200 μM ABA decreased endo-β-mannanase by 32% (Table 8.3).

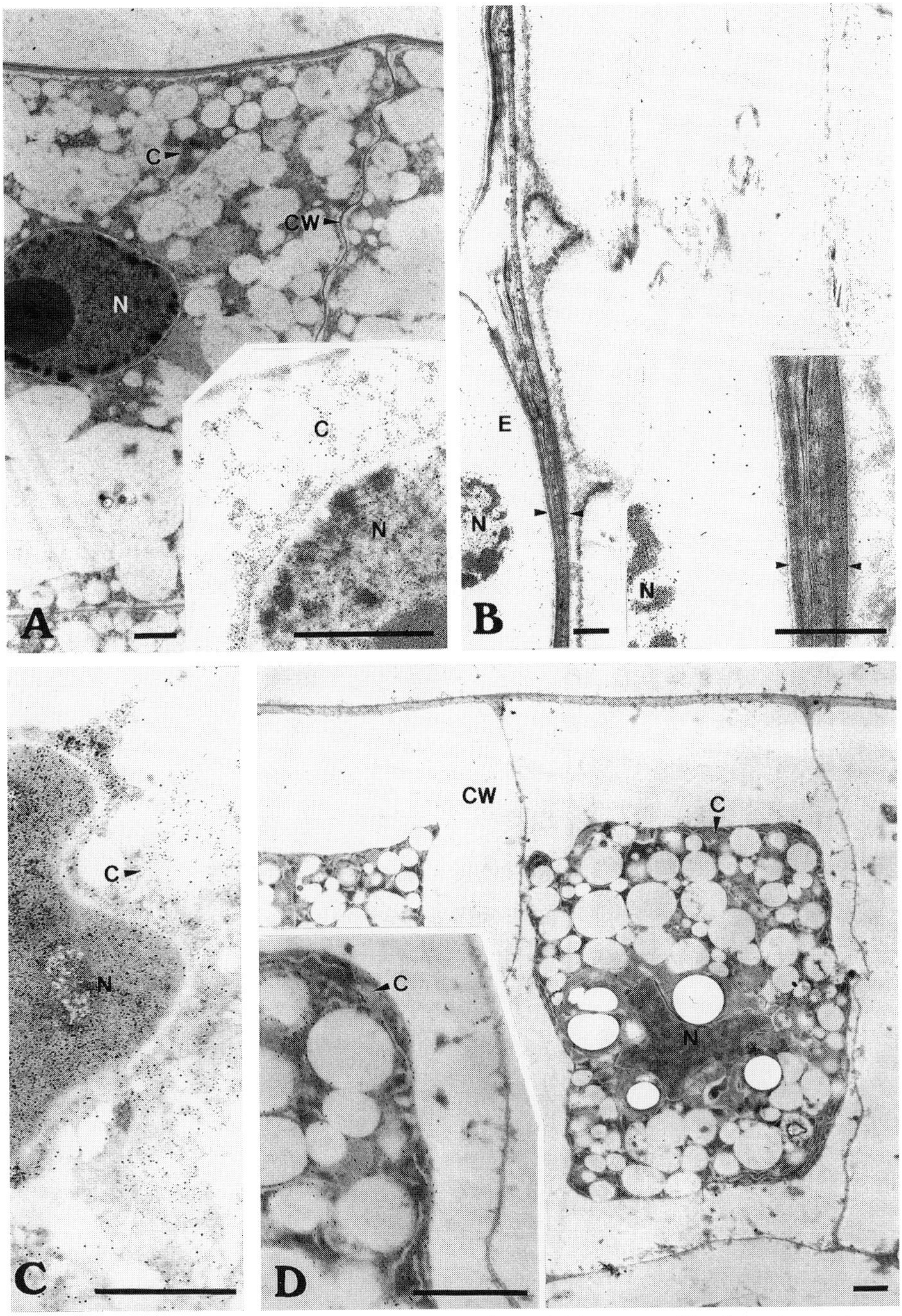

Fig. 8.3. A, 25 DAA embryonic axis tissue. Inset shows higher magnification of a gold-labelled nucleus. B, anticlinal section of 25 DAA endosperm tissue. Inset shows gold-labelling at a higher magnification. C, 25 DAA embryonic axis tissue that was not treated with EDC. D, periclinal section of 25 DAA endosperm tissue processed without EDC. Inset shows a higher magnification. c, cytoplasm; cw, cell wall (limits arrowed in B); E, endosperm; N, nucleus. Bars represent 1 μm.

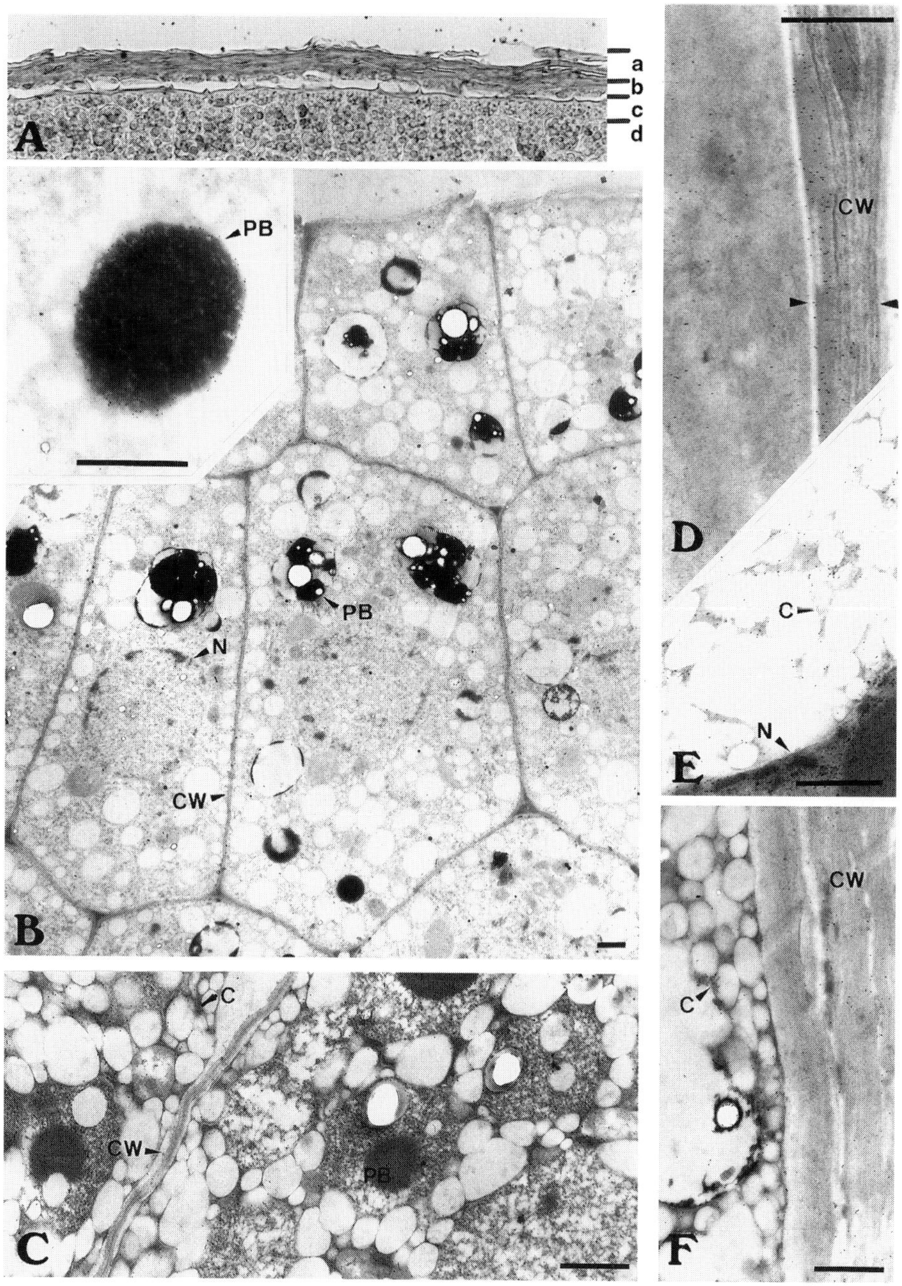

Fig. 8.4. A, transverse section of a 55 DAA decoated muskmelon seed stained with Sudan IV (227× magnification); a, callose layer; b, endosperm layer; c, epidermis of the cotyledon; d, cotyledon tissue. B, 55 DAA muskmelon embryonic axis tissue. Inset shows a higher magnification of a gold-labelled protein body. C, 40 DAA embryonic axis tissue after imbibition in 100 μM (±) *cis*-4 *trans*-ABA. D, transverse section of 55 DAA endosperm cell walls. E, endosperm tissue from 55 DAA seeds. F, 40 DAA endosperm tissue after imbibition in 100 μM ABA. c, cytoplasm; cw, cell wall (limits arrowed in D); N, nucleus; PB, protein body. Bars represent 1 μm.

Table 8.3. Comparison of gel diffusion mannanase activities from muskmelon endosperm tissue adjacent to the radicle.

Incubation period (h)	Initial water potential (MPa)	Mannanase activity (nKatals)	ABA concentration (μM)	Mannanase activity (nKatals)
20	0	6.6 ± 1.1	0	6.3 ± 0.5
20	−0.4	3.2 ± 0.7	100	4.3 ± 0.5
20	−0.6	2.6 ± 0.8	200	4.3 ± 0.5
36	−1.2	3.1 ± 0.6		

Samples were collected from 50 or 55 DAA intact seeds incubated at 25°C in ABA or PEG solutions of reduced water potential.

Discussion

Seed development in muskmelon cv. Top Mark was characterized previously (Welbaum and Bradford, 1988, 1989). Generally, maximum dry weight was obtained 35 DAA, but 100% viability was not obtained until 45 DAA (Welbaum and Bradford, 1988). Seeds at 25 DAA were rapidly growing and could not germinate unless embryos were rescued from the seed and incubated in water (Welbaum and Bradford, 1989). In contrast, 50 and 55 DAA intact seeds were fully viable and vigorous (Welbaum and Bradford, 1989).

Abscisic acid levels were highest early in development and declined dramatically as seeds matured (Figs 8.1, 8.2). The decline in endogenous ABA was correlated with a number of developmental processes widely associated with ABA. The development of viability, desiccation tolerance, dry weight accumulation, and loss in sensitivity to exogenous ABA all corresponded with changes in ABA concentrations as expected (cf. Fig. 8.1 and Welbaum and Bradford, 1989). However, changes in ABA concentration varied widely among individual seed tissues.

The large size of muskmelon seeds allowed for dissection and separate analysis of the embryonic axis and endosperm even though these tissues constitute only a small percentage of the total seed (Table 8.1). Abscisic acid was highest in endocarp, embryonic axis, and endosperm envelope tissue prior to maximum dry weight accumulation at 35 DAA (Figs 8.1, 8.2; Welbaum and Bradford, 1988, 1989). The sheath surrounding tomato seeds was similarly found to contain high concentrations of ABA (Berry and Bewley, 1992). Desiccation tolerance was attained after 30 DAA as ABA concentrations declined in muskmelon embryonic axis and cotyledon tissue (Welbaum and Bradford, 1989). The decline in ABA concentrations in the embryonic axis and endosperm were closely linked to an increase in the germinability of intact fresh seeds in water from 5% at 25 DAA to 98% 45 DAA in a previous report (Fig. 8.1; Welbaum *et al.*, 1990). Germination of mature *Arabidopsis* and tomato seeds is well correlated with endogenous seed ABA content although ABA concentrations in individual seed tissues were not reported (Hilhorst, 1995).

Immature muskmelon seeds were very sensitive to exogenous ABA early in development when endogenous ABA was high (Fig. 8.2). The decline in

ABA sensitivity in muskmelon seeds occurred later in development approximately 10 days after the sharp decline in endogenous ABA (Figs 8.1, 8.2; Welbaum *et al.*, 1990). Decreased sensitivity to exogenous ABA also corresponded with a drop in endogenous ABA in rape, castor bean, and wheat (Finkelstein and Crouch, 1986; Walker-Simmons, 1987; Kermode *et al.*, 1989). Early in alfalfa seed development, sensitivity to ABA decreased as endogenous ABA increased, but the two declined in concert later in development (Xu *et al.*, 1990; Xu and Bewley, 1991). Changes in sensitivity to ABA in seeds have been linked to tissue dehydration (Kermode, 1990). Muskmelon seeds dry to 35% water content on a fresh weight basis inside developing fruits (Welbaum and Bradford, 1989). This drop paralleled the decline in endogenous ABA but not the change in sensitivity to exogenous ABA (Welbaum *et al.*, 1990).

Muskmelon and tomato seeds do not germinate precociously during development because the low water potential inside the fruit prevents seeds from becoming sufficiently hydrated (Welbaum *et al.*, 1990; Berry and Bewley, 1992). However, the high endogenous ABA in the embryonic axis and endosperm coincided with the dormancy and weak germination exhibited by some seeds between 20 and 40 DAA (Welbaum and Bradford, 1988, 1989). Therefore, precocious germination is a very rare event in 'Top Mark' muskmelon because both low fruit water potential and high ABA concentrations in the embryonic axis and endosperm jointly inhibit germination of immature seeds. During the later stages of fruit development when seeds are fully developed, osmotic inhibition is the primary mechanism responsible for quiescence in muskmelon and tomato fruits (Welbaum *et al.*, 1990; Berry and Bewley, 1992).

Rapid drying increased germination percentages between 25 and 40 DAA as ABA concentrations in the cotyledons and, primarily, the embryonic axis declined (Fig. 8.1; Welbaum and Bradford, 1989). Slow drying of *Helianthus* embryos significantly decreased ABA amounts, while germination percentages increased (Bianco *et al.*, 1994). *In situ* ABA synthesis was also required to impose and maintain embryo dormancy in *Helianthus annus* (Le Page-Degivry and Garello, 1992).

Afterripening of muskmelon seeds was apparently not correlated with declining ABA because concentrations in all tissues were low before afterripening began. Rehydrated, afterripened 50 DAA seeds had ABA concentrations that were at least as high as at harvest (cf. Fig. 8.1B, D and Table 8.2). Dry storage of *Helianthus* seeds for 6 weeks did not further reduce ABA but increased germinability similar to muskmelon (Le Page-Degivry and Garello, 1992). However, in tomato seeds, ABA amounts declined by one half during one year of dry storage, and the decrease was correlated with a loss of dormancy upon reimbibition (Hilhorst, 1995).

Osmotic solutions and ABA act by similar mechanisms to inhibit germination (Welbaum *et al.*, 1990; Ni and Bradford, 1993). Water stress reduces turgor while ABA increases the yield threshold or the base or threshold water potential preventing radicle emergence (Bradford, 1995). To test whether reduced water potential may also stimulate ABA synthesis, seeds were incubated in osmotic solutions. In the embryonic axis, ABA concentrations increased at all water potentials and remained high after incubation for 120 h (Table 8.2).

Abscisic acid increased in response to water stress in the cotyledons, although this appeared to be a short-term response because concentrations were similar to initial values after 120 h (Table 8.2). Abscisic acid did not increase in tomato seeds incubated in osmotic solutions (Ni and Bradford, 1993).

While ABA is known to inhibit germination in many seeds by increasing the base or threshold water potential required for radicle emergence, the specific mechanism responsible for this inhibition is poorly understood (Schopfer and Plachy, 1985; Bradford, 1995). Endo-β-mannanase is a cell-wall hydrolase known to degrade barrier tissues adjacent to the radicle in germinating tomato, pepper, and muskmelons seeds (Welbaum *et al.*, 1998). Endo-β-mannanase activities decreased similarly after incubation in both ABA and osmotic solutions (Table 8.3). This suggests that reduced cell-wall hydrolase activity may be common to both osmotic and ABA inhibition of germination in muskmelon seeds. However, inhibitory concentrations of osmoticum and ABA either had no effect or enhanced endo-β-mannanase activity in tomato seeds (Still and Bradford, 1997).

The compartmentation of ABA changed during development. Nuclear labelling was apparent in immature but not mature seeds (cf. Figs 8.3 and 8.4). The cytoplasm was uniformly labelled in the embryonic axis and endosperm of immature tissue, and label was noticeably absent from vacuoles and other organelles at 25 DAA (Fig. 8.4A, D). At 55 DAA, labelling was less evident in all tissues, consistent with decreased measurements of soluble ABA by ELISA (Fig. 8.1). When 40 DAA seeds were imbibed in 100 μM ABA, label accumulated in the cytoplasm, protein bodies, and nuclei of the embryonic axis similar to the pattern associated with immature tissue (Fig. 8.4C).

The reason for ABA accumulation in the nucleus is unknown. Most models of ABA activity feature binding of ABA to plasma membrane receptors that initiate a second messenger cascade or penetrate the cell membrane to bind to internal targets (Hetherington and Quatrano, 1991). No distinctive membrane labelling was evident in the immunogold electron micrographs (Fig. 8.3B). It is unlikely that this technique could detect bound ABA because of the limited antigenic sites on bound ABA and the small target area in ultra-thin tissue sections.

A number of steps were taken to prevent non-specific antibody binding with cellular components. The Pab used for immunogold labelling detected (S)-(+)-ABA and (±)-ABA, and showed no cross reactivity with phaseic acid, (±)-abscisyl alcohol, (–)-ABA, (±)-abscisyl aldehyde, BSA, or EDC (Welbaum *et al.*, 1997). Plus and minus EDC controls tested the effectiveness of the fixative (Fig. 8.3D). The tissue was blocked before the addition of the secondary antibody and the Pab was preabsorbed on BSA. However, it is impossible to rule out the possibility that labelling was non-specific. However, gold labelling of specific organelles only at certain stages of development along with organelle-specific labelling after imbibition in ABA solutions suggests that these results were not artifactual. Immunogold labelling of ABA in the nucleus and cytoplasm increased in pea tissue after cold treatment as ABA concentrations and the synthesis of cold-induced proteins also increased (Welbaum *et al.*, 1997). Immunogold labelling was highest in nuclei of drought-stressed lavender leaves (Pastor *et al.*, 1999). In both studies, immunogold labelling occurred in the nuclei during periods of ABA-induced gene expression.

Acknowledgements

We thank Jennifer Mason for her assistance with ELISA. Supported in part by USDA/NRI grant 91-7100-6634 to R.L.G. and G.E.W. and USDA Regional Research Project W-168.

References

Barlow, E.W.R., Lee, J.W., Munns, R. and Smart M.G. (1980) Water relations of developing wheat grain. *Australian Journal of Plant Physiology* 7, 519–525.

Berry, T. and Bewley, J.D. (1992) A role for the surrounding fruit tissues in preventing the germination of tomato (*Lycopersicon esculentum*) seeds. *Plant Physiolology* 100, 951–957.

Berry, T. and Bewley, J.D. (1993) Comparisons between the roles of the fruit tissues, osmoticum and abscisic acid in maintaining tomato seed development and storage protein synthesis. *Seed Science Research* 3, 25–34.

Bertrand, S., Benhamou, N., Nadeau, P., Dostales, D. and Gosselin, A. (1992) Immunogold localization of free abscisic acid in tomato root cells. *Canadian Journal of Botany* 70, 1001–1011.

Bianco, J., Garello, G. and Le Page-Degivry, M.T. (1994) Release of dormancy in sunflower embryos by dry storage: involvement of gibberellins and abscisic acid. *Seed Science Research* 4, 57–62.

Bradford, K.J. (1995) Water relations in seed germination. In Kigel, J. and Galili, G. (eds) *Seed Development and Germination.* Marcel Dekker, Inc., New York, pp. 351–396.

Finkelstein, R.R. and Crouch, M.L. (1986) Rape embryo development in culture on high osmoticum is similar to that in seeds. *Plant Physiology* 81, 907–912.

Galau, G.A., Jakobsen, K.S. and Hughes, D.W. (1991) The controls of late dicot embryogenesis and early germination. *Physiologia Plantarum* 81, 280–288.

Groot, S.P.C. and Karssen, C.M. (1992) Dormancy and germination of abscisic acid-deficient tomato seeds. *Plant Physiology* 99, 952–958.

Groot, S.P.C., Van Yperen, I.I. and Karssen, C.M. (1991) Strongly reduced levels of endogenous abscisic acid in developing seeds of tomato mutant *sitiens* do not influence *in vivo* accumulation of dry matter and storage proteins. *Physiologia Plantarum* 81, 73–78.

Hetherington, A.M. and Quatrano, R.S. (1991) Mechanisms of action of abscisic acid at the cellular level. *New Phytologist* 119, 9–32.

Hilhorst, H.W.M. (1995) A critical update on seed dormancy I. Primary dormancy. *Seed Science Research* 5, 61–74.

Karssen, C.M., Brinkhorst-van der Swan, D.L.C., Breekland, A.E. and Koornneef, M. (1983) Induction of dormancy during seed development by endogenous abscisic acid: studies on abscisic acid deficient genotypes of *Arabidopsis thaliana. Planta* 157, 158–165.

Kermode, A.R. (1990) Regulatory mechanisms involved in the transition from seed development to germination. *Critical Reviews of Plant Science* 9, 155–196.

Kermode, A.R., Dumbroff, E.B. and Bewley, J.D. (1989) The role of maturation drying in the transition from seed development to germination. VII. Effects of partial and complete desiccation on abscisic acid levels and sensitivity in *Ricinus communis* L. seeds. *Journal of Experimental Botany* 40, 303–313.

Le Page-Degivry, M.T. and Garello, G. (1992) *In situ* abscisic acid synthesis. A requirement for induction of embryo dormancy in *Helianthus annuus. Plant Physiology* 98, 1386–1390.

Michel, B.E. (1983) Evaluation of the water potentials of solutions of polyethylene glycol 8000 both in the absence and presence of other solutes. *Plant Physiology* 72, 66–70.

Ni, B.R. and Bradford, K.J. (1993) Germination and dormancy of abscisic acid- and gibberellin-deficient mutant tomato seeds. Sensitivity of germination to abscisic acid, gibberellin, and water potential. *Plant Physiology* 101, 607–617.

Pastor, A., López-Carbonell, M. and Alegre, L. (1999) Abscisic acid immunolocalization and ultrastructural changes in water-stressed lavender (*Lavendula stoechas* L.) plants. *Physiologia Plantarum* 105, 272–279.

Quarrie, S.A. and Galfre, G. (1985) Use of different hapten-proteins conjugates immobilized on nitrocellulose to screen monoclonal antibodies to abscisic acid. *Analytical Biochemistry* 151, 389–399.

Schopfer, P. and Plachy, C. (1985) Control of seed germination by abscisic acid. III. Effect on embryo growth potential (minimum turgor pressure) and growth coefficient (cell wall extensibility) in *Brassica napus* L. *Plant Physiology* 77, 676–686.

Still, D.W. and Bradford, K.J. (1997) Endo-β-mannanase activity from individual tomato endosperm caps and radicle tips in relation to germination rates. *Plant Physiology* 113, 21–29.

Walbot, V., Clutter, M. and Sussex, I.M. (1972) Reproductive development and embryogeny in *Phaseolus. Phytomorphology* 22, 59–68.

Walker-Simmons, M. (1987) ABA levels and sensitivity in developing wheat embryos of sprouting resistant and susceptible cultivars. *Plant Physiology* 84, 61–66.

Welbaum, G.E. (1993) Water relations of seed development and germination in muskmelon (*Cucumis melo* L.). VIII. Development of osmotically distended seeds. *Journal of Experimental Botany* 44, 1245–1252.

Welbaum, G.E. and Bradford, K.J. (1988) Water relations of seed development and germination in muskmelon (*Cucumis melo* L.). I. Water relations of seed and fruit development. *Plant Physiology* 81, 406–411.

Welbaum, G.E. and Bradford, K.J. (1989) Water relations of seed development and germination in muskmelon (*Cucumis melo* L.). II. Development of germinability, vigour, and desiccation tolerance. *Journal of Experimental Botany* 40, 1355–1362.

Welbaum, G.E., Tissaoui, T. and Bradford, K.J. (1990) Water relations of seed development and germination in muskmelon (*Cucumis melo* L.). III. Sensitivity of germination to water potential and abscisic acid during development. *Plant Physiology* 92, 1029–1037.

Welbaum, G.E., Bian, D., Hill, D.R., Grayson, R.L. and Gunatilaka, M. (1997) Freezing tolerance, protein composition, and abscisic acid localization and content of pea epicotyl, shoot, and root tissue in response to temperature and water stress. *Journal of Experimental Botany* 48, 721–733.

Welbaum, G.E., Bradford, K.J., Yim, K-O., Booth, D.T. and Oluoch, M.O. (1998) Physiological and biochemical processes regulating seed germination. *Seed Science Research* 8, 123–135.

Xu, N. and Bewley, J.D. (1991) Sensitivity to abscisic acid and osmoticum changes during embryogenesis of alfalfa (*Medicago sativa*). *Journal of Experimental Botany* 42, 821–826.

Xu, N., Coulter, K.M. and Bewley, J.D. (1990) Abscisic acid and osmoticum prevent germination of developing alfalfa embryos, but only osmoticum maintains the synthesis of developmental proteins. *Planta* 182, 382–390.

9

Involvement of ABA and GAs in the Regulation of Dormancy in Developing Sorghum Seeds

R.L. Benech-Arnold[1], S. Enciso[1], R.A. Sánchez[1], F. Carrari[2], L. Perez-Flores[3], N. Iusem[2], H. Steinbach[1], D. Lijavetzky[4] and R. Bottini[5]

[1]*IFEVA, Facultad de Agronomía, Universidad de Buenos Aires, Argentina;* [2]*Laboratorio de Fisiología y Biología Molecular, FCEyN, Universidad de Buenos Aires, Argentina;* [3]*Departamento de Ciencias de la Salud, DCBS, Universidad Autónoma Metropolitana-Iztapalapa, D.F. Mexico;* [4]*Instituto de Biotecnología, CICV, INTA, Castelar, Argentina;* [5]*Laboratorio de Fisiologia Vegetal, Universidad Nacional de Río Cuarto, Argentina*

The participation of the hormone abscisic acid (ABA) in the control of dormancy of developing seeds is well documented for a number of species. It has recently been demonstrated that the degree of dormancy of developing sorghum seeds also depends on the extent to which ABA action is counterbalanced by the effect of gibberellins (GAs). The available evidence is discussed in relation to the possible role of: (i) a different endogenous concentration of either hormone; and (ii) a different sensitivity to the effects of ABA and GAs, in deciding the direction of the hormone balance in developing caryopses from varieties with contrasting dormancy. Although the endogenous concentration of ABA appeared to be similar in embryos from both varieties, embryos from the low-dormancy variety Redland B2 were found to be less sensitive to the inhibitory effect of ABA than embryos from the high-dormancy variety IS 9530. In addition, a genetic analysis using quantitative trait loci (QTLs) suggests the involvement of the transcription factor *Vp1* (whose participation in the control of embryo sensitivity to ABA is well documented for maize) in the determination of the dormancy level of these two varieties. On the other hand, neither GAs content nor embryo sensitivity to GAs appeared to be very different between high- and low-dormancy varieties. These results show that the only regulatory event so far capable of deciding the direction of the hormone balance and consequently determining the different degree of dormancy between varieties, is most likely established at the level of embryo sensitivity to ABA. However, other dormancy mechanisms not necessarily related to ABA action might be operating and further work is required to elucidate them.

Introduction

Sprouting resistance in sorghum is related to the maintenance of a sufficient dormancy level until late stages of seed development and maturation (Steinbach *et al.*, 1995). Indeed, although isolated embryos from both susceptible and resistant varieties can germinate equally well in water from early stages of development (i.e. 15 days after pollination (DAP) or earlier), the germination capacity of intact developing caryopses from resistant varieties is much less than that observed for susceptible genotypes and this higher dormancy is maintained until well after physiological maturity (Fig. 9.1). This coat-imposed dormancy is the barrier preventing untimely germination.

Research on the mechanisms of dormancy in the developing seeds of many species suggests a strong involvement of the phytohormone abscisic acid (ABA) (King, 1982; Fong *et al.*, 1983; Karssen *et al.*, 1983; Walker-Simmons, 1987; Black, 1991). Abscisic acid-deficient or -insensitive mutants of *Arabidopsis* and maize precociously germinate (Robichaud *et al.*, 1980; Karssen *et al.*, 1983) and application of the ABA-synthesis inhibitor fluridone has been shown to reduce dormancy in developing seeds of some species (Fong *et al.*, 1983; Xu *et al.*, 1990). In sorghum, the participation of ABA in the maintenance of dormancy is evidenced by the fact that inhibition of ABA

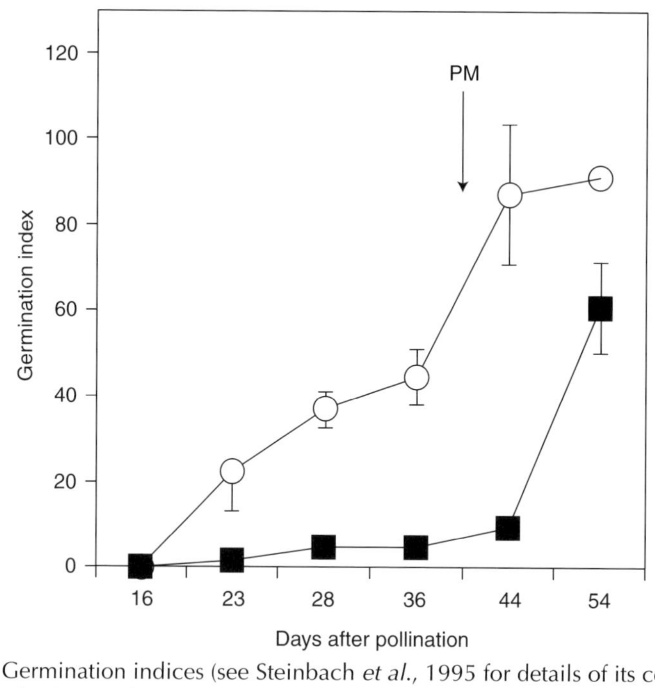

Fig. 9.1. Germination indices (see Steinbach *et al.*, 1995 for details of its construction) of developing sorghum caryopses harvested at different times after pollination and incubated at 25°C. Sprouting-susceptible variety Redland B2 (○) and sprouting-resistant IS 9530 (■). The arrow indicates the onset of physiological maturity (PM). Vertical bars, mean ± SE.

synthesis with fluridone applied at early developmental stages accelerates the termination of dormancy (Steinbach *et al.*, 1997).

The germination-promoting effect of gibberellins (GAs) is well documented for mature seeds of a number of species (Lona, 1956; Karssen *et al.*, 1989; Hilhorst, 1995; Karssen, 1995). It has been recently demonstrated that the degree of dormancy of developing sorghum seeds also depends on the extent to which ABA action is counterbalanced by the effect of GAs. Indeed, early application (i.e. 7 DAP) of the GAs synthesis inhibitor paclobutrazol, dramatically extends the period during which developing caryopses from a sprouting-susceptible variety show a high dormancy level (Steinbach *et al.*, 1997). On the other hand, applied GA_{4+7} stimulates germination of dormant, developing grains (Steinbach *et al.*, 1997). This suggests that the dormancy level of immature caryopses results from the balance between ABA and GA action. Hence, it could be expected that the regulatory event deciding the direction of the balance and consequently determining the different degree of dormancy between sprouting-susceptible and -resistant varieties, is established at one or more of the following levels: (i) a different endogenous concentration of either hormone; and/or (ii) a different sensitivity to the inhibitory effect of ABA and/or the promotive effect of GAs. Throughout this paper we discuss the available evidence, in relation to the possible role of one or more of the above-mentioned features in deciding the direction of the hormone balance, in developing caryopses from a sprouting-susceptible variety (Redland B2) and from a sprouting-resistant one (IS 9530).

Endogenous ABA Content throughout Development in Embryos from Sprouting-susceptible and -resistant Varieties

ABA content was measured in Redland B2 (susceptible) and IS 9530 (resistant) embryos excised from caryopses at various stages of development, using a radioimmunoassay as described elsewhere (Steinbach *et al.*, 1995). In spite of differences between varieties at some stages of development, it was clear that embryos from the more dormant variety IS 9530 did not have a constantly higher ABA content than embryos from the less dormant variety Redland B2 (Fig. 9.2). Therefore, although the possibility of a different ABA concentration at the site of action of the hormone cannot be ruled out, the direction of the hormone balance does not appear to be decided at the level of the inhibitor concentration.

The Role of Embryo Sensitivity to ABA in Deciding the Direction of Balance of the Hormone Effects

Sensitivity to ABA was assessed as the inability of the excised embryos to overcome the inhibitory effect of 50 μM ABA. Embryos from the less dormant Redland B2 were found to start losing sensitivity to this ABA concentration at very early stages of development (i.e. 23 DAP) since, from there on, embryo

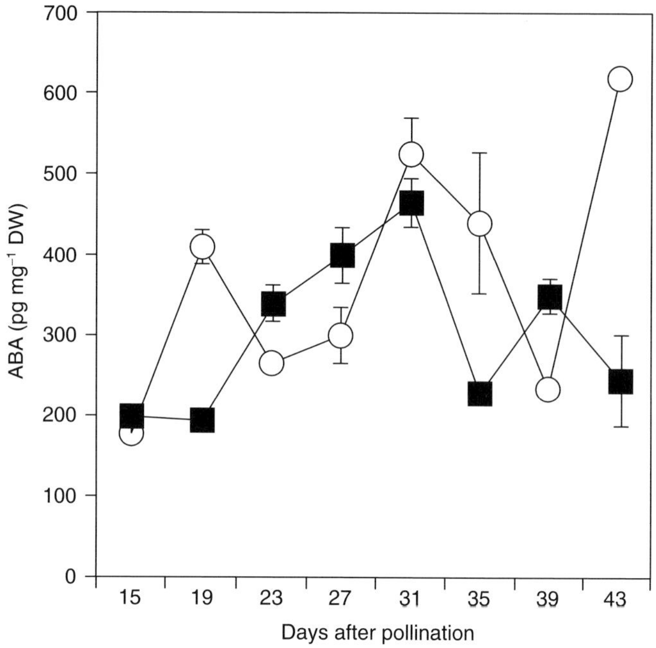

Fig. 9.2. Abscisic acid content in Redland B2 (○) and IS 9530 (■) embryos excised from the caryopses at different times after pollination. Vertical bars, mean ± SE.

germination was not effectively prevented (Fig. 9.3). In contrast, embryos from the dormant variety IS 9530 were not able to germinate when incubated in the presence of 50 μM ABA until well after physiological maturity (i.e. 44 DAP) (Fig. 9.3). These results clearly show that, although the endogenous concentration of ABA appears to be similar in embryos from both varieties, a higher concentration of the hormone is required to suppress germination of Redland B2 embryos during most of the maturation period.

In a genetic analysis of dormancy in sorghum, Lijavetzky *et al.* (1999) have shown that the character presents continuous variation in a segregating F2 population generated by crossing between IS 9530 and Redland B2, and found two unlinked quantitative trait loci (QTLs) that, together, explain more than 80% of the observed phenotypic variance. One of those QTLs was linked to the RFLP marker UMC3 that, in maize, is linked to the gene *Vp1*. This gene encodes a transcription factor which is involved in the control of embryo sensitivity to ABA. These results, together with those on embryo sensitivity to ABA, suggest the participation of *Vp1* in the control of sprouting resistance in sorghum. We cloned and partially sequenced *Vp1* from our sorghum varieties and found that the two alleles differ in about 2% in their sequence (F. Carrari *et al.*, unpublished observations). Analysis of gene expression throughout development in seeds from Redland B2 and IS 9530, however, showed a slightly higher *Vp1* expression in the former, in contrast to what could have been expected (Fig. 9.4). More interestingly, a different timing of expression was also observed between varieties; while *Vp1* expression in embryos from

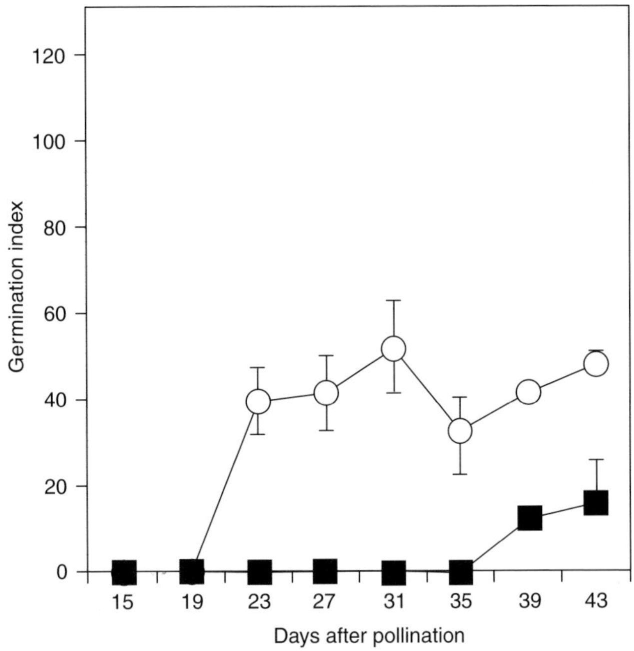

Fig. 9.3. Germination indices of Redland B2 (○) and IS 9530 (■) embryos excised at different times after pollination and incubated at 25°C in the presence of 50 μM ABA for 12 days. Vertical bars, mean ± SE.

the sprouting-susceptible Redland B2 keeps decreasing throughout development, expression in IS 9530 embryos, peaks at half way to harvest maturity (Fig. 9.4). Although much work is still required to understand the significance, if any, of this different pattern of gene expression, the following working hypothesis could be postulated: in IS 9530 embryos, sensitivity to ABA is not under *Vp1* control at early stages of development, but the late peak of *Vp1* expression in IS 9530 embryos is responsible for the maintenance of a high embryo sensitivity to ABA until late stages of maturation; in contrast, in Redland B2 embryos, sensitivity to ABA is under *Vp1* control from early stages of development and consequently keeps decreasing together with *Vp1* expression.

Both the physiological and the genetic studies strongly point to embryo sensitivity to ABA as a regulatory site of the hormone balance.

Endogenous GAs in Caryopses from Sprouting-resistant and -susceptible Varieties

Previous results had shown that inhibition of GA synthesis with paclobutrazol applied as early as 7 DAP, dramatically extended the duration of dormancy in the sprouting-susceptible variety Redland B2 (Steinbach *et al.*, 1997). These results not only were evidence for the participation of GAs in the regulation of dormancy but also suggested the possibility of a higher GAs content in sprouting-susceptible varieties. Therefore, IS 9530 and Redland B2 grains and excised embryos collected at different stages of development were analysed for GAs content. The procedure described by Gaskin *et al.* (1985) for extraction of plant material, purification of extracts and quantification using GC-MS, was essentially followed. Stable-isotope-labelled GA_1, GA_3 and GA_8 were used as internal standards.

In agreement with previous results in this species (Dewar *et al.*, 1998), GA_1 and GA_3 were detected in entire grains but not in isolated embryos. Although at some stages GA content was slightly higher in Redland B2 grains than in IS 9530 ones (Table 9.1), differences were not sufficiently high as to suggest GA content as a regulatory event in deciding the direction of the hormone balance.

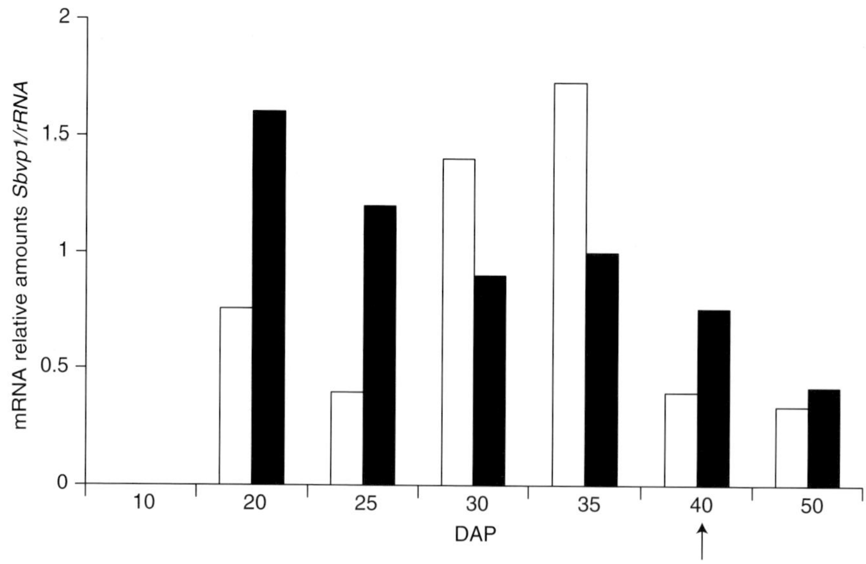

Fig. 9.4. Densitometric analysis of the expression of the Sb*Vp1* gene (*Sorghum bicolor Vp1*) in Redland B2 (■) and IS 9530 (□) embryos excised at different times after pollination (DAP). The bars indicate the level of expression corrected by the amount of total RNA used in each lane of the Northern blot experiments. The arrow indicates the time of physiological maturity.

Table 9.1. Endogenous gibberellin content (ng GA_{1+3} g^{-1} dry weight) in sorghum caryopses from varieties Redland B2 and IS 9530 harvested at different stages of development.

	Days after pollination			
	23	30	37	44
Redland B2	102.95	433.78	306.91	215.72
IS 9530	652.07	324.77	296.31	419.76

Sensitivity to GAs

Caryopsis sensitivity to GAs in sprouting-susceptible and -resistant varieties

Dose response curves to GAs were constructed by incubating caryopses harvested at different stages of development in the presence of increasing GA_3 concentrations. Caryopsis germination was recorded daily for 12 days and a germination index was constructed as described elsewhere (Steinbach *et al.*, 1995).

Caryopses from the less dormant variety Redland B2 were appreciably more sensitive to the promotive effect of GA_3 than those from the dormant IS 9530 (Fig. 9.5). Indeed, the slope of the line fitted to show the relationship between germination index and GA_3 concentration was, for most of the developmental stages considered, higher for Redland B2 caryopses than for IS 9530 ones (Fig. 9.5).

Embryo sensitivity to GAs in sprouting-susceptible and -resistant varieties

From the above-described results, and within the context of the hormone balance theory, it could be hypothesized that the inhibitory effect of ABA on embryo germination is more effectively counterbalanced by endogenous GAs, in Redland B2 embryos than in IS 9530 ones. To test this hypothesis, embryo sensitivity to GA_3 was determined at different stages of development as the effectiveness of this growth regulator to overcome the inhibitory effect imposed on embryo germination by 10 μM ABA. Thus, embryos from both varieties were incubated in 6 ml of 10 μM ABA alone, or 6 ml of 10 μM ABA increasing final concentrations of GA_3 at 5, 25, 50 or 500 μM. Embryo germination was daily recorded for 12 days and a germination index was constructed as described elsewhere (Steinbach *et al.*, 1995).

Differences in sensitivity should have been seen as a different slope of the line fitted to show the relationship between embryo germination index and GA_3 concentration in the incubation medium. Hence, Redland B2 embryos did not appear to present a higher sensitivity to GA_3 than IS 9530 ones (Fig. 9.6). The apparent higher sensitivity of 37 DAP IS 9530 embryos was in fact the result of the very low ABA sensitivity of Redland B2 embryos at this stage of development which elicited a response saturation with a very low GA_3 concentration (Fig. 9.6).

Overall, these results do not indicate that the inhibitory effect of ABA on embryo germination is more effectively counterbalanced by GA_3 in Redland B2 embryos than in IS 9530 ones. More likely, the higher sensitivity to GA_3 shown by developing Redland B2 caryopses (Fig. 9.5) appears to result from the fact that, in this variety, GAs are more effective than in IS 9530 in overcoming some other constraint for germination imposed on the embryo within the intact caryopses, which might not necessarily be related to ABA action.

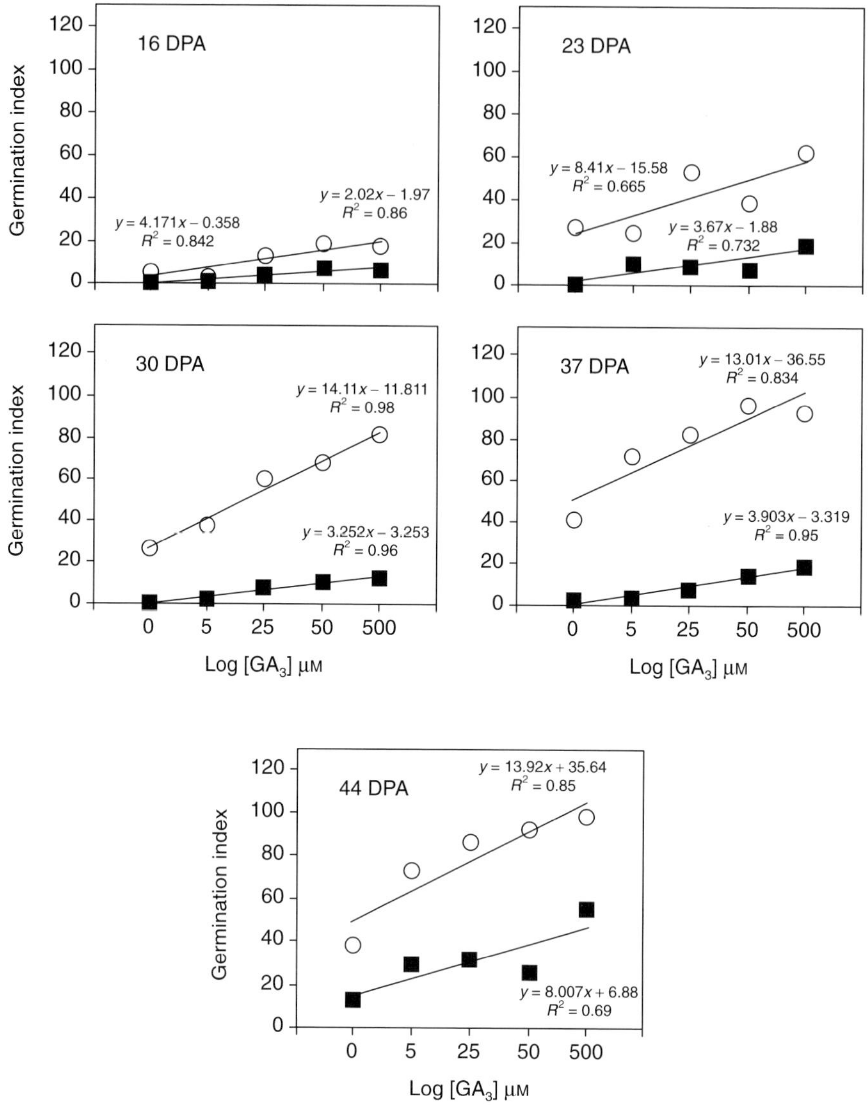

Fig. 9.5. Germination indices of Redland B2 (○) and IS 9530 (■) caryopses harvested at different times after pollination (16, 23, 30, 37 and 44 days post anthesis (DPA)) and incubated at 25°C in the presence of solutions containing various GA_3 concentrations. A linear model was fitted to each data set.

Conclusions

Throughout this paper we discussed the existence of one or more regulatory events deciding the direction of the hormone balance for determining the degree of dormancy of developing caryopses from a sprouting-susceptible variety (Redland B2) and from a sprouting-resistant one (IS 9530). ABA content during seed development did not differ between varieties; however, embryos

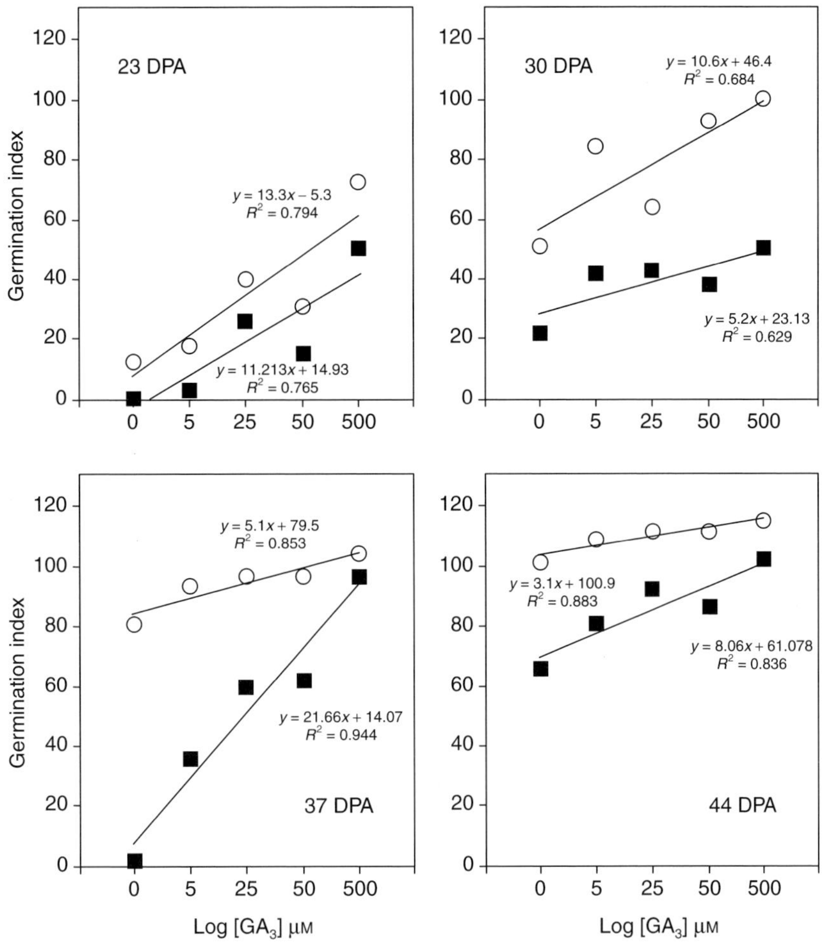

Fig. 9.6. Germination indices of Redland B2 (○) and IS 9530 (■) embryos excised at different times after pollination (23, 30, 37 and 44 days post anthesis (DPA)) and incubated at 25°C in the presence of 10 μM ABA together with various GA_3 concentrations. A linear model was fitted to each data set.

from the susceptible variety Redland B2 were found to have low sensitivity to ABA from early stages of development, whereas in IS 9530 embryos, sensitivity stayed high until well after physiological maturity. This points to embryo sensitivity to ABA as one regulatory element of the hormone balance. Furthermore, genetic analysis using QTLs suggests the participation of the gene *Vp1* in the control of dormancy: the involvement of this gene in the regulation of embryo sensitivity to ABA in maize is well documented (McCarty *et al.*, 1991). Expression analysis throughout development showed a later peak of expression of *Vp1* in IS 9530 embryos than in Redland B2 ones. The relationship between this late expression peak and the maintenance of a high sensitivity to ABA until late stages of development should be tested.

Previous results have demonstrated that the degree of dormancy of developing sorghum seeds also depends on the extent to which ABA action is counterbalanced by the effect of GAs. However, neither GA content nor embryo sensitivity to GAs appear to be different between dormant and non-dormant varieties, thus suggesting that differences in degree of dormancy between varieties are not controlled at this level.

In summary, the results from these experiments show that the only regulatory factor so far capable of deciding the direction of the hormone balance and consequently determining the different degree of dormancy between sprouting-susceptible and -resistant varieties, is most likely established at the level of embryo sensitivity to ABA. However, other dormancy mechanisms not necessarily related to ABA action might be operating and further work is required to elucidate them.

References

Black, M. (1991) Involvement of ABA in the physiology of developing and mature seeds. In: Davies, W.J. (ed.) *Abscisic Acid Physiology and Biochemistry.* Bios Scientific Publishers Limited, Oxford, pp. 99–124.

Dewar, J., Taylor, J.R.N. and Berjak, P. (1998) Changes in selected plant growth regulators during germination in sorghum. *Seed Science Research*, 8, 1–8.

Fong, F., Smith, J.D. and Koehler, D.E. (1983) Early events in maize seed development. *Plant Physiology* 73, 899–901.

Gaskin, P., Gilmour, S.J., MacMillan, J. and Sponsel, V.M. (1985) Gibberellins in immature seeds and dark-grown shoots of *Pisum sativum*. Gibberellins identified in the tall cultivar Alaska in comparison with those in the dwarf Progress No. 9. *Planta* 163, 283–289.

Hilhorst, H.W.M. (1995) A critical update on seed dormancy. I. Primary dormancy. *Seed Science Research*, 5, 61–73.

Karssen, C.M. (1995) Hormonal regulation of seed development, dormancy, and germination studied by genetic control. In: Kigel, J. and Galili, G. (eds) *Seed Development and Germination.* Marcel Dekker, New York, pp. 333–350.

Karssen, C.M., Brinkhorst-Van der Swan, D.L.C., Breekland, A.E. and Koorneef, M. (1983) Induction of dormancy during seed development by endogenous abscisic acid: studies on abscisic acid deficient genotypes of *Arabidopsis thaliana* (L.). *Planta* 157, 158–165.

Karssen, C.M., Zagorski, S., Kepczynski, J. and Groot, S.P.C. (1989) Key role for endogenous gibberellins in the control of seed germination. *Annals of Botany* 63, 71–80.

King, R.W. (1982) Abscisic acid in seed development. In: Kahn, A.A. (ed.) *The Physiology and Biochemistry of Seed Development, Dormancy and Germination.* Amsterdam, Elsevier Biomedical Press, pp. 157–181.

Lijavetzky, D., Martinez, C., Carrari, F. and Hopp, E. (1999) Pre-harvest sprouting resistance in sorghum is controlled by just two major QTLs showing epistatic effects. *Euphytica* (in press).

Lona, F. (1956) L´acido gibberéllico determina la germinazione del semi di *Lactuca scariola* in fase di scotoinhibizione. *Ateneo Parmense*, 27, 641–644.

McCarty, D.R., Hattori, T., Carson, C.B., Vasil, V., Lazar, M. and Vasil, I.K. (1991) The viviparous-1 developmental gene of maize encodes a novel transcriptional activator. *Cell* 66, 895–905.

Robichaud, C.S., Wong, J. and Sussex, I.M. (1980) Control of *in vitro* growth of viviparous embryo mutants of maize by abscisic acid. *Developmental Genetics* 1, 325–330.

Steinbach, H.S., Benech-Arnold, R.L., Kristof, G., Sánchez, R. A. and Marcucci Poltri, S. (1995) Physiological basis of pre-harvest sprouting resistance in *Sorghum bicolor* (L.) Moench. ABA levels and sensitivity in developing embryos of sprouting resistant and susceptible varieties. *Journal of Experimental Botany* 45, 701–709.

Steinbach, H.S., Benech-Arnold, R.L. and Sánchez, R.A. (1997) Hormonal regulation of dormancy in developing *Sorghum* seeds. *Plant Physiology* 113, 149–154.

Walker-Simmons, M.K. (1987) ABA levels and sensitivity in developing wheat embryos of sprouting-resistant and -susceptible cultivars. *Plant Physiology* 84, 61–66.

Xu, N., Coulter, K.M. and Bewley, J.D. (1990) Abscisic acid and osmoticum prevent germination of developing alfalfa embryos, but only osmoticum maintains the synthesis of developmental proteins. *Planta* 182, 382–390.

10 Irrigation and Seed Quality Development in Rapid-cycling Brassica

R.H. Ellis[1], U.R. Sinniah[1] and P. John[2]

[1]Department of Agriculture, University of Reading, Earley Gate, Reading RG6 6AT, UK; [2] School of Plant Sciences, University of Reading, Reading RG6 6AS, UK

Rapid-cycling brassica (*Brassica campestris* [*rapa*] L.) were irrigated for three different durations after pollination. The earlier irrigation ended, the earlier mass maturity, the lower the final seed dry weight, and the more rapidly seeds lost moisture. Desiccation tolerance developed soonest in seeds harvested from plants in which irrigation stopped earliest. Potential seed longevity (K_i) varied greatly with irrigation treatment and with duration from pollination. Maximum K_i values were attained 44, 36, and 32 days after pollination (DAP) (10, 6 and 7 days after mass maturity) for seeds from plants irrigated throughout, or until 24 DAP, or 16 DAP, respectively, and declined thereafter. Maximum K_i was greatest (4.61) where irrigation stopped 16 DAP and least (3.88) from plants irrigated throughout. Thus terminal drought resulted in more rapid seed quality development and also greater maximum seed quality. A simple multiple regression model of the relations between K_i and both the oligosaccharide:total sugar ratio and the relative content of a 58 kDa heat-stable protein content suggests that these sugars and proteins are equally likely to be required for seed quality development, that initially the sugars tend to accumulate at the greater rate, but that during maturation drying the heat-stable proteins accumulate at the greater rate.

Introduction

During the 1990s, the Seed Science Laboratory at Reading published the results of research in a wide range of contrasting species which showed that the quality of seeds on the mother plant continued to improve for some considerable time beyond the end of the seed-filling phase (e.g. Pieta Filho and Ellis, 1991; Demir and Ellis, 1992a,b, 1993; Ellis and Pieta Filho, 1992; Hong and Ellis, 1992; Zanakis *et al.*, 1994; Sanhewe and Ellis, 1996a). This has been a somewhat controversial conclusion, because it challenges the conventional view that seed quality is maximal at physiological maturity and that seeds begin to deteriorate immediately thereafter (Harrington, 1972), physiological maturity being defined as the end of the seed-filling phase (Shaw and Loomis, 1950). Indeed, TeKrony and Egli (1997) reported in the proceedings of the previous workshop in this series that maximum seed quality occurred at or before physiological maturity for species they harvested commercially as dry seed, but after this developmental stage for species whose seeds develop and mature in fleshy fruits. Although the latter agrees with the research at Reading in which seed quality improved after the end of the seed-filling period in seeds which maintain a high moisture content thereafter because they are within fleshy fruits (Demir and Ellis, 1992a,b, 1993), the former contrasts with the same conclusion for those which undergo maturation drying. Moreover, the benefits to seed quality from slow-drying treatments *ex planta* beginning at physiological maturity have mimicked the improvement in seed quality that occurs *in planta* during this period (Hong and Ellis, 1997). Given that seed quality is not necessarily maximal at the stage of development termed physiological maturity, the term mass maturity (Ellis and Pieta Filho, 1992) is used here to denote the end of seed-filling.

The research at Reading has also investigated the effect of several environmental variables on seed quality development, namely irradiance (Pieta Filho and Ellis, 1991), temperature (Ellis *et al.*, 1993; Sanhewe *et al.*, 1996c), and carbon dioxide concentration (Sanhewe *et al.*, 1996c). The initial objective of the current work was to determine what effect terminal drought on the mother plant has on seed quality development. We also sought to test further the hypothesis that seed quality is maximal at the end of the seed-filling phase (Harrington, 1972) in contrasting circumstances. Rapid-cycling brassica (*Brassica campestris* [*rapa*] L.) were chosen for study because: (i) seeds can be produced rapidly to a predictable time schedule of only a few weeks (Tomkins and Williams, 1990); and (ii) the compact plants can be grown in small containers and so (given the small rooting volume) withholding irrigation affects plant water status rapidly.

There has also been controversy concerning whether the accumulation of soluble sugars (Ooms *et al.*, 1993; Sun and Leopold, 1993; Bochicchio *et al.*, 1994; Lin and Huang, 1994; Ooms *et al.*, 1994; Still *et al.*, 1994; Brenac *et al.*, 1997) or heat-stable proteins (Blackman *et al.*, 1992, 1995; Gee *et al.*, 1994; Wechsberg *et al.*, 1994) are responsible for the improvement of seed quality during seed development and maturation.

Despite the recognition that desiccation tolerance is a complex phenomenon, much research to date has considered the role of either proteins or sugars separately. The present study examined the changes in both soluble carbohydrates and heat-stable proteins that occur during seed development and maturation. It addresses the question whether or not either is associated with improvement in potential seed longevity. Moreover, in order to reduce the inevitable associations between all aspects of seed development and maturation with duration from pollination, these topics were investigated in three seed production regimes in which, due to the differences in irrigation, seed development and maturation progressed at different rates at the same temperature.

Materials and Methods

Full details of experimental procedures have been provided elsewhere (Sinniah *et al.*, 1998a,b). Plants were grown in a growth chamber at 25°C, 75% relative humidity with 20 h day^{-1} of light until pollination, and at 20°C and 50% relative humidity thereafter. Embryo development was initiated by synchronous hand pollination of 19-day-old plants. All flowers open on this day were pollinated, such that a minimum of five uniformly-developed siliquae were subsequently harvested from each plant.

Plants were irrigated with 15 ml of water or nutrient solution per day per pot beginning 4 days after sowing until 15 DAP (days after pollination), but with only 5 ml per day thereafter (to prevent precocious germination). Three different irrigation treatments were imposed on plants: (i) irrigation ended at 16 DAP; (ii) irrigation ended at 24 DAP; (iii) plants irrigated throughout. Plants were harvested serially at frequent intervals and seed moisture content (%), mean seed dry weight, and ability to germinate determined. The remaining seeds in each sample were dried to below 10% moisture content in a drying cabinet maintained at 15–17°C with 15% relative humidity. After drying, seeds were tested for ability to germinate. Potential longevity was assessed by the value of the seed lot constant K_i in the equation

$$v = K_i - p/\sigma \qquad (10.1)$$

where v is probit percentage viability after p days in storage and σ is the standard deviation of the frequency distribution of seed deaths in time (days) (Ellis and Roberts, 1980) following experimental hermetic storage at 40°C with 15% moisture content. At each harvest date, four replicates of 100 fresh seeds each were frozen immediately in liquid nitrogen and then stored at −70°C for the subsequent determination of soluble carbohydrates (three of the replicates) and heat-stable proteins (the remaining replicate). The analytical procedures are described in detail in Sinniah *et al.* (1998b).

Results

Ending irrigation early resulted in earlier mass maturity and lighter seeds at maturity (Table 10.1). Desiccation tolerance developed soonest in seeds harvested from plants in which irrigation stopped earliest (Sinniah *et al.*, 1998a). Similarly, the duration from pollination to maximum potential longevity was reduced, but in all cases maximum potential longevity was not attained until 6–10 days after the end of the seed-filling phase (Table 10.1). This coincided with when maturation drying had reduced seed moisture content naturally to 6–7%. There was no evidence that maximum seed quality was reduced by

Table 10.1. Effect of irrigation treatments to rapid-cycling brassica plants on seed development and seed quality.

End irrigation (DAP)[a]	Durations: Mass maturity (DAP)[a]	Max. longevity (DAP)[a]	Max. longevity (DAMM)[b]	Mature seed dry wt (mg)	Max. K_i
Control	34.1 (2.2)	44	10	2.36 (0.026)	3.88
24	29.8 (3.2)	36	6	1.93 (0.086)	4.54
16	24.8 (2.3)	32	7	1.57 (0.017)	4.61

[a]Days after pollination.
[b]Days after mass maturity.

terminal drought. In fact, there was some suggestion that the two drought treatments enhanced K_i (Table 10.1).

Immature seeds of rapid-cycling brassica contained high amounts of reducing sugars (glucose and fructose), but the content of these monosaccharides declined substantially between 12 and 16 DAP, and at a lower rate thereafter such that by 45 DAP they were absent in the dry, mature seeds (Sinniah *et al.*, 1998b). The differences in glucose and fructose content between the three irrigation treatments were generally neglible.

Sucrose was the only soluble carbohydrate present at all stages of seed development. The accumulation of sucrose in seed tissues of rapid-cycling brassica followed a biphasic pattern. The high amounts of sucrose present in immature seeds were greatly reduced during the mid-developmental phase (15–20 DAP) but increased again thereafter. This latter effect was most clear and occurred earliest in the treatment where irrigation was stopped at 16 DAP.

The higher oligosaccharides, namely stachyose and raffinose, began to accumulate later during development relative to the monosaccharides and disaccharides. The accumulation of stachyose in plants irrigated throughout began at 24 DAP and stachyose content increased thereafter to a maximum at 44 DAP. Ending irrigation at 16 DAP resulted in stachyose accumulation beginning at 20 DAP (i.e. 4 days earlier than other treatments) and also resulted in more stachyose being present at 44 DAP. Ending irrigation at 24 DAP resulted in a more rapid increase in stachyose content compared with plants irrigated throughout.

Raffinose, another oligosaccharide implicated to play a role in protection against desiccation damage, began to accumulate earlier (at 16 DAP) than stachyose. In contrast to the trend observed for stachyose where the content either continued to increase or stabilized, raffinose content declined after attaining a maximum value. The content of this sugar reached a peak of almost 16 mg g^{-1} at 24 DAP for plants irrigated until 16 DAP, and a peak of just over 14 mg g^{-1} at 28 DAP for plants irrigated until 24 DAP or irrigated throughout.

Sucrose is an excellent glass former but may be incapable of protecting against desiccation damage unless in association with raffinose series oligosaccharides in order to prevent sucrose crystallization during water removal.

Consequently, mass ratios of sugars were calculated. The temporal patterns of changes in these mass ratios were generally similar for all three irrigation treatments but displaced depending upon when irrigation ended (Sinniah *et al.*, 1998b).

Correlations between individual sugar contents and sugar ratios with ability to tolerate desiccation (data from 12 to 44 DAP) and potential longevity (data from 20 or 24 to 44 DAP) were calculated. Sucrose was not correlated with ability to tolerate desiccation, but (among the later harvests) was correlated with potential longevity ($P < 0.01$). Glucose and fructose were negatively correlated with ability to tolerate desiccation ($P < 0.001$) and potential longevity ($P < 0.01$). Raffinose was correlated with ability to tolerate desiccation ($P < 0.01$) but not with potential longevity. Among the individual sugars, stachyose showed the strongest positive correlations with ability to tolerate desiccation ($r = 0.782$, 15 df, $P < 0.001$) and potential longevity ($r = 0.899$, 12 df, $P < 0.001$). Similarly, the ratio of oligosaccharide:total sugars showed very strong positive correlations with ability to tolerate desiccation ($r = 0.950$, 15 df, $P < 0.001$) and also with potential longevity ($r = 0.879$, 12 df, $P < 0.001$).

An SDS-gel electrophoretic profile of heat-stable proteins extracted from seeds of rapid-cycling brassica plants subjected to three different irrigation regimes during seed development and maturation identified several prominent bands (Sinniah *et al.*, 1998b). Ten of these (at 72, 58, 33, 31, 27, 26, 22.5, 21.5, 20, and 15 kDa) were subjected to densitometric analysis using a Whole Band Analyser programme (BioImage Applications, Millipore). Band intensity was quantified and expressed as a percentage of the maximum value obtained.

For seven of the bands (72, 58, 33, 31, 26, 22.5, and 21.5 kDa), significant positive correlations between band intensity and K_i were detected ($P < 0.05$), while in another (15 kDa) a significant negative correlation was detected ($P < 0.05$). The best correlation between K_i and protein band intensity was obtained for the 58 kDa protein ($r = 0.875$, 12 df, $P < 0.001$). The intensity of this 58 kDa protein band was positively correlated ($P < 0.05$) with those at 72, 33, 31, 27, 26, 22.5 and 21.5 kDa.

Discussion

The results summarized in Table 10.1 show that the widely-accepted hypothesis that seed quality is maximal at the end of the seed-filling phase and that viability and vigour decline thereafter (Harrington, 1972) is not correct. In all three plant irrigation treatments, the attainment of maximum potential longevity coincided with the reduction in seed moisture content by maturation drying on the mother plant to 6–7%. Since the timing of maturation drying was affected by the irrigation treatment, these results *in planta* would appear to provide evidence to support *ex planta* studies in which slow drying improved the quality of immature seeds (Adams *et al.*, 1983; Blackman *et al.*, 1992; Sanhewe and Ellis, 1996b; Hay *et al.*, 1997; Hong and Ellis, 1997).

Reducing irrigation to rapid-cycling brassica plants reduced final seed weight, altered the time course of maturation drying and seed quality development, and increased maximum seed quality. Clearly then, the results contradict the view that larger seeds are of better quality (Perry, 1980). Moreover, the demonstration here that irrigation can affect seed quality development substantially is not only of practical interest to commercial seedsmen, but is also of relevance to the elucidation of those factors responsible for seed quality.

Certain sugars and certain heat-stable proteins are both associated with the development of desiccation tolerance and potential longevity. The accumulation of both within the developing and maturing seeds was affected substantially by irrigation treatment. Since results from three different irrigation regimes were included in the analyses, the correlations between the contents of sugars and heat-stable proteins with desiccation tolerance and with potential longevity are not just a consequence of developmental time alone but include the effect of seed production environment. To test the possibility that the accumulation of both heat-stable proteins and sugars are required for the development of potential longevity, we selected the ratio of oligosaccharide:total sugars and the relative content of the heat-stable 58 kDa protein since each provided the best correlations with potential longevity. The following multiple regression model was obtained,

$$K_i = -1.27 + 10.5\,(OT) + 0.0194\,(P58) \quad (10.2)$$

where OT is the oligosaccharide:total sugars ratio, $P58$ is the relative content of the 58 kDa protein, and the respective standard errors are 0.924, 5.04, 0.0098. Both the oligosaccharide:total sugars ratio and the relative content of the 58 kDa protein provided similar values of t.

The model fitted is compared with the observations in Figure 10.1. The observations on this 3-D diagram track a curvi-linear pattern in time (from bottom centre to top centre) across the response surface. This shows that the oligosaccharides accumulated comparatively early in seed development (coinciding with the development of ability to tolerate desiccation), whereas the 58 kDa protein (and the other heat-stable proteins correlated with it) accumulated comparatively late, coinciding with the increase in potential longevity which continued once maximum ability to tolerate desiccation was achieved.

These results suggest that the accumulation of certain oligosaccharides and certain heat-stable proteins are equally likely (or unlikely) to be required for the development of high seed quality. However, since the heat-stable proteins accumulate comparatively late in seed development and maturation, we suspect that, in practice, differences in seed quality among different commercial seed lots are more likely to result from differences in heat-stable protein accumulation than in sugars since environmental effects, at least the deleterious effect of high temperature, on seed quality development tend not to be detected until comparatively late during seed development and maturation (Ellis *et al.*, 1993; Ellis and Hong, 1994). The research summarized above is described in more detail elsewhere (Sinniah *et al.*, 1998a,b), while Bettey *et al.* (1998) further show that three classes of stress proteins continue to accumulate after mass maturity at the same time that potential longevity is

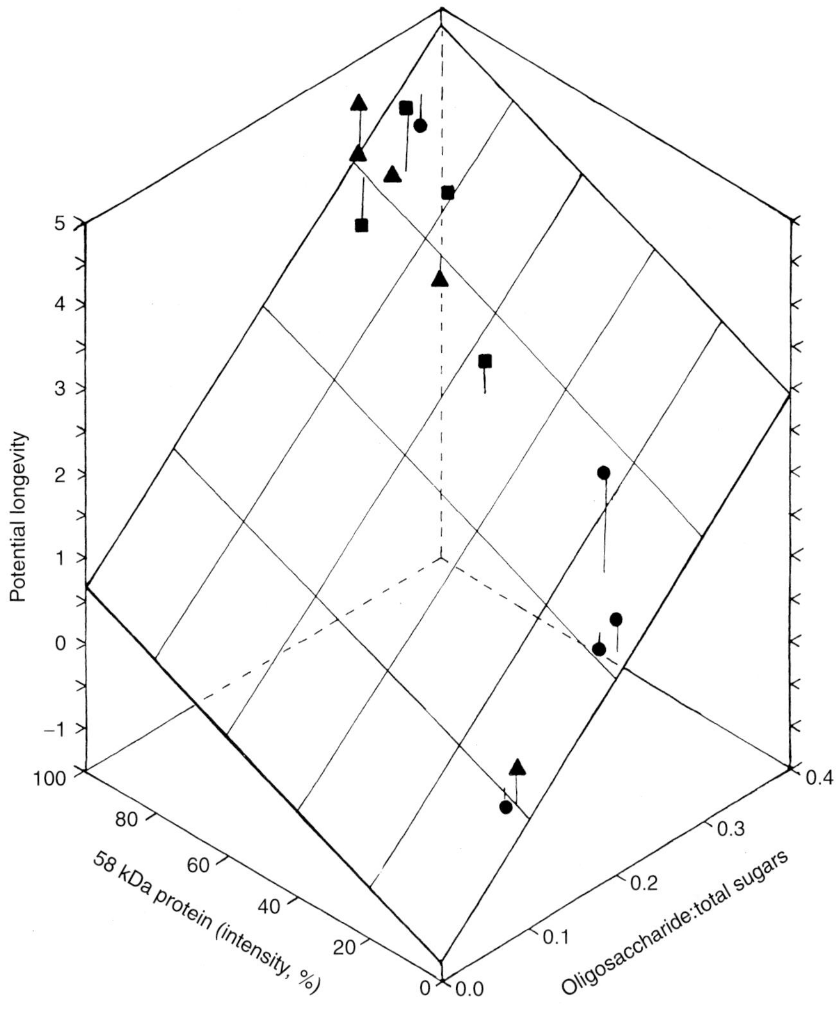

Fig. 10.1. Relations between the potential longevity (K_i of the seed viability equation) of seeds of rapid-cycling brassica collected after different durations of seed development and maturation from plants irrigated throughout (●), or until 24 DAP (■), or until 16 DAP (▲), and both the ratio of oligosaccharide:total sugars and the content (relative band intensity, %) of a 58 kDa heat-stable protein within the seeds. The fitted model ($R^2 = 0.782$, 11 d.f.) is described in the text: vertical lines show differences between observed and fitted values of K_i. (From Sinniah *et al.*, 1998b.)

increasing. Not only, then, does this research provide further evidence that seed quality continues to improve after mass maturity, but the concurrent accumulation of several low molecular weight heat-stable proteins after mass maturity also begins to suggest potential mechanisms for this vital aspect of seed physiology.

References

Adams, C.A., Fjerstad, M.C. and Rinne, R.W. (1983) Characteristics of soybean seed maturation: necessity for slow dehydration. *Crop Science* 23, 265–267.

Bettey, M., Sinniah, U.R., Finch-Savage, W.E. and Ellis, R.H. (1998) Irrigation and seed quality development in rapid-cycling brassica: accumulation of stress proteins. *Annals of Botany* 82, 657–663.

Blackman, S.A., Obendorf, R.L. and Leopold, A.C. (1992) Maturation proteins and sugar in desiccation tolerance of developing seeds. *Plant Physiology* 100, 225–230.

Blackman, S.A., Obendorf, R.L. and Leopold, A.C. (1995) Desiccation tolerance in developing soybean seeds: the role of stress proteins. *Physiologia Plantarum* 93, 630–638.

Bochicchio, A., Rizzi, E., Balconi, C., Vernieri, P. and Vazzana, C. (1994) Sucrose and raffinose contents and acquisition of desiccation tolerance in immature maize embryos. *Seed Science and Technology* 22, 361–370.

Brenac, P., Smith, M.E. and Obendorf, R.L. (1997) Raffinose series oligosaccharides and desiccation tolerance of developing viviparous maize embryos. In: Ellis, R.H., Black, M., Murdoch, A.J. and Hong, T.D. (eds) *Basic and Applied Aspects of Seed-Biology. Proceedings of the Fifth International Workshop on Seeds*, Reading, 1995. Dordrecht, Kluwer Academic Publishers, pp. 95–101.

Demir, I. and Ellis, R.H. (1992a) Changes in seed quality during seed development and maturation in tomato. *Seed Science Research* 2, 81–87.

Demir, I. and Ellis, R.H. (1992b) Development of pepper *(Capsicum annuum)* seed quality. *Annals of Applied Biology* 121, 385–399.

Demir, I. and Ellis, R.H. (1993) Changes in potential seed longevity and seedling growth during seed development and maturation in marrow. *Seed Science Research* 3, 247–257.

Ellis, R.H. and Hong, T.D. (1994) Desiccation tolerance and potential longevity of developing seeds of rice (*Oryza sativa* L.). *Annals of Botany* 73, 501–506.

Ellis, R.H. and Pieta Filho, C. (1992) The development of seed quality in spring and winter cultivars of barley and wheat. *Seed Science Research* 2, 9–15.

Ellis, R.H. and Roberts, E.H. (1980) Improved equations for the prediction of seed longevity. *Annals of Botany* 45, 13–30.

Ellis, R.H., Hong, T.D. and Jackson, M.J. (1993) Seed production environment, time of harvest and the potential longevity of seeds of three cultivars of rice (*Oryza sativa* L.). *Annals of Botany* 72, 583–590.

Gee, O.H., Probert, R.J. and Coomber, S.A. (1994) 'Dehydrin-like' proteins and desiccation tolerance in seeds. *Seed Science Research* 4, 135–141.

Harrington, J.K. (1972) Seed storage and longevity. In: Kozlowski, T.T. (ed.) *Seed Biology*, Volume III. New York, Academic Press, pp. 145–245.

Hay, F.R., Probert, R.J. and Coomber, S.A. (1997) Development of desiccation tolerance and longevity in seeds from detached capsules of foxglove (*Digitalis purpurea* L.). *Annals of Botany* 79, 419–417.

Hong, T.D. and Ellis, R.H. (1992) Development of desiccation tolerance in Norway maple (*Acer platanoides* L.) seeds during maturation drying. *Seed Science Research* 2, 169–172.

Hong, T.D. and Ellis, R.H. (1997) The effect of the initial rate of drying on the subsequent ability of immature seeds of Norway maple (*Acer platanoides* L.) to survive rapid desiccation. *Seed Science Research* 7, 41–45.

Lin, T.-P. and Huang, N.-H. (1994) The relationship between carbohydrate composition of some tree seeds and their longevity. *Journal of Experimental Botany* 45, 1289–1294.

Ooms, J.J., Leon-Kloosterziel, K.M., Bartels, D., Koorneef, M. and Karssen, C.M. (1993) Acquisition of desiccation tolerance and longevity in seeds of *Arabidopsis thaliana. Plant Physiology* 102, 1185–1191.

Ooms, J.J., Van der Veen, R. and Karssen, C.M. (1994) Abscisic acid and osmotic stress or slow drying independently induce desiccation tolerance in mutant seeds of *Arabidopsis thaliana. Physiologia Plantarum* 92, 506–510.

Perry, D.A. (1980) Deterioration of barley seed and its effect on field performance. In: Hebblethwaite P.D. (ed.) *Seed Production.* London, Butterworths, pp. 327–337.

Pieta Filho, C. and Ellis, R.H. (1991) The development of seed quality in spring barley in four environments. I. Germination and longevity. *Seed Science Research* 1, 163–177.

Sanhewe, A.J. and Ellis, R.H. (1996a) Seed development and maturation in *Phaseolus vulgaris.* II. Post-harvest longevity in air-dry storage. *Journal of Experimental Botany* 47, 959–965.

Sanhewe, A.J. and Ellis, R.H. (1996b) Seed development and maturation in *Phaseolus vulgaris.* I. Ability to germinate and to tolerate desiccation. *Journal of Experimental Botany* 47, 949–958.

Sanhewe, A.J., Ellis, R.H., Hong, T.D., Wheeler, T.R., Batts, G.R., Hadley, P. and Morison, J.I.L. (1996c) The effect of temperature and CO_2 on seed quality development in wheat (*Triticum aestivum* L.). *Journal of Experimental Botany* 47, 631–637.

Shaw, R.H. and Loomis, W.E. (1950) Bases for the prediction of corn yields. *Plant Physiology* 25, 225–244.

Sinniah, U.R., Ellis, R.H. and John, P. (1998a) Effect of irrigation on seed quality development in rapid-cycling brassica (*Brassica campestris* [*rapa*]. L.): ability to germinate, tolerate desiccation, and survive during air-dry storage. *Annals of Botany* 82, 309–314.

Sinniah, U.R., Ellis, R.H. and John, P. (1998b) Irrigation and seed quality development in rapid-cycling brassica: soluble carbohydrates and heat-stable proteins. *Annals of Botany* 82, 647–655.

Still, D.W., Kovach, D.A. and Bradford, K.J. (1994) Development of desiccation tolerance during embryogenesis in rice (*Oryza sativa*) and wild rice (*Zizania palustris*). *Plant Physiology* 104, 431–438.

Sun, W.Q.and Leopold, A.C. (1993) Acquisition of desiccation tolerance in soyabeans. *Physiologia Plantarum* 87, 403–409.

TeKrony, T.M.and Egli, D.B. (1997) Accummulation of seed vigour during development and maturation. In: Ellis, R.H., Black, M., Murdoch, A.J. and Hong, T.D. (eds) *Basic and Applied Aspects of Seed Biology. Proceedings of the Fifth International Workshop on Seeds,* Reading, 1995. Dordrecht, Kluwer Academic Publishers, pp. 369–384.

Tomkins, S.P. and Williams, P.H. (1990) Fast plants for finer science: an introduction to the biology of rapid-cycling *Brassica campestris* (*rapa*) L. *Journal of Biological Education* 24, 239–250.

Wechsberg, G.E., Bray, C.M. and Probert, R.J. (1994) Expression of 'Dehydrin-like' proteins in orthodox seeds of *Ranunculus sceleratus* during development and water stress. *Seed Science Research* 4, 241–246.

Zanakis, G.N., Ellis, R.H. and Summerfield, R.J. (1994) Seed quality in relation to seed development and maturation in three genotypes of soyabean *(Glycine max). Experimental Agriculture* 30, 139–156.

11 Molecular Genetic Analysis of *Arabidopsis* Seed Quality

S.P.C. GROOT[1], A.H.M. VAN DER GEEST[1], K. TESNIER[1], C. ALONSO-BLANCO[2], L. BENTSINK[2], H. DONKERS[1], M. KOORNNEEF[2], D. VREUGDENHIL[3] AND R.J. BINO[1]

[1]Plant Research International, PO Box 16, NL-6700 AA Wageningen, The Netherlands; [2]Department of Genetics, Wageningen University, Dreijenlaan 2, NL-6703 HA Wageningen, The Netherlands; [3]Department of Biomolecular Sciences, Wageningen University, Arboretumlaan 4, NL-6703 HA Wageningen, The Netherlands

Improvement of seed quality may greatly benefit from the characterization of genes involved in seed development and seed germination. To study the effect of genes on seed quality, a comparison was made among genetically different lines. Ideally, the genetic difference should be known at the DNA sequence level. For this purpose *Arabidopsis* offers a potentially powerful model system. The complete genome sequence of this plant species will soon become available, a large amount of genetic variation is present and daily new mutants are produced. To exploit *Arabidopsis* as a model system for genetic analysis of seed quality, appropriate seed quality assays were developed. Several mutants and ecotypes were compared for storability, using a controlled deterioration assay. Large variation was observed with regard to this trait. Seeds from abscisic acid deficient or insensitive mutants deteriorated faster compared to wild-type seeds. Seeds from the ecotype Landsberg *erecta* and the ecotype Cape Verde Island differ in dormancy, controlled deterioration tolerance and their content of raffinose and stachyose. A quantitative trait loci analysis of recombinant inbred lines derived from a cross between these two ecotypes showed that stachyose and raffinose content were linked to a single locus in the genome. The traits for dormancy, controlled deterioration tolerance and content of the sugars were not genetically linked. A very strong reduction in stachyose and raffinose content had no effect on the controlled deterioration tolerance of the seeds. It is concluded that *Arabidopsis* offers a powerful tool to study genetic variation in seed quality and the effect of modification of seed composition on seed quality.

Genetic Variation in Seed Quality

Seed quality is determined by events during the growth of the seed crop, as well as by conditions of harvest, treatments thereafter, storage and germination environment. Genetic variation for seed quality also exists. Sometimes the genetic basis of reduced quality is evident, such as for super sweet corn, seeds of which lack most of their starch due to a mutation in one of the enzymes involved in starch synthesis. Upon sowing, these shrunken sugary seeds are very vulnerable to attack by soil-borne microorganisms. Several other examples of genetic variation in seed quality have been published (e.g. Dahal and Bradford, 1990; Dahal *et al.*, 1990). However, in most cases the genetic and molecular basis of this variation is less clear.

Seed physiologists have exploited genetic variation for some time, especially in studying the roles of plant growth regulators in seed development and seed germination. Tomato (*Lycopersicon esculentum*) and *Arabidopsis* (*A. thaliana*) serve as model species to study the role of abscisic acid and gibberellins (Karssen *et al.*, 1983; Groot *et al.*, 1987; Groot and Karssen 1992; Ni and Bradford, 1993). It was demonstrated that during *Arabidopsis* and tomato seed development, abscisic acid is involved in the induction of seed dormancy, whereas during seed germination gibberellins are essential to obtain radicle protrusion.

Arabidopsis mutants with a severe *abi3* mutation and the *aba1-1/abi3-1* double mutant produce seeds without dormancy that fully germinate directly after harvest, but lose their vitality within a few weeks of dry storage (Koornneef *et al.*, 1989; Ooms *et al.*, 1993; Wolkers *et al.*, 1998). These studies indicate that in wild-type *Arabidopsis* seeds, abscisic acid is important for the induction of components involved in the acquisition of desiccation tolerance and longevity.

However, abscisic acid influences many other processes during plant development and it is not clear yet which of these induced processes or components is responsible for desiccation tolerance of the seeds. Late embryogenesis abundant (LEA) proteins are thought to be involved in the acquisition of stress tolerance during seed maturation. Indeed *aba1-1/abi3-1* double mutant seeds contain reduced levels of LEA proteins (Parcy *et al.*, 1994). This does not exclude that in wild-type seeds additional components are more crucial for seed longevity.

The galactosyl sucrose oligosaccharides, raffinose and stachyose, accumulate in many seeds during seed maturation. These oligosaccharides have been proposed to be important factors in the acquisition of desiccation tolerance and storability (Blackman *et al.*, 1992; Horbowicz and Obendorf, 1994; Black *et al.*, 1996; Obendorf, 1997). Indeed, desiccation-sensitive *aba1-1/abi3-1* double mutant seeds and poorly storable *abi3-5* mutant seeds do not accumulate these oligosaccharides in substantial amounts, but instead have high levels of sucrose (Ooms *et al.*, 1994; de Bruijn *et al.*, 1997).

Arabidopsis as a Model for Molecular Genetic Studies

Arabidopsis has many advantages for physiological, developmental and genetic studies of higher plants. These advantages include its relatively small genome, clear mutagenesis protocols and short generation cycles (Meinke *et al.*, 1998). Related to its plant size and the potential for high yields in seed numbers, the species provides a relatively inexpensive seed production system. Many *Arabidopsis* mutants are available, including those described above and other mutants involved in seed characteristics such as dormancy or morphology. Most of these mutations are mapped on the set of five chromosomes and detailed molecular marker maps have been produced for different ecotypes. Several gene-tagging systems are available for *Arabidopsis* and, within a few years, its complete genome sequence will be available (Meinke *et al.*, 1998). These factors make *Arabidopsis* an excellent model species for research on genetic effects on seed quality. Both mutagenesis (studying single gene effects) and quantitative trait loci (QTL) approaches (localizing multigenic traits, see later) have been used successfully in *Arabidopsis* and contributed to knowledge about many fundamental aspects of plant development.

The *Arabidopsis* Seed Quality Assay

Assays to test seed quality and to evaluate the effects of seed production conditions and seed treatments are available for most commercial crops (Hampton and TeKrony, 1995). These tests comprise evaluation of germination rate, total germination frequency and frequency of normal seedlings in germination tests at optimal conditions. Additional information can be obtained by germination tests performed at sub-optimal conditions, such as at low temperatures, in osmotica, at reduced oxygen levels or after a controlled deterioration treatment (Matthews and Powell, 1987). The latter tests aim to mimic loss of viability during storage. For *Arabidopsis*, no generally accepted seed quality test was available and had to be developed (K. Tesnier, in preparation).

For non-dormant *Arabidopsis* seeds the optimal temperature for germination is around 25°C. *Arabidopsis* seed lots exhibiting a low level of dormancy may germinate somewhat faster at temperatures between 15 and 20°C, due to the dormancy-breaking effect of these relatively lower temperatures (P. Toorop, personal communication). Some genotypes, e.g. those of the ecotype Cape Verde Island (Cvi), may produce seeds which are deeply dormant upon harvest and require several months of dry storage to lose their dormancy. As with many crop species, the germinability of seed lots expressing dormancy can be tested by a cold treatment and simultaneous application of exogenous gibberellins (GAs).

Placing black nylon membranes (Sleicher and Schuell, ref. no. 409712) between the filter paper and the seeds improves the evaluation of the germination rate of seed lots, since the higher contrast between the protruding radicle tip and the black membrane makes the recording of radicle protrusion more accurate and easy. When abnormal *Arabidopsis* seedlings with poor quality seed batches were observed, they closely resembled those observed with other cruciferous crops: blunt roots or yellow cotyledons. Application of 3 mM GA_3, in order to break dormancy, resulted in seedlings with relatively longer hypocotyls and shorter roots. However, when this was taken into account, the frequency of otherwise abnormal seedlings was not influenced by the GA_3 treatment.

In order to estimate differences in storability of seed lots, different controlled deterioration treatments were compared. Equilibration of the seeds at 85% relative humidity (RH), followed by hermetically sealed storage at 40°C for various durations (days), proved to be discriminative between different seed lots during the subsequent germination tests.

Genetic Variation in *Arabidopsis* Seed Quality

With the assays developed, seed lots from several *Arabidopsis* ecotypes and mutants were compared (K. Tesnier and co-workers, in preparation). As expected, differences were observed in the dormancy level of freshly harvested seeds. Abscisic acid-deficient seeds were able to germinate at more negative water potentials compared to wild type seeds.

Large differences were observed regarding tolerance to the controlled deterioration treatment. The *abi3-7* seeds were more sensitive to the controlled deterioration treatment compared to wild-type Landsberg *erecta* (L*er*) seeds and those of the *aba1-1* mutant. Variation was also observed among different ecotypes. Seeds from the ecotype Cvi were more tolerant compared to those from the ecotype L*er*. After about 5 days controlled deterioration treatment the frequency of normal L*er* seedlings was halved, whereas about 12 days of treatment were needed for a similar germination reduction of Cvi seeds.

Quantitative Trait Loci Analysis of *Arabidopsis* Seed Traits

In recent years, a combination of genetics, physiology and molecular biology has provided a powerful tool in the identification of genetic loci involved in quantitative traits as e.g. yield and flowering time. In order to identify these so-called quantitative trait loci (QTL), crosses are made between two genotypes differing genetically for a trait. Self-pollination of progeny plants for several generations will result in a population of lines, consisting of more or less homozygous plants, each of which is composed of different chromosome segments from both parent plants. The association of values for the trait that is studied with the molecular markers segregating in that material allows the location of multiple genes segregating for the trait (QTL mapping) (Fig. 11.1). Also, for *Arabidopsis*, such recombinant inbred line (RIL) populations have been produced (Alonso-Blanco *et al.*, 1998).

As mentioned above, Cvi and L*er* seeds also exhibit variation in sensitivity to the controlled deterioration treatment. Therefore a QTL analysis for this trait

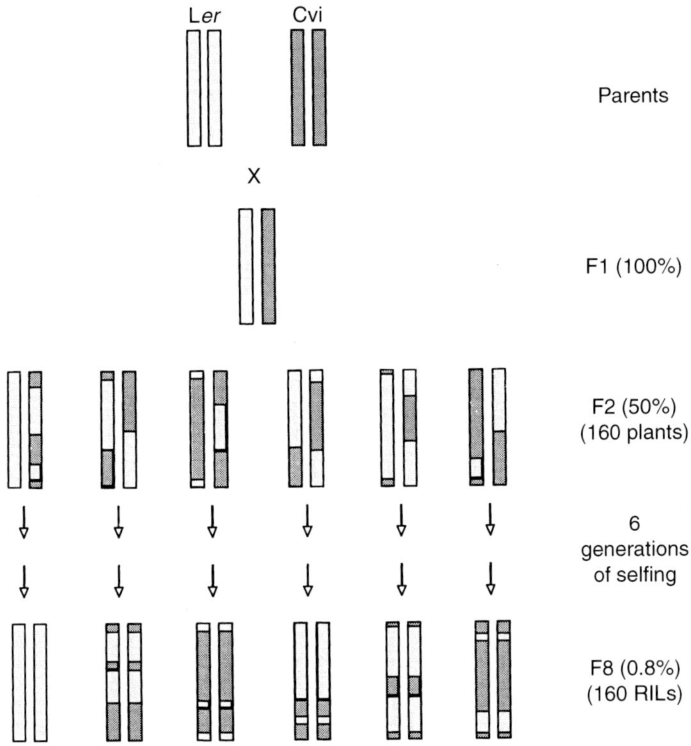

Fig. 11.1. Diagram showing production of recombinant inbred lines (RILs) for analysis of quantitative trait loci. The parent lines were the ecotypes Landsberg *erecta* (L*er*) and Cape Verde Island (Cvi). Self-pollination of the progeny lines for six generations results in near isogenic recombinant inbred lines.

was performed with the same Cvi/L*er* RIL populations as used for the analysis of dormancy (L. Bentsink and K. Tesnier, in preparation). The experiment was performed with seeds that had been stored for more than one year at ambient temperature, which had removed all dormancy from the seed lots. Seeds from the 162 RILs and the parent lines (Cvi and L*er*) were subjected to either zero or 4 days of controlled deterioration. Subsequently, all seed samples were dried at 32% RH and tested for germinability and frequency of resulting normal seedlings. Again, differences were observed between both parent lines, with L*er* seeds being more sensitive. The variation among the different RILs was even larger than between the parent lines. Based on the QTL analysis, two major loci were identified, both located on the top arm of chromosome 1, which significantly contributed to improved tolerance of the Cvi seeds.

Since dormancy and controlled deterioration tolerance were tested with the same Cvi/L*er* RILs, it is interesting to analyse whether any correlation exists between both traits. For each line, the level of dormancy after 6 weeks storage at ambient temperature was plotted against the differences in germination between zero and 4 days of controlled deterioration treatment (Fig. 11.2). No correlation was observed between the two parameters, which is confirmed by

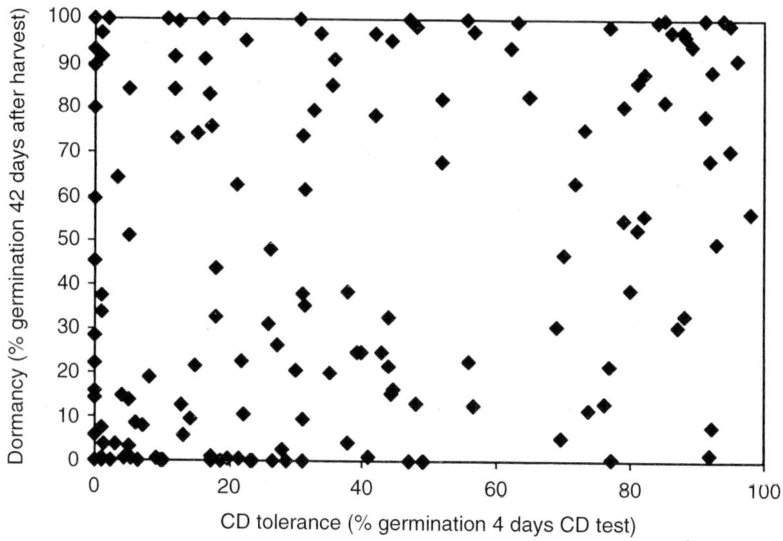

Fig. 11.2. Relation between controlled deterioration tolerance and dormancy for seeds from 160 recombinant inbred lines derived from a cross between the ecotypes Landsberg *erecta* (L*er*) and Cape Verde Island (Cvi). Controlled deterioration (CD) tolerance is expressed as the difference in percent germination between zero and 4 days hermetic storage at 40°C after equilibration at 85% relative humidity. Dormancy is expressed as the percent non-dormant) seed at 42 days after harvest.

the fact that the QTL observed for both traits are also located in different positions. This means that those loci governing dormancy in Cvi do not affect tolerance towards the controlled deterioration treatment and vice versa.

No Relation between Controlled Deterioration Tolerance and Stachyose Content

Stachyose and raffinose contents have also been analysed in Cvi and L*er* *Arabidopsis* seeds (L. Bentsink and co-workers, in preparation). Raffinose and stachyose contents were closely correlated (Fig. 11.3) and relatively low levels of stachyose and raffinose were found in Cvi seeds relative to L*er* seeds, without any negative effect on desiccation tolerance or storability of Cvi seeds.

Analysis of stachyose and raffinose contents in the RILs allowed the identification of a single major locus controlling the content of these oligosaccharides. This locus cosegregates with a member of the galactinol synthase gene family. The oligosaccharide locus maps at a different position as the QTLs identified for controlled deterioration tolerance. When, for each line, seed stachyose and raffinose contents are plotted against tolerance in the controlled deterioration test, no correlation could be observed between these two traits (Fig. 11.3).

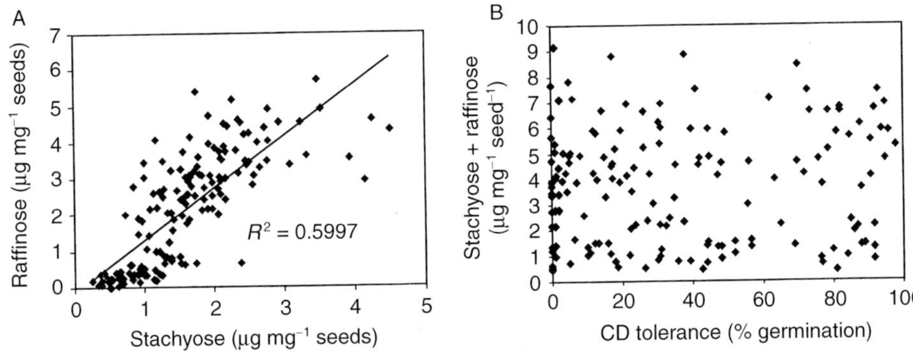

Fig. 11.3. Relation between raffinose content, stachyose content and controlled deterioration tolerance for seeds from 160 recombinant inbred lines derived from a cross between the ecotypes Landsberg *erecta* (L*er*) and Cape Verde island (Cvi). Raffinose and stachyose are presented as mg g^{-1} seed weight. Controlled deterioration (CD) tolerance is expressed as the difference in percent germination between zero and 4 days hermetic storage at 40°C after equilibration at 85% relative humidity. A, relation between raffinose and stachyose content; B, relation between CD tolerance and stachyose plus raffinose content.

These results showed that with *Arabidopsis* a strong reduction of stachyose and raffinose contents had no effect on desiccation tolerance, nor on tolerance of the seeds towards the deterioration treatment performed. One can question whether the deterioration test is indicative of seed longevity under more natural storage conditions. However, during the two-year storage at ambient temperature, Cvi seeds, with their low stachyose content, had survived and were still more tolerant towards the subsequently applied deterioration treatment compared to with L*er* seeds

Future Prospects

Besides the mutants already described, it is expected that many more seed quality mutants will be isolated in the near future. Mutagenized seeds can be obtained commercially and are used by several research groups to isolate mutants affected in seed development or seed germination. More recently, mutations are being created by T-DNA insertion (Dubreucq *et al.*, 1996). This technique, called T-DNA-tagging, is based on disruption of gene functioning through the integration of a foreign piece of DNA after transformation, and offers the great advantage that the gene involved in the mutation can be isolated through linkage with the T-DNA (the tag). A comparable approach is available, transposon-tagging, where the tag is a mobile piece of DNA that can integrate in a gene and thereby disrupt its function (Aarts *et al.*, 1995; Pereira and Aarts, 1998). Alternatively, by so-called reverse genetics, lines harbouring a T-DNA or transposon insertion in the gene of interest can be identified through two- or three-dimensional PCR-based screenings. With the progeny of these lines the phenotype of the mutation can be studied.

Other means to study the importance of certain genes are transgenic approaches. *Arabidopsis* can easily be genetically transformed. By over-expressing certain genes or their 'anti-sense' constructs, activity of the studied genes can be up- or down-regulated and the effect on seed viability or germination behaviour can be evaluated.

Quantitative trait loci analysis can also be performed for other seed quality traits. At present, QTL analyses for germination at low temperature or at low oxygen levels and with different RIL populations is under way. Further back-crossing of candidate lines and fine mapping of the loci involved will aid in identification and isolation of the gene(s) involved. This will especially be the case when the complete sequence of the *Arabidopsis* genome is made public.

Conclusions

Arabidopsis offers a very attractive model system to study the genetic variation in seed quality and the effect of modification of seed composition. Assays for *Arabidopsis* seed quality are now available. Combination of classic and molecular genetics with plant physiological research will provide, in the coming years, a wealth of mutants and information regarding genes involved in seed quality. This information may be used in plant breeding to improve seed quality of new varieties, for instance through the availability of molecular markers that can be used for indirect selection. *Arabidopsis* is a close relative of economically important cruciferous species such as cabbage and oilseed rape. Thus, gene sequences providing a positive influence on *Arabidopsis* seed quality might also be transferred to other crops, resulting in genetically modified crops with improved seed quality.

The information gained in the next years will also aid to a great extent in the fundamental understanding of processes and components involved in seed quality. The latter can be used to improve seed production and seed treatments, or to understand the ecology of species in natural habitats.

References

Aarts, M.G.M., Corzaan, P., Stiekema, W.J. and Pereira, A. (1995) A two-element *Enhancer-Inhibitor* transposon system in *Arabidopsis thaliana. Molecular and General Genetics* 247, 555–564.

Alonso-Blanco, C., El-Din El-Assal, S., Coupland, G. and Koornneef, M. (1998) Analysis of natural allelic variation at flowering time loci in the Landsberg *erecta* and Cape Verde Islands ecotypes of *Arabidopsis thaliana. Genetica* 199, 749–764.

Black, M., Corbineau, F., Grzesik, M., Guy, P. and Côme, D. (1996) Carbohydrate metabolism in developing and maturing wheat embryo in relation to its desiccation tolerance. *Journal of Experimental Botany* 47, 161–169.

Blackman, S.A., Obendorf, R.L. and Leopold, A.C. (1992) Maturation proteins and sugars in desiccation tolerance of developing soybean seeds. *Plant Physiology* 100, 225–230.

Dahal, P. and Bradford, K.J. (1990) Effects of priming and endosperm integrity on seed germination rates of tomato genotypes. II. Germination at reduced water potentials. *Journal of Experimental Botany* 41, 1441–1453.

Dahal, P., Bradford, K.J. and Jones R.A. (1990) Effects of priming and endosperm integrity on seed germination rates of tomato genotypes. I. Germination at sub-optimal temperature. *Journal of Experimental Botany* 41, 1431–1439.

De Bruijn, S.M., Ooms, J.J.J., Karssen, C.M. and Vreugdenhil, D. (1997) Effects of abscisic acid on reserve deposition in developing *Arabidopsis* seeds. *Acta Botanica Neerlandica* 46, 263–277.

Dubreucq, B., Grappin, P. and Caboche, M. (1996) A new method for the identification and isolation of genes essential for *Arabidopsis thaliana* seed germination. *Molecular and General Genetetics* 252, 42–50.

Groot, S.P.C. and Karssen, C.M. (1992) Dormancy and germination of abscisic acid-deficient tomato seeds: studies with the *sitiens* mutant. *Plant Physiology* 99, 952–958.

Groot, S.P.C., Bruinsma, J. and Karssen, C.M. (1987) The role of endogenous gibberellin in seed and fruit development of tomato: studies with a gibberellin-deficient mutant. *Physiologia Plantarum* 71, 184–190.

Hampton, J.G. and Te Krony, D.M. (1995) *Handbook of Vigour Test Methods*, 3rd edn. International Seed Testing Association, Zurich, 117 pp.

Horbowicz, M. and Obendorf, R.L. (1994) Seed desiccation tolerance and storability: dependence on flatulence producing oligosaccharides and cyclitols – review and survey. *Seed Science Research* 4, 385–405.

Karssen, C.M., Brinkhorst-van der Zwan, D.L.C., Breekland, A.E. and Koornneef, M. (1983) Induction of dormancy during seed development by endogenous abscisic acid: studies on abscisic acid deficient genotypes of *Arabidopsis thaliana* (L.) Heynh. *Planta* 157, 158–165.

Koornneef, M., Hanhart C.J., Hilhorst, H.W.M. and Karssen, C.M. (1989) *In vivo* inhibition of seed development and reserve protein accumulation in recombinants of abscisic acid biosynthesis and responsive mutants in *Arabidopsis thaliana. Plant Physiology* 90, 463–469.

Matthews, S. and Powell, A.A. (1987) Controlled deterioration tests. In: Fiala, F. (ed.) *Handbook of Vigour Test Methods*, 2nd edn. International Seed Testing Association, Zurich, pp. 49–56.

Meinke, D.W., Cherry, J.M., Dean, C., Rounsley, S.D. and Koornneef, M. (1998) *Arabidopsis thaliana*: a model plant for genome analysis. *Science* 282, 662–682.

Ni, B.R. and Bradford, K.J. (1993) Germination and dormancy of abscisic acid- and gibberellin-deficient mutant tomato (*Lycopersicon esculentum*) seeds. Sensitivity of germination to abscisic acid, gibberellin and water potential. *Plant Physiology* 101, 607–617.

Obendorf, R.L. (1997) Oligosacharides and galactosyl cyclitols in seed desiccation tolerance. *Seed Science Research* 7, 63–74.

Ooms, J.J.J., Leon-Kloosterziel, K.M., Bartels, D., Koornneef, M. and Karssen, C.M. (1993) Acquisition of desiccation tolerance and longevity in seeds of *Arabidopsis thaliana*: a comparative study using abscisic acid-insensitive *abi3* mutants. *Plant Physiology* 102, 1185–1191.

Ooms, J.J.J., Wilmer, J.A. and Karssen, C.M. (1994) Carbohydrates are not the sole factor determining desiccation tolerance in seeds of *Arabidopsis thaliana. Physiologia Plantarum* 90, 431–436.

Parcy, F, Valon, C., Raynal, M., Gaubier-Comella, P., Delseny, M. and Giraudat, J. (1994) Regulation of gene expression programs during *Arabidopsis* seed development: roles of the abi3 locus and of endogenous abscisic acid. *Plant Cell* 6, 1567–1582.

Pereira, A. and Aarts, G.M. (1998) Transposon tagging with the *En-I* system. In: Martinez–Zapater, J. and Salinas, J. (eds) *Methods in Molecular Biology*, Vol. 82, pp. 329–338.

Wolkers, W.F., Alberda, M., Koornneef, M., Leon-Kloosterziel, K.M. and Hoekstra, F.A. (1998) Properties of proteins and the glassy matrix in maturation-defective mutant seeds of *Arabidopsis thaliana*. *Plant Journal* 16, 133–143.

12 Analysis of the Cell Cycle in Sugarbeet Seed during Development, Maturation and Germination

E. Śliwińska

Department of Genetics and Plant Breeding, University of Technology and Agriculture, Kaliskiego 7, 85-796 Bydgoszcz, Poland

Cell cycle activity in sugarbeet seed was studied by flow cytometry. During embryo development the G_2/G_1 ratio decreases up to 24–28 days after pollination and then cell cycle activity is arrested (the G_2/G_1 ratio becomes constant). This is an indication that the embryo is already physiologically fully developed at this early stage. Under optimal conditions of maturation, drying and storage of the seeds, the G_2/G_1 ratio usually remains constant, decreasing slightly only after drying. It increases again at 24–48 h of imbibition, after completing DNA repair processes. It was found that environmental conditions during maturation of the seed, especially heavy rainfalls, could also increase the G_2/G_1 ratio of embryonic cells. When the seeds were collected at commercial harvest time (mature seeds) and 2 weeks before this (immature seeds), higher G_2/G_1 ratios were observed in the embryos of mature seeds, as compared with those from immature seeds. This phenomenon was most probably due to rainfalls and lower than optimal temperatures during the last 2 weeks of maturation, which induced germination while the seeds were still attached to the mother plant. A correlation was found between the G_2/G_1 ratio, vigour and laboratory and field test parameters, which suggests that this ratio can be used as an indicator of the physiological status of a seed. It is concluded that flow cytometry can be helpful in understanding and predicting sugarbeet seed quality.

Introduction

Flow cytometry is a fast and accurate method for measurement of DNA content. It also gives information about the DNA replication stages of the cells. This method makes it possible to study cell cycle activity in developing, maturing and germinating seeds. The aim of this study is to investigate if the changes in G_2/G_1 ratio in the embryo can be a valuable marker for determining the physiological status of the seed which, in turn, determines its germination

 Seed Biology: Advances and Applications (eds M. Black, K.J. Bradford and J. Vázquez-Ramos)

ability. Since seed quality (vigour, germination capacity) is one of the most important problems in sugarbeet breeding and seed production, such a marker could be helpful to provide seed lots of high standard to the growers.

Cell Cycle Activity in Developing Seeds

In the process of triploid ($2x \times 4x$) seed formation one of the sperm nuclei fuses with the egg cell, which results in triploid embryo formation. The second sperm nucleus fuses with the polar nuclei giving rise to a tetraploid endosperm. The first division of the primary endosperm nucleus occurs between 16 and 24 h after pollination. Approximately 7 days after pollination (DAP) the embryo becomes baton-shaped and the endosperm passes from the nuclear into the cellular phase (Jassem, 1972). At that time, however, the embryo is too small to be detected by flow cytometry (Fig. 12.1). Only the cells of the diploid ovule (2C and 4C) and of the endosperm (4C and 8C) are visible (Śliwińska, 1998). The embryo cells (3C and 6C) can be identified first at about 14 DAP.

With the growth of the embryo, the proportion of endosperm cells decreases rapidly (from 70% of the total seed cells at 14 DAP to 17% at 28 DAP; Śliwińska, 1998). During later stages of development, endosperm is resorbed and only a single-cell-layer remains in the mature seed, surrounding the radicle, and its share does not exceed 10% of the true seed cells.

At the beginning of the seed development, high cell cycle activity is observed in the embryo (G_2/G_1 ratio as high as 0.8 at 14 DAP). The G_2/G_1 ratio decreases up to 24–28 DAP and then becomes stabilized at the level of about 0.1 (Fig. 12.2). At that time the proportion of particular DNA replication stages in the seed becomes fixed. This suggests that cell cycle activity is already arrested in such young sugarbeet seeds. The results from flow cytometric analysis confirm observations of Jassem (1972), who found that triploid sugarbeet embryos are morphologically fully developed at 24 DAP.

Since seeds of different developing stages exist on branched sugarbeet plants even as late as 8–9 weeks after the beginning of flowering, it is very difficult to predict the optimal harvest time. Analysis of the cell cycle can help seed producers to fix the most economic harvest date, when the share of fully developed seeds is the highest and the oldest seeds still do not start to shatter.

Effect of Harvest Time on the G_2/G_1 Ratio

High quality sugarbeet seed production occurs when the period of seed development and ripening is warm and dry (Longden, 1986). Under such conditions cell cycle activity does not change much after being arrested at 24–28 DAP. Only during seed drying after harvest does the G_2/G_1 ratio decrease slightly (E. Śliwińska, unpublished data). However, it is suggested that the external conditions during seed maturation as well as harvest time affect DNA synthesis in embryonic cells (Bino *et al.*, 1993; Śliwińska *et al.*, 1999).

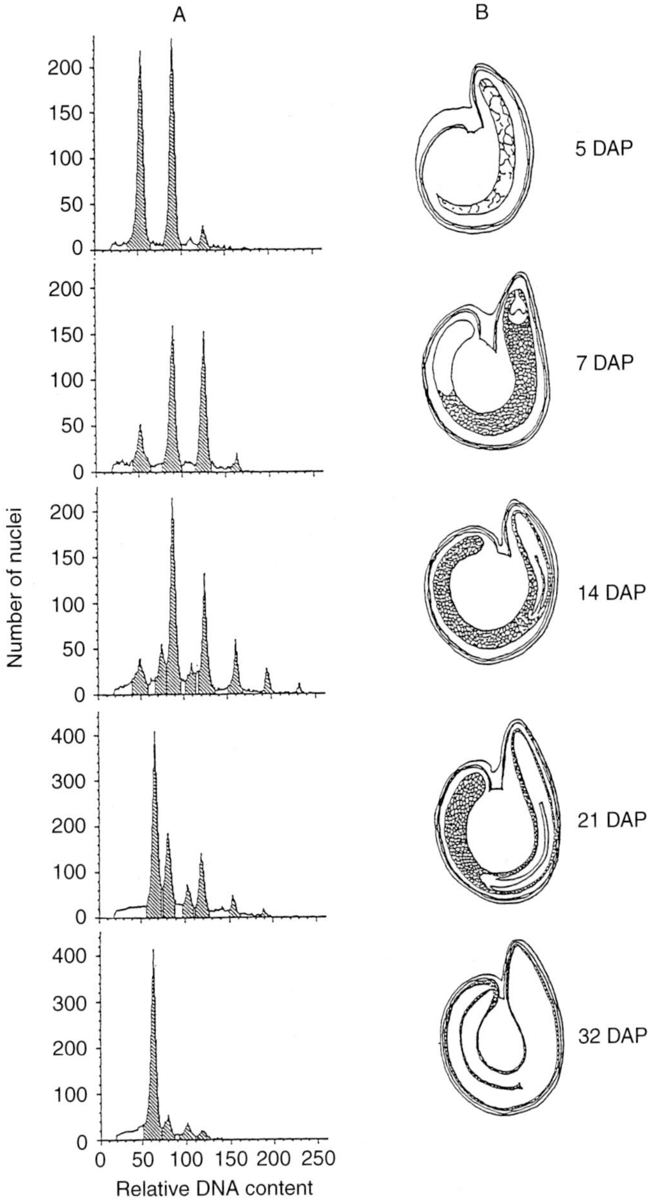

Fig. 12.1. Development of triploid ($2x \times 4x$) sugarbeet seed. A, DNA-histograms (after Śliwińska, 1998); channels: 50 – 2C, 68 – 3C, 86 – 4C, 104 – 6C, 122 – 8C; B, cross section of the seed (based on Jassem, 1972); DAP, days after pollination.

Very important factors during seed production, influencing seed characteristics, are temperature and rain. Rainy and cold periods cause delayed maturation (Battle and Whittington, 1969). On the other hand, some rainfall on ripening seeds could be beneficial to seed germinability (Lexander, 1981;

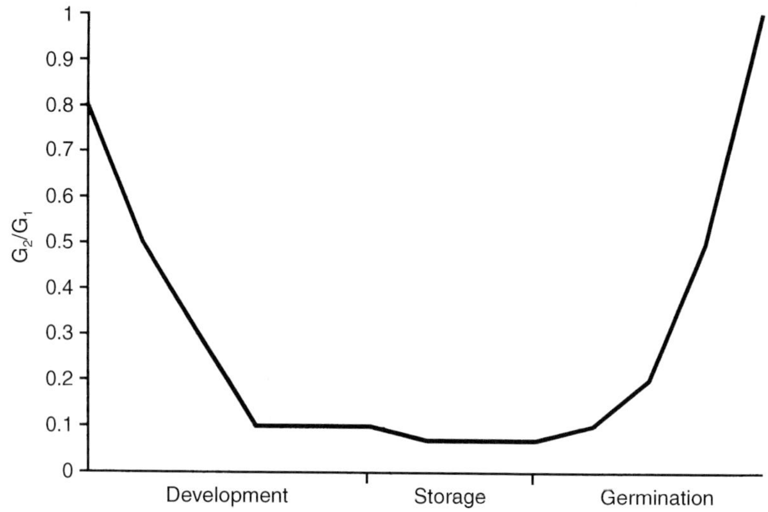

Fig. 12.2. Changes in G_2/G_1 ratios during development, storage and germination of sugarbeet seeds (under optimal conditions).

Longden, 1986). Besides leaching out germination inhibitors from the pericarp it can directly influence the embryo (Śliwińska *et al.*, 1999).

Since cells in the G_2 phase of the cell cycle are primarily located in the radicle tip (Bino *et al.*, 1992, 1993; Śliwińska, 1997), changes in cell cycle activity can be detected first in that part of the seed. Śliwińska *et al.* (1999) stated that the G_2/G_1 ratio was higher in the radicles of seeds collected at commercial harvest time than in seeds harvested 2 weeks earlier. This indicates that cell cycle activity has resumed during that period. Rain could increase seed moisture level and induce DNA replication, which would cause the augmentation of G_2 signals. Also, in the late-collected seeds a higher amount of the soluble B-subunit of the 11S globulin was evident as compared to that present in the early harvested ones. Normally, B-chain solubilization occurs during seed priming and early germination (Job *et al.*, 1997). Thus, these results demonstrate that germination processes can start in seeds still attached to the mother plant.

Relationship between the G_2/G_1 Ratio and Seed Vigour

One of the major factors influencing vigour and viability is physiological maturity of the seeds at harvest. As a marker of seed development stages, the G_2/G_1 ratio proved to be a good indicator for recognizing seed lots of low quality (Śliwińska, 1997). Such lots usually contain high percentages of undeveloped seeds with increased numbers of G_2 cells in the embryo and a relatively high proportion of endosperm cells.

An enhanced G_2/G_1 ratio does not always indicate unripeness, however. It was found that rainfall during seed ripening can increase cell cycle activity in the seeds (Śliwińska *et al.*, 1999). In this case, the G_2/G_1 ratio correlated positively with vigour, germination capacity and field emergence. The above reported results indicate that for predicting vigour of particular seed lots based on the G_2/G_1 ratio, it has to be considered that the high ratio could be due to lack of seed development or to commencement of germination (Fig. 12.2). The difference between undeveloped and fully developed germinating seeds, having the same augmented G_2/G_1 ratio in the embryo, can be detected easily by flow cytometry. A proportion of endosperm cells higher than 10% of the total seed cells is suggestive of poor development of the tested seed (Fig. 12.3A), while in mature germinating seeds, the proportion of endosperm cells is much lower (Fig. 12.3B).

Changes in the Proportion of DNA Replication Stages during Germination

Flow cytometry can be used to observe the progress of germination. Since cell cycle activity increases before the radicle emerges from the pericarp, this method gives information on seedling growth capacity much earlier than does the standard laboratory germination test.

In dry, mature sugarbeet seeds about 90% of the embryo cells are arrested in the G_0/G_1 phase of the cell cycle (Table 12.1; 2C DNA content). During the first 24 h of germination under optimal conditions (plated filter paper, 65% relative substrate moisture, darkness, temperature 20°C) no DNA replication is observed (Śliwińska, 1996; Śliwińska *et al.*, 1999). This is probably due to DNA repair processes which take place within the first hours of imbibition

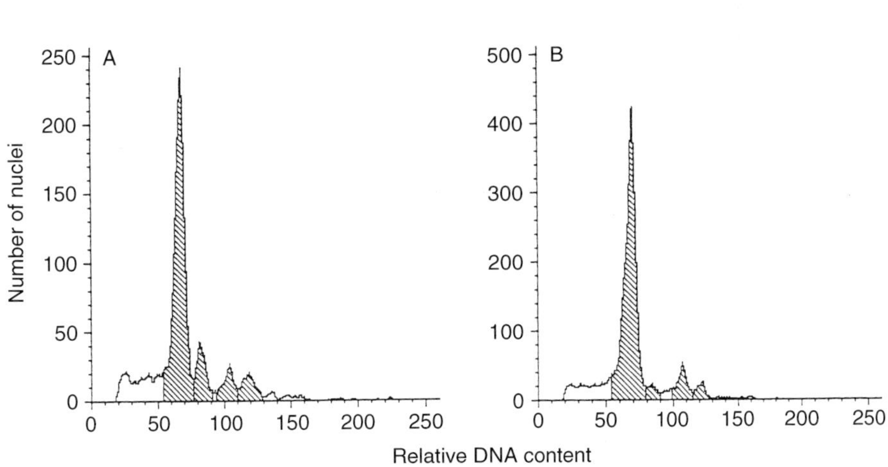

Fig. 12.3. DNA-histograms of triploid sugarbeet seeds. A, undeveloped (G_2/G_1 ratio in the embryo, 0.12, proportion of the endosperm cells, 19%); B, fully developed at the beginning of germination (G_2/G_1 ratio in the embryo, 0.12, proportion of the endosperm, 9%); channels 68 and 104, embryo cells of 3C and 6C DNA content, respectively; channels 86 and 122, endosperm cells of 4C and 8C DNA content, respectively.

Table 12.1. The proportions of cells with different DNA content, and the G_2/G_1 ratio, in dry and germinating diploid sugarbeet seeds.

Time of germination (h)	Nuclear DNA content (peak's area %)					G_2/G_1 ratio
	2C	4C	8C	16C	32C	
0	92.1	7.9	0	0	0	0.09 a[1]
12	90.6	9.4	0	0	0	0.11 a
24	88.2	10.2	1.6	0	0	0.15 a
36	75.1	21.3	3.7	0	0	0.30 ab
48	63.5	27.6	7.4	1.5	0	0.53 b
60	60.7	30.4	7.5	1.4	0	0.69 c
72	42.0	38.5	14.4	4.4	0.7	1.14 d
84	23.3	47.0	19.2	8.9	1.6	2.01 e
96	24.4	47.2	18.2	8.7	1.5	1.94 e

[1]Values followed by the same letter are not significantly different at $P = 0.05$ (Duncan's test).
After Śliwińska, 1996.

(Osborne, 1982). Later the percentage of G_2 cells increases from below 10% in dry seeds to about 20% at 36 h and 50% at 96 h of germination (Śliwińska, 1996). Apparently, the G_2/G_1 ratio can be a marker of the onset of germination. A rapid increase of G_2 cells occurs when the primary root becomes visible (about 60–72 h of germination). At that time the G_2/G_1 ratio reaches 1, and when the radicle/hypocotyl axis is 1 cm long an augmentation of the ratio to 2 is observed. During the first 96 h of germination and seedling growth endoreplication up to 32C is evident (Table 12.1).

It is well known that radicle elongation characterizes the transition from a reversible physiological state to an irreversible state, which means that dehydration at this later gemination stage kills the seed (Côme and Thévenot, 1982). Observations of cell cycle activity by flow cytometry allow us to recognize a transition from germination *sensu stricto* to growth. It provides useful information for establishing the proper harvest time of seeds when there is a rainy ripening period.

Conclusions

Flow cytometric evaluation of DNA replication stages gives valuable information about the physiological status of the seeds. Observations of changes in proportions between embryo and endosperm cells as well as in the G_2/G_1 ratio can be helpful in predicting the most economic harvest time. It also offers possibilities for the recognition of seed lots of low vigour. Estimation of the G_2/G_1 ratio in the embryo allows detection of the onset and progress of germination. Thus, a study of cell cycle activity is useful in the production of high quality sugarbeet seed lots.

Acknowledgements

Professor Marek Jassem (University of Technology and Agriculture, Bydgoszcz, Poland) and Professor J. Derek Bewley (University of Guelph, Canada) are kindly acknowledged for critically reading the manuscript.

References

Battle, J.P. and Whittington, W.J. (1969) The influence of genetic and environmental factors on the germination of sugar beet. *Journal of Agricultural Science* 73, 329–335.

Bino, R.J., de Vries, J.N., Kraak, H.L. and van Pijlen, J.G. (1992) Flow cytometric determination of nuclear replication stages in tomato seeds during priming and germination. *Annals of Botany* 69, 231–236.

Bino, R.J., Lanteri, S., Verhoeven, H.A. and Kraak, H.L. (1993) Flow cytometric determination of nuclear replication stages in seed tissues. *Annals of Botany* 72, 181–187.

Côme, D. and Thévenot, C. (1982) Environmental control of embryo dormancy and germination. In: Khan, A.A. (ed.) *The Physiology and Biochemistry of Seed Development, Dormancy and Germination.* Elsevier Biomedical Press, New York, pp. 271–298.

Jassem, M. (1972) Embryological investigations into the development of monogerm diploid, triploid and tetraploid sugarbeet seeds (in Polish). *Bydgoskie Towarzystwo Naukowe, Prace Wydzialu Nauk Przyrodniczych* Seria B 15, 69–121.

Job, C., Kersulec, A., Ravasio, L., Chareyre, S., Pepin, R. and Job, D. (1997) The solubilization of the basic subunit of sugarbeet seed 11-S globulin during priming. *Seed Science Research* 7, 225–243.

Lexander, K. (1981) Physical and physiological seed characteristics influencing field emergence of sugar beet. *Proceedings of 44th Winter Congress of IIRB*, pp. 21–36.

Longden, P.C. (1986) Influence of the seed crop environment on the quality of sugar beet seed. *Proceedings of 49th Winter Congress of IIRB*, pp. 1–16.

Osborne, D.J. (1982) DNA integrity in plant embryos and the importance of DNA repair. In: *Embryonic Development, Part L: Cellular Aspects*, A.R. Liss, Inc., New York, pp. 577–592.

Śliwińska, E. (1996) Flow cytometric analysis of the cell cycle of sugar beet seed during germination. *Journal of Applied Genetics* 37A, 254–257.

Śliwińska, E. (1997) Flow cytometric analysis of sugar beet seeds different in vigour. In: Ellis, R.H., Black, M. and Hong, T.D. (eds) *Basic and Applied Aspects of Seed Biology.* Kluwer Academic Publishers, Dordrecht, pp. 577–584.

Śliwińska, E. (1998) Cell cycle activity during development of sugar beet seed. In: Maluszynska, J. (ed.) *Plant Cytogenetics.* Silesian University Publishers (Prace Naukowe Uniwersytetu Slaskiego 1696), Katowice, pp. 51–59.

Śliwińska, E., Jing Hai-Chun, Job, C., Job, D., Bergervoet, J.H.W., Bino, R.J. and Groot, S.P.C. (1999) Effect of harvest time and soaking treatment on cell cycle activity in sugar-beet seeds. *Seed Science Research* 9, 91–99.

13 Possible Roles of Phosphoenolpyruvate Carboxylase and Pyruvate Kinase in Assimilate Partitioning in Developing Maize Embryos

R. RODRÍGUEZ-SOTRES, A. LARA-NUÑEZ AND M. RODRÍGUEZ-PENAGOS

Departamento de Bioquímica, Facultad de Química, Universidad Nacional Autónoma de México, 04510 México D.F., México

During their development, seed tissues convert the nutrients coming from the mother plant into proteins, lipids, starch and energy. Thus, they must regulate carefully the use of the available carbon. When isolated developing maize embryos were incubated in a diluted medium without abscisic acid, their lipid synthesis rate was strongly reduced; this metabolic change was accompanied by a small but significant change in the affinity of phosphoenolpyruvate carboxylase for phosphoenolpyruvate, with no changes in the extractable pyruvate kinase activity. The antibiotic phosphomycin was found to activate phosphoenolpyruvate carboxylase and to inhibit pyruvate kinase in extracts from immature embryos. When the isolated maize embryos were incubated in the presence of phosphomycin, the relative incorporation of [^{14}C]-acetate into proteins increased while that into lipids was reduced. These results suggest a central role for phosphoenolpyruvate metabolism in carbon partitioning during the development of maize embryos.

Introduction

Developing seeds must convert the carbon they receive from the mother plant into energy, cell components, and storage compounds in order to complete their growth and development (Bewley and Black, 1994). Since the most abundant source of carbon for the developing seeds are soluble sugars (Weber *et al.*, 1997), glycolysis must play a central role in carbon partitioning. Thus, understanding the regulation of this pathway in seed tissues will help us to identify the key points in the control of carbon utilization.

The regulation of plant glycolysis is complex since many steps can take place in both the cytosol and the plastid, while others are exclusive to one or

(eds M. Black, K.J. Bradford and J. Vázquez-Ramos)

the other compartment (Plaxton, 1996). In addition, hexose-phosphates, triose phosphates, phosphoenolpyruvate (PEP), pyruvate and perhaps other metabolites can be transported through the plastidial envelope (Emes and Neuhaus, 1997; Fischer *et al.*, 1997). The glycolytic reactions occurring in the plastid depend on the tissue and its developmental stage, since the available evidence indicates that chloroplasts and leucoplasts do not contain identical glycolytic machinery (Emes and Neuhaus, 1997; Fischer *et al.*, 1997); and to further complicate the matter, some glycolytic steps can be bypassed by accessory reactions. Such is the case of the PEP to pyruvate conversion catalysed by pyruvate kinase (PK), that can be performed by the anapleurotic step catalysed by PEP carboxylase, followed by the action of the malate dehydrogenase and the malic enzyme (Plaxton, 1996). In fact, this pathway has been proposed to feed carbon into fatty acid synthesis during seed development, because isolated leucoplasts from *Brassica napus* incorporate malate more efficiently than pyruvate into fatty acids (Smith *et al.*, 1992), because PEP carboxylase is highly active in some oily seeds (Sangwan *et al.*, 1992), and because a novel malate/Pi antiporter has been identified in these organelles (Eastmond *et al.*, 1997). However, PEP carboxylase has been more often associated with a high amino acid synthesis rate in other plant tissues (Rodríguez-Penagos and Muñoz-Clares, 1999), and oily seeds also accumulate significant amounts of protein (Bewley and Black, 1994).

Developing embryos of cereal are very active in both fatty acid and protein synthesis and the fatty acid synthesis can be virtually abolished by incubation of the embryos in dilute buffer (Rodríguez-Sotres and Black, 1994; Pacheco-Moisés *et al.*, 1997). Since changes in regulatory enzymes are expected to accompany all large changes in metabolic fluxes, we studied the possible changes in PK and PEP carboxylase associated with the incubation of isolated developing embryos of maize in dilute buffer (treatment A), as compared with those embryos that were incubated in buffer in the presence of 60 mM sucrose, 500 mM mannitol and 10 μM abscisic acid (ABA) (BSMA, treatment B), since these latter conditions have been shown to induce and maintain high rates of fatty acids and TAG synthesis in this model system (Pacheco-Moisés *et al.*, 1997). Based on our findings, we propose that these two enzymes, together with the exchange of metabolites through the leucoplast envelope are key points for the regulation of carbon partitioning in maize embryos during their development.

Methods, Results and Discussion

Maize embryos were isolated from developing seeds at 25–35 days after pollination. When these embryos were homogenized in buffer TEA-HCl 100 mM, pH 7.4 with 50 μg ml^{-1} of chymostatin and 10 μg ml^{-1} of leupeptin as protease inhibitors, high levels of PEP carboxylase (Table 13.1) and PK (Table 13.2) activities were observed in the soluble fraction. Since PK isozymes have been reported to be present in the cytosolic and plastidial fractions (Plaxton, 1996), we took advantage of the fact that the plastidial isoenzyme (PKp) is more heat-labile than the cytoplasmic one (PKc; Gottlob McHugh *et al.*, 1992). The extract was heated for 10 min at 55°C, cooled in ice for 5 min and centrifuged in a microfuge for 10 min. The remaining PK activity was considered to be

Table 13.1. Activity and malate sensitivity of PEP carboxylase in developing embryos isolated from maize seeds at 25–35 days after pollination.

Incubation condition	V_{max}[a] (nmol min^{-1} mg^{-1})	Apparent K_m[a] (µM PEP)	*V/K*	Malate sensitivity (mM)[b]
None	62 ± 8	41 ± 18	1.51	0.286 ± 0.038
A	41 ± 2	11 ± 2	3.72	0.467 ± 0.042
B	54 ± 2	42 ± 5	1.28	0.295 ± 0.078

[a]Estimated by non-linear fit of the PEP saturation curves to the Michaelis–Menten equation.
[b]Concentration of malate that produces 50% inhibition at 0.15 mM PEP and 2 mM $MgCl_2$.
The isolated embryos were incubated as described in Pacheco-Moisés *et al.* (1997), in media containing 10 mM morpholinoethanesulphonic-NaOH pH 5.5 without (treatment A), or with 10 µM ABA, 500 mM mannitol and 60 mM sucrose (treatment B). After 24 h at 25°C with gentle shaking, the embryos were ground in 2 volumes of 100 mM triethanolamine-HCl, with 2 mM $MgCl_2$, 10 µg ml^{-1} leupeptin, 100 µg ml^{-1} chymostatin and 10% glycerol. The activity and malate sensitivity of PEP carboxylase was determined as described by Rodríguez-Penagos and Muñoz-Clares (1999). As a control, some embryos were extracted fresh (condition, none).

Table 13.2. Pyruvate kinase activity in developing embryos isolated from maize seeds at 25–35 days after pollination.

Incubation treatment	Temperature treatment	V_{max}[a] (nmol min^{-1} mg^{-1})	Apparent K_m[a] (µM PEP)
None	None	42 ± 6	21 ± 13
A	None	26 ± 2	18 ± 7
B	None	27 ± 2	14 ± 6
None	55°C	38 ± 5	17 ± 9
A	55°C	20 ± 3	10 ± 5
B	55°C	30 ± 4	14 ± 8
None	–	4 (9.5)[b]	nd
A	–	6 (23)[b]	nd
B	–	3 (11)[b]	nd

[a]Estimated by non-linear fit of the PEP saturation curves to the Michaelis–Menten equation.
[b]Estimated by difference. The values as percentage of the total PK activity are given in parenthesis.
Embryos isolated and treated as described in Table 13.1. The activity of pyruvate kinase was determined before and after heating the extracts for 5 min at 55°C, as described by Gottlob McHugh *et al.* (1992). The activity remaining after the heat treatment was designated as PKc and the heat-sensitive activity was determined by difference.

the PKc and the PKp activity was calculated by difference of the the total PK activity minus the PKc fraction (Table 13.2). Our results showed that the PKc activity recovered after the previous treatment was very stable and further heating, cooling and centrifugation cycles had only a small effect on the activity levels (not shown), indicating that this activity was indeed the heat-stable isozyme and not the result of a complex thermal inactivation kinetics.

Both PKc and PKp putative fractions were present in the crude extract from developing maize embryos, although the plastidial component was no more than 25% of the total PK activity. Tables 13.1 and 13.2 also show that the maximum specific activity of PEP carboxylase and PKc or PKp were not significantly different in the freshly isolated (not incubated) embryos or after the A and the B treatments. Desalting these extracts through a G-25 column had no effect in the above findings (not shown), indicating that metabolites in the extract were not masking differences in activity.

This last result suggests that PEP carboxylase has to compete with PKc for the cytoplasmic pool of PEP, unless two different pools are available to these enzymes; which, to the best of our knowledge, has not been documented.

Many regulatory enzymes are thought to be working at subsaturating levels of substrate *in vivo*, as seems to be the case for PEP carboxylase (Tovar-Méndez *et al.*, 1998). Therefore, we examined the affinity and the specificity constants of these enzymes for their common substrate PEP, because even relatively small changes in their kinetic constant can result in a shift in the PEP consumption balance by the two enzymes. Data in Table 13.1 show that the embryos from treatment A have a PEP carboxylase activity with a lower K_m value and a higher specificity constant (V/K) than the one present in the BSMA treated or the freshly isolated embryos (To). In contrast, we did not find large differences in any of the PK activity components (Table 13.2). This change is enough to duplicate the PEP used by PEP carboxylase if both enzymes are working below their K_m values *in vivo*. We could not find evidence of the molecular origin of this change, because the sensitivity of PEP carboxylase to its inhibitor malate, that may give an indication of the enzyme phosphorylation state (Chollet *et al.*, 1996), did not change significantly. The enzyme molecular weight in native gel electrophoresis stained for PEP carboxylase activity, or in SDS-PAGE followed by Western blot (detected with polyclonal antibodies against the C_4 maize isoform) was not different in both extracts. Yet, it is possible that the change in affinity for PEP of PEP carboxylase is related to the *de novo* synthesis of a similar, but not identical PEP carboxylase isozyme, because the plant PEP carboxylases reported to date are very similar to each other in their amino acid sequences (Lepiniec *et al.*, 1994).

The above results suggest that less PEP is converted by PEP carboxylase in the embryos with an active TAG synthesis. If this is true, and if PEP carboxylase and PK regulate carbon partitioning, any external factor able to increase the ratio of PEP carboxylase to PK activity *in vivo* should reduce fatty acid synthesis, and perhaps, increase protein synthesis. We decided to take advantage of the uncommon activation of PEP carboxylase by the PEP analog phosphomycin (Mujica-Jiménez *et al.*, 1998); but because the effect of this antibiotic on the non-photosynthetic maize PEP carboxylase and on PK has not

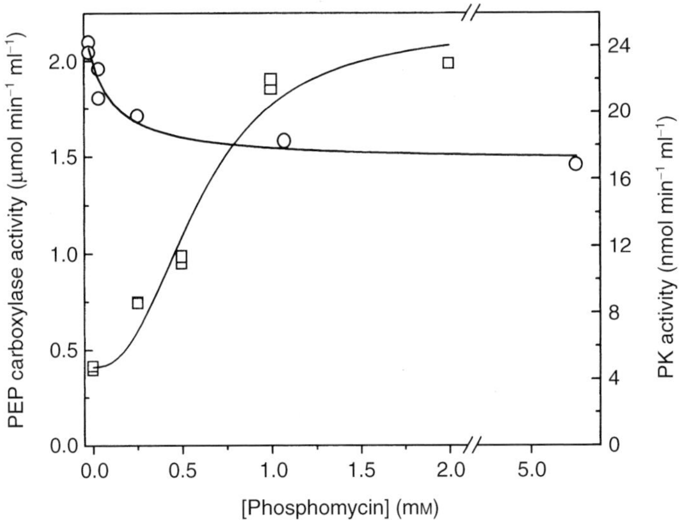

Fig. 13.1. Developing maize embryos were isolated and extracted as described in Table 13.1, and PEP carboxylase or total PK activity were determined at the indicated phosphomycin concentrations. Only the ADP-dependent pyruvate formation was considered as true PK activity. Curves were generated by non-linear fit of the experimental points to an equation for sigmoidal activation or hyperbolic partial inhibition, respectively. The A_{50} for PEP carboxylase activation was 0.61 ± 0.09 mM, with a Hill number of 2.40 ± 0.61. The I_{50} for PK inhibition was 0.22 ± 0.11 mM phosphomycin, and 70.8% of the activity was apparently insensitive to the inhibitor.

been reported, we first analysed their sensitivity to phosphomycin *in vitro*. Figure 13.1 shows that while PEP carboxylase activity in crude extracts from maize embryos was strongly activated by phosphomycin, PK (total) showed only partial inhibition, i.e. phosphomycin did have a differential effect on these enzymes.

We then incubated developing embryos of maize in A or B media for 24 h followed by incubation with or without 25 mM phosphomycin. At the end of these incubations the embryos were fed with [^{14}C]-acetate, and finally the radioactivity incorporated into the protein, the lipid and the triacylglycerol fractions was determined. The results shown in Fig. 13.2 indicate that phosphomycin produced a clear reduction in [^{14}C]-lipids, while increasing the radioactivity in the protein fraction. This was true whether the synthesis of triacylglycerol in the embryos was rapid (Fig. 13.2B) or slow (Fig. 13.2A). However, phosphomycin had almost no effect on the amount of [^{14}C]-triacylglycerol at the lower rate of synthesis (Fig. 13.2A), which means that reduction in lipid synthesis did not affect the same lipid fractions in both conditions. The reason for this last unexpected effect of phosphomycin is unclear, but reflects the complexity behind the regulation of seed oil synthesis. The effect of phosphomycin on pure PKc and pure PKp, on the PEP transport across the leucoplast envelope, and other side-effects of this drug have not been studied in detail. Our results oppose the view that PEP carboxylase feeds

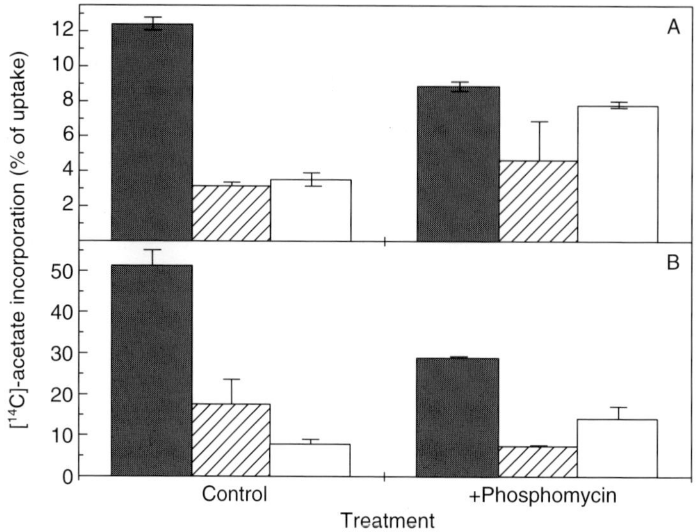

Fig. 13.2. Developing maize embryos were isolated and incubated as described in Table 13.1. At the end of the first period of incubation, 25 mM phosphomycin was added to half of the samples and the incubation was extended for 12 h. Finally, the embryos were fed with 0.5 μCi of [^{14}C]-acetate, and incubated for a further 4 h and the radioactivity in the protein (□), total lipids (■) and triacylglycerol (▨) fractions was determined as described by Pacheco-Moisés *et al.* (1997). A, medium A; B, medium B.

carbon into the seed oils. In fact, our data suggest that this last enzyme drives the carbon away from oil synthesis and into the biosynthesis of amino acids and protein. Experiments are being carried out in our laboratory to verify this last possibility.

Acknowledgements

We want to thank Dr Rosario A. Muñoz-Clares and Professor Michael Black for their helpful suggestions throughout the development of this project. The research reported here was funded by the grant DGAPA-UNAM IN210295. A.L.-N. received a DGAPA-UNAM scholarship.

References

Bewley, D. and Black, M. (1994) *Seeds. Physiology of Development and Germination*, 2nd edn. Plenum Press, New York, London, 445 pp.

Chollet, R., Vidal, J. and O'Leary, M.H. (1996) Phosphoenolpyruvate carboxylase: an ubiquitous, highly regulated enzyme in plants. *Annual Review of Plant Physiology and Plant Molecular Biology* 47, 273–298.

Eastmond, P.J., Dennis, D.T. and Rawsthorne, S. (1997) Evidence that a malate inorganic phosphate exchange translocator imports carbon across the leucoplast envelope for fatty acid synthesis in developing castor seed endosperm. *Plant Physiology* 114, 851–856.

Emes, M.J. and Neuhaus, H. E. (1997) Metabolism and transport in non-photosynthetic plastids. *Journal of Experimental Botany* 48, 1995–2005.

Fischer, K., Kammerer, B., Gutensohn, M., Arbinger, B., Weber, A., Haeusler, R.E. and Fluegge, U.I. (1997) A new class of plastidic phosphate translocators: a putative link between primary and secondary metabolism by the phospho*enol*pyruvate/phosphate antiporter. *Plant Cell* 9, 453–462.

Gottlob McHugh, S., Rajender, S.S., Blakeley, S.D., Vanlergerhe, G.C., Ko, K., Turpin, D.H., Plaxton, W.C., Miki, B.L. and Dennis, D.T. (1992) Normal growth of transgenic tobacco plants in the absence of cytosolic pyruvate kinase. *Plant Physiology* 99, 952–958.

Lepiniec, L., Vidal, J., Chollet, R., Gadal, P. and Cretin, C. (1994) Phospho*enol*pyruvate carboxylase: structure, regulation, and evolution. *Plant Science* 99, 111–124.

Mujica-Jiménez, C., Castellanos-Martínez, A. and Muñoz-Clares, R. (1998) Studies of the allosteric properties of maize leaf phospho*enol*pyruvate carboxylase with the phospho*enol*pyruvate analog phosphomycin as activator. *Biochimica Biophysica Acta* 1386, 132–144.

Pacheco-Moisés, F., Valencia-Turcotte, L., Altuzar-Martínez, M. and Rodríguez-Sotres, R. (1997) Regulation of acyltransferase activity in immature maize embryos by abscisic acid and the osmotic environment. *Plant Physiology* 114, 1095–1101.

Plaxton, W.C. (1996) The organization and regulation of plant glycolysis. *Annual Review of Plant Physiology and Plant Molecular Biology* 47, 185–214.

Rodríguez-Penagos, M. and Muñoz-Clares, R. (1999) Response of phospho*enol*pyruvate carboxylase from maize leaves to moderate water deficit. *Journal of Plant Physiology* 155, 631–638.

Rodríguez-Sotres, R. and Black, M. (1994). Osmotic potential and abscisic acid regulate triacylglycerol synthesis in developing wheat embryos. *Planta* 192, 9–15.

Sangwan, R.S., Singh, N. and Plaxton, W.C. (1992) Phospho*enol*pyruvate carboxylase activity and concentration in the endosperm of developing and germinating castor oil seeds. *Plant Physiology* 99, 445–449.

Smith, R.G., Gauthier, D.A., Dennis, D.T. and Turpin, D.H. (1992) Malate- and pyruvate-dependent fatty acid synthesis in leucoplasts from developing castor endosperm. *Plant Physiology* 96, 1233–1236.

Tovar-Méndez, A., Rodríguez-Sotres, R., López-Valentín, D.M. and Muñoz-Clares, R.A. (1998) Re-examination of the roles of PEP and Mg^{2+} in the reaction catalyzed by the phosphorylated and non-phosphorylated forms of phospho*enol*pyruvate carboxylase from leaves of *Zea mays*. Effects of the activators glucose-6-phosphate and glycine. *Biochemical Journal* 332, 633–642.

Weber, H., Borisjuk, L. and Wobus, U. (1997) Sugar import and metabolism during seed development. *Trends in Plant Science* 2, 169–174.

III Storage and Vigour

14 Effects of Seed Ageing on the Enzymic Antioxidant System of Maize Cultivars

I. BERNAL-LUGO[1], A. CAMACHO[1] AND A. CARBALLO[2]

[1]*Departamento de Bioquímica, Facultad de Química, Universidad Nacional Autónoma de México, 04510 México D.F., México;* [2]*Centro de Genética, Colegio de Posgraduados, Montecillo, 56230 México*

Seed ageing includes a wide range of degenerative events that accumulate over time, causing loss of viability. These detrimental changes are due in part to ageing-induced oxidative reactions. This study was undertaken to evaluate the role of the enzymatic oxidative mechanism in limiting oxidative damage and maintaining seed viability during dry storage. To test this hypothesis the effect of ageing on the enzymatic antioxidative capacity of two maize cultivars showing differences in seed storage performance has been measured. There is a positive association between the hydrogen peroxide detoxification capacity retained in the aged germinated seed and storage performance in these two cultivars.

Introduction

Loss of vigour and viability during dry storage includes a wide range of degenerative events that accumulate over time, causing loss of viability (for review see Priestley, 1986; Smith and Berjak, 1995). These can be grouped as physiological and biochemical events. Among the physiological changes induced by seed ageing are: decreased rates of germination and seedling growth (Heydecker, 1972), increased number of morphologically abnormal seedlings (Mackay, 1972), decreased ability to emerge when sown under stressful conditions (Mackay, 1972), increased metabolite and ion leakiness (Roberts and Ellis, 1982), and greater susceptibility of seedlings to pathogens (Christensen, 1972). At the biochemical level, seed ageing is accompanied by: decline in metabolic activity upon germination (Ferguson *et al.*, 1990; Dreyer and Van de Venter, 1992), changes (in most cases a decrease) in enzymatic activities (Ganguli and Sen-Mandi, 1993; Bernal-Lugo *et al.*, 1994; Aung and McDonald, 1995), and a decline in protein and nucleic acid biosynthesis (Bray and Chow, 1976; Dell'Aquila, 1994; Cruz-Garcia *et al.*, 1995). Lesions in DNA (reviewed by

Roberts, 1988) and loss of membrane integrity have also been reported (Senaratna *et al.*, 1988; Basavarajappa *et al.*, 1991; Dawidowicz-Grzegorzewska and Podstolski, 1992). While this information describes the symptoms of seed ageing, the biochemical and molecular basis for loss of vigour and longevity remains enigmatic.

In much of the literature, there is the assumption that oxidative reactions contribute to the detrimental changes observed in aged seeds. These include free-radical oxidations (Dobretsov *et al.*, 1977; Wilson and McDonald, 1986), enzymic dehydrogenation (Dapron, 1985), aldehyde oxidation of proteins (Stadtman, 1992) and protein glycation (Maillard) reactions (Wettlaufer and Leopold, 1991; Sun and Leopold, 1995).

Recently, it has been suggested that the glassy state contributes to seed longevity in dry storage, particularly during the initial period of relative stability (Leopold *et al.*, 1994; Bernal-Lugo and Leopold, 1998). Because of the high viscosity of the glass, deteriorative reactions are suppressed. Some detrimental events may take place under environmental conditions in which the glass has been partially or totally melted; nevertheless, the seed mortality curve still shows an initial period of relative stability preceding a dynamic rate of cumulative mortality (Bernal-Lugo and Leopold, 1998). The above suggests that besides the glassy state the seed may have other mechanisms that could serve to protect against oxidative reactions, either during storage, or during germination. Since it is well accepted that enzymatic and non-enzymatic oxidations contribute to the reactions regulating seed ageing, it is reasonable to hypothesize that the cell defence mechanisms against oxidative damage may be also involved in maintaining seed vigour and viability. To test this hypothesis, we have measured the effect of ageing on the enzymatic anti-oxidative capacity of two maize cultivars showing differences in seed storage performance. The results showed that there is a positive association between the hydroperoxide detoxification capacity retained in the aged germinated seed and storage performance.

Materials and Methods

Plant materials

Maize seeds (*Zea mays* L.) of accession T_{100} were obtained from the Maize Germplasm Bank at CIMMYT. Dr Aquiles Carballo (Centro de Genética, Colegio de Posgraduados) provided A_6.

Seed storability

Seed storage characteristics were determined by accelerated ageing at 30°C and 75% RH using a saturated NaCl solution to buffer the environmental humidity. Seeds were germinated for 7 days in wet paper rolls in the dark at 25°C and were scored as having germinated if the axes had broken through the seed coat. Each germination test comprised 20 seeds and was repeated three times. Seed survival curves were plotted on a probit scale using the origin program. A probit of value 5 (equivalent to 50% germination) identifies the time for 50% viability (P_{50}).

Enzyme preparation and assays

Embryonic axes (30) were excised from control and aged seeds, surface sterilized and plated in Petri dishes containing sterile fiter paper wetted with 5 ml of 2% sucrose. At different times of imbibition, protein was extracted by homogenizing axes in 3 ml of 5 mM Tris/HCl, pH 7.5 containing 1 mM EDTA, 4 mM cysteine and 0.1% Triton. The homogenate was filtered through three layers of cheese cloth and centrifuged at 18,000 *g* for 15 min. The supernatant obtained was used for enzyme assays. All operations were carried out at 4°C. All enzyme activities were measured in a final volume of 1 ml at 25°C.

Activities of catalase and guaiacol peroxidase were measured by modifications of the method of Chance and Maehly (1955). The reaction mixture for catalase contained 25 mM phosphate buffer (pH 7.0), 10 mM H_2O_2 and an enzyme aliquot. The decomposition of H_2O_2 was measured by following the decrease in absorbance at 240 nm (E = 39.4 mM^{-1} cm^{-1}). For guaiacol peroxidase, the reaction mixture contained 25 mM phosphate buffer (pH 7.0), 10 mM H_2O_2, 0.05% guaiacol and the enzyme aliquot. The oxidation of guaiacol was measured at 470 nm (E = 26.6 mM^{-1} cm^{-1}). Superoxide dismutase activity was assayed by autooxidation of epinephrine to adrenochrome at pH 10.2 based on the method described by Misra and Fridovich (1972). The oxidation of epinephrine was measured at 480 nm (E = 4.02 mM^{-1} cm^{-1}). Total protein content was determined by the method of Lowry *et al.* (1951) using BSA as a standard.

H_2O_2 assay

The amount of hydrogen peroxide was assayed in the diffusates from the axes by the loss of scopoletin fluorescence at 460 nm following excitation at 350 nm (Hildebrandt and Roots, 1975). Scopoletin was obtained from Sigma Co. The diffusates were prepared by soaking 30 germinated or non-germinated axes in 3 ml of 30 mM potassium phosphate buffer pH 7.0 at 25°C.

Results

Storage performance was expressed as the time for loss of 50% viability during storage (P_{50} Table 14.1). The poorest storage performance was found for A_6, whereas T_{100} showed better storability (Table 14.1).

To know whether the storage performance was associated with the level of the enzymatic antioxidant capacity of the seed or with its stability, we compared the activity of the main enzymes of hydroperoxide metabolism, superoxide dismutase (SOD), catalase (CAT) and guaiacol peroxidases (POX) upon imbibition of embryonic axes isolated from control and aged seeds of maize cultivars exhibiting different storage performance.

In control dry seeds of either A_6 or T_{100}, CAT showed low activity whereas SOD and POX showed higher enzymatic activities. With the exception of POX, which showed higher activity in T_{100} than in A_6, the activities of the enzymes were similar in both cultivars (Table 14.2).

During imbibition of control seeds the SOD activity increased during the first 5 h. Thereafter, this enzyme remained at a constant level (Fig. 14.1A). After 24 h of imbibition, the time at which the cultivars germinated, the amount of SOD was similar between cultivars.

The activities of the enzymes involved in hydroperoxide detoxification (catalase and peroxidase) were also measured during imbibition. The activity of catalase remained constant throughout in both cultivars (Fig. 14.1B). However, for peroxidase the activity increased rapidly up to 24 h of imbibition. During the whole period the activity of guaiacol peroxidase was similar for both cultivars (Fig. 14.1C). Peroxidase showed the most significant increase in activity during imbibition (Fig. 14.1). The amount of hydroperoxide detoxified after 24 h of germination, calculated as the difference between the hydroperoxide quantified at 0 and 24 h of imbibition, was about 317 nmol $axis^{-1}$ (Table 14.3) for both cultivars. From the above results we can conclude that the enzymatic antioxidant capacity was similar among cultivars that show differences in storage performance.

Table 14.1. Effect of storage[a] on seed characteristics.

Cultivar	P_{50} (months)[b]	Viability (%)[c]
A_6	2.1	20
T_{100}	5.4	90

[a]75% RH, 30°C.
[b]Storage time in which seed lot germination decreases 50%.
[c]After 4 months of storage at 75% RH, 30°C.

Table 14.2. Activities of the enzymes involved in $^{\cdot}O_2^-$ and H_2O_2 consumption in control and aged maize axes.

	Enzyme activity (μmol H_2O_2 min^{-1} $axis^{-1}$)		
Lot	SOD	Catalase	Peroxidase
Cultivar A_6			
Control	2.1 ± 0.22[a]	1.3 ± 0.06	3.1 ± 0.11
Aged	1.1 ± 0.07	0.4 ± 0.03	1.7 ± 0.08
Cultivar T_{100}			
Control	2.6 ± 0.28	1.1 ± 0.06	8.2 ± 0.71
Aged	2.4 ± 0.19	0.9 ± 0.04	8.3 ± 0.81

[a]Values are expressed as the mean of three independent experiments ± the standard deviation.

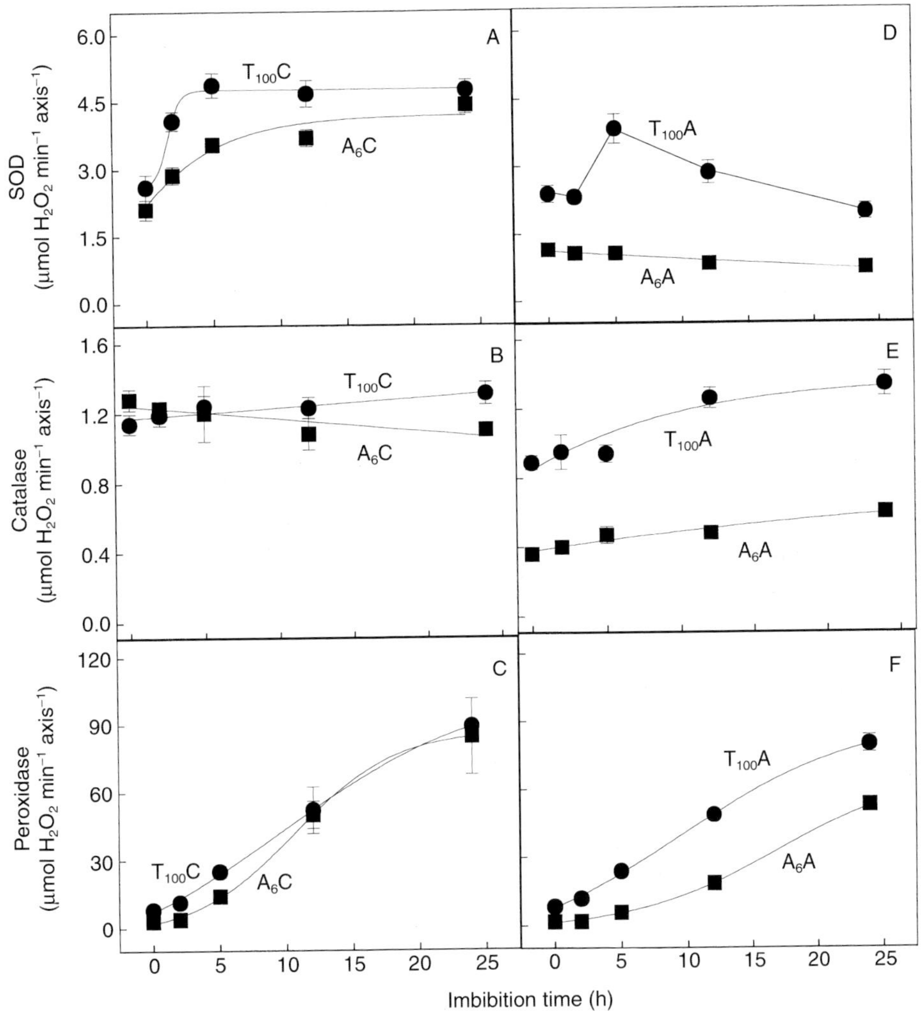

Fig. 14.1. Changes in superoxide dismutase (A, D), catalase (B, E) and peroxidase (C, F) before and during germination of control and aged seeds of two maize cultivars. Values presented are the mean of three determinations. Bars indicate the standard deviation.

We then proceeded to evaluate if in the cultivars the enzymatic antioxidant system could withstand storage. With the exception of POX, ageing decreased the activity of the antioxidant enzymes in dry seeds of both cultivars (Table 14.2). Such impairment was also observed during imbibition. In aged seeds of A_6, SOD activity remained low and no increase could be observed during imbibition (Fig. 14.1D), whereas in T_{100}, the well-storing cultivar, an increase in SOD activity was observed during early imbibition (Fig. 14.1D). This increment was transient and was observed 2–6 h after the addition of water, and then gradually decreased to reach the level present in the dry seed

Table 14.3. Hydrogen peroxide (nmol $axis^{-1}$) accumulated in control and aged maize axes during imbibition.

	Imbibition time (h)	
Lot	0	24
Cultivar A_6		
Control	407^a	90^b
Aged	380^a	250^c
Cultivar T_{100}		
Control	403^a	110^b
Aged	430^a	120^b

Values are expressed as the mean of three independent experiments. Numbers followed by the same letter are not significantly different at $\alpha = 0.05$.

(Fig. 14.1D). Catalase activity in aged seeds of A_6 remained low whereas in aged T_{100} the time-course pattern of CAT activity was similar to that exhibited by the control (Fig. 14.1E). With regard to peroxidase an increase in activity was observed for both cultivars, but in A_6 the amount of the increment was lower than in the respective control (Fig. 14.1F). In T_{100} POX activity did not seem to be modified.

Since ageing induced an impairment in the activity of the enzymes for hydroperoxide degradation, there was the possibility that this metabolite may accumulate. In order to know if this was the case, we determined the hydroperoxide detoxification capacity of aged seeds (Table 14.3). The hydroperoxide detoxification capacity of aged T_{100} was similar to that shown by control seeds, whereas in A_6 ageing decreased significantly this capacity (Table 14.3).

Discussion

Recent findings have led to the suggestion that reactive oxygen species, O_2^- and H_2O_2, play an important role in seed deterioration during ageing (Puntarulo and Boveris, 1990; Simontacchi *et al.*, 1993; Sun and Leopold, 1995; Sung, 1996). These oxygen radicals are derived from mitochondrial oxidative metabolism during germination (Puntarulo *et al.*, 1991). In fresh seeds, these metabolites are kept at low steady-state levels by the action of superoxide dismutase, catalase and peroxidases that use diverse electron donors. The activity of these enzymes also increases during germination (Puntarulo *et al.*, 1991; Cakmak *et al.*, 1993; Gidrol *et al.*, 1994). Therefore, deterioration of the seed during ageing may depend on its seed efficiency to maintain sufficient enzymic systems as protection against the oxidative stress.

In this paper changes in the main enzymatic activities of hydroperoxide metabolism have been related to seed storage performance. The reported results indicate that control maize lots with different storage performance have a similar capacity for hydroperoxide metabolism (Fig. 14.1 and Table 14.3). However, the effect of ageing on the activity of the antioxidant enzymatic system did relate to storage performance. A decrease in SOD activity was detected in both maize lots during germination (Fig. 14.1A vs. Fig. 14.1D), whereas the activity of H_2O_2-scavenging enzymes only decreased in A_6, the badly storing lot (Fig. 14.1B, C, vs. E, F). SOD protects cells against oxidative damage caused by oxygen (Scandalios, 1993) and during seed germination a high production of toxic O_2 species can be expected in view of the high O_2 consumption and respiratory activity following imbibition (Bewley and Black, 1994). Therefore, the decreased synthesis or accelerated inactivation of SOD occurring upon ageing may lead to a less protected condition in the emerging axis.

Germinating maize seeds of both lots showed a significant increase in guaiacol peroxidase activity, whereas catalase activity increased only slightly. Puntarulo *et al.* (1991), Cakmak *et al.* (1993) and Gidrol *et al.* (1994) showed that catalase activity did not change before the onset of germination, and as in this study with maize, the activity of guaiacol peroxidase in wheat and soybean was higher than that of catalase (Gaspar *et al.*, 1985; Cakmak *et al.*, 1993). In all the above systems it has been proposed that during germination, catalase is the predominant enzyme using H_2O_2 although peroxidases are also expected to contribute to the H_2O_2-scavenging system.

Interestingly, only in aged embryonic axes of A_6 was the activity of the H_2O_2-consuming enzymes lower than in the control counterparts, whereas in T_{100}, the good storer, ageing did not significantly affect this system. In A_6, this decline might lead to the accumulation of H_2O_2 which is likely to impair further germination either directly or via the formation of hydroxyl radicals (HO) by a Haber-Weiss reaction. The finding of this study that aged A_6, the poorly storing lot, accumulates higher levels of H_2O_2 is in agreement with the above proposition (Table 14.3). Moreover, after 4 months of ageing A_6 showed lower viability than T_{100} (Table 14.1), suggesting that the accumulation of H_2O_2 was detrimental to radicle growth. The results of this paper show that in aged seeds of a poor-storage lot, whether germinated or non-germinated, catalase activity was more susceptible to loss during ageing although peroxidase activity was also impaired. At the moment we do not know if the accumulation of H_2O_2 observed in the aged poorly-storing lot is due to a decline in catalase activity or to a decline in peroxidase activity or to both. Thus, apparently, ageing impairs the induction of the antioxidant defence enzymatic system that is necessary for protecting the germinating embryo from oxidative stress injury. The level of this impairment is related to seed storage performance.

In conclusion, these experiments demonstrate two important points. First, the stability of the enzymatic antioxidant system varies between maize cultivars and may be associated with storage performance and ageing stability of the seed. Second, during seed ageing, deteriorative reactions may occur in part during storage, and in part during the early stages of germination. The mechanisms involved in maintaining seed vigour and viability during dry storage and during early stages of germination, may include the efficiency of the enzymatic system for hydroperoxide detoxification.

Acknowledgements

We thank Professor A.C. Leopold and Professor E. Sánchez for critical review of the manuscript.

This work was supported by a grant to I.B.-L. from the Mexican Research and Technology Council (Project 4798-N9406).

References

Aung, U.T. and McDonald, M.B. (1995) Changes in esterase activity associated with peanut (*Arachis hypogea* L.) seed deterioration. *Seed Science and Technology* 23, 101–111.

Basavarajappa, B.S., Shetty, H.S. and Prakash, H.S. (1991) Membrane deterioration and other biochemical changes, associated with accelerated ageing of maize seeds. *Seed Science and Technology* 19, 279–286.

Bernal-Lugo, I. and Leopold, A.C. (1998) The dynamics of seed mortality. *Journal of Experimental Botany* 49, 1455–1461.

Bernal-Lugo, I., Parra, C., Carballo, A. and Hamabata, A. (1994) Enzymic systems altered by accelerated aging of seeds. *Plant Physiology* (Life Sci. Adv.) 13, 287–294.

Bewley, J.D. and Black, M. (1994) Cellular events during germination and seedling growth. In: *Seeds: Physiology of Development and Germination*, 2nd edn. Plenum Press, pp. 147–197.

Bray, C.M. and Chow, T.Y. (1976) Lesions in ribosomes of non-viable pea (*Pisum arvense*) embryonic axis tissue. *Biochimica et Biophysica Acta* 422, 14–23.

Cakmak, I., Dragan, S. and Marschner, H. (1993) Activities of hydrogen peroxide-scavenging enzymes in germinating wheat seeds. *Journal of Experimental Botany* 44, 127–133.

Chance, B. and Maehly, A.C. (1955) Assay of catalase and peroxidase. *Methods in Enzymology* 2, 764–765.

Christensen, C.M. (1972) Microflora and seed deterioration. In: Roberts, E.H. (ed.) *Viability of Seeds*. Syracuse University Press, pp. 59–93.

Cruz-Garcia, F., Gonzalez-Hernandez, V.A., Molina-Moreno, J. and Vazquez Ramos, J.M. (1995) Seed deterioration and respiration as related to DNA metabolism in germinating maize. *Seed Science and Technology* 23, 477–486.

Dapron, R. (1985) Enzyme activity as a function of water activity. In: Simatos, D. and Multon J.L. (eds) *Properties of Water in Foods*. Martinus Nijhoff, The Hague, The Netherlands, pp. 171–190.

Dawidowicz-Grzegorzewska, A. and Podstolski, A. (1992) Age-related changes in the ultrastructure and membrane properties of *Brassica napus* L. *Annals of Botany* 69, 39–46.

Dell'Aquila, A. (1994) Wheat seed ageing and embryo protein degradation. *Seed Science Research* 4, 293–298.

Dobretsov, G.E., Borschevskaya, T.A., Petrov, V.A. and Vladimorov, Y.A. (1977) The increase of phospholipid bilayer rigidity after lipid peroxidation. *FEBS Letters* 84, 125–128.

Dreyer, M. and Van de Venter, H.A. (1992) Differential effect of temperature on mitochondrial activity in shoots from freshly and moderately aged kernels of maize (*Zea mays* L.). *Plant Growth Regulation* 11, 267–271.

Ferguson, J.M., Tekrony, D.M. and Egli, D.B. (1990) Changes during early soybean seed and axes deterioration: lipids. *Crop Science* 30, 179–182.

Ganguli, S. and Sen-Mandi, S. (1993) Effects of ageing on amylase activity and scutellar cell structure during imbibition in wheat seed. *Annals of Botany* 71, 411–416.

Gaspar, T., Penel, C., Castillo, J.F. and Greppin, H. (1985) A two step control of basic and acidic peroxidases and its significance for growth and development. *Physiologia Plantarum* 64, 418–423.

Gidrol, X., Lin, W.S., Degousee, N., Yip, S.F. and Kush, A. (1994) Accumulation of reactive oxygen species and oxidation of cytokinin in germinating soybean seeds. *European Journal of Biochemistry* 224, 21–28.

Heydecker, W. (1972) Vigour. In: Roberts, E.H. (ed.) *Viability of Seeds*. Syracuse University Press, pp. 209–252.

Hildebrandt, A.G. and Roots, I. (1975) H_2O_2 and liver mixed function oxidation. *Archives of Biochemistry and Biophysics* 171, 385–397.

Leopold, A.C., Sun, W.Q. and Bernal-Lugo, I. (1994) The glassy state in seeds. *Seed Science Research* 4, 267–274.

Lowry, O.H., Rosenbrough, N.J., Farr, A.L. and Randall, R.L. (1951) Protein measurement with the Folin phenol reagent. *Journal of Biological Chemistry* 193, 265–275.

Mackay, D.B. (1972) The measurement of viability. In: Roberts, E.H. (ed.) *Viability of Seeds*. Syracuse University Press, pp. 172–208.

Misra, H.P. and Fridovich, I. (1972) The role of superoxide anion in the autoxidation of epinephrine and a simple assay for superoxide dismutase. *Journal of Biological Chemistry* 247, 3170–3175.

Priestley, D.A. (1986) *Seed Aging*. Comstock Publishing Associates, Ithaca, New York, London.

Puntarulo, S. and Boveris, A. (1990) Effect of natural and accelerated aging on the hydroperoxide metabolism of soybean embryonic axes. *Plant Science* 69, 27–32.

Puntarulo, S., Galleano, M., Sanchez, R.A. and Boveris, A. (1991) Superoxide anion and hydrogen peroxide metabolism in soybean embryonic axes during germination. *Biochimica et Biophysica Acta* 1074, 277–283.

Roberts, E.H. (1988) Seed aging: the genome and its expression. In: Nooden, L.D. and Leopold, A.C. (eds) *Senescence and Aging in Plants*. Academic Press, pp. 465–498.

Roberts, E.H. and Ellis, R.H. (1982) Physiological, ultrastructural and metabolic aspects of seed viability. In: Khan, A.A. (ed.) *The Physiology and Biochemistry of Seed Development, Dormancy and Germination*. Elsevier Biomedical Press, pp. 465–485.

Scandalios, J.G. (1993) Oxygen stress and superoxide dismutases. *Plant Physiology* 101, 7–12.

Senaratna, T., Gusse, J.F. and McKersie, B.D. (1988) Ageing-induced changes in cellular membranes of imbibed soybean seed axes. *Physiologia Plantarum* 73, 85–91.

Simontacchi, M., Caro, A., Fraga, C.G. and Puntarulo, S. (1993) Oxidative stress affects α-tocopherol content in soybean embryonic axes during germination. *Plant Phyisiology* 103, 949–953.

Smith, M.T. and Berjak, P. (1995) Deteriorative changes associated with the loss of viability of stored dessication-tolerant and dessication-sensitive seeds. In: Kigel, J. and Galili, G. (eds) *Seed Development and Germination*. Marcel Dekker, Inc., New York, pp. 701–746.

Stadtman, K.J. (1992) Protein oxidation and ageing. *Science* 257, 1220–1224.

Sun, W.Q. and Leopold, A.C. (1995) The Maillard reaction and oxidative stress during ageing of soybean seeds. *Physiologia Plantarum* 94, 94–104.

Sung, J.M. (1996) Lipid peroxidation-scavenging in soybean seeds during aging. *Physiologia Plantarum* 97, 85–89.

Wettlauffe, S.H. and Leopold, A.C. (1991) Relevance of Amadori and Maillard reactions to seed deterioration. *Plant Physiology* 97, 165–169.

Wilson, D.O., Jr and McDonald, M.B., Jr (1986) The lipid peroxidation model of seed aging. *Seed Science and Technology* 14, 269–300.

15 Towards the *ex situ* Conservation of Aquatic Angiosperms: a Review of Seed Storage Behaviour

F. Hay, R. Probert, J. Marro and M. Dawson

Royal Botanic Gardens, Kew, Wakehurst Place, Ardingly, West Sussex RH17 6TN, UK

Many aquatic plants are threatened due to a variety of reasons, for example, eutrophication, competition by algal and vascular weeds, global warming and loss and degradation of wetlands. Whilst *in situ* conservation measures are the priority, *ex situ* conservation techniques are urgently needed as a back-up and to provide propagating material for investigation and re-introduction. Seed storage is a valuable method of *ex situ* conservation. There is a lack of information on the seed biology of aquatic plants. Often, flowering and/or fruit development occurs below the water surface and the seeds are never exposed to air drying. Consequently, it has been suggested that seeds of aquatic plants may not tolerate rapid enforced desiccation and would thus be unsuited to long-term storage in seed banks. In this paper, the seed storage physiology of 87 aquatic species is described and discussed. Only six (6.9%) of these species can be described as having recalcitrant seed storage behaviour, three (3.4%) have intermediate seed storage behaviour and a further 13 (14.9%) are unclassified; the remaining 65 species (74.7%) have orthodox seed storage behaviour. These figures suggest that, contrary to the prediction that aquatic plants are likely to have recalcitrant seed storage physiology, the relative proportions of seed physiology types in aquatic plants appears to be closer to those predicted for the world's total spermatophyte flora. Conventional seed banking methodologies would therefore be a reliable method of conserving many aquatic plant species whose existence may become threatened in the future.

Introduction

Vascular aquatic plants have been defined as those plants (fern, fern-allies, and seed-bearing plants) whose photosynthetically active parts are permanently or, at least, for several months each year submerged in water or float on the surface of water (Cook *et al.*, 1974). According to this definition, between

1 and 2% of all angiosperms are aquatic but there are no aquatic gymnosperms (Cook, 1990).

Aquatic angiosperms can be broadly divided into two categories: those which live in the marine environment and those which live in freshwater. Marine aquatic angiosperms are collectively known as seagrasses due to their outward resemblance to terrestrial grasses. At present, there are 57 recorded seagrass species from 13 genera within four families of monocotyledons (Larkum *et al.*, 1989). They often form highly productive underwater meadows in open coastal waters or on intertidal flats. Freshwater aquatic angiosperms are found in almost every natural or man-made water course, including lakes, rivers, fens, bogs, swamps, and floodplains. In total, there are freshwater aquatic species present in 75 families in the angiospermae, including both mono- and di-cotyledonous families (derived from Cook, 1990). Aquatic angiosperms are primary producers and therefore support a variety of fauna including echinoderms, fishes, larger herbivores such as manatees, and waterfowl. They are also important for sediment stabilization and for improving water quality by oxygenation, absorption of minerals, and filtration.

However, aquatic plants are threatened due to the effects of, for example, eutrophication, pollution, algal and vascular weeds, and global warming. They may also be under threat due to the loss of wetland habitats as a result of in-filling and land drainage, dredging and channelling, and groundwater abstraction. For example, the US has lost an estimated 54% of its original wetland coverage (Maltby, 1986).

The designation of wetlands and coastal waters as protected areas is one way in which aquatic plant species can be conserved. However, without the back up of *ex situ* conservation measures, species may still be lost since the factors that are detrimental to these habitats cannot be easily controlled and because the realization and implementation of successful management practices is extremely complex. Seed banking is a useful tool in the *ex situ* conservation of plant species (Miller *et al.*, 1995); seeds are dried to a low moisture content (typically 3–7%, fresh weight basis), sealed inside air-tight containers and placed at low temperature (–20°C). Under such conditions it is predicted that the seeds from many species will remain viable for many decades. Unfortunately, not all species produce seeds that tolerate drying to such low moisture contents. Seeds which are desiccation tolerant and which can therefore be placed in long-term storage have been termed 'orthodox' (Roberts, 1973). By contrast, 'recalcitrant' seeds are desiccation intolerant (desiccation sensitive) and do not survive drying below a relatively high moisture content, typically 40–50% fresh weight. More recently, it has been found that not all species produce seeds which conform to either of these two categories. Seeds of, for example, coffee (*Coffea arabica*) and papaya (*Carica papaya*) have been described as having 'intermediate' storage behaviour since they tolerate a greater degree of drying than recalcitrant seeds but are less desiccation tolerant than orthodox seeds (Ellis *et al.*, 1990, 1991a,b). The longevity of intermediate seeds, in contrast with orthodox seeds, is impaired at low temperatures; this impairment of longevity may occur at higher temperatures in intermediate seeds from tropical species (i.e. longevity is impaired at temperatures ≤ 10°C) compared with intermediate seeds from temperate species.

Hanelt (1977) suggested that there is an ecological connection between seed storage characteristics and the habitat in which the species is found. In particular, recalcitrant seeds tend to be produced by species growing in moist habitats, in tropical rainforests, temperate gallery forests, and in aquatic environments (elaborated by Roberts and King, 1980). Several studies have sought to identify further what makes recalcitrant and intermediate seeds distinct from orthodox seeds, particularly in view of the fact that even closely related species can have contrasting seed storage behaviour. For example, seeds of sycamore (*Acer pseudoplatanus*) do not tolerate drying below about 45% moisture content and are therefore described as having recalcitrant seed storage physiology whereas seeds of the closely related Norway maple (*Acer platanoides*) have orthodox storage physiology (Dickie *et al.*, 1991).

In general, recalcitrant seeds tend to be larger than intermediate seeds which in turn tend to be larger than orthodox seeds; particular fruit types may be associated with seed storage behaviour (species which produce many-seeded fruits tend to have orthodox seeds whilst species which have fruits with just one or a few seeds may show recalcitrant, intermediate or orthodox storage physiology); recalcitrant seeds also tend to be shed at higher moisture contents than intermediate and orthodox seeds (Hong *et al.*, 1998).

The suggestion that many aquatic plants are recalcitrant arose in part from the observation that, of the few aquatic species that had been considered, the seeds from the aquatic species, *Glyceria striata*, *Trapa natans*, and *Zizania aquatica* did not germinate after desiccation (Muenscher, 1936a, cited in King and Roberts, 1980). Starting from this point, the conclusion that seeds from many aquatic plants are likely to be recalcitrant appears to be justified, not least because many of these species flower and/or set fruit under water and the seeds are never naturally exposed to air drying. In this paper, this hypothesis will be re-evaluated in the light of more recent studies carried out on seeds from aquatic plants. In particular, evidence produced in recent years from research in this laboratory, predominantly on aquatic plants native to the UK, will indicate that orthodox seed storage behaviour is more common in aquatic plants than expected.

Seed Storage Behaviour of Aquatic Angiosperms

Historical review

There is a lack of information on the seed biology of aquatic plants, probably a reflection of the fact that the reproductive structures of most aquatic plant species are not readily accessible. In particular, very few studies on only a small number of species have considered seed storage characteristics. However, there have been rather more studies which have looked at the population dynamics of aquatic plants; some of these studies give insights to the degree of desiccation tolerated by the seeds. The following section will give a review of the literature specifically relating to the desiccation-tolerance/desiccation-intolerance of seeds from aquatic plants.

Guppy (1897) was one of the first workers to publish results on the seed biology of aquatic plants. The focus of this study was to look at the germination requirements of the seeds, mainly shortly after harvest or following wet storage. However, it was reported that seeds of *Nuphar luteum* (= *Nuphar lutea*) and *Nymphaea alba* rapidly decayed when placed to germinate after drying (conditions of drying not reported) for periods ranging from 2 to 15 months (it is perhaps interesting to note, however, that seeds did germinate after being frozen in mud for 2 weeks). In contrast, fruits of *Myriophyllum spicatum*, *M. alterniflorum*, *Ceratophyllum demersum*, *Sparganium ramosum*, *S. simplex*, *Calla palustris*, *Damasonium stellatum*, *Alisma plantago* (= *Alisma plantago-aquatica*), *A. ranunculoides*, *Zannichellia palustris*, *Callitriche aquatica*, *Ranunculus hederaceus*, and *Limnanthemum nymphoides* (= *Nymphoides peltata*) all germinated following a period of dry storage (2–30 months). Similarly, fruits of *Potamogeton densus* (= *Groenlandia densa*), *P. natans* and *P. crispus* germinated after drying although prolonged storage was detrimental in the case of *G. densa* and *P. natans*. Clearly, although the experiments might have been somewhat rudimentary, these results suggest that all but two of these 18 species (*Nuphar lutea* and *Nymphaea alba*) show orthodox storage characteristics, at least in terms of their response to drying (although, since drying conditions were not given, we cannot rule out intermediate seed storage behaviour).

Muenscher (1936b) investigated the germination (in water in glass jars placed under diffuse light in a glasshouse for 8 months) of 21 species of *Potamogeton* following dry storage for 2–3, 5–6, or 12 months. The only germination recorded was 1% germination for seeds of *P. americanus* and 0.5% germination for seeds of *P. gramineus* after 12 months storage. In a second study, Muenscher (1936a) described the results of storage and germination experiments carried out on 43 aquatic species representing 30 genera and 20 families (including five species of *Potamogeton* which were examined in the first study). Seeds were stored under four different conditions: (i) in water at 1–3°C in the dark; (ii) in water at room temperature (18–20°C) with diffuse light; (iii) air dry in manila envelopes in the dark at 1–3°C; and (iv) air dry in manila envelopes at room temperature. Of these 43 species, only seeds from 12 species germinated following dry storage (at 1–3°C and/or room temperature) of up to 7 months (one of which was *P. pectinatus* which failed to germinate following dry storage in the first study). Muenscher concluded that seeds from aquatic plants must be kept moist to maintain their viability and ensure prompt germination following planting. However, he suggested that this was more likely to be due to the induction of dormancy and that drying resulted in actual embryo death in only four species, *Orontium aquaticum*, *Peltandra virginica*, *Vallisneria americana* (although there was 1% germination after 5 months dry storage at 1–3°C) and *Zizania aquatica*.

There has been some debate regarding the storage behaviour of seeds from *Zizania palustris*. Since Muenscher (1936a) found that seeds of *Z. aquatica* did not appear to tolerate drying, Probert and Longley (1989) proposed that it was likely that seeds from *Z. palustris* would also be desiccation intolerant (these species are either very closely related or are not distinct). Their results supported this proposal; there was a linear relationship between embryo moisture content and probit germination with only 50% viability following drying to ~30% moisture content and less than 5% viability after drying to ~10% moisture content. These results were obtained when seeds were dried by immersion in polyethylene glycol at 1°C (at −10 MPa, 92.4% RH), over saturated NH_4Cl at 16°C (82% RH), or in a dry-room (15% RH) at 15°C. However, Kovach and Bradford (1992) subsequently showed that seeds of *Z. palustris* could tolerate drying to 6–8% moisture content depending on the temperature during drying and subsequent rehydration. High levels of viability (>80%) were obtained if seeds were dried at temperatures ≥ 25°C and then rehydrated over an extended period (3 or 4 weeks) at temperatures between 10°C and 25°C. In a study of the properties of water and the survival of embryos following drying, Vertucci *et al.* (1994, 1995) concluded that desiccation tolerance of *Z. palustris* embryos increased during seed development but never achieved the orthodox condition. In a synthesis of the data available for this species, Hong *et al.* (1998) concluded that *Z. palustris* seeds should be categorized as intermediate.

Seeds from two other species of *Zizania*, *Z. texana* and *Z. latifolia* are also thought to have intermediate storage physiology (Hong *et al.*, 1998). Vertucci *et al.* (1994, 1995) described results for *Z. texana* which were similar to the results obtained for *Z. palustris*. Furthermore, studies carried out in this laboratory showed that there was a gradual reduction in the germination of seeds of *Z. latifolia* dried to equilibrium at 15°C at different relative humidities (~80%, ~60%, ~30% and ~15%) with just 20% surviving desiccation to equilibrium with 15% RH (N.J. Bowhay and R. Probert, personal communication).

Studies on seeds from other aquatic grasses have indicated recalcitrant seed storage behaviour. There was less than 20% germination of *Spartina anglica* seeds dried to 20% embryo moisture content at 6°C either over saturated NaI (42% RH) or over silica gel; similarly, seeds from the tropical aquatic grass *Porteresia coarctata* lost viability during drying at 15% RH and 15°C or at ~80% RH and 6°C (Probert and Longley, 1989). Seeds from *Spartina alterniflora* also show desiccation sensitivity, although recalcitrant storage behaviour has not been confirmed (Mooring *et al.*, 1971 cited in Probert and Longley, 1989).

Seeds from the seagrasses *Zostera marina* and *Z. capricorni* dried at a range of temperatures (6, 16 or 26°C) and relative humidities (~15 to ~94%) did not appear to tolerate drying below a moisture content in the region of 25–30% fresh weight (J. Brenchley and R. Probert, personal communication), supporting other reports indicating recalcitrant storage physiology for these two species (Hootsmans *et al.*, 1987; Conacher *et al.*, 1994).

Najas marina, the holly-leaved naiad, is an aquatic annual herb which is included on the Schedule 8 list of threatened British plant species. Work carried out at Wakehurst Place on seeds harvested from a population in the east of England indicated that whilst ~50% of the population survived drying to

~20% moisture content, drying below a critical moisture content of 13–14% (approximately 80% RH at 25°C) quickly resulted in loss of viability (C. Bone and R. Probert, unpublished report). Mild water stress or drying under anaerobic conditions significantly increased the proportion of seeds able to withstand partial drying (i.e. to 80% RH at 25°C). Seeds which had survived partial drying could also withstand storage at −20°C and −196°C. These results indicate that, in terms of their response to drying, seeds from *N. marina* should be classified as having intermediate seed storage physiology (Hong *et al.*, 1998). However, Agami and Waisel (1984) reported ~62% viability following 4 years storage under dry, dark conditions. C. Bone and R. Probert suggest (personal communication) that seed immaturity may have contributed to the desiccation sensitivity which they observed.

Drying reduced the germination percentage of seeds of *Nymphaea odorata* (Else and Riemer, 1984). Germination fell from approximately 45% to less than 20% after drying in air for 96 h (seed moisture contents/equilibrium RHs were not determined), suggesting that seeds from this species have some desiccation sensitivity. In a study of seed dispersal in three other nymphaeid species, Smits *et al.* (1989) considered the desiccation tolerance of seeds as evidence for/against epizoochory (dispersal of seeds on the outside of animals). Seeds of *Nuphar lutea*, *Nymphaea alba*, and *Nymphoides peltata* were dried on filter paper in a desiccator at room temperature (~20°C). Under these conditions, the viability of seeds of *Nuphar lutea* had declined to nearly 0% after 7 days; the viability of seeds of *Nymphaea alba* steadily declined from 100% to ~0% over 21 days; seeds of *Nymphoides peltata* maintained their viability over 28 days and were still viable after 15 months storage in the desiccator. Germination results were supported by staining tests using tetrazolium solution (viable tissues stain deep red where the insoluble formazan is deposited). No information was given relating to drying rates or final moisture content of the seeds. None the less, it seems justified to conclude that the seeds of *Nuphar lutea* are likely to be recalcitrant and those of *Nymphoides peltata* are orthodox (Smits *et al.* (1989) concluded that only *Nymphoides peltata* was likely to be dispersed by epizoochory.) The longer period of survival of the *Nymphaea alba* seeds compared with those of *Nuphar lutea* may have been due to a slower rate of drying (although this seems unlikely since *Nymphaea alba* seeds are smaller than those of *Nuphar lutea*) or may indicate some desiccation tolerance.

Although the desiccation tolerance of seeds of *Amphibolis antartica*, *A. griffithii*, *Thalassodendron ciliatum*, and *T. pachyrhizum* has not been studied *per se*, these species have viviparous seedlings (Kuo and Kirkman, 1990) and it is therefore unlikely that desiccation-tolerant seeds could be collected from these species.

There are a number of other studies which have indicated orthodox seed storage behaviour for a range of aquatic species: *Nymphaea gigantea* (Ewart, 1908); *Nelumbo lutea* and *N. nucifera* (Dent, 1942); *Eichhornia crassipes* (Barton and Hotchkiss, 1951); *Myriophyllum spicatum* (Patten, 1955); *Ruppia maritima* (van Vierssen *et al.*, 1984); *Lobelia dortmanna* (Farmer and Spence, 1987); *Griffithella hookeriana* (Vidyashankari and Mohan Ram, 1987);

Hottonia palustris (Brock *et al.*, 1989); *Potamogeton pectinatus* (van Wijk, 1989); *Littorella uniflora* (Arts and van der Heijden, 1990); *Zannichellia obtusifolia* and *Z. pedunculata* (Grillas *et al.*, 1991); *Hydrilla verticillata* (Lal and Gopal, 1993); *Marathrum haenkeanum, M. rubrum, Oserya coulteriana, Tristicha trifaria,* and *Vanroyenella plumosa* (Philbrick and Novelo, 1994); many duckweed species (Lemnaceae), e.g. *Lemna aequinoctialis, L. disperma, L. gibba, Wolffiella repanda* and *W. rotunda* (Landolt, 1997).

Current studies on UK species

As part of the Millennium Seed Bank Project at the Royal Botanic Gardens, Kew, Wakehurst Place, a study commenced in 1997 which aimed to investigate the desiccation tolerance of seeds from 60–70 freshwater aquatic plant species native to the UK. Viability was assessed before and after drying to equilibrium at 15% RH and 15°C by carrying out a staining test with tetrazolium solution and/or a germination test after a rehydration period over distilled water. For the germination tests, a 2 × 2 × 2 × 2 × 2 factoral design was used to assess the effects of mean temperature, temperature regime, light, oxygen, and cold stratification; where germination levels were still low, a small incision was made to the fruit and/or seed coat. Of the 38 species examined to date, non-orthodox behaviour is only indicated for three species: *Najas flexilis, Nuphar lutea* and *Nymphaea alba.*

There was total loss of viability in seeds of *Najas flexilis* dried for just 2 h at 15% RH, 15°C (moisture content 22.5%, equilibrium RH not determined), indicating recalcitrant storage behaviour. Seeds from *Nuphar lutea* and *Nymphaea alba* also showed no signs of tolerating desiccation. The embryos from fresh seeds of *Nymphaea alba* stained well in the tetrazolium test; however, there was no staining after drying below ~25% moisture content (seeds dried at 15% RH, 15°C or at ~79% RH, 15°C (over a saturated solution of NH_4Cl) followed by a 2 day rehydration period at 21°C). In further tests, dried seeds of both *Nuphar lutea* and *Nymphaea alba* were rehydrated at a range of temperatures between 6° and 26°C for up to 20 days before exposing the embryos to tetrazolium solution; no staining was visible. Thus, together with the results of Smits *et al.* (1989), we may conclude that these two species probably possess recalcitrant seed storage physiology. For the other 35 species, some viability was retained after drying to moisture contents in equilibrium with 15% RH and 15°C (moisture contents ranged from 3.2 to 11.3% fresh weight), although in some cases, only low percentages of germination were recorded after drying (Table 15.1). For example, the viability of seeds of *Hydrocharis morsus-ranae* was reduced from 100% to ~8% following drying at 15% RH and 15°C from a harvest moisture content of 27.8% down to 5.7% moisture content. However, those seeds of *H. morsus-ranae* which were desiccation tolerant also survived subsequent storage at −20°C. The viability of six other species was also maintained during 1 month storage at −20°C. These species appear to show orthodox seed storage physiology.

Table 15.1. Evidence for orthodox and non-orthodox seed storage behaviour in 38 UK freshwater aquatic species (survey carried out on collections made in 1997 and 1998) (Hay, Marro, Dawson and Probert, previously unpublished data).

Family	Species	Common name	Initial mc (mean)	Initial viability (%)* (tz)	Initial viability (%)* (germ)	mc after drying (mean)	Viability after drying (%)* (tz)	Viability after drying (%)* (germ)
Orthodox								
Alismataceae	*Sagittaria sagittifolia*	Arrowhead	83.0	100	76	9.1	100	82[b]
Apiaceae	*Apium inundatum*	Lesser marshwort	–	–	–		100	93
Brassicaceae	*Subularia aquatica*	Awlwort	–	–	–	–	–	35[b]
Callitricaceae	*Callitriche brutia*	Pedunculate water-starwort	57.0	–	–	5.5	10	93
	C. hermaphroditica	Autumnal water-starwort	80.4	–	–	–	0	65
	C. stagnalis	Common water-starwort	67.6	–	–	5.4	50	87
Campanulaceae	*Lobelia dortmanna*	Water lobelia	–	–	–	–	–	83
Ceratophyllaceae	*Ceratophyllum demersum*	Rigid hornwort	72.0	100	–	5.1	66.7	–
Cyperaceae	*Eleogiton fluitans*	Floating club-rush	–	–	–	4.9	100	–
Haloragaceae	*Myriophyllum spicatum*	Spiked water-milfoil	59.1	80	60	4.4	100	100[b]
Hippuridaceae	*Hippuris vulgaris*	Mare's tail	–	–	–	3.2	40	34
Hydrocharitaceae	*Hydrocharis morsus-ranae*	Frogbit	27.8	?[a]	100	5.7	?[a]	8[b]
Lentibulariaceae	*Utricularia vulgaris*	Greater bladderwort	–	–	–	–	–	33
Menyanthaceae	*Nymphoides peltata*	Fringed water-lily	52.8	100	–	3.8	70[b]	–
Plantaginaceae	*Littorella uniflora*	Shoreweed	42.2	100	–	4.8	70	100[b]
Potamogetonaceae	*Groenlandia densa*	Opposite-leaved pondweed	83.0	100	–	6.2	30	61
	Potamogeton acutifolius	Sharp-leaved pondweed	–	–	–	–	60?	40
	P. berchtoldii	Small pondweed	–	–	100	87	100	90
	P. filiformis	Slender-leaved pondweed	–	–	–	–	100	75

	P. gramineus	Various-leaved pondweed	76.6	40	–	–	80	57
	P. lucens	Shining pondweed	–	100	–	10.6	90	–
	P. natans	Broad-leaved pondweed	50.8	100	100	5.6	100	100[b]
	P. obtusifolius	Blunt-leaved pondweed	–	100	–	11.3	62.5	–
	P. pectinatus	Fennel pondweed	78.2	100	100	6.6	70[b]	–
	P. perfoliatus	Perfoliate pondweed	–	100	–	10.6	87.5	–
	P. polygonifolius	Bog pondweed	–	–	–	6.2	0	13
Ranunculaceae	*P. pusillus*	Lesser pondweed	–	–	–	–	80	21
	P. trichoides	Hairlike pondweed	–	–	–	–	100	33
Ruppiaceae	*Ranunculus baudotii*	Brackish water-crowfoot	–	–	–	3.7	100	97
Sparganiaceae	*R. circinatus*	Fan-leaved water-crowfoot	–	–	–	–	–	75
	Ruppia maritima	Beaked tasselweed	56.2	90	–	8.7	80	–
	Sparganium angustifolium	Floating bur-reed	–	–	–	5.3	10	–
Zannichelliaceae	*S. emersum*	Unbranched bur-reed	–	–	–	5.5	100	–
Non-orthodox?	*S. erectum*	Branched bur-reed				5.3	100	–
Hydrocharitaceae	*Zannichellia palustris*	Horned pondweed	67.9	100	–	8.7	90	85
Nymphaceae								
	Najas flexilis	Slender naiad	38.9	100	–	7.3	0	0
	Nymphaea alba	White water-lily	43.5	70	–	7.9	0	–
	Nuphar lutea	Yellow water-lily	47.8	100	–	9.2	0	–

*Sample sizes: 5–20 embryos in the case of tetrazolium tests, 15–100 fruits/seeds in the case of germination tests (at every factor combination).

[a]Tetrazolium test inconclusive.

[b]Viability retained during 1 month storage at −20°C.

Discussion

Seed storage categories

According to the *Compendium of Information on Seed Storage Behaviour* which describes the seed storage behaviour of 6919 species, 88.6% of those species are listed as having orthodox seed storage behaviour, 1.9% are intermediate, 7.4% are recalcitrant, and the remaining 2.1% are unclassified (Hong *et al.*, 1998). This is probably the best estimate that is currently available on the relative proportions of seed storage types in the spermatophyta. However, the species which have been studied to date are strongly biased towards crop and other cultivated, predominantly temperate species and therefore the value of 9.3–11.4% is probably a minimum estimate for the overall percentage of the world spermatophyte flora with non-orthodox seed storage behaviour. Within particular groups of plants, for example those found in wet habitats such as tropical rainforests (Tompsett and Kemp, 1996), temperate gallery forests, and aquatic environments where seeds may not be naturally exposed to desiccation, the proportion is expected to be considerably higher (Hanelt, 1977 elaborated by Roberts and King, 1980). However, this is not corroborated by the data summarized in this paper which shows that the relative proportions of seed physiology types in aquatic plants are likely to be closer to the estimates for the spermatophyte flora in general. Of the 87 species for which there are data, only six (6.9%) can be defined as being recalcitrant, three (3.4%) have intermediate storage physiology and a further 13 (14.9%) may prove to be recalcitrant, intermediate, or orthodox; the remaining 65 species (74.7%) have orthodox seed storage physiology (Table 15.2). Indeed, in compiling this review, it became clear that the question to be asked is not 'which aquatic plants have orthodox seed storage physiology?' but 'which aquatic plants don't have orthodox seed storage physiology?'

It is also clear that there is relatively little information on the storage behaviour of seeds from aquatic plants, particularly tropical aquatic plant species. Only a few detailed studies have been carried out and in some studies, particularly earlier work, some of the results are unreliable. Furthermore, it is often not easy to categorize seeds to a particular storage type. For example, in the case of *Najas marina*, should the seeds be defined according to the results of C. Bone and R. Probert (unpublished data) which suggest intermediate seed storage physiology or according to those of Agami and Waisel (1984) where 64% survival following 4 years dry-storage clearly indicates orthodox storage behaviour? Similarly, in the case of *Hydrocharis morsus-ranae*, does 8% survival of drying and sub-zero temperatures indicate intermediate storage physiology or possible orthodox storage physiology? Most of the species listed as 'unclassified (non-orthodox?)' (Table 15.2) require further study since they rely on just a few basic observations (e.g. 'dry seeds failed to germinate').

Table 15.2. Provisional list of aquatic angiosperms showing non-orthodox seed storage behaviour.

Family	Species	Reference
Recalcitrant		
Gramineae	*Porteresia coarctata*	Probert and Longley, 1989
	Spartina anglica	Probert and Longley, 1989
Nymphaceae	*Nuphar lutea*	Guppy, 1897; Smits *et al.*, 1989; unpublished data[a]
	Nymphaea alba	Guppy, 1897; Smits *et al.*, 1989; unpublished data[a]
Zosteraceae	*Zostera capricorni*	Conacher *et al.*, 1994; unpublished data[b]
	Z. marina	Hootsman *et al.*, 1987; unpublished data[b]
Intermediate		
Gramineae	*Zizania latifolia*	Hong *et al.*, 1998
	Z. palustris	Hong *et al.*, 1998
	Z. texana	Hong *et al.*, 1998
Unclassified (non-orthodox)?		
Araceae	*Orontium aquaticum*	Muenscher, 1936a
	Peltandra virginica	Muenscher, 1936a
Cymodoceaceae	*Amphibolis antarctica*	Kuo and Kirkam, 1990
	A. griffithii	Kuo and Kirkam, 1990
	Thallassodendron ciliatum	Kuo and Kirkam, 1990
	T. pachyrhizum	Kuo and Kirkam, 1990
Gramineae	*Spartina alterniflora*	Mooring *et al.*, 1971
	Zizania aquatica	Muenscher, 1936a
Hydrocharitaceae	*Najas flexilis*	Unpublished data[a]
	N. marina	Unpublished data[c]
	Vallisneria americana	Muenscher, 1936a; Catling *et al.*, 1994
Nymphaceae	*Nymphaea odorata*	Else and Riemer, 1984
Trapaceae	*Trapa natans*	Muenscher, 1936a

[a]F. Hay, J. Marro, M. Dawson and R. Probert, 1997–1998.
[b]J. Brenchley and R. Probert, 1996–1997.
[c]C. Bone and R. Probert, 1993–1995.

Problems of classification

The confirmation of recalcitrant storage physiology relies upon a negative result, namely the loss of viability following drying. However, there are several reasons why a seed lot of a particular species may appear to be non-viable following desiccation when in fact the species does have orthodox seed storage physiology (see Roberts *et al.*, 1984). In the case of seeds from aquatic plants, the most likely reasons why seeds may be misclassified are immaturity, improper drying methods (leading to ageing-induced viability loss), and short-duration viability tests. These factors could also explain why some species, which we have described as orthodox, show low levels of viability following drying.

The fruits of many aquatic plants ripen below, at, or just above the water surface. Consequently, as soon as the fruits are ripe, they are likely to be dislodged by water currents or by activity at the water surface (wind and waves, animals, floating or waving vegetation). There is often a very short period in which it is possible to make a collection of ripe or near-ripe seeds before they are dispersed (perhaps considerably narrower than for terrestrial species). As a result, some of the seeds within the collection are likely to be relatively immature, especially since most of these aquatic plants are wild species with indeterminate flowering habits. Desiccation tolerance is acquired during seed development and consequently, even in terrestrial orthodox species, a collection may contain immature seeds which are not desiccation tolerant i.e. they appear recalcitrant. Indeed, a collection containing seeds at a range of maturities may include individuals displaying the full spectrum of storage categories.

Problems of seed immaturity may be expected when plants are at the extremes of their range of distribution. C. Bone and R. Probert (personal communication) suggested that this might have been why their seeds of *Najas marina* were more desiccation sensitive than those seeds used by Agami and Waisel (1984); Bone and Probert were working on seeds collected in the east of England whereas Agami and Waisel worked on seeds from Israel. Similarly, Vertucci *et al.* (1994, 1995) showed that maturity affected the desiccation tolerance of seeds of *Zizania palustris*, although orthodox seed storage behaviour was not shown in even the most mature seeds. Immaturity could also be a factor for some of the other UK species which have been studied in our laboratory. For example, British populations of *Hydrocharis morsus-ranae* only appear to set fruit in warm summers (Preston and March, 1996) and even in warm summers, a high proportion of fruits may fail to develop and ripen sufficiently for the seeds to acquire tolerance of desiccation to seed bank drying standards.

The dormancy-breaking requirements of seeds from aquatic plants are often complex and/or unknown. Seeds may therefore fail to germinate following drying because these requirements have not been met or treatments are not continued for a sufficient length of time. For a review of the germination requirements for a range of aquatic plant species, see Baskin and Baskin (1998). For the UK species which have been studied in our laboratory over the last 2 years, a range of germination conditions has been tested; many species respond favourably to a period of cold stratification; some species achieve higher levels of germination at constant temperatures whilst others germinate to higher levels in alternating temperature regimes; some species germinate only under anaerobic conditions, some under aerobic conditions, and yet others will germinate whether or not oxygen is present; for some of the species, for example, many of the *Potamogeton* species, high levels of germination are only achieved when mechanical dormancy is removed by chipping into the fruit to expose the seed (F. Hay, J. Marro, M. Dawson and R. Probert, unpublished data).

It is therefore with a degree of caution that the terms recalcitrant or intermediate are assigned for just a handful of species: *Porteresia coarctata*, *Spartina anglica*, *Nuphar lutea*, *Nymphaea alba*, *Zostera capricorni*, *Z. marina*, *Zizania latifolia*, *Z. palustris*, and *Z. texana*. The other species listed as non-orthodox (Table 15.2) should be subjected to further tests before confirmation of seed storage category, ideally on a number of seed-lots from different populations covering the whole of the species' distribution range.

Characteristics associated with seed storage behaviour

Determining the seed storage category of aquatic species may help our understanding of seed storage behaviour. However, trends which have been suggested from studies on seeds of terrestrial species (e.g. von Teichman and van Wyk, 1994; Hong *et al.*, 1998) may not be upheld. For example, contrary to the suggestion that, in general, recalcitrant seeds are larger than orthodox seeds (Roberts and King, 1980), many of the aquatic species included in our study have relatively small seeds and yet there is evidence of recalcitrant, intermediate, and orthodox seed storage behaviour. Furthermore, of those UK species, the one with the largest seed is *Ceratophyllum demersum* (length ~8 mm, diameter ~6 mm), which has orthodox seed storage behaviour. It has also been noted that orthodox species tend to shed their seeds after an on-plant drying phase whereas recalcitrant seeds are, necessarily, shed at high moisture content. In the case of aquatic plants, seeds of all types are likely to be shed at high moisture contents (excepting those emergent species whose fruits remain well above the water surface), typically greater than 50% fresh weight (Table 15.1). Attempts have also been made to associate particular fruit types and morphologies with seed storage characteristics; many recalcitrant terrestrial species have single-seeded fruits. However, amongst the aquatics, many single-seeded fruits (e.g. *Potamogeton* spp.) contain orthodox seeds whilst there are a few many-seeded fruits (e.g. *Nymphaea* and *Nuphar*) with recalcitrant seeds.

Ecologically, desiccation tolerance would be a selective advantage for those species whose seeds might be threatened by desiccation as a result of reductions in water levels and/or initial dispersal to a dry environment. However, amongst aquatic species, seed storage type does not appear to be associated with niche preference or any other obvious trait. Furthermore, as has been seen with some terrestrial species (e.g. *Acer platanoides* and *A. pseudoplatanus* (Dickie *et al.*, 1991)), closely related species can have contrasting seed storage physiology; *Nymphaea alba* and *Nuphar lutea* both appear to have recalcitrant seed storage physiology whereas, according to Harrington (1972), seeds of *Nymphaea gigantea* have been dry-stored successfully (Ewart, 1908). Thus, whilst we might predict that for example, all *Potamogeton* species are likely to have orthodox seed storage physiology (based on our work on 12 species (Table 15.1)), we strongly recommend carrying out a desiccation experiment before making a large collection for seed banking.

Recalcitrance is considered to be a relatively primitive characteristic (von Teichman and van Wyk, 1994). In more advanced plant groups, the production of desiccation-tolerant seeds is thought to have evolved relatively early; those species with recalcitrant seeds may represent isolated lineages within those groups. Alternatively, von Teichman and van Wyk (1994) suggest that recalcitrance may have evolved secondarily (i.e. as a reversal away from orthodox seed production) in some taxa, such as certain aquatics. Aquatic angiosperms are found in groups of plants throughout the phylogenetic tree and it is believed that they have evolved from terrestrial plants which have 'returned' to the water. This return to the aquatic environment could have occurred before or after seed desiccation tolerance had evolved. However, it seems more likely that, in some groups at least, it occurred, as von Teichman and van Wyk (1994) suggest, after seed desiccation tolerance had arisen. For example the *Ceratophyllaceae* are thought to be a relatively ancient group of plants and yet there is some evidence of orthodox seed storage behaviour in this family. It seems unlikely that desiccation tolerance would have been acquired after returning to the aquatic environment (this group of angiosperms is somewhat unusual in that it is placed as a phylogenetically-distinct group of angiosperms (Chase *et al.*, 1993)). In contrast, the *Nymphaceae* are also thought to be relatively primitive and there is evidence of both recalcitrant and orthodox physiology in this group (Table 15.1; Ewart, 1908); the recalcitrant species could represent relicts or reversions.

Recommendations

In conclusion, the evidence reviewed in this paper shows that, contrary to the prediction that many aquatic plants are likely to have recalcitrant seed storage physiology, the relative proportions of seed physiology types are likely to be closer to the proportions within the spermatophyte flora in general. Conventional seed banking methodologies would therefore be a reliable method of conserving many aquatic plant species whose existence may become threatened in the future. For non-orthodox species, some success may be achieved through 'alternative' storage techniques. In the short term, seeds from aquatic plants can be stored in water at low temperatures and, given that many of these species do not germinate readily, they may remain viable for a few years (although some species may germinate at temperatures ≤ 10°C). For example seeds of *Zizania palustris* have been successfully stored at between 9 and 11.5% moisture content at −2°C and 3°C for a year (Oelke and Stanwood, 1988; Oelke, 1990). For seeds which tolerate some drying, seeds may remain viable for a longer period of time; an 'intermediate' seed lot of *Najas marina* dried to equilibrium at 80% RH (~15% moisture content) is being stored in the Royal Botanic Gardens Kew Seed Bank where it is predicted that the time for viability to fall by one normal equivalent deviate (or probit value) (for example from 84% down to 50%) is ~7 years (C. Bone and R. Probert, personal communication). Alternatively, it may be possible to develop successful cryopreservation methodologies for seeds from many of the species which cannot be conserved using conventional seed banking techniques.

Acknowledgements

This work was carried out as part of the Millennium Seed Bank Project. We would like to acknowledge the support given to the Millennium Seed Bank by our three major funders in the UK – The Millennium Commission, the Wellcome Trust, and Orange plc. The UK Flora Programme received further financial support from English Nature. Financial support was also given by English Nature for the work carried out on *Najas marina*, and by the Leverhulme Trust for the work on *Zostera marina* and *Z. capricorni*. We gratefully acknowledge the work carried out by Chris Bone, Leon Terry, Nicholas Bowhay, and Jennie Brenchley. We also thank Dr C. Preston and Mrs J. Croft at the Institute of Terrestrial Ecology – Monks Wood, Ruth Wingfield, English Nature (Norfolk), Mr R. Lansdown, Mrs M. Briggs, Mr A. Knapp, Mr J. Wood, and Miss E. Lerges for help and advice with the seed collecting, and colleagues at Wakehurst for comments on the manuscript.

References

Agami, M. and Waisel, Y. (1984) Germination of *Najas marina* L. *Aquatic Botany* 19, 37–44.

Arts, G.H.P. and van der Heijden, R.A.J.M. (1990) Germination ecology of *Littorella uniflora* (L.) Aschers. *Aquatic Botany* 37, 139–151.

Barton, L.V. and Hotchkiss, J.E. (1951) Germination of seeds of *Eichhornia crassipes* Solms. *Contributions of the Boyce Thompson Institute* 16, 215–220.

Baskin, C.C. and Baskin, J.M. (1998) *Seeds. Ecology, Biogeography, and Evolution of Dormancy and Germination*, Academic Press.

Brock, T.C.M., Mielo, H. and Oostermeijer, G. (1989) On the lifecycle and germination of *Hottonia palustris* L. in a wetland forest. *Aquatic Botany* 35, 153–166.

Catling, P.M., Spicer, K.W., Biernacki, M. and Lovett Doust, J. (1994) The biology of Canadian weeds. 103. *Vallisneria americana* Michx. *Canadian Journal of Plant Science*, 74, 883–897.

Chase, M.W. *et al.* (41 other authors) (1993) Phylogenetics of seed plants: an analysis of nucleotide sequences from the plastid gene *rbc*L. *Annals of the Missouri Botanical Garden* 80, 528–580.

Conacher, C.A., Poiner, I.R., Butler, J., Pun, S. and Tree, D.J. (1994) Germination, storage and viability testing of seeds of *Zostera capricorni* Aschers. from a tropical bay in Australia. *Aquatic Botany* 49, 47–58.

Cook, C.D.K. (1990) *Aquatic Plant Book*. SPB Academic Publishing, The Hague, The Netherlands.

Cook, C.D.K., Gut, B.J., Rix, E.M., Schneller, J. and Seitz, M. (1974) *Water Plants of the World: a Manual for the Identification of the Genera of Freshwater Macrophytes*. Junk, The Hague, The Netherlands, 561 + viii pp.

Dent, T.V. (1942) Some records of extreme longevity of seeds of Indian forest plants. *Indian Forester* 68, 617–631.

Dickie, J.D., May, K., Morris, S.V.A. and Titley, S.E. (1991) The effects of desiccation on seed survival in *Acer platanoides* L. and *A. pseudoplatanus* L. *Seed Science Research* 1, 149–162.

Ellis, R.H., Hong, T.D. and Roberts, E.H. (1990) An intermediate category of seed storage behaviour? I. Coffee. *Journal of Experimental Botany* 41, 1167–1174.

Ellis, R.H., Hong, T.D. and Roberts, E.H. (1991a) An intermediate category of seed storage behaviour? II. Effects of provenance, immaturity and imbibition on desiccation-tolerance in coffee. *Journal of Experimental Botany* 42, 653–657.

Ellis, R.H., Hong, T.D. and Roberts, E.H. (1991b) Effect of storage temperature and moisture on the germination of papaya seeds. *Seed Science Research* 1, 69–72.

Else, M.J. and Riemer, D.N. (1984) Factors affecting germination of seeds of fragrant waterlily (*Nymphaea odorata*). *Journal of Aquatic Plant Management* 22, 22–25.

Ewart, A.J. (1908) On the longevity of seeds. *Proceedings of the Royal Society of Victoria* 21, 1–120.

Farmer, A.M. and Spence, D.H.N. (1987) Flowering, germination and zonation of the submerged aquatic plant *Lobelia dortmanna* L. *Journal of Ecology* 75, 1965–1076.

Grillas, P., van Wijck, C. and Bonis, A. (1991) Life history traits: a possible cause for the higher frequency of occurrence of *Zannichellia pedunculata* than of *Zannichellia obtusifolia* in temporary marshes. *Aquatic Botany* 42, 1–13.

Guppy, H.B. (1897) On the postponement of the germination of the seeds of aquatic plants. *Proceedings of the Royal Physiological Society xxvi*, Edinburgh.

Hanelt, P. (1977) Okologische und systematische Aspekte der Lebensdauer von Samen. *Biologische Rundschau* 15, 81–91.

Harrington, J.F. (1972) Seed storage longevity. In: Kozlowski, T.T. (ed.) *Seed Biology, Volume III.* Academic Press, New York, pp. 145–245.

Hong, T.D., Linington, S. and Ellis, R.H. (1998) *Compendium of Information on Seed Storage Behaviour, Volumes 1 and II.* Royal Botanic Gardens, Kew, UK.

Hootsmans, M.J.M., Vermaat, J.E. and van Vierssen, W. (1987) Seed-bank development, germination and early seedling survival of two seagrass species from the Netherlands: *Zostera marina* L. and *Zostera noltii* Hornem. *Aquatic Botany* 28, 275–285.

King, M.W. and Roberts, E.H. (1980) Maintenance of recalcitrant seeds in storage. In: Chin, H.F. and Roberts, E.H. (eds) *Recalcitrant Crop Seeds.* Tropical Press, Malaysia, pp. 53–89.

Kovach, D.A. and Bradford, K.J. (1992) Imbibitional damage and desiccation tolerance of wild rice (*Zizania palustris*) seeds. *Journal of Experimental Botany* 43, 747–757.

Kuo, J. and Kirkman, H. (1990) Anatomy of viviparous seagrass seedlings of *Amphibolis* and *Thalassodendron* and their nutrient supply. *Botanica Marina* 33, 117–126.

Lal, C. and Gopal, B. (1993) Production and germination of seeds in *Hydrilla verticllata. Aquatic Botany* 45, 257–261.

Landolt, E. (1997) How do Lemnacea (duckweed family) survive dry conditions? *Bulletin of the Geobotanical Institute ETH* 63, 25–31.

Larkum, A.W.D., McComb, A.J. and Shepard, S.A. (1989) *Biology of Seagrasses: A Treatise on the Biology of Seagrasses with Special Reference to the Australian Region.* Elsevier, Amsterdam, 841 pp.

Maltby, E. (1986) *Waterlogged Wealth: Why Waste the World's Wet Places?* Earthscan, London.

Miller, K., Allegretti, M.H., Johnson, N. and Jonsson, B. (1995) Measures for conservation of biodiversity and sustainable use of its components. In: Heywood, V.H. (ed.) *Global Biodiversity Assessment.* Cambridge University Press, Cambridge.

Mooring, M.T., Cooper, A.W. and Seneca, E.D. (1971) Seed germination response and evidence for height ecophenes in *Spartina alterniflora* from North Carolina. *American Journal of Botany* 58, 48–55.

Muenscher, W.C. (1936a) Storage and germination of seeds of aquatic plants. Bulletin 652, Cornell University Agricultural Experiment Station, Ithaca, New York.

Muenscher, W.C. (1936b) The germination of seeds of *Potamogeton*. *Annals of Botany* 50, 805–821.

Oelke, E.A. and Stanwood, P.C. (1988) Wild rice seed moisture content and viability. *Agronomy Abstracts*. American Society of Agronomy, Madison, Wisconsin, 146 pp.

Oelke, E.A. McClellan, M. and Leif, J. (1990) Wild rice production research. In: *Minnesota Wild Rice Research 1989, Miscellaneous Publication 64*. Minnesota Agricultural Experiment Station, University of Minnesota, St Paul, pp. 1–15.

Patten, B.C. (1955) Germination of the seed of *Myriophyllum spicatum* L. *Bulletin of the Torrey Botanical Club* 82, 50–56.

Philbrick, C.T. and Novelo, A.R. (1994) Seed germination of Mexican Podostemaceae. *Aquatic Botany* 48, 145–151.

Preston, C.D. and March, M.D. (1996) *Hydrocharis morsus-ranae* L. (Hydrocharitaceae) fruited in Britain in 1995. *Watsonia* 21, 206–208.

Probert, R.J. and Longley, P.L. (1989) Recalcitrant seed storage physiology in three aquatic grasses (*Zizania palustris*, *Spartina anglica* and *Porteresia coarctata*). *Annals of Botany* 63, 53–63.

Roberts, E.H. (1973) Predicting the storage life of seeds. *Seed Science and Technology* 1, 499–514.

Roberts, E.H. and King, M.W. (1980) The characteristics of recalcitrant seeds. In: Chin, H.F. and Roberts, E.H. (eds) *Recalcitrant Crop Seeds*. Tropical Press, Malaysia, pp. 1–5.

Roberts, E.H., King, M.W. and Ellis, R.H. (1984) Recalcitrant seeds: their recognition and storage. In: Holden, J.H.W. and Williams, J.T. (eds) *Crop Genetic Resources: Conservation and Evaluation*. George Allen and Unwin, London, pp. 38–52.

Smits, A.J.M., van Ruremonde, R. and van der Velde, G. (1989) Seed dispersal of three nymphaeid macrophytes. *Aquatic Botany* 35, 167–180.

Tompsett, P.B. and Kemp, R.J. (1996) *Database of Tropical Tree Seed Research (DABATTS)*. Royal Botanic Gardens, Kew, Surrey, UK.

van Vierssen, W.V., van Kessel, C.M. and van der Zee, J.R. (1984) On the germination of *Ruppia* taxa in western Europe. *Aquatic Botany* 19, 381–393.

van Wijk, R.J. (1989) Ecological studies on *Potamogeton pectinatus* L. III. Reproductive strategies and germination ecology. *Aquatic Botany* 33, 271–299.

Vertucci, C.W., Crane, J., Porter, R.A. and Oelke, E.A. (1994) Physical properties of water in *Zizania* embryos in relation to maturity status, water content and temperature. *Seed Science Research* 4, 211–224.

Vertucci, C.W., Crane, J., Porter, R.A. and Oelke, E.A. (1995) Survival of *Zizania* embryos in relation to water content, temperature and maturity status. *Seed Science Research* 5, 31–40.

Vidyashankari, B. and Mohan Ram, H.Y. (1987) *In vitro* germination and origin of thallus in *Griffithella hookeriana* (Podostemaceae). *Aquatic Botany* 28, 161–169.

von Teichman, I. and van Wyk, A.E. (1994) Structural aspects and trends in the evolution of recalcitrant seeds in dicotyledons. *Seed Science Research* 4, 225–239.

16 Treatment of Immature Embryos of Maize with Water Reduces their Storability and the Desiccation Tolerance of the Scutellum

A. Bochicchio[1], B. Mariotti[1], P. Berjak[2] and N.W. Pammenter[2]

[1]*Department of Agronomy and Land Management, University of Florence, Italy;* [2]*School of Life and Environmental Sciences, University of Natal, Durban 4041, South Africa*

Embryos of seeds exposed to a hydration treatment may show reduced longevity. The aim of the study was to determine whether this effect also occurs in immature, but desiccation-tolerant embryos of maize, and whether it is related to reduced oligosaccharide content. Embryos isolated 25 days after pollination (DAP) were selected because: (i) they are desiccation tolerant; (ii) they have no raffinose, but raffinose is produced on slow drying; (iii) treatment with water prior to drying decreases the raffinose produced during drying; and (iv) the scutellum of the dried embryo can be easily and completely separated from the endosperm. Excised embryos were dried slowly either directly or after exposure to water and were then subjected to artificial ageing. Results showed: (i) embryos treated with water had lower vigour initially; (ii) during artificial ageing, embryos with less raffinose (treated with water) accumulated damage earlier than those with higher raffinose contents (dried directly); (iii) in both treatments, raffinose content in the scutellum was lower than in the axis; and (iv) electron microscopy showed that scutellum was the primary site of damage on drying of embryos exposed to water. This decrease in desiccation tolerance of the scutellum might be responsible for the earlier accumulation of damage in these embryos. It cannot be concluded whether the earlier accumulation of damage is a result of low raffinose or low initial vigour, and whether these two effects are related.

Introduction

Techniques like hydration and osmopriming are widely used to increase the speed and synchrony of germination and improve the vigour of aged seeds.

However, it has also been reported that hydration and/or osmopriming reduce the storability of seeds (e.g. Liu *et al.*, 1996). Powell and Yule (1998) showed that this effect depended on the initial vigour: storability of low vigour seeds was increased by a prior hydration, whereas that of high vigour seeds was decreased.

In dry seeds, the glassy state has been suggested to serve as a physical stabilizer and protector against deteriorative reactions (reviewed by Bernal-Lugo and Leopold, 1998). Vitrification is favoured by the oligosaccharides of the raffinose family which inhibit the crystallization of sucrose. Consistent with this observation, the storability of maize seeds of different cultivars was found to be correlated with the extent of vitrification, and the greatest vitification was associated with the highest raffinose content (Bernal-Lugo and Leopold, 1995). Moreover, the mass ratio of oligosaccharides:sucrose was found to be positively correlated with longevity of several orthodox seeds (Lin and Huang, 1994). Oligosaccharides of the raffinose family have been found to decrease in mature seeds dried after osmopriming or hydration treatments (Hoekstra *et al.*, 1994; Lin *et al.*, 1998), and immature embryos of maize, dried after incubation on water, produced less raffinose than immature embryos dried upon excision (A. Bochicchio, unpublished data). The question arises whether the reduced storability of hydrated and osmoprimed seeds might be related to their reduced content of oligosaccharides of the raffinose family.

The aim of our study was to determine: (i) whether a treatment with water prior to dehydration affects storability of immature but desiccation-tolerant maize embryos; (ii) whether the reduced storability is related to a reduced raffinose content; and (iii) whether scutellum and axis are differently affected by dehydration after the water treatment.

Materials and Methods

Embryos of maize (inbred line Lo904) excised at 25 days after pollination (DAP) were chosen because: (i) they are desiccation tolerant; (ii) they have no raffinose but raffinose is produced on slow drying (Wolkers *et al.*, 1997); (iii) a water treatment prior to drying decreases the raffinose produced during drying (A. Bochicchio, unpublished data); and (iv) the scutellum of the dried embryo can be easily and completely isolated from the endosperm. Material was grown in Durban, South Africa, and Florence, Italy, in January and September, 1997, respectively.

Embryos excised at 25 DAP were dried slowly either directly (treatment D) or after a treatment with sterile distilled water for 8 h in Petri dishes on two layers of filter paper (treatment WD). The moisture content (MC) of newly-excised embryos was 2.13 ± 0.11 g g^{-1} (dry mass basis) and increased by 11.5% after 8 h of water treatment. Slow drying was achieved as reported in Bochicchio *et al.* (1997). The dried embryos were artificially aged at 27°C and 65% relative humidity in the dark. After slow drying the MC of the embryos was lower than 0.08 g g^{-1} and did not change during ageing.

During the ageing treatment embryos were tested for germination and vigour. This was assessed by the length of whole normal seedlings at 7 days from sowing and as the production of abnormal seedlings. Sucrose and raffinose were quantified separately in both axes and scutella by HPLC. Embryos of the Florence harvest were not exposed to artificial ageing and were tested only for seedling length after dehydration.

The ultrastructure of the radicle tip and of the peripheral part of the scutellum was observed in material that had been routinely prepared for electron microscopy.

Results

Total germination (normal plus abnormal seedlings) did not decrease during artificial ageing in either embryos dried directly or in those dried after the treatment with water (Fig. 16.1A), and WD embryos did not show a germination percentage lower than D embryos. However, the vigour of WD embryos declined during artificial ageing earlier and to a greater extent than D embryos. The length of seedlings produced by WD embryos decreased steadily after the initial 40 days of ageing, while the length of seedlings from D embryos did not decrease until after 127 days of ageing (Fig. 16.1B). The percentage of abnormal seedlings increased during ageing in both treatments, but earlier and more so for WD embryos. In this case, abnormal seedlings appeared much earlier during ageing (Fig. 16.1C). These data indicate that the embryos treated with water prior to dehydration accumulated damage earlier during artificial ageing than those dried directly. The WD embryos also showed a lower initial vigour: prior to any ageing they produced seedlings significant shorter than those developed from D embryos. This occurred with embryos harvested from plants grown in both locations (Fig. 16.2).

Sucrose content was higher in WD embryos than in D embryos in both axes and scutella, and did not decline during ageing in either treatment (Fig. 16.3A). The results for raffinose content showed the opposite: WD embryos always had a lower raffinose content than D embryos, in both axes and scutella (Fig. 16.3B and C). In the scutellum (except for one instance) the difference between the treatments was always statistically significant (Fig. 16.3C). At any time and for any treatment, the raffinose content of the scutellum was lower than that in the corresponding axis (compare Fig. 16.3B and C). After a decline during the initial 40 days, raffinose did not decrease in either treatment during ageing. In summary, embryos with less raffinose (treated with water) accumulated damage earlier and to a greater extent than those with higher raffinose (dried directly) (cf. Figs 16.1 and 16.3).

Ultrastructural observations showed that cells of freshly-excised axes (not illustrated) presented an appearance of relative inactivity, with a general scattering of lipid bodies. However, after slow drying (D), the lipid bodies were orientated peripherally (as is typical of the fully-mature condition after maturation drying) and there were few signs of ultrastructural damage (Fig. 16.4A). During exposure to water, there was differentiation of organelles,

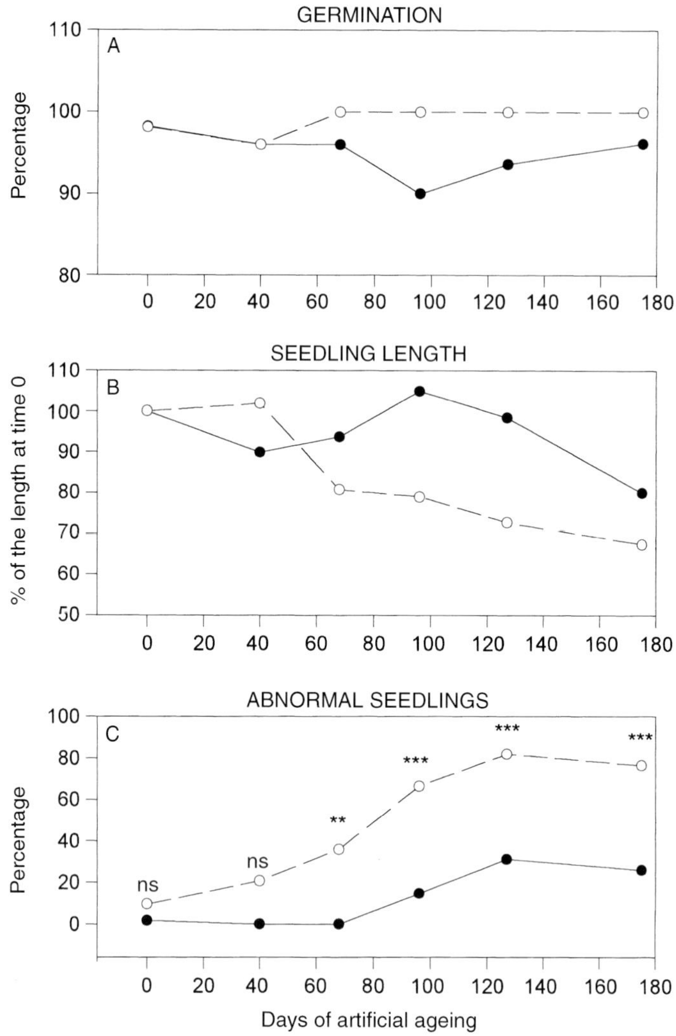

Fig. 16.1. Response to artificial ageing of immature (25 DAP) embryos of maize dried either directly (●) or after an 8 h treatment with water (○). (A) Germination (normal plus abnormal seedlings; minimum 25 seeds per datum point). (B) Mean length of normal seedlings at 7 days (minimum 12 replicates) relative to seedlings from unaged seeds. (C) Abnormal seedlings as percentage of the germinated seeds (minimum 24 seedlings). Significance of difference between frequencies of abnormal seedlings in D and WD embryos: ns, not significant; $^{**}P \leq 0.01$; $^{***}P \leq 0.001$ (χ^2 test).

suggesting that axis cells had become more metabolically active but, after these axes were dried (WD), the cells generally retained an orderly, undamaged appearance (Fig. 16.4B). The scutellar cells from newly-excised embryos were packed with storage material having the appearance of lipid containing granular or crystalline cores (Fig. 16.4C), interspersed with occasional, large starch grains. During water exposure, depletion of the stored reserves and

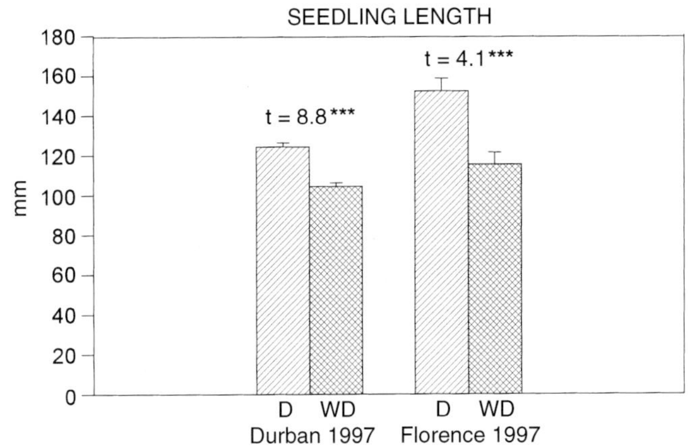

Fig. 16.2. Lengths of whole seedlings 7 days after sowing of unaged 25 DAP embryos harvested in 1997 in Durban or Florence. Embryos were either dried directly (D) or after an 8 h water treatment (WD). Bars represent one standard error of the mean; t, Student's t statistic for the difference of the means within the same harvest; $^{***}P \leq 0.001$.

development of organelles were among marked changes in many scutellar cells and are indicative of metabolic activation (Fig. 16.4D). Some cells of the scutellum were damaged in material dried immediately after excision (D), but when dehydration followed water exposure (WD), marked ultrastructural damage could be seen in many scutellar cells (Fig. 16.4E). It thus appears that drying of excised axes, whether or not they had been exposed to water, caused considerably less damage than did dehydration of the scutella. Furthermore, during the period of water exposure, although ultrastructural activity was enhanced in cells of both axes and scutella, in the latter, intracellular development in conjunction with depletion of stored reserves appeared to dispose the cells to severe damage when they were subsequently dehydrated.

Discussion and Conclusions

The treatment with water prior to drying reduced storability of immature embryos of maize even though it did not affect final germination percentage within the time scale of the ageing treatment. Reduced storability was always associated with a lower raffinose content, which is consistent with the hypothesis of a positive role of raffinose in storability of immature embryos of maize. In our experiments, raffinose did not decrease during artificial ageing, contrary to the case for mature embryos of maize (Bernal-Lugo and Leopold, 1995). The discrepancy is probably because our embryos, besides being immature, were equilibrated to a lower relative humidity and showed a low MC during ageing, and it is possible that under these temperature and MC conditions the glassy state would exist (see Williams and Leopold, 1989). However, they emphasize that 'the glassy state is a metastable state and the rate at which glasses

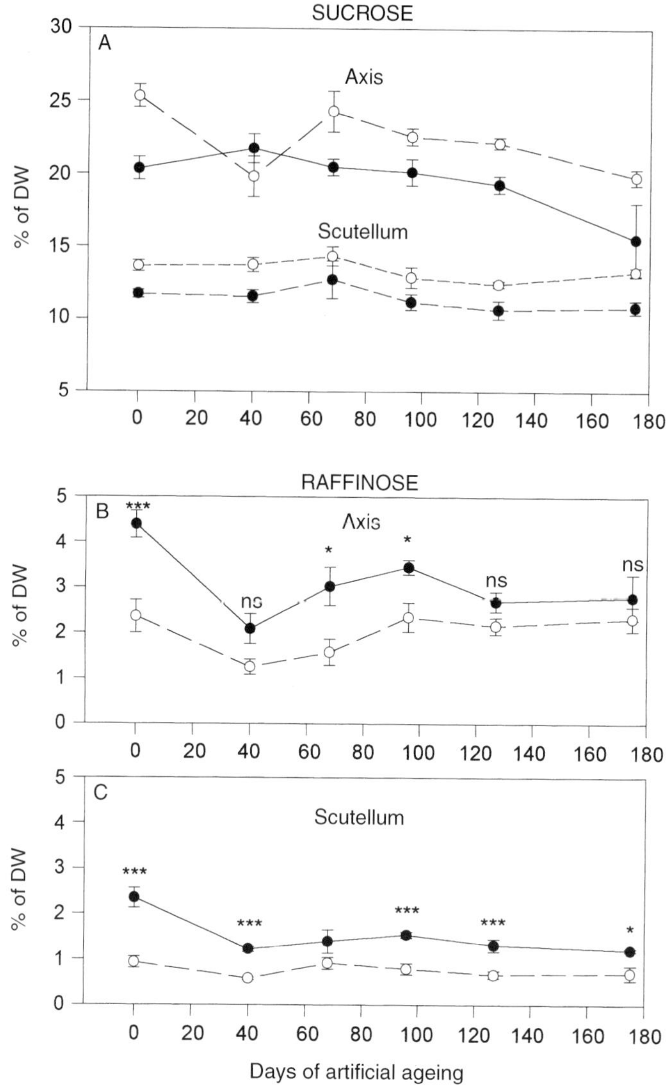

Fig. 16.3. Changes of sucrose and raffinose contents (percentage dry weight) during artificial ageing in the scutella and axes of embryos dried directly (●) or after an 8 h water treatment (○); means of at least three samples, vertical bars represent one standard error. (A) Sucrose content of axis and scutellum. (B) Raffinose content of axis. (C) Raffinose content of scutellum. Significance of differences in raffinose content between treatments tested by the Student's t test: ns, not significant; $*P \leq 0.05$; $***P \leq 0.001$.

decompose varies widely depending upon composition and temperature'. In our D embryos, with a higher raffinose content, the glassy state domains may have been more extensive and more stable than in WD embryos. Raffinose did not appear to be hydrolysed during the 6 months of ageing, the raffinose

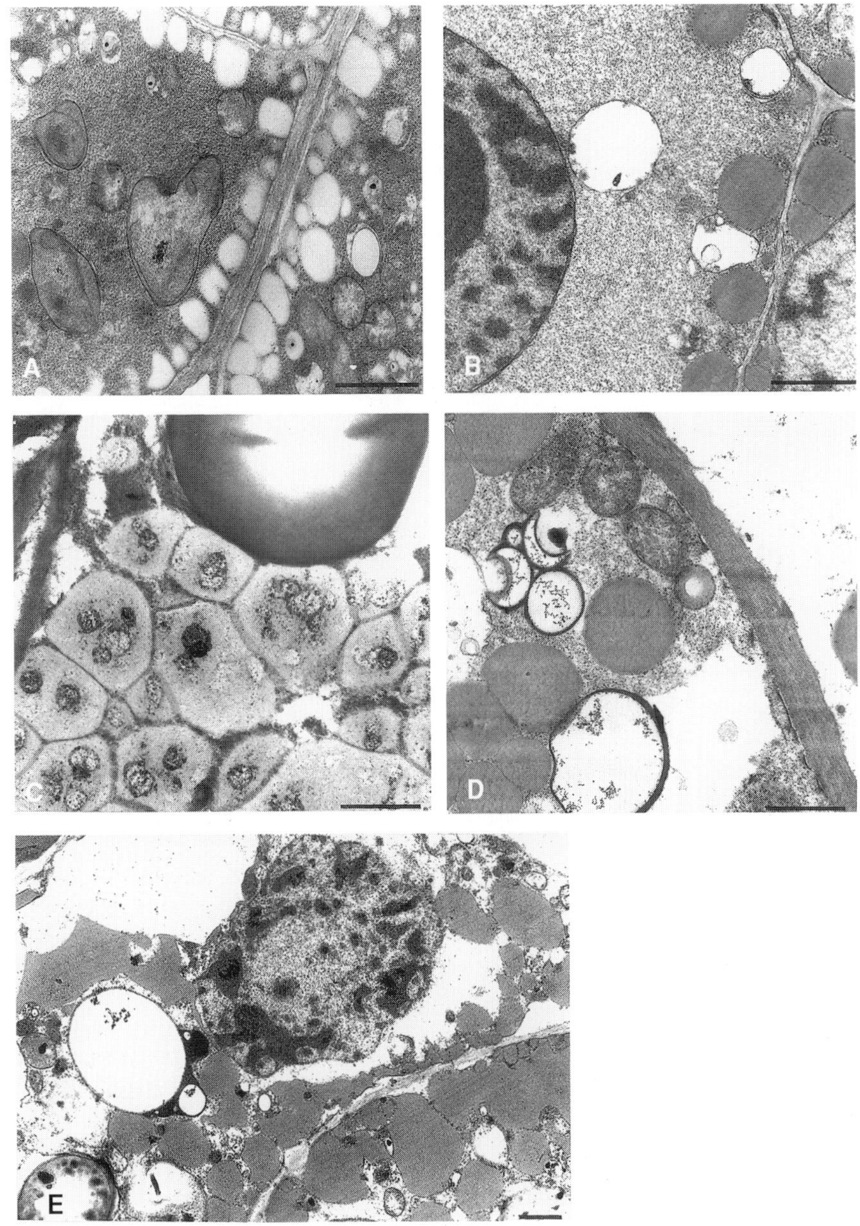

Fig. 16.4. Ultrastructural characteristics of 25 DAP embryos of maize. In each micrograph the bar represents 1 µm. (A) Axis cells after direct slow drying, and (B) drying after exposure to water. Little damage is apparent in either case. (C) Scutellar cells from undried embryos, showing storage material. (D) Similar cells after exposure to water, showing reserve depletion and organelle development. (E) Scutellar cells dried after exposure to water, exhibiting considerable subcellular damage.

content remaining constant except for a decrease during the initial 40 days of ageing. The glassy state may have slowly decomposed and decomposition may have occurred earlier in WD embryos than in D embryos (because of the lower raffinose content in the WD embryos) leading to an earlier accumulation of damage.

This possible involvement of raffinose and vitrification does not rule out other processes which could contribute to the observed effects. Particularly, it must be considered that WD embryos had lower vigour initially, and low vigour usually leads to early deterioration during storage. Thus, the lower storability of WD embryos may be because of their lower initial vigour which, in turn, may be or may be not related to their lower raffinose content. Ultrastructural results indicate that treatment of the embryos with water enhanced subcellular activity, particularly in the scutella where there was considerable reserve depletion. Subsequent drying led to considerable ultrastructural damage to the scutella, as is common for metabolically active tissue. The lower initial vigour of WD embryos could have been a consequence of this damage to the scutella.

References

Bernal-Lugo, I. and Leopold, A.C. (1995) Seed stability during storage: raffinose content and seed glassy state. *Seed Science Research* 5, 75–80.

Bernal-Lugo, I. and Leopold, A.C. (1998) The dynamics of seed mortality. *Journal of Experimental Botany* 49, 1455–1461.

Bochicchio, A., Vernieri, P., Puliga, S., Murelli, C. and Vazzana, C. (1997) Desiccation tolerance in immature embryos of maize: sucrose, raffinose and the ABA-sucrose relation. In: Ellis, R.H., Black, M., Murdoch, A.J. and Hong, T.D. (eds) *Basic and Applied Aspects of Seed Biology*. Kluwer Academic Publishers, Dordrecht, pp. 13–22.

Hoekstra, F.A., Haigh, A.M., Tetteroo, F.A.A. and van Roekel, T. (1994) Changes in soluble sugars in relation to desiccation tolerance in cauliflower seeds. *Seed Science Research* 4, 143–147.

Lin, T.P. and Huang, N.H. (1994) The relationship between carbohydrate composition of some tree seeds and their longevity. *Journal of Experimental Botany* 45, 1289–1294.

Lin, T.P., Yen, W.L. and Chien, C.T. (1998) Disappearance of desiccation tolerance of imbibed crop seeds is not associated with the decline of oligosaccharides. *Journal of Experimental Botany* 49, 1203–1212.

Liu, Y., Bino, R.J., van der Burg, W.J., Groot, S.P.C. and Hilhorst, H.W.M. (1996) Effects of osmotic priming on dormancy and storability of tomato (*Lycopersicon esculentum Mill.*) seeds. *Seed Science Research* 6, 49–55.

Powell, A.A. and Yule, J. (1998) Influence of the aerated hydration seed invigoration treatment on the response of cauliflower (*Brassica oleracea* var. *botrytis*) seed to storage. *Abstracts of the 25th Congress of the International Seed Testing Association*, Pretoria, South Africa, p. 47.

Williams, R.J. and Leopold, A.C. (1989) The glassy state in corn embryos. *Plant Physiology* 89, 977–981.

Wolkers, W.F., Bochicchio, A., Selvaggi, G. and Hoekstra, F.A. (1997) Fourier transform infrared microspectroscopy detects changes in protein secondary structure associated with desiccation tolerance in developing maize embryos. *Plant Physiology* 116, 1169–1177.

17 Maillard Reactions Cause Browning in Bean Seed Coats during Ageing: Inhibition by Aminoguanidine

A.G. TAYLOR[1], T-G. MIN[1,2] AND D.H. PAINE[1]

[1]*Department of Horticultural Sciences, New York State Agricultural Experiment Station, Cornell University, Geneva, New York, USA;*
[2]*Department of Agronomy, Taegu University, South Korea*

Physiological ageing was studied in a white-seeded cultivar of snap bean (*Phaseolus vulgaris*). Seeds were aged by first equilibrating them at 70% relative humidity followed by incubation at 40°C, removing samples at one-week increments up to 6 weeks. Standard germination tests revealed that non-aged seeds had 95% germination, and no seeds germinated after 6 weeks. Changes in seed coat colour were observed with ageing, and browning increased with duration of ageing. Light reflectance measurements were made with a Hunter Colorimeter, and 'L' readings (white = 100 and black = 0) were obtained from a sample of intact seeds. A slight decrease in L values was measured during the first 4 weeks of ageing, followed by a significant decline in the 5- and 6-week samples. Maillard reactions were proposed to be responsible for seed coat browning. Since chemical analysis of advanced glycosylation endproducts is problematic, a specific chemical inhibitor of Maillard chemistry was utilized. Application of this inhibitor, aminoguanidine, prior to ageing reduced the usual browning, and thus supported the proposition that Maillard reactions may be responsible for seed coat browning in snap beans.

Introduction

Morphological changes have been observed with seed ageing, may be specific to a particular species, and may occur at the seedling or seed stage (reviewed by Priestley, 1986). Physiological necrosis is an example of a seedling morphological disorder in lettuce (*Lactuca sativa* L.). Seedling cotyledons exhibit a reddish-brown coloration on either side of the midrib, which increases with ageing (Tomas *et al.*, 1992). However, morphological changes that can be detected prior to germination are of primary interest. Commercial lots of white-seeded beans have been shown to have a proportion of seeds with tan and brown-coloured seed coats. These off-colour seeds were attributed to physiological quality of the lot rather than to a mixture of genotypes. Selected tan and brown seeds had lower germination potential than white seeds (Lee *et al.*, 1998). The cause of the seed coat browning was not determined.

Seed coat browning may be attributed to at least two different chemical reactions in biological systems. The seed coats of many species contain soluble tannins (leuco-anthocyanins) (Bate-Smith and Ribereau-Gayon, 1959). Polymerization of soluble tannins can form high molecular weight (MW) polymers, which are brown-coloured, condensed tannins (Nozzolillo and De Bezada, 1984). The Maillard reaction is an alternative mechanism for browning, and a brief review of Maillard chemistry follows (see Fig. 17.1). Reducing sugars (glucose or fructose) can non-enzymatically react with the epsilon amino group of basic amino acids, such as lysine, present in proteins (Hodge, 1953). The condensation product loses a molecule of water to form a Schiff base. The reaction proceeds to form an Amadori product; at this point, the protein is glycosylated. Several subsequent steps follow, and in the final stage, brown melanoidin pigments are produced from the polymerization of many reactive compounds formed during the advanced Maillard reaction through several interacting routes (Hayashi and Namiki, 1986; Hodge, 1953). In the medical field, the final products of the Maillard reactions have been termed 'advanced glycosylation endproducts' (AGEs) (Brownlee *et al.*, 1986). The reaction can be inhibited by aminoguanidine, which has been proposed to react with the Amadori rearrangement product, or a post-Amadori product that is an intermediate in AGE formation.

Early work on Maillard reactions and seeds was performed by cereal chemists on wheat to be used as grain. A malady known as 'sick wheat syndrome' occurred when the germ or embryo darkened in colour, reducing the

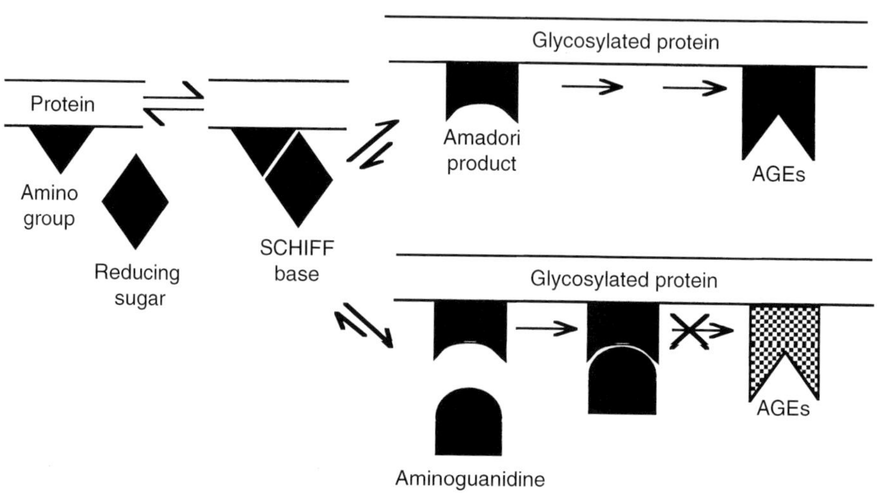

Fig. 17.1. Non-enzymatic reactions of reducing sugars with proteins to form a Schiff base and then form an Amadori product. The Amadori product can undergo several more steps to form AGEs, or can be blocked with aminoguanidine (adapted from Cerami *et al.*, 1987).

commercial grade of the grain. The brown pigment isolated from the embryos was characteristic of Maillard products, increased as storage temperature and moisture content increased, and was not attributed to mould growth (McDonald and Milner, 1954). It was later shown that an increase in reducing sugars occurred at the expense of non-reducing sugars, and this change preceded visual browning (Linko *et al.*, 1960).

The role of Maillard products was studied in soybean (*Glycine max* L. Merr.) by Wettlaufer and Leopold (1991). Seeds were aged either at 40°C and 100% relative humidity or 30°C and 75% RH. Amadori products increased dramatically after one week of ageing and then declined in the 40°C treatment, while at the lower temperature, Amadori products continued to increase over a 4-week ageing period. Maillard products were only found to increase in the 40°C treatment and were accompanied by a loss in germination. No loss of germination was measured in the 30°C treatment. In contrast, the accumulation of Maillard products was not related to seed viability in several small-seeded vegetable crops (Baker and Bradford, 1994). Sun and Leopold (1995) also found that Maillard products accumulated in soybean axes and cotyledons after ageing at 36°C and 75% RH. The increase in Maillard products was correlated with the loss of seed germinability; however, this was less obvious under long-term storage conditions at low temperatures.

The objective of this study was to quantify seed coat browning in white-seeded beans aged in a controlled environment. The effect of aminoguanidine in inhibiting Maillard reactions and subsequently decreasing browning was tested.

Materials and Methods

A seed lot of the white-seeded cultivar 'Labrador' snap beans was provided by Asgrow Seed Co., Twin Falls, Idaho, and was not commercially treated with chemical or biological treatments. Seed ageing was performed under controlled environmental conditions in the laboratory. Seeds were first equilibrated to 70% relative humidity, which was maintained by a solution of water and glycerol in custom-designed Plexiglas chambers (Forney and Brandl, 1992). The corresponding moisture content was 15% (fresh weight basis). Samples were heat-sealed in aluminium-foil plastic-laminate packets and aged at 40°C for up to 6 weeks. A single packet was removed at weekly intervals. After each ageing period, seed quality was determined using the standard germination test in which normal and abnormal seedlings were classified (AOSA, 1992, 1993).

The effect of aminoguanidine on browning was tested. Aminoguanidine hydrochloride was provided from Alteon Corp., Ramsey, New Jersey, and pH of the solutions was adjusted to pH 7.4 with potassium phosphate buffer. A known volume of solution was added to a sample of seeds to increase the seed moisture content to 40% (Taylor *et al.*, 1998). Seeds were brought to an initial seed moisture content of 15% to avoid hydration damage during treatment (Taylor *et al.*, 1992). A distilled water check was included. Seeds imbibed in the solutions for 24 h and then were dried under ambient conditions back to the original seed moisture content of 15%. Seeds were aged at 70% RH and 50°C for 2 weeks, and a non-aged control was included.

Seed coat browning was measured with a Hunter Colorimeter, D25 L optical sensor, Reston, VA. The 'L' readings (white = 100 and black = 0) were obtained from a sample of intact seeds. Several replicates were measured and means with standard errors were calculated.

Results and Discussion

Snap bean seeds were aged in one-week increments up to 6 weeks. The percentage of normal seedlings was 95% from non-aged seeds, and the decline in normal seedlings followed a sigmoidal-shaped curve (Fig. 17.2). The percentage abnormal seedlings increased with ageing duration and then decreased after 4 weeks of ageing. Thus the seed quality ranged from high quality in non-aged samples to a very low quality from which no normal seedlings were produced after 6 weeks of ageing.

Light reflectance from aged and off-coloured bean seeds has been determined (Lee *et al.*, 1998). Both aged and naturally off-coloured seeds had a lower percentage of light reflectance over a broad wavelength region in comparison with non-aged and seeds with white-coloured seed coats, respectively. Thus, instrumentation that quantifies light reflectance with a white–black scale should provide a measure of browning. The Hunter Colorimeter is commonly used in food technology from which the 'L' scale can be used to measure

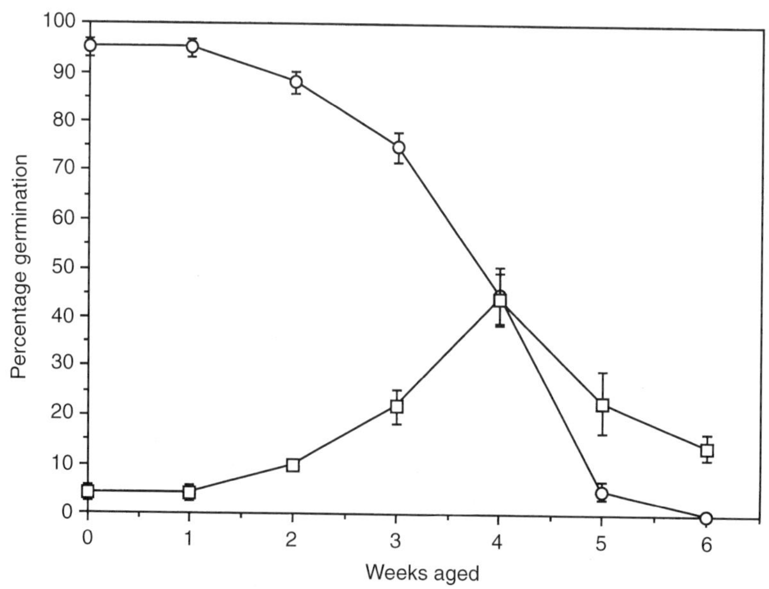

Fig. 17.2. Snap beans (*Phaseolus vulgaris*) seeds aged at 40°C and 15% moisture content for 0 to 6 weeks. ○, Normal seedlings; □, abnormal seedlings.

balance between white and black. A slight decrease in L values was measured during the first 4 weeks of ageing, followed by a significant decline in the 5- and 6-week samples (Fig. 17.3). There was a highly significant relationship between L values and the percentage normal seedlings ($r = 0.978^{**}$). Thus, light reflectance from a sample of intact seeds can provide a rapid and non-destructive method to assess seed quality in white-seeded snap beans.

Seed coat browning may be attributed to condensation of tannins or Maillard reactions. Browning in lentil (*Lens culinaris* Medic.) was shown to result by polymerization of soluble tannins to form condensed tannins (Nozzolillo and De Bezada, 1984). However, most white-seeded snap bean cultivars have the recessive p gene (Dickson and Petzoldt, 1988). White-seeded cultivars lack soluble tannins as precursors (Bate-Smith and Ribereau-Gayon, 1959), and therefore cannot form condensed tannins. Precursors of Maillard reactions are reducing sugars and amino groups of proteins (Fig. 17.1). Reducing sugars have been found in higher concentration in seed coats of soybean than in axes or cotyledons (Kuo *et al.*, 1997). Research in our laboratory performed in collaboration with R. Obendorf's laboratory has shown similar results in snap beans (unpublished data). Snap bean is considered a starch-storing seed; however, the protein content is also high (22%) (cited by Taylor, 1997). Micro-Kjeldahl analysis of seed coat tissue revealed 0.64% N. Using a multiplication factor of 6.25 yields an estimated protein content of the seed coat of 4%. Collectively, seed coats not only contain higher concentrations of reducing sugars than embryos but also contain some protein.

Confirmation of Maillard reactions could be pursued by analysis of AGEs; however, the chemistry of the final Maillard reactions is complex (Hodge,

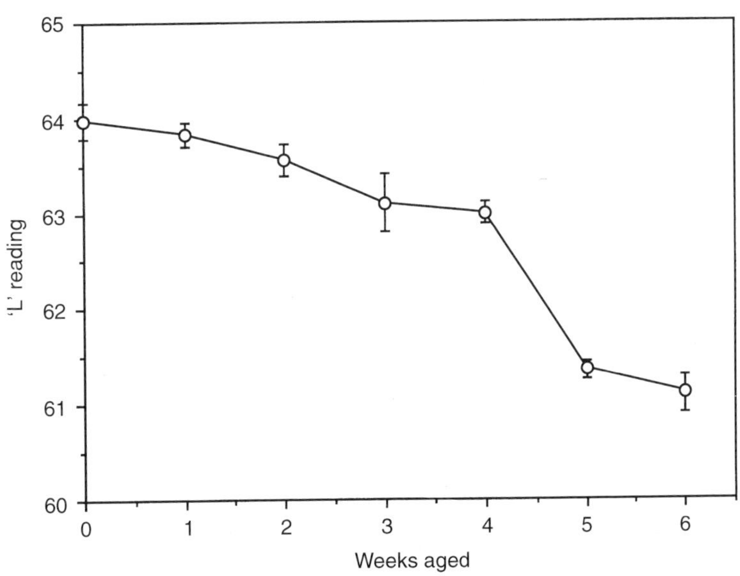

Fig. 17.3. Hunter Colorimeter 'L' readings from aged bean seeds aged for 0 to 6 weeks.

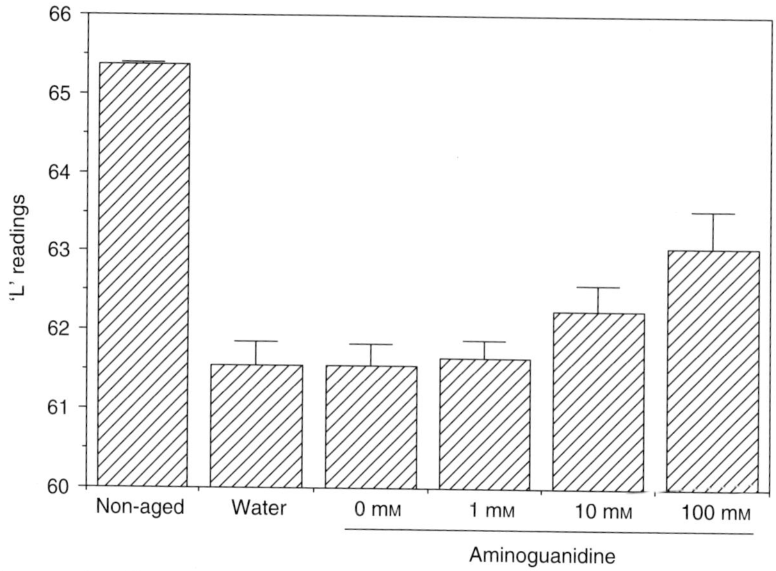

Fig. 17.4. The effect of different concentrations of aminoguanidine on seed coat browning.

1953). Chemical analysis using fluorescence measurements may be confounded as seeds of many species exhibit autofluorescence. Other chemical methods used to assay AGEs also have limitations and have been reviewed by Furth (1988). Therefore, an alternative approach was taken. Rather than analysing or attempting to analyse the endproducts of Maillard reactions, research was focused on using specific inhibitors of Maillard chemistry.

Aminoguanidine has been proposed to react with Amadori products, thus blocking downstream reactions (Cerami *et al.*, 1987; Fig. 17.1). Aminoguanidine has found applications in the medical field to arrest deleterious protein cross-linking caused by elevated glucose levels (Brownlee *et al.*, 1986; Ulrich and Cerami, 1992). This compound may be useful in the study of Maillard reactions in seeds. Aminoguanidine is water-soluble and can be applied by soaking seeds in solutions. Increasing the concentration of aminoguanidine from 1 to 100 mM increased 'L' readings (Fig. 17.4). Non-aged seeds had the highest 'L' readings indicating that browning was not completely inhibited even with 100 mM treatment. In conclusion, aminoguanidine was able to retard browning in a log-linear relationship. The reduction of browning by aminoguanidine provided further evidence that Maillard reactions are responsible for seed coat browning in white-seeded beans.

Acknowledgements

The authors wish to thank Dr Alka Bhargava of the Tropical Forest Research Institute (Indian Council of Forestry Research and Education) for her technical expertise on tannin chemistry. Dr Bhargava was an International Scholar at Cornell University funded by the World Bank. Tai-Gi Min's sabbatical to Cornell University was partially supported by a grant funded by the National Instrumentation Center for Environmental Management (NICEM) in Seoul National University, Korea. The aminoguanidine was supplied as a gift from Alteon Inc. to conduct this research project.

References

Association of Official Seed Analysts (1992) *Seedling Evaluation Handbook*. Association of Official Seed Analysts, Contribution No. 35.

Association of Official Seed Analysts (1993) Rules for testing seeds. *Journal of Seed Technology* 16(3), 1–113.

Baker, E.H. and Bradford, K.J. (1994) The fluorescence assay for Maillard product accumulation does not correlate with seed viability. *Seed Science Research* 4, 103–106.

Bate-Smith, E.C. and Ribereau-Gayon, P. (1959) Leuco-anthocyanins in seeds. *Qualitas Plantarum et Materiae Vegetabiles* 8, 189–198.

Brownlee, M., Vlassara, H., Kooney, A., Ulrich, P. and Cerami, A. (1986) Aminoguanidine prevents diabetes-induced arterial wall protein cross-linking. *Science* 232, 1629–1632.

Cerami, A., Vlassara, H. and Brownlee, M. (1987) Glucose and aging. *Scientific American* 256, 90–96.

Dickson, M.H. and Petzoldt, R. (1988) Deleterious effects of white seed due to p gene in beans. *Journal of the American Society for Horticultural Sciences* 113, 111–114.

Forney, C.F. and Brandl, D.G. (1992) Control of humidity in small controlled-environment chambers using glycerol-water solutions. *HortTechnology* 2, 52–54.

Furth, A.J. (1988) Methods for assaying nonenzymatic glycosylation. *Analytical Biochemistry* 175, 347–360.

Hayashi, T. and Namiki, M. (1986) Role of sugar fragmentation in an early stage browning of amino-carbonyl reaction of sugar with amino acid. *Agricultural and Biological Chemistry* 50, 1965–1970.

Hodge, J. (1953) Chemistry of browning reactions in model systems. *Journal of Agricultural Food Chemistry* 1, 928–943.

Kuo, T.M., Lowell, C.A. and Smith, P.T. (1997) Changes in soluble carbohydrates and enzymic activities in maturing soybean seed tissues. *Plant Science* 125, 1–11.

Lee, P.C., Paine, D.H. and Taylor, A.G. (1998) Detection and removal of off-colored bean seeds by color sorting. *Seed Technology* 20, 43–55.

Linko, P., Cheng, Y.-Y. and Milner, M. (1960) Changes in the soluble carbohydrates during browning of wheat embryos. *Cereal Chemistry* 37, 548–556.

McDonald, C.E. and Milner, M. (1954) The browning reaction in wheat germ in relation to 'sick' wheat. *Cereal Chemistry* 31, 279–295.

Nozzolillo, C. and De Bezada, M. (1984) Browning of lentil seeds, concomitant loss of viability, and the possible role of soluble tannins in both phenomena. *Canadian Journal of Plant Science* 64, 815–824.

Priestley, D.A. (1986) *Seed Aging Implications for Seed Storage and Persistence in the Soil.* Comstock Publishing Associates, Ithaca, New York.

Sun, W.Q. and Leopold, A.C. (1995) The Maillard reaction and oxidative stress during aging of soybean seeds. *Physiologia Plantarum* 94, 94–104.

Taylor, A.G. (1997) Seed storage, germination and quality. In: Wien, H.C. (ed.) *The Physiology of Vegetable Crops.* CAB International, Wallingford, UK, pp. 1–36.

Taylor, A.G., Prusinski, J., Hill, H.J. and Dickson, M.D. (1992) Influence of seed hydration on seedling performance. *HortTechnology* 2, 336–344.

Taylor, A.G., Allen, P.S., Bennett, M.A., Bradford, K.J., Burris, J.S. and Misra, M.K. (1998) Seed enhancements. *Seed Science Research* 8, 245–256.

Tomas, T.N., Taylor, A.G., Ellerbrock, L.A. and Chirco, E.M. (1992) Lettuce seed necrosis. *Seed Science and Technology* 20, 539–546.

Ulrich, P.C. and Cerami, A. (1992) Inhibitors of nonenzymatic cross-linking. *United States Patent #* 5140048.

Wettlaufer, S.H. and Leopold, A.C. (1991) Relevance of Amadori and Maillard products to seed deterioration. *Plant Physiology* 97, 165–169.

18 Effects of Desiccation on the Subcellular Matrix of the Embryonic Axes of *Quercus robur*

D.J. MYCOCK[1], P. BERJAK[2] AND W.E. FINCH-SAVAGE[3]

[1]Department of Botany, University of the Witwatersrand, Johannesburg 2050, South Africa; [2]School of Life and Environmental Sciences, University of Natal, Durban 4041, South Africa; [3]Horticulture Research International, Wellesbourne, Warwick CV35 9EF, UK

The effects of desiccation on the ultrastructure of *Quercus robur* embryonic axes isolated from intact seeds that had been dried to various nominal water contents was investigated. Root meristem tissues were prepared for conventional transmission electron microscopy and the microtubule and microfilament systems were visualized using specific probes and fluorescence and confocal microscopy. The axes of this species were able to withstand some desiccation in that drying from 55% to 40% water content had little effect on viability. This was supported by the overall ultrastructure, and microfilament and microtubules systems of this material appearing similar to the control. Drying to 35% caused a 40% decline in viability. At the ultrastructural level such axes were characterized by plasmalemma withdrawal from the cell wall, cytoplasmic clearing between the endoplasmic reticulum sheets and nuclear lobing. At this water content, the number of cells in which the microfilament and microtubule systems could be demonstrated declined, and this important cytoskeletal system did not reconstitute to control levels on imbibition. Drying to below 35% caused drastic losses in viability and ultrastructural disruption.

Introduction

Quercus robur is one of several important temperate tree species that produce seeds that do not develop desiccation tolerance during their development (Grange and Finch-Savage, 1992). As a consequence, these seeds, which are shed at high water contents and are desiccation sensitive, cannot be stored for protracted periods using conventional low temperature, low relative humidity storage regimes applicable to orthodox seeds (Roberts, 1973). The effects of desiccation on the tissues of recalcitrant seeds has been the focus of much research over the past few years. The present contribution forms part of that research effort and is an investigation into the ultrastructural effects, in particular cytoskeletal effects, of drying on the root meristem of *Q. robur* seeds.

Materials and Methods

Seeds

Seeds were collected at Wellesbourne in the United Kingdom and air freighted to the laboratories in South Africa. On receipt, the seeds were tested for germinability and water content. For the viability assessments intact seeds were imbibed for 24 h wrapped in moistened paper, the pericarp was then removed and the basal third of the cotyledons excised. The seeds were then buried, cut surface down, in moistened vermiculite and maintained at 20°C in an environmental chamber. Percentage radicle extension, complete germination and normal seedling development were recorded. Axis and cotyledon moisture content were determined gravimetrically after drying at 80°C, and are expressed on a wet mass basis.

Drying

Using the initial seed moisture content, the seed weight estimated to be equivalent to 40, 35, 27 and 20% water content were calculated. The pericarp was removed and the seeds were mixed with an equal volume of activated silica gel and dried, at 20°C, to each target water content. Viability and actual water content were assessed for each drying treatment.

Transmission electron microscopy (TEM)

The ultrastructure of the root meristem was examined before and after drying. The terminal 3 mm of the radicle were excised and processed for TEM using a standard glutaraldehyde-osmium method. Sections were stained with lead citrate and uranyl acetate, then viewed and photographed with a Jeol 1010 transmission electron microscope.

Microtubule and microfilament studies

An antibody to β-tubulin with an attached CY3 fluorescent tag was used to label the microtubules and FITC-labelled phalloidin was used to detect the microfilaments. Vibratome sections (20–25 μm thickness) cut through the root meristem were collected on glass slides and stained for 1 h. The antibody to β-tubulin was carried in 0.01 M PBS, pH 7.4, containing 1% BSA, 1% methanol, 1% Triton X and 1% DMSO. The FITC phalloidin was carried in 0.01 M phosphate buffer, pH 7.2, containing 10 mM EGTA, 0.1 M $MgSO_4$, 2% methanol,

1% Triton X and 1% DMSO. After staining, the sections were washed in buffer devoid of the stain and mounted in a drop of antifade agent (Citifluor). The sections were then viewed and photographed with a Zeis Axiovert 100 LSM confocal microscope or a Zeiss Axiophot fluorescence microscope using the appropriate excitation wavelength. The number of cells exhibiting microfilament fluorescence was quantified using image analysis.

Results and Discussion

As was expected, the actual water contents of the axes and cotyledons isolated from the dried seeds were different from the target water contents, with the larger cotyledonary tissues having water contents closer to the nominal water contents (Table 18.1). As is typical of recalcitrant seeds, there was a tremendous variation in water content (Table 18.1).

The drying of the oak seeds had a deleterious effect on their germinability (Fig. 18.1). Below water contents of 35% the number of seeds showing radicle

Table 18.1. Average actual water contents of seeds dried to various nominal water contents. The range of water contents is given in parentheses.

	Actual	
Nominal	Cotyledons	Axes
Control	46.9 (42.3–51.7)	56.5 (53.8–59.9)
40%	39.5 (33.8–43.6)	43.6 (38.2–46.7)
35%	31.6 (27.8–33.1)	40.4 (37.6–43.6)
27%	24.9 (20.2–27.2)	33.5 (27.1–43.1)
20%	21.5 (19.7–26.8)	26.8 (20.1–32.6)

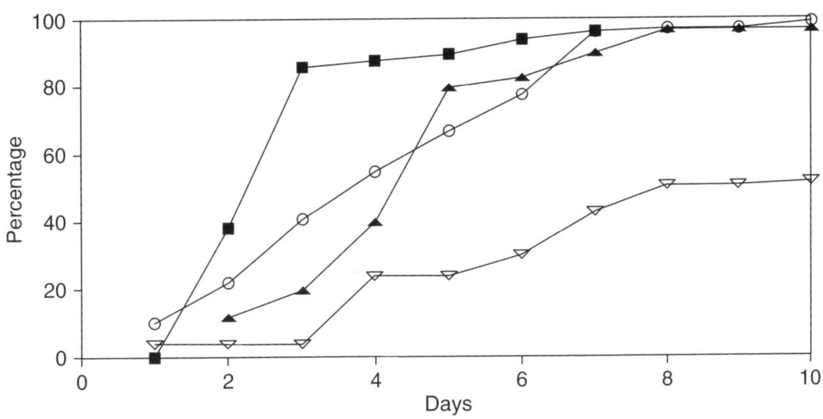

Fig. 18.1. Total germination of *Quercus robur* seeds after drying to various nominal water contents (■, Control; ○, 40%; ▲, 35%; ▽, 27%). No germination was recorded for material dried to 20%.

extension, proceeding to complete germination and exhibiting normal seedling growth was drastically reduced, whereas at water contents above 35%, these features were similar to the control (data not shown).

These results were not unexpected in terms of our current understanding of the effects of drying on recalcitrant seeds (e.g. Berjak *et al.*, 1989). That is, the seeds were able to withstand a certain amount of drying, but there was a limit after which vigour and viability were adversely affected.

The cells of the root meristem of the control undried axes were characterized by a central and spherical nucleus (Fig. 18.2). The nuclear membrane was clearly definable and the heterochromatin, nucleoli were frequently observed. Numerous mitochondria and strands of endoplasmic reticulum (ER) were observed in the dense cytoplasm and there was some evidence of polysome formation. As is typical of meristem tissues, the cells were not highly vacuolated. Collectively, these observations were indicative of active metabolism. Cells of the root meristem of seeds dried to 40% exhibited an ultrastructural profile similar to the control. When the seeds were dried to 35%, the most obvious feature was an increase in the extent of vacuolation. The ER was dilated and the sheets of membrane had separated (Fig. 18.3). The extent of ER dilation varied among the cells and it is possible that these expanded sheets of ER ultimately formed the vacuoles (Fig. 18.4). Although the cytoplasm remained dense there was not much evidence of polysome formation. There were also some aggregates of microfilaments. Material dried to 27% exhibited a range of damage. Some cells did not appear much more damaged than cells from the 35% treatment. There was however, some evidence of clearing of the mitochondrial matrices. In other cells, the desiccation had caused the plasmalemma to withdraw from the cell wall although the membrane did often remain attached at the plasmodesmata. In these cells, vacuolation was greatly increased and the aggregates of the cytoskeletal components were more frequently observed (Fig. 18.5). In other cells, the membranes were ruptured and the cell contents dispersed in the confines of the walls. In this destruction the nucleus was generally the only clearly definable body. As in the previous cases, there was a range of ultrastructural damage in material dried to the lowest water content. The great majority of the cells of the meristem were extensively damaged and were presumably dead, thereby supporting the germination data.

The increase in vacuolation in response to drying has been previously reported in recalcitrant seed tissues and in hydrated orthodox seed and vegetative tissues (e.g. Smith and Berjak, 1995). The aggregation of the cytoskeletal components has also been reported as a response to drying in other recalcitrant species, for example *Hevea brasiliensis* and *Camellia sinensis* (Berjak, 1989; Wesley-Smith *et al.*, 1992).

The cytoskeleton is a filamentous system which orders and supports the membranes, organelles and ground substance of the cell. The principal components of the cytoskeleton are the microtubules, the microfilaments and intermediate filaments, although in plants, intermediate filaments appear to be represented only by the lamins of the nuclear matrix and lamina (Wolfe, 1993). The system is dynamic and responds to changes in both the intra- and

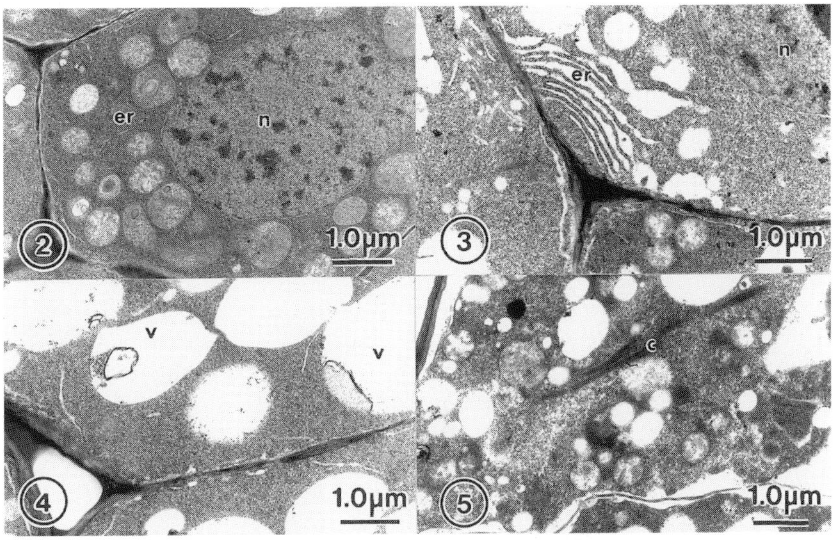

Figs 18.2–18.5. Ultrastructural detail of control (undried) and dried root meristem. The ultrastructure of the control was indicative of active metabolism (Fig. 18.2). Drying the material to 35% water content caused dilation of the ER (Fig. 18.3) and extensive vacuolation (Fig. 18.4). Further drying resulted in a range of damage, in particular the cytoskeletal elements appeared to aggregate (Fig. 18.5). c, Cytoskeletal elements; er, endoplasmic reticulum; m, mitochondrion; n, nucleus; v, vacuole.

extra-cellular environments. The cytoskeleton has been shown to regulate a wide array of activities including: the positioning and movement of organelles (Cole and Lippincott-Schwartz, 1995); the localization and regulation of biochemical pathways e.g. glycolysis (Masters, 1984); and both the microfilament and microtubule systems are involved in karyo- and cytokinesis (Staiger and Lloyd, 1991).

All of the cells of the control undried tissue exhibited actin fluorescence. Drying the seeds to 40% resulted in a drop in the number of cells exhibiting the characteristic yellow-green fluorescence. However, reimbibition after drying resulted in a reconstitution of the microfilament system as exhibited by an increase in the number of cells showing fluorescence (Fig. 18.6). Drying to 35% and below resulted in a further reduction in the number of cells containing actin microfilaments and although this situation was improved after reimbibition (Fig. 18.6) reconstitution did not return to the level of the control.

In the control, the CY3 signal was so strong that it was difficult to visualize individual microtubules. After drying to 40% water content, the strength of the CY3 signal was reduced and it was possible to distinguish the microtubule arrays within the cells (Fig. 18.7). For example the cortical array around the cell periphery and the pre-prophase band which surrounds the nucleus and extends into the cortex prior to mitosis were frequently observed. Reimbibing the seeds for 24 h after drying increased the strength of the fluorescence signal and the number of cells exhibiting this signal, thus implying that the microtubule system was able to reconstitute after some drying. However, if the seeds were dried to the nominal water content of 35% the number of cells showing fluorescence was greatly reduced and this situation was not improved after reimbibition. Drying to 27 and 20% was correlated with a total breakdown of the microtubule system and little staining was observed (Fig. 18.8).

It therefore appears that drying *Q. robur* seeds below the nominal water content of 35%, under the conditions presently used (Pammenter *et al.*, 1998), causes disassembly of the microtubule and microfilament systems.

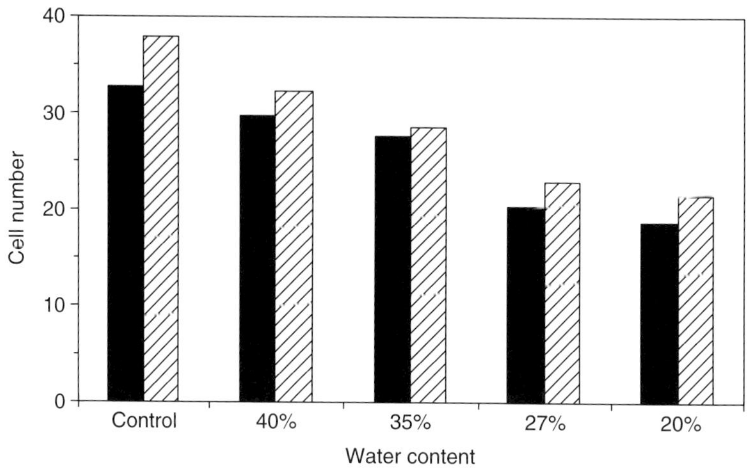

Fig. 18.6. Average number of cells per 200 μm^2 showing actin fluorescence in seeds of different water contents. Hatching indicates the value after re-imbibition.

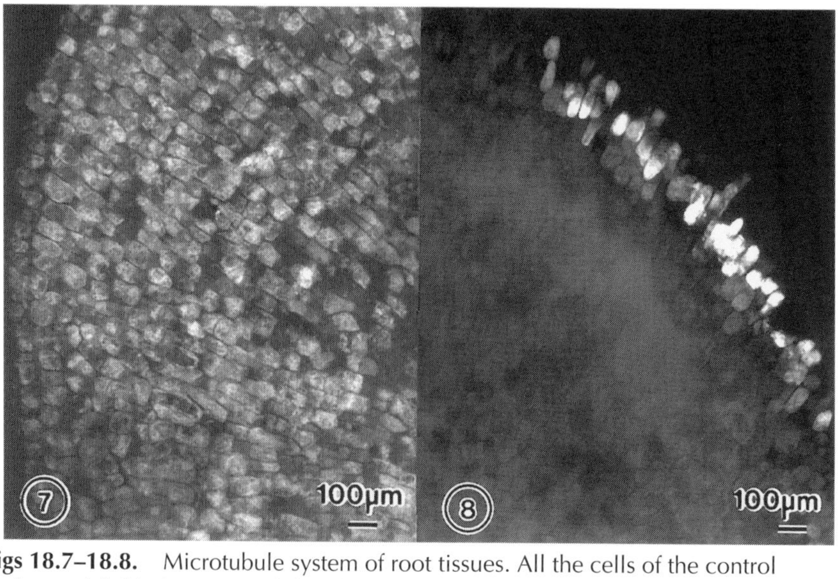

Figs 18.7–18.8. Microtubule system of root tissues. All the cells of the control and material dried to 40% exhibited CY3 fluorescence indicative of microtubules (Fig. 18.7). Drying the seeds below 35% resulted in a reduction in the number of cells with microtubules (Fig. 18.8).

It can be theorized that the reductions in the rate of germination after drying are due, at least in part, to the requirement for the reconstitution of the cytoskeleton, and, in particular, the reconstitution of the functional biochemical pathways associated with the cytoskeleton. A reduction in ATP synthesis, due to the inefficient functioning of glycolysis (a microfilament-associated biochemical pathway (Master, 1984)), due to the drying would slow the rate of growth. Microtubules have been shown to move and hold in position the Golgi bodies and the ER cisternae (Cole and Lippincott-Schwartz, 1995). Disruption of the microtubules by drying could be responsible for these membrane sheets separating. This could help to explain our TEM observations of dilated ER and increased vacuolation.

References

Berjak, P. (1989) Storage behaviour of seeds of *Hevea brasiliensis*. *Journal of Natural Rubber Research* 4, 195–203.

Berjak, P., Farrant, J.M. and Pammenter, N.W. (1989) The basis of recalcitrant seed behaviour. In: Taylorson, R.B. (ed.) *Recent Advances in the Development and Germination of Seeds*. Plenum Press, New York, pp. 89–108.

Cole, N.B. and Lippincott-Schwartz, J. (1995) Organisation of organelles and membrane traffic by microtubules. *Current Opinions in Cell Biology* 7, 55–64.

Grange, R.I. and Finch-Savage, W.E. (1992) Embryo water status and development of the recalcitrant species *Quercus robur* L.: determination of water relations parameters by pressure-volume analysis. *Journal of Experimental Botany* 43, 657–662.

Masters, C. (1984) Interactions between glycolytic enzymes and components of the cytomatrix. *Journal of Cell Biology* 99, 222–225.

Pammenter, N.W., Greggains, V., Kioko, J.I., Wesley-Smith, J., Berjak, P. and Finch-Savage, W.E. (1998) Effects of differential drying rates on viability retention of recalcitrant seeds of *Ekebergia capensis*. *Seed Science Research*, 8, 463–471.

Roberts, E.H. (1973) Predicting the storage life of seeds. *Seed Science and Technology* 1, 499–514.

Smith, M.T. and Berjak, P. (1995) Deteriorative changes associated with the loss of viability of stored desiccation-tolerant and desiccation-sensitive seeds. In: Kigel, J. and Galili, G. (eds) *Seed Development and Germination*. Marcel Dekker, Inc., New York, pp. 701–746.

Staiger, C.J. and Lloyd, C.W. (1991) The plant cytoskeleton. *Current Opinions in Cell Biology* 3, 33–42.

Vertucci, C.W. and Farrant, J.M. (1995) Acquisition and loss of desiccation tolerance. In: Kigel, J. and Galili, G. (eds) *Seed Development and Germination*. Marcel Dekker, Inc., New York, pp. 237–271.

Wesley-Smith, J., Vertucci, C.W., Berjak, P., Pammenter, N.W. and Crane, J. (1992) Cryopreservation of desiccation-sensitive axes of *Camellia sinensis* in relation to dehydration, freezing rate and the thermal properies of tissue water. *Journal of Plant Physiology* 140, 596–604.

Wolfe, S.L. (1993) *Molecular and Cellular Biology*. Wadsworth Publishing Company, Belmont, California, pp. 496–52.

19 Loss of Viability in Rye Embryos at Different Levels of Hydration: Senescence with Apoptotic Nucleosome Cleavage or Death with Random DNA Fragmentation

I. BOUBRIAK[1,2], V. NAUMENKO[2], L. LYNE[1] AND D.J. OSBORNE[1]

[1]*Oxford Research Unit, The Open University, Boars Hill, Oxford OX1 5HR, UK;* [2]*Institute of Cell Biology and Genetic Engineering, Kiev 252143, Ukraine*

In experiments with embryos of rye seeds which have been held at different moisture contents, different fragmentation patterns of DNA occur as a result of DNAase activities operating at different levels of water activity. Embryos of seed held in the dry state (*c.* 9% moisture content) show a progressive accumulation of random DNA fragments with time associated with a progressive loss of viability and death within 4 years. In contrast, embryos of seeds held under accelerated ageing conditions (75% relative humidity, 40°C – embryos reaching *c.* 14% moisture content) remain viable for less than 15 days. These embryos show an accumulation of DNA nucleosome multimers rather than random fragmentation. The evidence suggests that different specific nucleases are activated at different levels of embryo hydration. This means that the results of accelerated ageing of seeds can not be used as a model system to represent a fast achievement of natural ageing. The relevance of these results to priming is discussed.

Introduction

Plant cells can die from many causes; desiccation, overheating, senescence or pathogen attack. They can also die *in situ* amongst their living neighbour cells as part of a programmed and differentiation-determined terminal function. But the timing and the causes of death of the embryos of dry seeds have long been a debated question. Also, the ways in which the loss of viability takes place may not always be the same and could well be determined by the level of cell hydration. Although each species has an approximate maximum life span under the best conditions of storage, we do not know if all the cells die at the

(eds M. Black, K.J. Bradford and J. Vázquez-Ramos)

same rate. The embryos of the desert-dry *Canna* lily seeds are possibly the longest lived and authentically carbon-dated dry seeds at 620 ± 60 years that have germinated (Lerman and Cigliano, 1971), whilst those of the air-dispersed willow are said to live less than 2 weeks. A Sacred Lotus embedded in the mud of a Manchurian lake and more recently dated at 700+ years is known to have survived and germinated (Shen-Miller *et al.*, 1983).

We have known for a long time that the DNA of an embryo becomes progressively cleaved as dry seeds age and lose their ability to germinate (Cheah and Osborne, 1978) and that one of the first events to occur in the viable embryo when the seed imbibes water is an active repair of the breaks and lesions, with the restoration of genomic integrity (Osborne *et al.*, 1981). We also know that on imbibition, breaks in the DNA following natural ageing are less readily repaired as the seeds become older mainly because a number of metabolic enzymes and at least two of the DNA repair cohort present within the dry embryo slowly lose function in the dry-stored state. One enzyme, DNA ligase, is not sufficiently synthesized on imbibition of low viability material to restore genomic integrity before other cell degenerative processes dominate. The failure to synthesize the enzyme on imbibition is seen as being due to the loss in activity of DNA ligase in the dry state and to breaks in the ligase DNA coding sequences present when water again becomes available (Elder *et al.*, 1987). DNA polymerases also lose activity with age in dry seeds (Yamaguchi *et al.*, 1978; Coello and Vazquez-Ramos, 1996) so, in this sense, death can be facilitated by the loss of the DNA repair capability. What is not known is the pattern of cleavage fragmentation of the DNA during ageing processes, nor is it known if the fragmentation that proceeds in the dry dead embryo continues with a similar pattern of fragmentation when the cells are hydrated. In this respect, the similarity in loss of viability by seeds aged dry and those subjected to enhanced moisture levels and increased temperatures has been a subject of much discussion and controversy over the years (Priestley, 1986).

In studies of loss of viability, the holding of seeds at enhanced hydration levels and at elevated temperatures that preclude germination (but are permissive to protease, nuclease and other hydrolytic enzyme activities) leads to significant accelerations of embryo death. Seeds that are ‘accelerated-aged’ in this way have often been used experimentally as model systems for studying the ageing of relatively long-lived species such as wheat (Dell’Aquila, 1994). In this way, the rate of loss of viability can be speeded and the time to lose viability can be reduced from 30 to 40 years in the dry state to less than 28 days under the high temperature, high humidity conditions of accelerated ageing.

The supply of limited amounts of water to a dry seed is not however necessarily harmful and under controlled conditions can actually improve subsequent germination performance. The concept that the embryo can be held to a water potential that precludes germination, but permits certain metabolic processes to proceed unhindered is the basis of ‘priming’. It is also the basis of longevity of the dormant seed under imbibed conditions in the soil. Both the imposed (controlled) and the natural homeostasis that this entails, permits the protein, RNA and lipid synthesis of the maintenance cell metabolism to proceed (Cuming and Osborne, 1978), but blocks the onset of cell cycling, DNA replication and growth.

Of the cell maintenance that operates under these conditions, we would see the enzymes of the DNA repair complex as critical to embryo survival. They certainly operate during priming of leek (Ashraf and Bray, 1993) and in dormant embryos of *Avena fatua* (Elder and Osborne, 1993). It seems more than likely that they also operate for survival during the many (over 90) wetting and drying periods experienced by desert species such as *Blepharis* that alternate each day between morning dew and extreme noonday heat (Gutterman, 1993).

So far, changes in enzyme activities (other than those of the DNA repair system) have not been found consistently for embryo cells dying in the dry or under accelerated ageing conditions; water binding, as measured by spin-echo NMR appears unaffected; values for the levels of polyunsaturation of lipids are contradictory between different researchers using different species and results for a role for free radicals, including singlet oxygen and excited carbonyl groups, have also proved equivocal (Priestley, 1986).

The present study described here set out to determine whether the loss in DNA integrity that occurs consistently during loss of viability in embryos of seeds of rye held dry or under the enhanced hydration conditions of accelerated ageing is attributable to an apoptopic-type senescence with a formation of nucleosome oligomers as break-down products of chromosomal DNA (as in animal systems) or is, by definition, the result of a non-apoptotic and mainly random fragment length catastrophic DNA cleavage. Both DNA-degrading mechanisms are potentially capable of destroying genome integrity and bringing about cell death if the DNA repair process does not function.

Our results show that both mechanisms can operate in living, non-germinating embryos but the type of nuclease activity that occurs is determined by the degree of the hydration to which the embryo cells are exposed.

Material and Methods

Plant source

Rye (*Secale cereale* var. Rheidol) grains of 96% viability were a gift from Dr P.I. Payne of Plant Breeding International, Cambridge, UK. The 7-year-dead rye had 0% germination since 1991 after a 4-year period of dry storage at ambient temperature. Isolated embryos were hand dissected from the dry grain and used at once for each experiment.

Accelerated ageing of seeds

Accelerated ageing was achieved by suspending the seeds over a saturated NaCl solution (giving 75% relative humidity) in closed Kilner jars maintained in an incubator at 40°C. Isolated embryos were weighed before and at different times after hydration for calculation of water uptake at each time.

Irradiation

Nucleosome lysis solution and isolated embryos from dry seed were gamma-irradiated with 500 and 750 Gy respectively (dose rate 0.14 Gy s^{-1}) from a ^{137}Cs-source of a Gravitron RX 30/55 Irradiator (Gravatom, UK).

DNA extraction

For DNA isolation, a commercial Genomic DNA kit (InViSorb™, Bioline, UK) was used. The content of DNA for each isolated sample was quantified at 260/280 nm in a Spectrophotometer CE-4400 (Cecil Instruments, UK) or at 550 nm in a Luminescence Spectrometer L550B (Perkin Elmer, UK) using the DNA-specific dye PicoGreen™ (Molecular Probes Europe, The Netherlands).

Nuclear DNA contents

Embryos were fixed overnight in ethanol:glacial acetic acid (3:1 v/v), then hydrolysed in 5 M HCl for 30 min at room temperature, rinsed in water, stained in Feulgen reagent (BDH, UK) for 60 min at 26°C, rinsed with 0.5% potassium metabisulphite in 0.05 M HCl, then with water, followed by 10 min in 45% acetic acid. After rinsing in water, embryos were squashed on slides, air dried and mounted in Canada Balsam (Sigma, UK). The density and area of at least 200 nuclei for five embryos was measured on a M85 Scanning Microdensitometer (Vickers, UK) at 550 nm, mask size 4, against adjacent non-nuclear cytoplasmic areas. Values are expressed in relative DNA units.

Nucleosome analysis

Nucleosome contents were measured using a Nucleosome ELISA kit (Oncogene Research Products, Calbiochem, USA). For this, 55 mg of embryos were ground in dry ice and lysed in 750 μl of the buffer supplied. Samples were diluted 1:8 and 1:16 for immunological analysis. Data are presented as units of nucleosomes per 1 ml of original lysis solution calculated from kit standards. DNA content per ml did not differ by more than 2% between samples.

Results

The moisture contents of the embryos held dry were below 9% for both the viable and the 7-year-dead material. On transferring seed from the dry to accelerated ageing conditions at 40°C for 15 to 19 days, the moisture contents increased to 13–14% for both the rye embryo samples (Table 19.1) Feulgen analysis of the 2C DNA values of these embryos either dry, after accelerated ageing or after transfer of embryos from accelerated aged seeds to water for 20 h, shows that the total DNA per nucleus remained unchanged – there was no loss of DNA per nucleus resulting from the treatments (Table 19.2).

When DNA was extracted from dry viable and dry 7-year-dead rye embryos and subjected to electrophoretic fractionation (Fig. 19.1A, lanes 2 and 3 respectively) the enhanced random fragmentation in dead embryos is clearly visible. Imbibition of viable embryos for 3 h leads to immediate DNA repair of DNA breaks and restitution to high molecular weight DNA (Elder and Osborne, 1993), but when 7-year-dead embryos were imbibed for 3 h (lane 4) a visible accumulation of DNA fragment multimers is seen superimposed upon the random fragment background. These multimers are not seen in DNA of viable embryos when they receive water because of the immediate operation of a DNA repair process.

When seeds of dry viable (Fig. 19.1B, lane 2) and dry, 7-year-dead (lane 3) rye were put under conditions of accelerated ageing for 19 days and the DNA then extracted and run on gels we see that viable (now dead) embryos acquired multimers of fixed-sized fragment breaks in their DNA (lane 4) which were stable for at least 70 days when these embryos were dried back from *c.* 14% to less than 9% moisture content (lane 5). No multimers were formed when the 7-year-dead embryos were accelerated aged (lane 6) – only further random-sized fragmentation took place (compare lanes 2 and 6). In the originally alive (viable) embryos held at *c.* 14% moisture content, therefore, an additional nuclease activity was initiated, which did not arise in 7-year-dead material unless these embryos were transferred to water (Fig. 19.1A, lane 4).

Since there is no evidence for protein synthesis in live embryos at *c.* 14% moisture content and none is possible in the 7-year-dead embryos, the

Table 19.1. Moisture content (% H_2O) of dry and accelerated aged viable and 7-year-dead embryos of rye.

Sample	Air dry	Accelerated aged
Viable	8.6 ± 0.1	13.5 ± 0.5
7-year-dead	5.9 ± 0.2	12.7 ± 0.9

Table 19.2. DNA content per 2C nucleus in different samples of rye embryos.

Sample	Conditions	DNA content (Feulgen units)
Viable	Dry	22.5 ± 5.0
Viable	Accelerated aged, 19 days	22.5 ± 6.7
Viable	Accelerated aged, 19 days, Imbibed, 20 h	25.8 ± 6.4
7-year-dead	Dry	27.1 ± 4.4
7-year-dead	Accelerated aged, 19 days, Imbibed, 20 h	23.3 ± 5.6

formation of the DNA multimers in originally viable embryos has to be by some metabolic activation process that can be shut down again when the dead embryos are returned to the dry state of < 9% moisture content.

We already know that rye DNA has a 200 bp nucleosome structure with a 140 bp core and *c.* 60 bp H1 linker region which is, in fact, indistinguishable

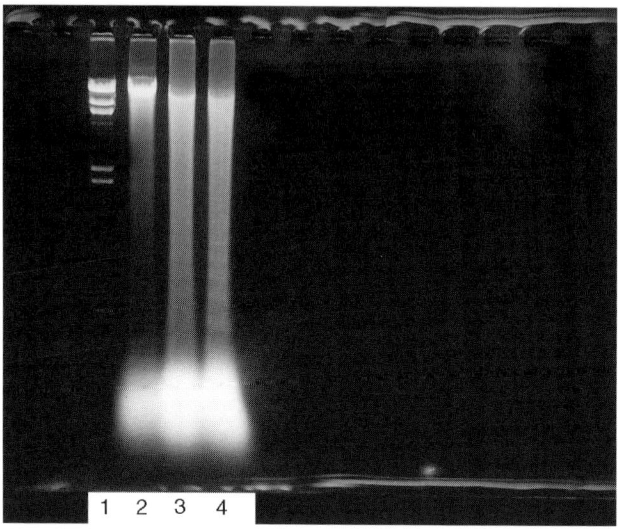

Fig. 19.1A. Electrophoretic fractionation of DNA from 7-year-dead, dry, and imbibed embryos (neutral agarose gel). Lane 1, *Hind*III digest of λ DNA; lane 2, viable dry (control); lane 3, 7-year-dead, dry; lane 4, 7-year-dead, imbibed, 3 h.

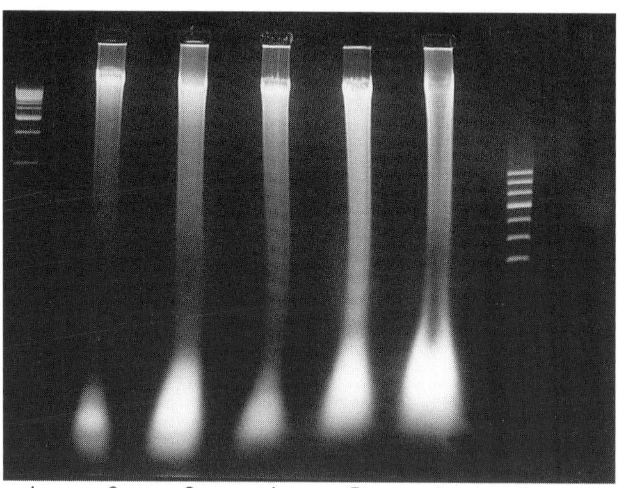

Fig. 19.1B. Electrophoretic fractionation of DNA from viable and 7-year-dead embryos after accelerated aged treatment (neutral agarose gel). Lane 1, Bioladder 1 kb (1,000–12,000 bp); lane 2, viable, dry; lane 3, 7-year-dead, dry; lane 4, viable, accelerated aged; lane 5, viable, accelerated aged, stored dry, 70 days; lane 6, 7-year-dead, accelerated aged; lane 7, Bioladder 200 (800–2000 bp).

from that of rat. The nucleosome profiles, prepared from isolated nuclei of viable and 7-year-dead embryos by monococcal nuclease digestion, are also indistinguishable from each other (Cheah and Osborne, 1977).

By comparing nucleosome contents from lysates of embryos of dry and accelerated aged seeds we have now shown that the appearance of DNA multimers in the embryos of viable seeds when accelerated aged (Table 19.3) can be accounted for by the metabolic activation of a H1 cleaving nuclease with the liberation of nucleosome fragments. This activation does not occur in the embryos of seeds held under the dry conditions (Table 19.4), nor does it occur under accelerated ageing condition in the embryos of dead seeds (Table 19.5). Only live embryos activate a nuclease-generated fragmentation of DNA under accelerated ageing conditions (*c.* 14% moisture content) which, we suggest, is then accompanied by an apoptotic-type cell death.

The nucleosome fragmentation observed in water after 3 h (Fig. 19.1) or 20 h (Tables 19.3, 19.4) with the 7-year-dead embryos must represent either a non-metabolic H1 nuclease activation or a free-water-induced change in chromatin conformation that renders the DNA accessible to nucleosome fragmentation. There is no formation of nucleosome multimers when either the lysis embryo solution or the dry embryos of viable seeds are subjected to gamma-irradiation levels that induce random DNA fragmentation (Table 19.4).

Table 19.3. Nucleosome contents in viable embryos of rye seed under accelerated aged conditions.

Seed treatment	Nucleosome content (U ml^{-1}*)
Dry	3.6 ± 1.4
Accelerated aged, 15 days	11.6 ± 2.4
Accelerated aged, 19 days	14.4 ± 3.4
Accelerated aged, 19 days, Stored dry, 70 days	16.0 ± 2.0
Accelerated aged, 19 days, Imbibed, 20 h	26.8 ± 4.8

*Threshold for significant nucleosome detection 5.7 U ml^{-1}.

Table 19.4. Comparison of nucleosome contents in embryos of viable and 7-year-dead rye seed.

Sample	Treatment	Nucleosome content (U ml^{-1}*)
Viable	Dry	3.7 ± 1.8
Viable	Irradiation of nucleosome lysis solution, 500 Gy	4.6 ± 0.9
Viable	Irradiation of dry embryos, 750 Gy	4.5 ± 2.0
Viable	Imbibed, 20 h	5.1 ± 1.7
7-year-dead	Dry	5.9 ± 1.2
7-year-dead	Imbibed, 20 h	19.0 ± 2.3

*Threshold for significant nucleosome detection 5.7 U ml^{-1}.

Discussion

One outcome of these experiments is the clear evidence that the type of DNA degradation that takes place as the cells of the embryos die is dictated by the extent of hydration of the cells. The dry state (< 9% moisture content) permits at least one nuclease action, which with time leads to progressive random DNA cleavage giving fragmentation to varying sized poly- and oligo-nucleotides. In contrast, an accelerated ageing regime (75% RH at 40°C) which increases the moisture content of the embryo to *c.* 14% (a value insufficient for germination of viable embryos) activates another form of nuclease that must pre-exist in the cells from the time of maturation upon the mother plant; this activity leads to a 3–4 times increase in the nucleosome content and the early death of the embryo (Table 19.3). When embryos that are already dead for 7 years are subjected to similar accelerated ageing conditions, no nucleosomes are generated even after 19 days (Table 19.5). This indicates that an H1 cleaving nuclease is activated at 14% moisture content only in live embryos, which implies some metabolic involvement within viable embryos. However, the 7-year-dead rye, which shows only random fragmentation of DNA when dry or accelerated aged, does produce multiple nucleosomes when exposed to free water for 20 h (Table 19.4). In the complete absence of metabolic activity in such embryos (and the impossibility of any new protein synthesis), the H1 cleaving process can be caused to operate at high levels of hydration.

Since nucleosome generation can result only from a specific nuclease function that must pre-exist in the dry embryo, it would seem that the hydration of an accelerated aged condition is sufficient to activate the specific nucleosome-forming nuclease only in live embryos. Two possibilities may explain this. Either the conformation of DNA in the dry embryo makes it inaccessible to the nucleosome-generating nuclease, or the nuclease itself is rendered inactive at low levels of moisture (< 9%). The latter perhaps seems the more likely, from the evidence that accelerated aged material continues to show random DNA cleavage when these dead embryos are transferred back from accelerated aged to the dry state whilst the total nucleosome level achieved during accelerated ageing remains essentially unchanged. Only on imbibition in water will the random molecular weight DNA fragments in dead embryos be degraded further to nucleosome multimers.

Nucleases are remarkably stable enzymes, particularly in dry plant material. We know, for example, that the 103-year-old caryopses of wheat,

Table 19.5. Nucleosome content in 7-year-dead embryos of rye seed under accelerated aged conditions.

Seed treatment	Nucleosome content (U ml^{-1}*)
Dry	4.8 ± 1.6
Accelerated aged, 19 days	3.1 ± 1.9
Accelerated aged,19 days, Imbibed, 20 h	21.3 ± 4.4

*Threshold for significant nucleosome detection 5.7 U ml^{-1}.

which must have been non-viable for at least 60 years, yielded active nucleases (Osborne *et al.*, 1974). Since there is no new synthesis in dead embryos, hydration changes to the chromatin supercoiling and conformation requiring free water may be the permissive event that allows the nucleosome nuclease to function.

The transition of cytoplasmic water from a tenaciously bound glassy state at < 9% moisture content to that of weakly bound (or free) water of accelerated aged conditions (*c.* 14%) may be conducive to such change (Seewaldt *et al.*, 1981; Bruni and Leopold, 1991).

Whether or not the cells of plant embryos operate caspase activated events, which lead to apoptotic death and nucleosome production remains to be discovered. Clearly, however, the cells of embryos can proceed in two ways to cell death; the dry-regulated process of loss of DNA integrity which is slow (in rye, 3–4 years) or the rapid and apparently apoptotic accelerated ageing pathway that can be complete in 15–19 days. It seems clear that both a catastrophic and an apoptotic-like death are part of the repertoire of rye embryos and accelerated aged seeds cannot act as a model system for the normal loss of viability in seeds held dry.

This now raises the question: is there a commonality between priming and accelerated ageing? We suggest that once the water activity in the cell reaches that permissive for DNA repair, for however short a period (30 min is sufficient), an embryo will start to restore its genomic integrity. But if this restoration is then overtaken by an enhanced nucleosome DNAase activity a decline in viability will then follow and the cells will become progressively apoptotically aged.

The borderline between these two activities will depend upon the intimate water relations and temperatures that each of the embryo cells experiences. The continued maintenance of conditions that favour activation of nucleosome cleavage will, we suggest, lead to a final apoptotic embryo cell death.

Conclusions

Seeds held at a moisture content *c.* 14% are at risk of a multiple nuclease degradation of their DNA which is a fast process of H1 fragmentation compared with a slow DNAase nicking seen in the random single strand cleavage of DNA in embryos held at less than 9%. This means that the results of accelerated ageing of seeds cannot be used as a model system to represent a fast achievement of natural ageing. Once accelerated aged seeds are dried back to the moisture content of the normal dry state (< 9%) the nucleosome-generating DNAase activity is again restricted. The nucleosome content then remains essentially unchanged in the presence of a continued random cleavage of high molecular weight DNA by the less aqueous-demanding DNAase that functions in the dry state.

We suggest that there is a critical interplay between the hydration states of these two nuclease activities, when the priming of seed can become directed to accelerated ageing.

References

Ashraf, M. and Bray, C.M. (1993) DNA synthesis in osmoprimed leek (*Allium porrum* L.) seeds and evidence for repair and replication. *Seed Science Research* 3, 15–23.

Bruni, F. and Leopold, A.C. (1991) Cytoplasmic glass formation in maize embryos. *Seed Science Research* 2, 251–253.

Cheah, K.S.E. and Osborne, D.J. (1977) Analysis of nucleosomal deoxyribonucleic acid in a higher plant. *Biochemical Journal* 163, 141–144.

Cheah, K.S.E. and Osborne, D.J. (1978) DNA lesions occur with loss of viability in embryos of ageing rye seed. *Nature* 272, 593–599.

Coello, P. and Vazquez-Ramos, J.M. (1996) Maize DNA polymerase 2 (an alpha-type enzyme) suffers major damage after seed deterioration. *Seed Science Research* 6, 1–7.

Cuming, A.C. and Osborne, D.J. (1978) Membrane turnover in imbibed and dormant embryos of the wild oat (*Avena fatua* L.) I. Protein turnover and membrane replacement. *Planta* 139, 208–217.

Dell'Aquila, A. (1994) Wheat seed aging and embryo protein degradation. *Seed Science Research* 4, 293–298.

Elder, R.H. and Osborne, D.J. (1993) Function of DNA synthesis and DNA repair in the survival of embryos during early germination and in dormancy. *Seed Science Research* 3, 43–53.

Elder, R.H., Dell'Aquila, A., Mezzina, M., Sarasin, A. and Osborne, D.J. (1987) DNA ligase in repair and replication in the embryos of rye, *Secale cereale*. *Mutation Research* 181, 61–71.

Gutterman, Y. (1993) *Seed Germination in Desert Plants.* Springer-Verlag, Berlin, Heidelberg, Germany, 253 pp.

Lerman, J.C. and Cigliano, E.M. (1971) New carbon-14 evidence for six hundred years old *Canna compacta* seed. *Nature* 252, 568–570.

Osborne, D.J., Roberts, B.E., Payne, P.I. and Sen, S. (1974) Protein synthesis and viability in rye embryos. In: Bieleski, R.L., Ferguson, A.R. and Cresswell, M.M. (eds) *The Royal Society of New Zealand, Bulletin 12.* Wellington, pp. 805–812.

Osborne, D.J., Sharon, R. and Ben-Ishai, R. (1981) Studies on DNA integrity and DNA repair in germinating embryos of rye (*Secale cereale*). *Israel Journal of Botany* 29, 259–272.

Priestley, D.A. (1986) *Seed Aging.* Comstock Publishing Associates, Cornell University Press, Ithaca, New York, 304 pp.

Shen-Miller, J., Schopf, J.W. and Berger, R. (1983) Germination of a ca.700 year old lotus seed from China: evidence of exceptional longevity of seed viability. *American Journal of Botany* 70, 78–80.

Seewaldt, V., Priestley, D.A., Leopold, A.C. and Feigenson, G.W. (1981) Membrane organisation in soybean seeds during hydration. *Planta* 152, 19–23.

Yamaguchi, H., Naito, T. and Tatur, A. (1978) Decreased activity of DNA polymerase in seeds of barley during storage. *Japanese Journal of Genetics* 53, 133–135.

20 The Effect of Drying Rate on Recalcitrant Seeds: 'Lethal Water Contents', Causes of Damage, and Quantification of Recalcitrance

N.W. PAMMENTER[1], P. BERJAK[1] AND C. WALTERS[2]

[1]*School of Life and Environmental Sciences, University of Natal, Durban 4041, South Africa;* [2]*USDA-ARS National Seed Storage Laboratory, 1111 South Mason Street, Fort Collins, CO 80521, USA*

The rate of drying influences the response of recalcitrant seeds to desiccation. Seeds or excised axes that are dried rapidly can survive to lower water contents than those dried slowly. It has been suggested that at intermediate water contents metabolism becomes uncoordinated, leading to damaging oxidative chemical process. At low water contents, damage is biophysical, directly affecting macromolecular structures. Slowly dried material is exposed to high water contents for a period adequate to permit aqueous-based chemical oxidative processes to occur, and so loses viability at a high water content. Rapidly dried material passes through this water content range before excessive oxidative damage accumulates and is killed by biophysical processes occurring at lower water contents. The time component of drying cannot be ignored, and it is not possible to identify a specific or damaging water content to describe the desiccation tolerance of a seed species.

Introduction

Recalcitrant seeds are shed at high water contents, are metabolically active at shedding (Farrant *et al.*, 1985; Berjak *et al.*, 1993; Salmen Espindola *et al.*, 1994) and are desiccation-sensitive. There is, however, considerable variation in the post-harvest physiology of these seeds, with the 'lethal' or 'injurious' water contents reported in the literature varying widely within and among species. This variation in post-harvest physiology led to the concept of a continuum of recalcitrant seed behaviour (Farrant *et al.*, 1988), ranging from extremely to moderately sensitive to desiccation.

It is now apparent that a further factor contributing to the observed variability in the response of desiccation-sensitive seeds to dehydration is the rate of drying. Generally, the faster the drying rate, the lower the water content that can be tolerated. This is particularly noticeable when isolated axes are

 Seed Biology: Advances and Applications (eds M. Black, K.J. Bradford and J. Vázquez-Ramos)

dried (Normah *et al.*, 1986; Fu *et al.*, 1990; Pammenter *et al.*, 1991; Berjak *et al.*, 1993) and it has also been observed, although to a much lesser extent, in whole seeds (Farrant *et al.*, 1985; Pritchard, 1991; Pammenter *et al.*, 1998), but it is not always apparent (Finch-Savage, 1992).

This paper presents data pertinent to the questions of whether the effect of drying rate is widespread, and if so, what could be the cause of the effect and what this means in terms of quantification of seed recalcitrance.

Materials and Methods

Axes excised from seeds of *Trichilia dregeana* Sond., *Castanospermum australe* A. Cunn. and *Camellia sinensis* (L.) O. Kuntze (tea) and whole seeds of *Ekebergia capensis* Sparrm. were dehydrated at a range of drying rates at room temperature. After various drying treatments viability was assessed and water contents (on a dry mass basis (g water (g dry mass)$^{-1}$, g g^{-1})) were determined.

Results

For both *T. dregeana* and *C. australe* the faster the isolated axes were dried, the lower the water content that was reached before viability was completely lost (Table 20.1). From these 'lethal' water content data it might appear that *T. dregeana* is more desiccation-tolerant than *C. australe.* Similar responses to drying rate were observed with isolated axes of tea and with whole seeds of *E. capensis* dried at two rates. In the case of tea (Fig. 20.1) axes dried at the rapid initial rate of 1.29 g water (g dry mass)$^{-1}$ h^{-1} survived to water contents of about 0.4 g g^{-1} before significant loss of viability occurred. At slower drying rates, equivalent viability loss occurred at water contents around 1.0 g g^{-1}, and the slower the drying rate the more deleterious was further dehydration. Whole seeds of *E. capensis* (Fig. 20.2) survived to axis water contents of about 0.7 g g^{-1} when dried rapidly (over 24 h), whereas with seeds dried over 10 days viability declined at axis water contents below 1.3 g g^{-1}.

Table 20.1. The water content (WC) at the stage at which germination had fallen to zero in axes of *Trichilia dregeana* and *Castanospermum australe* dried at different rates under different relative humidities. The time required to reach zero germination is also indicated.

	Trichilia dregeana		*Castanospermum australe*	
Drying treatment	WC (g g^{-1})	Time (day)	WC (g g^{-1})	Time (day)
Silica gel	0.09	< 1	0.19	0.33
75% RH	0.11	1	0.28	4
92% RH	0.22	2		
96% RH	0.60	9	0.98	17

The time factor in drying experiments becomes particularly apparent if the viability as well as the water content is plotted on a conventional drying time course. Figure 20.3 shows these data for axes of *T. dregeana* dried over silica gel or at 96% RH. For both drying rates, axis water content dropped too close to its equilibrium value before viability showed a marked decline. For axes dried over silica gel, the relative water content (relative to freshly shed material) declined to 0.17 after 0.17 days (4 h), and thereafter declined only slightly. At 0.17 days viability was 80–90% and did not decline until after 0.25 days (6 h). When axes were dried at 96% RH equilibrium water content (RWC 0.4)

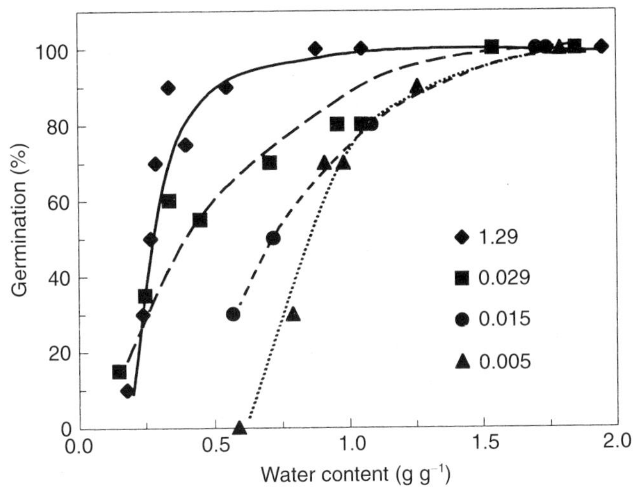

Fig. 20.1. Viability of isolated axes of tea seeds dried at different rates. Initial drying rates (in g water g^{-1} dry matter h^{-1}) are indicated.

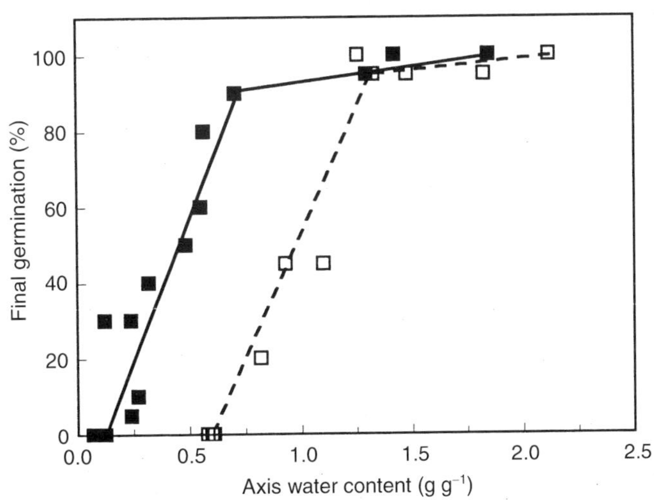

Fig. 20.2. The effect of drying rate (□, slow; ■, fast) on the response to dehydration of whole seeds of *Ekebergia capensis*.

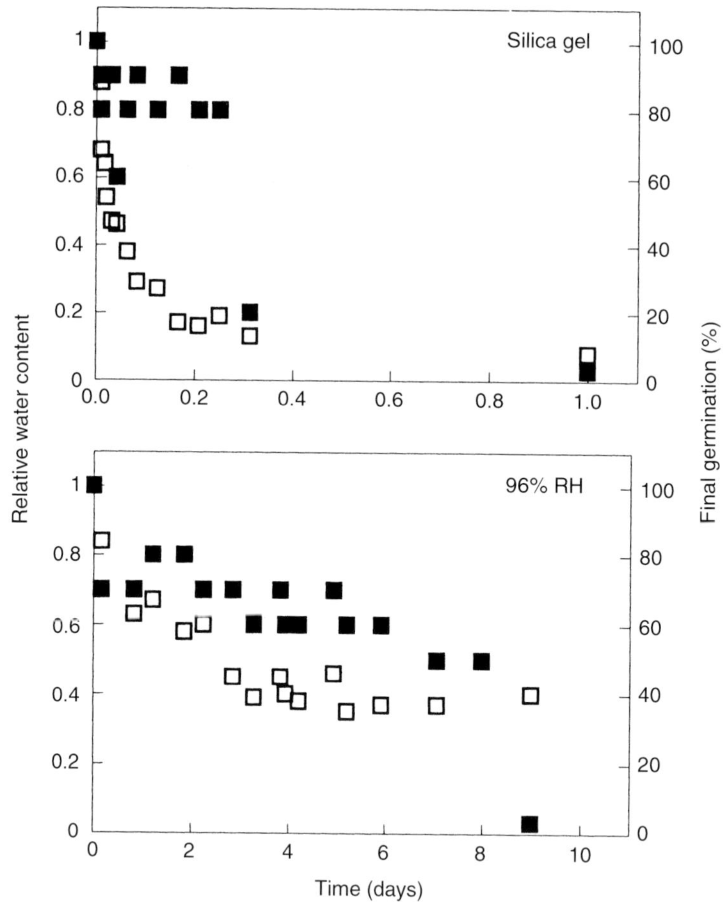

Fig. 20.3. The relative water content (□) and viability (■) of isolated axes of seeds of *Trichilia dregeana* dried either over silica gel (upper panel) or at 96% relative humidity (lower panel). Relative water content is calculated relative to the water content at shedding.

was achieved after about 3.5 days. Viability showed a very different pattern, declining gradually for about 8 days, and then dropping precipitously.

The data presented in Fig. 20.3 effectively show a short-term storage response of partially dehydrated axes. This phenomenon was more closely examined with tea axes. Isolated axes were rapidly dried to a range of water contents and then stored under conditions precluding further water loss for 10 days. At each water content, viability was assessed before and after storage, and the rate of viability loss during storage (% day^{-1}) calculated (Fig. 20.4). There appear to be three water content ranges where the pattern of viability loss as function of water content differed. In the high water content range, from about 0.9 g g^{-1} upwards, as water content increased the rate of deterioration declined. In an intermediate water content range, from about 0.5 to 0.8 g g^{-1}, the reverse occurred: as water content increased the rate of deterioration

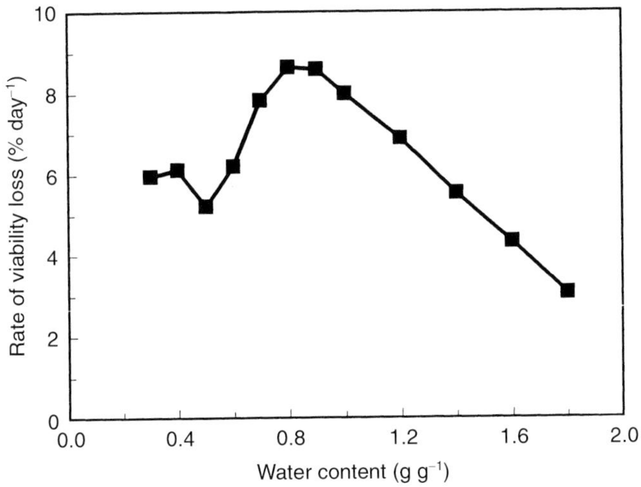

Fig. 20.4. The rate of viability of loss of isolated axes of tea seeds dried to a range of water contents and stored for 10 days.

increased. At water contents below 0.5 g g^{-1} the pattern is not as clear; it was difficult to calculate rates of viability loss during storage because dehydration to these water contents reduced viability to low levels prior to storage. These different patterns of viability loss in the different water content ranges suggest that different deleterious processes may be leading to the loss of viability in the different water content ranges.

Discussion

The data for four recalcitrant seed species confirm that the rate of drying influences the response to dehydration. The effect was shown with whole seeds of *E. capensis* as well as isolated axes of the other three species, and so is not an artefact of removing the embryonic axis from the rest of the seed, as was suggested by Finch-Savage (1992).

A number of processes have been suggested to be involved in the loss of viability of desiccation-sensitive material, and different processes may predominate in different water content ranges (reviewed by Vertucci and Farrant, 1995; Pammenter and Berjak, 1999). At intermediate water contents, uncontrolled oxidative reactions, dependent upon metabolism, can occur, leading to desiccation damage (Leprince *et al.*, 1996). At lower hydration levels, the remaining water is non-freezable and is associated with macromolecular structures. Its removal can lead to conformational changes that may be irreversible and damaging. Thus, at least two types of damage that can occur on drying recalcitrant seeds are envisaged: strict desiccation damage that occurs when sufficient water is removed to lead to damage to macromolecular structures, and aqueous-based oxidative damage that occurs at intermediate water contents and is a consequence of unregulated metabolism. The different patterns of viability loss when tea axes at different water contents are subjected to short-term storage (Fig. 20.4) is entirely consistent with the hypothesis that different deleterious processes occur in different water content ranges. With respect to the effect of the rate of drying, rapidly dried tissue can survive in the short term to lower water contents because the tissue spends insufficient time at intermediate water contents for damage consequent upon the deleterious aqueous-based reactions to accumulate; the lethal process that kills rapidly dried isolated axes that have successfully passed through the intermediate water contents is the removal of structure-associated water.

Although different species having recalcitrant seeds are considered to constitute a continuum in terms of post-harvest physiology (Farrant *et al.*, 1998), the influence of drying rate on the response to dehydration means that it is not possible to identify a specific water content (or water potential) corresponding to desiccation damage. Indeed, the data from Fig. 20.3 indicate that if material is dried under conditions such that an equilibrium water content is reached, the loss of viability recorded will depend on how long the material has been at that water content. Thus, during dehydration experiments, the time component of the experiment cannot be ignored, and this makes it difficult to quantify desiccation sensitivity and compare this parameter among species.

There is a possible alternative approach to quantifying the degree of recalcitrance that does not depend upon an absolute specific water content corresponding to desiccation damage. When axes of *T. dregeana* were dried at 96% RH, they reached equilibrium water content after about 3.5 days, at which point the axes were 60% viable, but total viability loss did not occur until 9 days (Fig. 20.3). By contrast, axes of *C. australe* reached the equilibrium water content at 96% RH after 8 days, at which point viability was 70%. Total viability loss occurred after 17 days (data not shown). Thus *T. dregeana* axes survived at the water content in equilibrium with 96% RH for about 5.5 days, whereas those of *C. australe* survived similar conditions for 9 days. On this basis, *C. australe* could be considered to be less recalcitrant than *T. dregeana*, although the water contents at which viability was lost under the different drying treatments might indicate the opposite (Table 20.1). If aqueous-based oxidative processes constitute the major cause of damage under 'normal' drying conditions, the rate at which these proceed may be a better measure of desiccation sensitivity than the water content at which viability loss occurred (which is drying rate dependent). Perhaps recalcitrance could be quantified in terms of the rate of viability loss of material dehydrated to specified water potentials, rather than in terms of an apparent critical water content corresponding to desiccation damage.

References

Berjak, P., Vertucci, C.W. and Pammenter, N.W. (1993) Effects of developmental status and dehydration rate on characteristics of water and desiccation-sensitivity in recalcitrant seeds of *Camellia sinensis*. *Seed Science Research* 3, 155–166.

Farrant, J.M., Berjak, P. and Pammenter, N.W. (1985) The effect of drying rate on viability retention of propagules of *Avicennia marina. South African Journal of Botany* 51, 432–438.

Farrant, J.M., Pammenter, N.W. and Berjak, P. (1988) Recalcitrance – a current assessment. *Seed Science and Technology* 16, 155–166.

Finch-Savage, W.E. (1992) Embryo water status and survival in the recalcitrant species *Quercus robur* L.: evidence for a critical moisture content. *Journal of Experimental Botany* 43, 663–669.

Fu, J.R., Zhang, B.Z., Wang, X.P., Qiao, Y.Z. and Huang, X.L. (1990) Physiological studies on desiccation, wet storage and cryopreservation of recalcitrant seeds of three fruit species and their excised embryonic axes. *Seed Science and Technology* 18, 743–754.

Leprince, O., Hendry, G.A.F., Atherton, N.M. and Vertucci-Walters, C. (1996) Free radicals and metabolism associated with acquisition and loss of desiccation tolerance in developing seeds. *Biochemical Society Transactions* 24, 451–455.

Normah, M.N., Chin, H.F. and Hor, Y.L. (1986) Desiccation and cryostorage of embryonic axes of *Hevea brasiliensis* Muell.-Arg. *Pertanika* 9, 299–303.

Pammenter, N.W. and Berjak, P. (1999) A review of recalcitrant seed physiology in relation to desiccation-tolerance mechanisms. *Seed Science Research* 9, 13–37.

Pammenter, N.W., Vertucci, C.W. and Berjak, P. (1991) Homeohydrous (recalcitrant) seeds: dehydration, the state of water and viability characteristics in *Landolphia kirkii. Plant Physiology* 96, 1093–1098.

Pammenter, N.W., Greggains, V., Kioko, J.L., Wesley-Smith, J., Berjak, P. and Finch-Savage, W.E. (1998) Effects of differential drying rates on viability retention of seeds of *Ekebergia capensis. Seed Science Research* 8, 463–471.

Pritchard, H.W. (1991) Water potential and embryonic axis viability in recalcitrant seeds of *Quercus rubra. Annals of Botany* 67, 43–49.

Salmen Espindola, L., Noin, M., Corbineau, F. and Côme, D. (1994) Cellular and metabolic damage induced by desiccation in recalcitrant *Araucaria angustifolia* embryos. *Seed Science Research* 4, 193–201.

Vertucci, C.W. and Farrant, J.M. (1995) Acquisition and loss of desiccation tolerance. In: Kigel, J. and Galili, G. (eds) *Seed Development and Germination.* Marcel Dekker, Inc., New York, pp. 237–271.

21 Conservation of Genetic Resources Naturally Occurring as Recalcitrant Seeds

P. BERJAK[1], D.J. MYCOCK[3], M. WALKER[1], J.I. KIOKO[1], N.W. PAMMENTER[1] AND J. WESLEY-SMITH[1,2]

[1]School of Life and Environmental Sciences, [2]Electron Microscope Unit, University of Natal, Durban 4041, South Africa; [3]Botany Department, University of the Witwatersrand, Johannesburg 2050, South Africa

Conservation of the genetic resources of many species is effectively precluded by their production of seeds that are hydrated, desiccation-sensitive and storable only for the short-term. The only means of conservation of such species has been, and largely still is, in the form of the growing plants in field genebanks. In the relatively recent past, however, efforts have been made to cryopreserve the germplasm of many species, but, with only a few exceptions, these have met with either limited, or no, success. Reasons for this are broadly examined, and particularized in the case of the pedunculate oak, for which the bases of lethal vs. successful handling of excised zygotic axes are analysed.

Introduction

Intensive discussions that took place 20 years ago had already noted that conservation of the genetic resources of recalcitrant seeds by storage of these propagules would probably be possible only for the short-term, and that cryostorage (in, e.g. liquid nitrogen) of suitable explants appeared to be the only alternative (Chin and Roberts, 1980; Withers and Williams, 1980). At that time, little was known about the basis of recalcitrance in seeds, other than that they were all more or less desiccation-sensitive, but since then our understanding of the phenomenon, while certainly not yet complete, has increased substantially (Pammenter and Berjak, 1999). Based on the discovery that although it was practically impossible even to consider cryostorage of intact recalcitrant seeds of most species, excised zygotic axes would retain viability to water contents sufficiently low to facilitate their being frozen, there have been concerted efforts to cryopreserve these explants.

Success has, however, been far from widespread, and in many cases difficult to assess from the reported results. The only real measure of successful

 Seed Biology: Advances and Applications (eds M. Black, K.J. Bradford and J. Vázquez-Ramos)

cryopreservation is the ultimate production of vigorous, normal seedlings rooted in soil. However, much reported survival is recorded as root development and/or greening in culture and, even when shoot development is recorded as being initiated under *in vitro* conditions, the ultimate test of the production of hardened-off plants is not. We have found for a variety of species that while onwards growth of the radicle might occur, shoot development does not follow (unpublished observations). In a few cases, however, variably successful plantlet formation following axis cryopreservation has been reported, e.g. for *Hevea brasiliensis* (Normah *et al.*, 1986); *Euphoria longan* (Fu *et al.*, 1990); *Cocos nucifera* (Assy-Bah and Engelmann, 1992); *Calamus manan* (Krishnapillay *et al.*, 1992); *Artocarpus heterophyllus* (Chandel *et al.*, 1995; Thammasiri, 1999); and *Camellia japonica* (Janeiro *et al.*, 1996).

One of the persistent inadequacies of much of the work on cryopreservation of recalcitrant zygotic axes is the lack of analysis of factors underlying relative success or failure of particular protocols applied to individual species. In the case of *Quercus* spp., while some callus production after cryopreservation of zygotic axes might be recorded, there have been no reports to date of seedling formation (Pence, 1990; Poulsen, 1992; Chmielarz, 1997). In view of encouraging results that we obtained for *Q. robur* L., we examined the root and shoot apices microscopically at all the critical stages during axis manipulation for, and retrieval from, cryostorage, and compared these observations with parallel results obtained for axes treated according to two reported protocols (Poulsen, 1992; Chmielarz, 1997) for which those authors reported little or no survival.

Materials and Methods

Seeds of *Q. robur* were obtained both locally and imported from the UK. In both cases, the seeds were collected soon after shedding and consigned to Durban in the minimum time, under conditions not permitting water loss. As *Q. robur* seeds can be successfully stored for some months if they are maintained hydrated and under refrigeration (Chin and Roberts, 1980), the seeds were stored for short periods at 6°C, although the experiments were performed as soon as possible.

Details of surface-sterilization, dehydration, cooling (freezing) and retrieval from cryostorage appear elsewhere (Berjak *et al.*, 1999), hence are only summarized here. The three protocols used are referred to as N (Natal (our protocol), P (Poulsen, 1992) or C (Chmielarz, 1997). Excised axes were briefly surface sterilized with 1% NaOCl (N) or 0.1% $HgCl_2$ (C) or subjected to a multi-step procedure, including anti-oxidant treatments, following preliminary treatment of the entire embryo (P). Thereafter, manipulation involved fast flash-drying of axes for 2–4 h (N), laminar air-flow dehydration for 8 h (P) or cryoprotectant treatment and encapsulation followed by dehydration for 21 h (C). Axes prepared to this point by all three protocols were then subjected to cooling (freezing) ultra-rapidly in isopentane held in a liquid nitrogen reservoir (N; Wesley-Smith *et al.*, 1999), or in batches of five enclosed in cryotubes which were plunged into liquid nitrogen, or by slow/stepwise cooling (1°C min^{-1} to −38°C (P) or −20°C (C)). In all cases, initial cooling was immediately followed by immersion in liquid nitrogen. Axes were retrieved from cryostorage and briefly immersed in distilled water at 40°C, as is common practice, and, in a pilot experiment with those processed according to the N protocol, also into a 1:1 solution of 1 μM $CaCl_2$ and 1 mM $MgCl_2$ for 2–40 min at room temperature. After each step in the protocols, survival was monitored over 30 days in culture, while in all cases axes for light and electron microscopy were sampled after a 6 day recovery period *in vitro*.

Results and Discussion

Surface sterilization is a vital step for any material that is ultimately to be cultured *in vitro*. While all three procedures used eliminated the fungi that were invariably located between the shoot apex and leaf bases in newly-excised axes, treatment with $HgCl_2$ (C) proved least injurious, while considerably more damage to both the shoot apical meristem and the root cap was associated with use of the rigorous procedure of the P protocol. As full germinability of the axes was retained notwithstanding any of the surface-sterilization procedures, these treatments alone did not compromise viability: however, damage, particularly to the protective root cap, caused during the protracted P procedure, is held to have pre-disposed the axes to lethal injury when introduced into the cryogen (see below). In principle, therefore, in cryopreservation trials the most gentle, but effective, surface-sterilant possible should be identified before any further manipulations of the axes are attempted.

There were considerable differences in the duration and rate of dehydration of the axes among the three protocols. Fast flash-drying (N), using the apparatus described by Wesley-Smith *et al.* (1999), achieved axis water contents averaging 0.27 g g^{-1} (dmb) after 4 h, while axes subjected to laminar air-flow drying for double this time (P) reached an average water content of 0.32 g g^{-1}. Following initial dehydrating pre-treatment (below), axes dried using the C protocol (laminar air-flow, 1 h; silica gel exposure, 20 h) were at an average of 0.19 g g^{-1}. Even excepting the cryoprotectant pre-treatment used by Chmielarz (1997), involving stepwise passage through solutions of increasing sucrose concentration and finally use of 1 M glycerol followed by encapsulation in calcium alginate, the longer and more slowly the axes were dehydrated, the more they were damaged. Neither the root cap nor the root or shoot apical meristems of material that had been fast flash-dried showed any marked signs of microscopically-visible abnormality when axes were sampled following the 6 day recovery period, whereas those that had been subjected to the 8 h laminar air-flow dehydration period (P) exhibited persisting abnormality. The deleterious effects of prolonged and excessive dehydration were, however, most vividly displayed by axes manipulated by the C protocol. Extensive degradation of the root cap had occurred, and frequent foci of necrosis were apparent in both root and shoot apical meristems. Nevertheless, ultimate axis viability was not compromised irrespective of the dehydration regime, although the effects of dehydration stress were clearly manifested by decreasing vigour (measured as time taken until shoot development) that was inversely proportional to the drying period. This is in agreement with the findings that during relatively slow dehydration of recalcitrant seeds (or excised axes), the time during which water levels are relatively high affords a period during which injurious effects of unbalanced metabolism are manifested (Pammenter *et al.*, 1998). It follows therefore, that if this period is minimized, then such metabolism-related damage is similarly minimized. Ultimate desiccation damage, which occurs at low water contents (no matter how rapidly these are achieved (Pammenter *et al.*, 1998)) is a further complicating factor that might well have contributed to the damage seen in axes dehydrated down to 0.19 g g^{-1} (C).

In view of the persistent damage and decreased vigour of axes following preparation by the P, and especially the C, protocols, it was not surprising that none survived cryopreservation, no matter how this was achieved. When liquid nitrogen was used as the cryogen, complete devastation of both root and shoot apices occurred, despite preliminary stepwise cooling (P and C). Although not exhibiting such extreme damage, neither did axes prepared by the N protocol survive, which is in agreement with the poor cooling properties of liquid nitrogen (Wesley-Smith *et al.*, 1999).

In contrast, when axes that had been fast flash-dried were plunged into isopentane held in a liquid nitrogen reservoir (N; Wesley-Smith *et al.*, 1999) and subsequently thawed in distilled water, 66% produced vigorous roots in culture. However, two problems persisted: firstly, these roots remained horizontal, and secondly, the shoots failed to develop and ultimately became necrotic. Ultrastructural examination particularly of shoot apices after the 6 day recovery period proved illuminating: while, at this stage, there was evidence of considerable activity, numbers of cells of the apical meristem showed a displacement of organelles to the peri-nuclear area, and frequently the nuclei themselves were atypically featureless and somewhat mis-shapen. These features suggested to us the possibility that elements of the cyto- and nucleo-skeletons had not become reorganized in a manner appropriate to normal onwards development of the shoots. While cells of the root cap and apex did not show similar aberrations, roots nonetheless did not respond normally in terms of gravitropism. Both this, and the intracellular situation of the shoots, suggested to us that use of a thawing medium incorporating the divalent cations Ca^{2+} and Mg^{2+} in concentrations known to promote normal assembly of the intracellular skeletons, might be efficacious. Modifying a procedure developed for the recovery of cryopreserved somatic and zygotic embryos (Mycock, 1999), we instituted a thawing regime using a 1:1 mixture of 1 μM $CaCl_2$ and 1 mM $MgCl_2$.

Immersion of the axes in this solution for only 2 min resulted in significant improvement: 100% of the roots grew vigorously, and all showed strong gravitropic curvature, while for 70% of the axes, the shoots greened, developed normally, and relatively soon produced leaves. At the ultrastructural level, it was apparent that provision of the divalent cations had stimulated the formation of large, spherical or oval amyloplasts in the central columella cells (the statocytes) of the root cap: these organelles – and their cytoskeleton-based orientation – are known to be central to gravity perception and it is also well-established that calcium particularly is implicated (e.g. Baluška *et al.*, 1997). In the shoot apical meristem cells, instead of the disorientation of organelles and often featureless nuclei seen after water thawing, the short exposure to the cation solution promoted organization that was already manifested in intense mitotic activity by 6 days after axis retrieval from cryopreservation. Germinated plantlets have now been rooted in soil, and give the appearance of developing into normal plants.

We are presently experimenting with the post-freezing treatment with the calcium/magnesium solution, and have found that immersion of the *Q. robur* axes for 40 min is optimal. Investigations are also currently in progress to assess the responses of the cytoskeleton (and hopefully also those of the nucleo-skeleton) to various manipulations, and to explain other intracellular phenomena that may be responsive to the post-freezing treatment with the divalent cation solution, and which collectively contribute to the success of cryopreservation of zygotic axes by the N protocol. The technology has also been applied to axes of two other recalcitrant species – this time of tropical origin – with similarly promising results.

References

Assy-Bah, B. and Engelmann, F. (1992) Cryopreservation of mature embryos of coconut (*Cocos nucifera* L.) and subsequent regeneration of plantlets. *CryoLetters* 13, 117–126.

Baluška, F., Kreibaum, A., Vitha, S., Parker, J.S., Barlow, P.W. and Sievers, A. (1997) Central root cap cells are depleted of endoplasmic microtubules and actin microfilament bundles: implications for their role as gravity-sensing statocytes. *Protoplasma* 196, 212–223.

Berjak, P., Walker, M., Watt, M.P. and Mycock, D.J. (1999) Experimental parameters underlying failure or success in plant germplasm cryopreservation: a case study on zygotic axes of *Quercus robur* L. *CryoLetters* (in press).

Chandel, K.P.S., Chaudhury, R., Radhamani, J. and Malik, S.K. (1995) Desiccation and freezing sensitivity in recalcitrant seeds of tea, cocoa and jackfruit. *Annals of Botany* 76, 443–450.

Chin, H.F. and Roberts, E.H. (1980) *Recalcitrant Crop Seeds.* Tropical Press SDN.BHD, Kuala Lumpur, Malaysia.

Chmielarz, P. (1997) Preservation of *Quercus robur* L. embryonic axes in liquid nitrogen. In: Ellis, R.M., Black, M., Murdoch, A.J. and Hong, T.D. (eds) *Basic and Applied Aspects of Seed Biology.* Kluwer Academic Publishers, Dordrecht, pp. 765–769.

Fu, J.-R., Zhang, B.Z., Wang, X.P., Qiao, Y.Z. and Huang, X.L. (1990) Physiological studies on desiccation, wet storage and cryopreservation of recalcitrant seeds of three fruit species and their excised embryonic axes. *Seed Science and Technology* 18, 743–754.

Janeiro, L.V., Vieitez, A.M. and Ballester, A. (1996) Cryopreservation of somatic embryos and embryonic axes of *Camellia japonica. Plant Cell Reports* 15, 699–703.

Krishnapillay, B., Marzalina, M. and Aziah, M.Y. (1992) Cryopreservation of excised embryos of rattan manau. In: *Proceedings of the 3rd National Biotechnological Research Conference of National Universities of Malaysia.* Kuala Lumpur, Malaysia, pp. 325–330.

Mycock, D.J. (1999) Addition of calcium and magnesium to a glycerol and sucrose cryoprotectant solution improves the quality of plant embryo recovery from cryostorage. *CryoLetters* 20, 77–82.

Normah, M.N., Chin, H.F. and Hor, Y.L. (1986) Desiccation and cryopreservation of embryonic axes of *Hevea brasiliensis* Muell.-Arg. *Pertanika* 9, 299–303.

Pammenter, N.W. and Berjak, P. (1999) A review of recalcitrant seed physiology in relation to desiccation-tolerance mechanisms. *Seed Science Research* 9, 13–37.

Pammenter, N.W., Greggains, V., Kioko, J.I., Wesley-Smith, J., Berjak, P. and Finch-Savage, W.E. (1998) Effects of differential drying rates on viability of recalcitrant seeds of *Ekebergia capensis. Seed Science Research* 8, 463–471.

Pence, V.C. (1990) Cryostorage of embryonic axes of several large-seeded temperate tree species. *Cryobiology* 27, 212–218.

Poulsen, K.M. (1992) Sensitivity to desiccation and low temperature (–196°C) of embryo axes from acorns of the pedunculate oak (*Quercus robur* L.). *CryoLetters* 13, 75–82.

Thammasiri, K. (1999) Cryopreservation of embryonic axes of jackfruit. *CryoLetters* 20, 21–28.

Wesley-Smith, J., Walters, C., Berjak, P. and Pammenter, N.W. (1999) A method for the cryopreservation of embryonic axes at ultra-rapid cooling rates. *Proceedings of the IUFRO Symposium, 'Recalcitrant Seeds',* Kuala Lumpur, Malaysia, 132–139.

Withers, L.A. and Williams, J.T. (1980) Crop genetic resources. The conservation of difficult material. *Proceedings of an International Workshop, University of Reading, UK.* IUBS Série B 42.

IV Germination

22 Gene Expression Prior to Radicle Emergence in Imbibed Tomato Seeds

K.J. Bradford, F. Chen, M.B. Cooley, P. Dahal, B. Downie, K.K. Fukunaga, O.H. Gee, S. Gurusinghe, R.A. Mella, H. Nonogaki, C-T. Wu, H. Yang and K-O. Yim

Department of Vegetable Crops, One Shields Avenue, University of California, Davis, CA 95616-8631, USA

In tomato (*Lycopersicon esculentum* Mill.) seeds, the primary control of germination resides in the tissues of the radicle tip and the endosperm cap enclosing it. It is a reasonable assumption that germination is regulated ultimately by specific gene expression in these tissues. We are therefore interested in identifying the genes that are expressed in these tissues during imbibition and prior to radicle emergence and in understanding how their expression is influenced by environmental and hormonal factors that also modify the germination rate or percentage. Two approaches were utilized to identify genes specifically associated with the events initiated by seed imbibition and culminating in radicle emergence. The genes for proteins or enzymes that were expected to play a role in germination have been cloned using polymerase chain reaction (PCR) or library-screening methods, including genes for several cell wall hydrolases (endo-β-mannanase, cellulase, polygalacturonase, xyloglucan endotransglycosylase, β-1,3-glucanase, and chitinase), expansins and galactinol synthase. In addition, mRNA populations from both radicle tips and endosperm caps of the gibberellin-deficient (*gib-1*) tomato mutant were compared by differential cDNA display following imbibition in the presence or absence of gibberellin (GA). Additional mRNAs identified by this approach that are either expressed or repressed in these tissues in response to GA include ones coding for an arabinosidase, a vacuolar H^+-translocating ATPase subunit, a GA-responsive transcript and the activating subunit of a protein kinase involved in sugar sensing. These studies are beginning to reveal the wide array of genes and biochemical processes that are involved in the transition from seed development to germination.

Introduction

The time between seed imbibition and radicle emergence is the period of germination in the strict sense. Following initial water uptake, this phase of development is characterized by relatively little change in seed water content until it is terminated by the initiation of embryo growth. During this time, energy metabolism resumes, repair processes are activated, and the cell cycle may be initiated, while events associated with seed maturation are suppressed (Bray, 1995; Hilhorst *et al.*, 1998). These changes are reflected in patterns of gene expression, which quickly switch from a developmental mode to a germinative mode (Kermode, 1995). Although the comprehensive transition in gene expression patterns following imbibition has been well documented, there is remarkably little information about the identities of the genes that are responsible for the initiation of embryo growth. Detailed information about gene expression in germinating seeds is largely concerned with enzymes involved in reserve mobilization, the majority of which occurs post-germination (Bewley and Black, 1994). The phenomenon of dormancy, during which metabolism is active but germination is not completed, indicates that there must be additional checkpoints controlling the transition from potential to actual growth. It is likely that thousands of genes are turning on or off as the seed shifts from its prior life as a maturing seed to its new role as a growing seedling. Which of these genes are specifically associated with the biochemical and physiological processes leading to (or preventing) the initiation of embryo growth?

Genes associated with cell enlargement are good candidates, as the initial protrusion of the embryo from the seed is due to water uptake and cell expansion. In seeds where the tissues covering the embryo are weak or are split during imbibition, this may be the only process required for the completion of germination (e.g. Schopfer and Plachy, 1985). In many seeds, on the other hand, the embryo is enclosed within endosperm and/or testa tissues that present a mechanical barrier to radicle emergence. In these seeds, in addition to the expansive force of the embryo, weakening of the external tissues is required to allow radicle emergence to occur (Welbaum *et al.*, 1998). The biochemical processes involved in tissue weakening are likely to involve modifications to the cell walls of the restraining tissue, and may share common mechanisms with other developmental processes where cell wall disassembly or cell separation occur, such as fruit ripening, abscission, or lateral root emergence (Osborne and Jackson, 1989).

These considerations have led us to pursue two approaches to identify genes involved in the regulation of radicle emergence. The first is to use cDNAs or sequences of genes known or thought to be involved in cell growth or cell wall disassembly processes to determine whether homologues of these genes are expressed in imbibed seeds prior to radicle emergence. The second is to use differential cDNA display (DCD) (Liang and Pardee, 1992) to compare the mRNA populations of seeds that will or will not complete germination. The latter technique does not require prior information about gene identity or function, and can therefore potentially identify novel genes involved in germination. We will first describe the characteristics of the tomato seed system that we have employed and then summarize progress in identifying and characterizing genes expressed in tomato seeds prior to radicle emergence.

Tomato Seed Germination

Tomato (*Lycopersicon esculentum* Mill.) seeds are comprised of an embryo completely surrounded by a relatively hard and brittle endosperm and a thin testa (Fig. 22.1). Tomato seeds have been particularly well studied as a model system for the interaction between embryo expansive force and weakening of enclosing tissues in controlling germination (reviewed in Hilhorst *et al.*, 1998; Welbaum *et al.*, 1998). The endosperm tissue enclosing the radicle tip (called the endosperm cap or the micropylar endosperm) can be excised (Fig. 22.1) and its strength determined by puncture force measurements using a mechanical force analyser (Groot and Karssen, 1987). These experiments showed that weakening of the endosperm cap is required for radicle emergence and that the weakening process is primarily controlled by gibberellin (GA). The testa also contributes to the restraint that must be overcome by the radicle (Hilhorst and Downie, 1995), but the testa is non-living at maturity, and its resistance to puncture remains relatively constant during germination. The endosperm tissue opposite the radicle tip can be distinguished anatomically by its having smaller cells and thinner walls than the remainder of the endosperm (called the lateral endosperm) (Haigh, 1988), and, as will become evident below, it is also physiologically differentiated from the adjacent lateral endosperm.

Some advantages of tomato seeds as an experimental system include their relatively large size, allowing dissection and separation of specific seed tissues (Fig. 22.1). GA-deficient, abscisic acid (ABA)-deficient, auxin-insensitive and

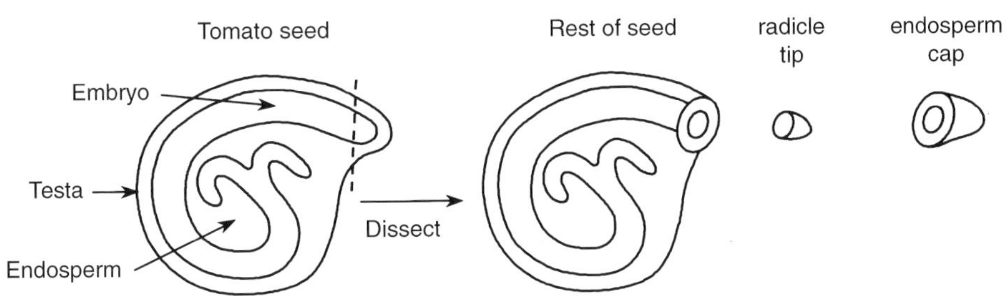

Fig. 22.1. Tomato seed anatomy. The endosperm and testa surround the tomato embryo. The micropylar region of the endosperm, or endosperm cap, is physiologically and anatomically differentiated from the remaining lateral endosperm. To examine the tissue localization of gene expression, the micropylar tip can be excised and the radicle tip can be removed. This results in the endosperm cap, radicle tip, and the rest of the seed (remainder of embryo and lateral endosperm). The strength of the endosperm cap can also be determined by inserting a metal probe into the cavity of the cap and measuring the force required to penetrate the tissue.

ethylene-insensitive mutants are available for studying the roles of hormones in regulating germination (Kelly and Bradford, 1986; Groot and Karssen, 1987, 1992; Ni and Bradford, 1993; Lashbrook *et al.*, 1998). Although generally not dormant, tomato seeds can exhibit dormancy under some conditions, and germination is sensitive to far-red light (Downie *et al.*, 1999) and water potential (Dahal and Bradford, 1990). The effects of environmental or hormonal manipulation on gene expression can therefore be compared to the corresponding effects on germination. Tomato is also easily transformable and has a large and growing database of genetic and DNA sequence information.

Cell Wall Hydrolases and Expansins

Since weakening of the endosperm cap is required for radicle emergence, cell wall hydrolases are likely to be involved. We will summarize here the current situation with respect to the cloning and expression of cell wall hydrolases and related proteins (e.g. expansins) in tomato seeds prior to radicle emergence.

Mannanase

The endosperm cap cell walls of tomato contain 60–70% mannose, suggesting that they are primarily composed of mannan polymers, most likely galactomannans (Groot *et al.*, 1988; Dahal *et al.*, 1997b). Groot *et al.* (1988) showed that endo-β-mannanase, mannosidase, and galactosidase activities were all present in GA-treated gibberellin-deficient (*gib-1*) tomato seeds, with endo-β-mannanase showing the most dramatic increase prior to radicle emergence. Much subsequent work has therefore focused on the role of endo-β-mannanase in endosperm cap weakening. Multiple electrophoretic isoforms of endo-β-mannanase are present in tomato seeds, with different isoforms being expressed first in the endosperm cap, then in the embryo and the lateral endosperm (Nonogaki and Morohashi, 1996; Toorop *et al.*, 1996; Voigt and Bewley, 1996; Nonogaki *et al.*, 1998). In addition, most reports have noted much greater activity in the endosperm cap than in the embryo prior to radicle emergence (Nonogaki and Morohashi, 1996; Dahal *et al.*, 1997b; Still *et al.*, 1997; Nonogaki *et al.*, 1998), although some have found just the reverse (Toorop *et al.*, 1996; Toorop, 1998). While endo-β-mannanase activity is consistently associated with tomato seed germination, the exact details of its involvement remain unclear (reviewed by Bewley, 1997).

The cloning of an endo-β-mannanase cDNA from germinated tomato seeds (Bewley *et al.*, 1997) should help resolve this situation. Southern blots indicated that there could be four or more endo-β-mannanase genes in tomato. We recently isolated a second endo-β-mannanase cDNA from a pre-germinative cDNA library (H. Nonogaki, O.H. Gee and K.J. Bradford, unpublished results) and M. Banik and J.D. Bewley (Guelph, 1999, personal communication) have isolated a genomic clone having a sequence distinct from the two cDNAs. Thus, at least three endo-β-mannanase genes are known in tomato. It seems likely that different mannanase genes might be expressed in the different seed tissues, with one of them accounting for the pregerminative endosperm cap isoform and others accounting for the embryo and lateral endosperm isoforms. Alternatively (or in addition), post-transcriptional or post-translational modifications could result in the diverse electrophoretic isoforms that have been detected. Further work with gene-specific probes and isoform-specific antibodies will be needed to decipher the expression patterns of endo-β-mannanases in germinating tomato seeds. Antisense experiments to selectively eliminate specific gene products will be particularly valuable in determining whether endo-β-mannanases are essential for endosperm weakening and radicle emergence.

Cellulase

Cellulase (β-1,4-endoglucanase) activity has been detected in tomato seeds, but the measured activity was not closely linked with germination timing (Leviatov *et al.*, 1995; Toorop, 1998). However, there was a closer correlation between cellulase activity and phytochrome-dependent germination in *Datura* seeds (Sánchez *et al.*, 1985), which also express endo-β-mannanase and mannosidase activities prior to radicle emergence (Sánchez and de Miguel, 1997). We have isolated a cDNA (termed *Cel55*) having high homology to known β-1,4-endoglucanases whose corresponding mRNA increases in abundance in tomato radicles and endosperm caps prior to radicle emergence (B. Downie and K.J. Bradford, unpublished results). Accumulation of the mRNA is reduced by low water potential, far-red light or in dormant seeds (Table 22.1). Thus, at least one unique β-1,4-endoglucanase gene is expressed following imbibition of tomato seeds and a functional role for cellulases in germination cannot be ruled out. As for mannanase, it is possible that multiple genes/isoforms having differential expression patterns or substrate specificities will be present that will complicate the interpretation of extracted enzyme activity data in relation to germination.

Polygalacturonase

Tomato endosperm cell walls become more 'porous' in appearance and are visibly degraded prior to radicle emergence (Nonogaki *et al.*, 1998; Toorop, 1998). In addition, it has been observed that the cap cells separate between adjacent cell walls, rather than by tears through individual cells (Haigh, 1988). As polygalacturonases (PGs) are involved in cell separation or cell wall disassembly in other developmental processes, including abscission and fruit ripening (Hadfield and Bennett, 1998), we have looked for expression of PGs in tomato endosperm caps. Polygalacturonase activity in tomato seeds increased following imbibition, and we have isolated a cDNA encoding a PG whose mRNA is expressed in the endosperm cap and particularly in the vascular

Table 22.1. Expression patterns of cloned genes in imbibed tomato seeds.

Gene	cDNA	Tissue localization	Regulation of expression: GA	ABA	Low ψ	FR	Dormancy
Endo-β-mannanase*	*LeMAN*	CAP, RT, EMB, LAT	+	O	O		–
Cellulase	*Cel55*	CAP, RT, ROS	+	O	–	–	–
Polygalacturonase	*LeXPG1*	CAP, RT		O		–	
Arabinosidase	*LeARA1*	CAP, LAT	+	O	–	–	
Xyloglucan endotransglycosylase	*LeXET*		+				
β-1,3-Glucanase	*GluB*	CAP	+	–			
Chitinase	*Chi9*	CAP, RT	+	O			
Expansin	*LeEXP4*	CAP	+	O	–	–	
SNF4 kinase-activating subunit	*LeSNF4*	CAP, RT	–	+	+	+	+
SNF1 kinase subunit	*LeSNF1*			O	O	O	O
Galactinol synthase	*LeGalS*	CAP, RT	–	O	+	+	+
Vacuolar H^+-ATPase	*LVA-P1*	CAP, RT	+	–			
GA-stimulated transcript	*LeGAS2*	CAP, RT	+	–	–		–

*Localization and regulation based upon enzyme activity rather than mRNA detection.
Genes that have been cloned from imbibed tomato seeds prior to radicle emergence are listed on the left with the corresponding cDNA. If known, the tissue localization of expression is indicated (CAP, endosperm cap; EMB, embryo; LAT, lateral endosperm; RT, radicle tip; ROS, rest of seed except micropylar tip). The qualitative effects of GA, ABA, low water potential (Low ψ), far-red light (FR) and primary dormancy, if known, are also indicated (+, promotes expression; O, no effect on expression; –, inhibits expression; blank, information not available). See text for details.

tissue of the expanding tissue of the radicle (Sitrit *et al.*, 1999). Extracts from seeds exhibited only exo-PG activity and no endo-PG activity, so we assume that the cloned gene (termed *LeXPG1*) encodes the former enzyme, although this has not been conclusively demonstrated. Exo-PGs are particularly well represented in pollen (Hadfield and Bennett, 1998), and PGs are expressed specifically near the developing meristems of lateral root primordia (Peretto *et al.*, 1992). Pollen tubes grow through the style and lateral roots grow through enclosing cortical and epidermal tissues, much like an emerging radicle grows through the endosperm cap. De-esterification of pectin is also associated with cell separation (Stevenson and Hawes, 1994), and pectin methylesterase activity increases in tomato radicle tips prior to emergence (Downie *et al.*, 1998a). Thus, degradation or modification of pectins may be involved in the endosperm-weakening process and/or in the initial expansion of the embryo associated with radicle protrusion.

Arabinosidase

While the genes described above were isolated by PCR or library screening with known sequences or cDNAs, another putative hydrolase cDNA (*LeARA1*) was isolated from a differential cDNA display screen of *gib-1* mutant tomato seeds imbibed in the presence or absence of GA (Dahal *et al.*, 1997a). The predicted amino acid sequence showed significant homology to bacterial and fungal arabinosidases, and seed extracts exhibited arabinosidase activity (P. Dahal and K.J. Bradford, unpublished results). The corresponding mRNA was initially expressed only in the endosperm caps in response to GA, and subsequently expression spread through the remainder of the endosperm, but not in the embryo. Interestingly, accumulation of *LeARA1* mRNA was prevented by imbibition in low water potential or in far-red light, but not by ABA (Table 22.1), although all three conditions prevent germination.

Xyloglucan endotransglycosylase

Another hydrolase that could be involved in germination is xyloglucan endotransglycosylase (XET). It is believed that xyloglucans, the principal hemicellulosic components in the primary cell walls of dicots, can form a tightly bound, non-covalent association with cellulose. Xyloglucan endotransglycosylase catalyses both the endo-type splitting of a xyloglucan molecule and the linking of the newly generated reducing end to a nonreducing end of another xyloglucan molecule or an oligosaccharide (Fry *et al.*, 1992; Nishitani and Tominaga, 1992). This lengthening and rearrangement of xyloglucans may release tension and accommodate wall expansion during cell growth. However, XET activity also increases during fruit ripening, suggesting that XET may be involved in tissue softening or wall disassembly (Maclachlan and Brady, 1994). Transglycosylation can reduce the length of polysaccharides if the acceptor molecule is a small oligosaccharide, and the kiwifruit XET was shown to have xyloglucan hydrolase as well as endotransglycosylase activities (Schröder *et al.*, 1998). While xylose and glucose constitute significant fractions (15% and 30%, respectively) of tomato embryo cell walls, they represent much smaller fractions of the sugars in endosperm cap cell walls (2% and 20%, respectively) (Dahal *et al.*, 1997b). However, xylose and glucose constituted larger fractions of the soluble sugars released from isolated endosperm caps (10% and 20–30%, respectively) (Dahal *et al.*, 1997b). Thus, the xyloglucans in the endosperm cap may be preferentially hydrolysed during imbibition and germination. We have recently isolated two different cDNAs having sequence homology to known XETs from our imbibed tomato seed cDNA library (F. Chen and K.J. Bradford, unpublished results), but do not yet know whether the corresponding mRNAs are expressed in the embryo, the endosperm cap or the lateral endosperm or how their expression is regulated (Table 22.1). Further characterization of these genes and their expression patterns will reveal whether they could be involved in the endosperm weakening process or in embryo growth.

β-1,3-Glucanase and chitinase

Two additional cell wall hydrolases expressed in tomato endosperm caps present an intriguing situation. In general, the pathogenesis-related (PR) proteins β-1,3-glucanase and chitinase are expressed in response to pathogens or wounding and in abscission zones or other places where interior plant tissues are (or will be) exposed (Linthorst, 1991; González-Carranza *et al.*, 1998). Together, these enzymes can degrade the cell walls of many fungi and can enhance plant resistance to infection (Mauch *et al.*, 1988). Using the GUS reporter driven by the promoter of β-1,3-glucanase, Meins and coworkers (Vögeli-Lange *et al.*, 1994) demonstrated that this gene is expressed specifically in the endosperm cap tissue of tobacco seeds. Subsequent experiments confirmed that β-1,3-glucanase mRNA, protein and enzyme activity were present prior to radicle emergence and were regulated in a manner consistent with a role in germination (Leubner-Metzger *et al.*, 1995, 1996). Curiously, chitinase was not expressed in germinating tobacco seeds, leading to the hypothesis that β-1,3-glucanase might be involved in cell wall degradation and weakening of the endosperm (Vögeli-Lange *et al.*, 1994).

In cooperation with Meins' group, we have now performed similar experiments in tomato and find that both β-1,3-glucanase and chitinase (mRNAs, proteins, and activities) are expressed specifically in the endosperm cap prior to radicle emergence (Wu *et al.*, 1998). Gibberellin induces both genes in *gib-1* seeds, while ABA can inhibit expression of β-1,3-glucanase but not of chitinase in wild type seeds (Table 22.1). Both expressed genes are of the Class I (basic) isoforms that are localized primarily in the vacuole, rather than of the acidic isoforms that are found in the cell wall or apoplast. Furthermore, callose is not detectable in tomato endosperm caps (Beresniewicz *et al.*, 1995; our unpublished results), and plant cell walls are not thought to contain chitins. Thus, rather than being involved in cell wall hydrolysis, these enzymes could constitute a prophylactic defence mechanism expressed in the tissue through which the radicle will emerge, or they may be induced in response to elicitors released by wall degradation. Further work will be required to determine whether these enzymes are involved in endosperm cell wall degradation during germination or whether they contribute to a pathogen defence mechanism.

Expansins

Although cell wall hydrolases are almost certainly involved in cell expansion, hydrolytic enzymes alone, including β-1,4-endoglucanases and XET, are unable to cause wall extension in *in vitro* assays using killed hypocotyl segments held under tension (McQueen-Mason *et al.*, 1992). However, using such an assay, proteins termed 'expansins' have been identified that can cause extension of plant cell walls (McQueen-Mason *et al.*, 1992; Shcherban *et al.*, 1995). As purified expansin protein had little or no hydrolytic activity, it was proposed that expansins disrupt the non-covalent linkages (e.g. hydrogen bonds) at the cellulose–hemicellulose interface, allowing the polymers to creep under tension and the wall to expand (McQueen-Mason and Cosgrove, 1995). Cell wall hydrolases are proposed to work in conjunction with expansin to modify the hemicellulosic matrix by cleaving and reforming bonds to alter the strength or plasticity of the wall (Cosgrove, 1998).

Expansins have been highly conserved throughout plant evolution, as homologous genes have been identified in gymnosperms and in both monocots and dicots among the angiosperms (Cosgrove, 1998). Expansins occur as multi-gene families in *Arabidopsis*, rice, cucumber, tomato and other species where they have been examined in detail. The large number of expansin-like genes (e.g. at least 22 in *Arabidopsis*) suggests multiple developmental or tissue-specific roles for these proteins, possibly in addition to vegetative growth *per se*. Expansins are expressed in the shoot meristem during the early stages of leaf initiation (Fleming *et al.*, 1997; Reinhardt *et al.*, 1998) and in developing tomato fruits (Brummell *et al.*, 1999). Extensive cell wall degradation and solubilization of wall components occurs during ripening, resulting in tissue softening and cell separation without cell enlargement. Thus, expansins appear to be involved in diverse gene-specific roles in developmental processes related to cell formation and cell wall expansion, modification or disassembly (Cosgrove, 1997). Of particular interest here is the report (Cosgrove *et al.*, 1998) that the promoter of the *AtEXP8* expansin gene of *Arabidopsis* is active specifically in: (i) endosperm cells directly opposite the radicle tip; (ii) a ring of cells at the point where lateral roots emerge through the root cortex; and (iii) the outer cell layers of the root cap. This expression pattern suggests that this expansin participates in cell separation processes associated with the penetration of roots through other tissues.

We hypothesized that GA might induce expression of expansins in tomato endosperm cap tissue, resulting in tissue weakening or cell separation and allowing emergence of the radicle to occur. Reverse-transcription PCR (RT-PCR) using primers to conserved expansin sequences amplified six expansin-like fragments present in mRNA from imbibed tomato seeds. Full-length cDNAs corresponding to these fragments were isolated and characterized (Chen and Bradford, 1998). Transcripts for one of these genes (*LeEXP4*, Brummell *et al.*, 1999) were not detectable in GA-deficient *gib-1* tomato seeds imbibed on water, but accumulated within 12 h of imbibition in 100 μM GA_{4+7} (Fig. 22.2). A similar time course of expression occurred in wild type seeds imbibed on water. In both wild type and *gib-1* seeds, expression was localized specifically to the endosperm cap tissue prior to radicle emergence (Fig. 22.2). Weakening of the endosperm cap tissue paralleled *LeEXP4* expression (data not shown). Low water potential or far-red light also delayed or prevented both expression of *LeEXP4* (Table 22.1) and tissue weakening while inhibiting germination of wild type seeds. Somewhat surprisingly, ABA had no effect on either *LeEXP4* expression (Table 22.1) or on endosperm cap weakening, although it did inhibit radicle emergence. Thus, ABA must affect the force generated by the radicle, or possibly a second phase of endosperm weakening (Toorop, 1998). Nonetheless, expression of *LeEXP4* is consistently associated with endosperm cap weakening. Two additional expansin genes have been identified as being expressed in imbibed tomato seeds, one of which is

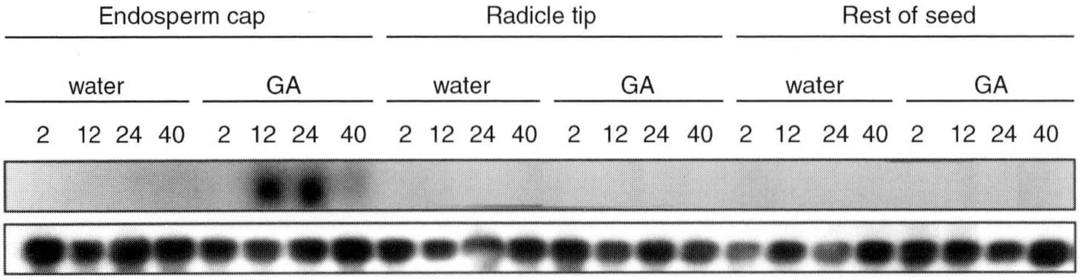

Fig. 22.2. Expression of the expansin *LeEXP4* in the endosperm cap, radicle tip and rest of the tomato seed (see Fig. 22.1). Gibberellin-deficient (*gib-1*) tomato seeds were imbibed in either water or 100 μM GA_{4+7} for the indicated times before dissection of the tissues and extraction of RNA. Total RNA was separated on an agarose gel and hybridized with a riboprobe complementary to *LeEXP4* cDNA. *LeEXP4* mRNA was not detectable at the time of imbibition and remained so in the absence of GA. In the presence of GA, *LeEXP4* mRNA appeared within 12 h only in the endosperm cap. The lower panel shows hybridization with a cDNA for a constitutive ribosomal protein to indicate RNA loading of each lane.

localized to the embryo (F. Chen and K.J. Bradford, unpublished results). Specific expansin genes may therefore be involved in endosperm weakening and in expansion of the embryo associated with radicle emergence.

Energy Metabolism, Stress Responses and Other Genes

In addition to the genes and proteins described above, which may be directly involved in tissue weakening or expansion, a number of other genes have been identified that are either expressed or repressed in association with tomato seed germination. They may have roles in energy metabolism, stress responses, reserve mobilization, or signalling pathways in the imbibed seed.

LeSNF1/LeSNF4 protein kinase complex

The yeast SNF (*sucrose non-fermenting*) protein kinase and mammalian AMP-activated protein kinase (AMPK) complexes are central components of conserved kinase cascades that act as metabolic sensors of changes in glucose availability and of ATP/AMP ratios, respectively (Hardie *et al.*, 1998). Protein sequence and functional homology exists between the kinase subunits (SNF1/AMPK-α), activating subunits (SNF4/AMPK-γ) and bridging or docking subunits (SIP/GAL83/AMPK-β) that constitute the functional kinase complexes in yeast and mammals (Fig. 22.3). In plants, a large family of SNF1-related kinases (SnRK1s) has been identified that shares sequence and functional homology with the kinase subunit (Halford and Hardie, 1998), and a homologue of the yeast GAL83 docking subunit has been identified in potato (Lakatos *et al.*, 1999). Abe *et al.* (1996) reported the sequence of a cDNA (*Pv42*) isolated from developing bean (*Phaseolus vulgaris*) seeds that had

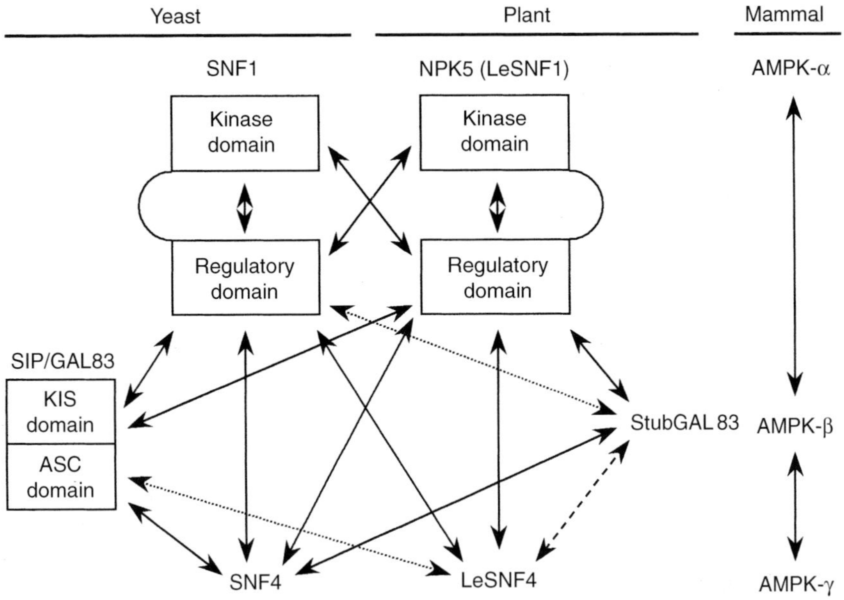

Fig. 22.3. The yeast SNF1/SNF4 complex and homologues in plants and mammals. The yeast SNF1 protein is composed of a kinase domain (KD) and a regulatory domain (RD). The RD can interact with the KD of the same protein and with the SNF4 protein. In addition, the SNF1 and SNF4 proteins form a complex with one of a family of proteins (SIP1/SIP2/GAL83). Each of these latter proteins can independently interact with SNF1 at a kinase interacting sequence (KIS) and with SNF4 at an association (ASC) domain. When glucose is available, interaction between the kinase and regulatory domains of SNF1 autoinhibits its kinase activity. When glucose is low, SNF4 interacts with the RD of SNF1, activating the kinase activity. The NPK5 protein is a tobacco homologue of SNF1; homologous cDNAs have been isolated from several species, including LeSNF1 from tomato described in this paper. We have also identified the tomato SNF4 homologue (LeSNF4) described herein, which mutant complementation experiments show can interact with yeast SNF1, and yeast complementation and two-hybrid assays indicate can interact with LeSNF1 also. A homologue of the GAL83 protein was recently identified in potato (StubGAL83) that can bind to yeast SNF1 (Lakatos *et al.*, 1999). The α-subunit of the mammalian AMP-activated protein kinase (AMPK) is a homologue of SNF1, the β-subunit is a homologue of SIP/GAL83, and the γ-subunit is a homologue of SNF4. The protein–protein interactions indicated by the solid arrows have been shown to occur *in vivo* by mutant complementation or two-hybrid assays, while those with dashed arrows are assumed to occur but have not been demonstrated experimentally as yet. (Figure modified from Jiang and Carlson, 1996 and Hardie *et al.*, 1998 to incorporate new data.)

significant homology to the SNF4/AMPK-γ activating subunit of the protein kinase complex.

In our DCD screen of water- or GA-imbibed *gib-1* tomato seeds, we identified a mRNA that was present in seeds imbibed in water but which disappeared rapidly in response to GA. We subsequently cloned a full-length cDNA corresponding to this mRNA (Yang *et al.*, 1997) and found that it had significant sequence homology to SNF4/AMPK-γ and to the bean *Pv42* cDNA. We termed this gene *LeSNF4* (*Lycopersicon esculentum SNF4*) and have demonstrated that it functionally complements a yeast *SNF4* deletion mutant. We also cloned a tomato SNF1 kinase subunit homologue (*LeSNF1*) that complements the corresponding *SNF1* deletion mutant in yeast. When expressed together, the tomato *LeSNF1* and *LeSNF4* genes complement a yeast strain having deletions in both the *SNF1* and *SNF4* genes. In yeast, the SNF1 and SNF4 proteins physically interact when the SNF4 protein activates the SNF1 kinase in response to low glucose concentrations, as can be demonstrated by the two-hybrid assay (Chien *et al.*, 1991; Jiang and Carlson, 1996). We used this method to show that the tomato homologue proteins, LeSNF1 and LeSNF4, also physically interact (P. Dahal and K.J. Bradford, unpublished results; Fig. 22.3).

Thus, tomato contains functional homologues of both the kinase and the activating subunits of the yeast SNF1/SNF4 complex, and given the close evolutionary relationship between tomato and potato, we can predict that tomato will contain the GAL83 homologue as well. It has been proposed that in plants this kinase complex may be the central system by which sugars (particularly sucrose) are sensed and carbon partitioning is regulated (Halford *et al.*, 1999). For example, antisense suppression of expression of the potato SNF1 homologue reduced sucrose synthase activity in the tubers and prevented sucrose induction of sucrose synthase transcripts in leaves (Purcell *et al.*, 1998).

As mentioned above, *LeSNF4* mRNA rapidly disappears upon imbibition of wild type seeds or of *gib-1* seeds in GA. However, conditions that prevent the completion of germination, including low water potential, ABA, far-red light or natural dormancy, all maintain *LeSNF4* mRNA abundance (Table 22.1). In seeds that have been imbibed until *LeSNF4* mRNA has disappeared, transfer to low water potential or dehydration will re-induce accumulation of the mRNA. This is also the case with leaves, where either water loss or ABA causes rapid accumulation of *LeSNF4* mRNA. This expression pattern is similar to that of an ABA-responsive protein kinase from wheat (PKABA1) (Holappa and Walker-Simmons, 1995) that was recently demonstrated to be involved in the regulation of ABA-suppressed genes in aleurone cells (Gomez-Cadenas *et al.*, 1998; Walker-Simmons, Chapter 25, this volume). On the other hand, *LeSNF1* mRNA amounts are relatively unresponsive to developmental transitions or stress conditions (Table 22.1), as is the case with the yeast and animal kinase complexes. In the latter cases, biochemical regulation of activity (as by phosphorylation/dephosphorylation and protein/protein interactions) is thought to be more important than transcriptional regulation. Our results suggest that in plants, transcriptional regulation of the activating subunit (SNF4 homologue) may be an additional mechanism by which kinase cascades are altered to shift entire metabolic pathways in response to developmental or environmental cues. For example, high *LeSNF4* expression during seed development could be associated with sink activity and the synthesis of storage compounds from transported sugars, while low expression during germination could be related to the mobilization of storage reserves and the export of sucrose to the growing seedling.

Galactinol synthase

Another gene exhibiting an expression pattern similar in some respects to that of *LeSNF4* and possibly also involved in stress adaptation encodes galactinol synthase (GalS). GalS catalyses the formation of galactinol from UDP-galactose and *myo*-inositol, and galactinol is then used as the galactose donor to form raffinose by the addition of galactose to sucrose (Liu *et al.*, 1995). The synthesis of galactinol is the rate-limiting, committed step in the biosynthesis of the raffinose family oligosaccharides (RFOs; Smith *et al.*, 1991). Seeds of diverse species accumulate these oligosaccharides in varying proportions during late maturation (Black *et al.*, 1996; Brenac *et al.*, 1997), and they are proposed to be involved in desiccation tolerance and seed longevity (Horbowicz and Obendorf, 1994; Obendorf, 1997). Using the sequence of an *Arabidopsis GalS* gene (Liu *et al.*, 1998), we cloned a tomato *GalS* homologue from tomato seed mRNA by RT-PCR (Downie *et al.*, 1998b). Based on Northern blots, *GalS* mRNA is present in immature tomato seeds and increases in abundance late in seed development but declines rapidly upon imbibition of mature seeds. Dehydrating imbibed seeds or transferring them to reduced water potentials caused the mRNA to reappear. *GalS* mRNA was maintained in imbibed dormant seeds or seeds imbibed in far-red light, which did not germinate. Similarly, *GalS* mRNA remained abundant in *gib-1* tomato seeds imbibed in water but decreased rapidly upon exposure to GA. Wild type seeds imbibed in 100 μM ABA did not germinate, but *GalS* mRNA declined with a time course similar to that for seeds imbibed on water. Thus, while both *GalS* and *LeSNF4* mRNA are present in developing seeds, decline rapidly upon imbibition or exposure to GA, and are induced by dehydration and far-red light, *GalS* is not responsive to ABA (Table 22.1; Downie *et al.*, 1998b; Liu *et al.*, 1998). Whether *GalS* expression is required for stress adaptation and/or desiccation tolerance is currently under investigation by several groups.

Vacuolar H^+-translocating ATPase

Among the mRNAs identified by DCD as being up-regulated by GA in *gib-1* seeds was one encoding the 16 kDa proteolipid subunit of vacuolar H^+-translocating ATPase (V-ATPase), termed *LVA-P1* (Cooley *et al.*, 1999). *LVA-P1* mRNA expression was particularly abundant in the micropylar region prior to radicle emergence. Detection of the 57 and 68 kDa regulatory and catalytic subunits of the V-ATPase using specific antisera confirmed that the enzyme complex was induced by GA in *gib-1* seeds and was primarily located in the micropylar region. V-ATPases establish and maintain an acidic pH in the vacuole, which enhances the activity of acid hydrolases involved in the breakdown of protein reserves and the transformation of protein bodies into vacuoles during germination (Maeshima *et al.*, 1994; Sze *et al.*, 1995). The mobilization of protein reserves, protein-body breakdown, and vacuolization are initiated in the micropylar endosperm in tomato seeds prior to radicle emergence (Nonogaki *et al.*, 1998). Similar results were reported previously for *Datura ferox* seeds (Mella *et al.*, 1995; Sánchez and de Miguel, 1997), which have a structure and germination physiology similar to those of tomato. Swanson and Jones (1996) demonstrated that GA induces vacuolar acidification in barley aleurone cells, but did not detect significant differences in V-ATPase protein content among control, GA- or ABA-treated aleurone cells. They suggested that other mechanisms, such as cytosolic pH or redox state, might regulate the activity of the V-ATPase. Although expression of *LVA-P1* in *gib-1* tomato seeds is dependent upon GA, expression in wild type seeds is essentially constitutive during development and germination, with endogenous GA apparently being sufficient to maintain *LVA-P1* mRNA abundance (Cooley *et al.*, 1999). Thus, additional post-transcriptional mechanisms are undoubtedly involved in regulating V-ATPase activity, as the biochemistry of vacuoles that are becoming protein bodies in developing seeds will undoubtedly be quite different from those undergoing the reverse process during germination.

GA-stimulated transcript

Our DCD screen also identified a transcript induced by GA in *gib-1* seeds (termed *LeGAS2*) whose predicted amino acid sequence shares significant homology with a family of GA-stimulated and ABA-repressed transcripts first identified in tomato as *GAST1* (Shi *et al.*, 1992). In wild type tomato seeds, *LeGAS2* mRNA was detected at low amounts during seed development, but accumulated to high amounts during imbibition and germination (Dahal *et al.*, 1998). *LeGAS2* mRNA accumulation was localized primarily in the micropylar tip region of the seed prior to radicle emergence. Primary dormancy, reduced water potential or ABA inhibited or delayed germination, and also blocked the increase in *LeGAS2* mRNA (Table 22.1). A family of related *GAS* genes has been identified in *Arabidopsis* (*GASA*), two of which were expressed exclusively in seeds and siliques (Herzog *et al.*, 1995). A related gene from tomato (*RSI-1*) is expressed in lateral root primordia induced by auxin, but disappears shortly after the lateral roots emerge from the parent tissue (Taylor and Scheuring, 1994). The similarity to the emergence of the radicle from the enclosing endosperm tissue is apparent. The function of the small 11 kDa protein encoded by these genes remains unknown, but it may be a useful marker for GA and ABA signalling pathways during seed development and germination.

Summary and Future Prospects

Although much remains to be done, considerable progress has been made in identifying genes that may be involved in the biochemical mechanisms of germination or in regulating those mechanisms in response to hormonal and environmental signals (Table 22.1). As might be expected, mRNAs for a number of cell wall hydrolases are up regulated following imbibition specifically in the micropylar tissues, including polygalacturonase, arabinosidase, β-1,3-glucanase and chitinase. The latter two enzymes may contribute to defence against fungal invasion of the lateral endosperm through the opening made by the radicle in the endosperm cap. Further work is likely to confirm that unique isoforms of many of these hydrolases, like endo-β-mannanase, are expressed in different tissues and stages of germination. The discovery that a specific expansin gene is expressed in the endosperm cap in association with weakening of this tissue supports the hypothesis that these proteins play tissue-specific roles in cell wall modification during plant growth and development. The expression of functional homologues of the SNF1/SNF4 protein kinase complex during seed development and their differential regulation during germination or stress provides an exciting new avenue for exploring how carbon and energy metabolism are regulated in sink and source tissues and in dormant seeds. Study of galactinol synthase expression may provide clues to the role of raffinose family oligosaccharides in seed desiccation tolerance and longevity. The genes identified thus far undoubtedly represent only a small sample of those that are involved in the germination process, but each new one provides additional insight into the biochemical and regulatory pathways active in imbibed seeds.

While protein identities and mRNA expression patterns are suggestive of function, additional studies using transgenic plants are required to confirm whether a given gene plays an essential role in germination. Studies are in progress using constitutive promoters to antisense or over-express several of the genes described here to determine the consequences for seed biology. In addition, we have isolated genomic clones and promoters for many of these genes, allowing targeted expression of transgenes in specific seed tissues and at certain times. Use of these promoters to develop non-destructive reporters allowing *in vivo* monitoring of gene expression patterns is an exciting next step, particularly for studying dormancy and other phenomena where physiological variation among individual seeds is involved. These studies are also in progress in our laboratory, and should provide a wealth of new tools for the investigation of both basic and applied aspects of seed biology.

Acknowledgements

This work was supported by National Science Foundation grants IBN-9407264 and IBN-9722978 to K.J.B. and by Natural Sciences and Engineering Research Council of Canada Post-doctoral Scholarships to B.D.

References

Abe, H., Kamiya, Y. and Sakurai, A. (1996) A cDNA clone encoding yeast SNF4-like protein (Accession No. U40713) from *Phaseolus vulgaris* (PGR95–126). *Plant Physiology* 110, 336.

Beresniewicz, M.M., Taylor, A.G., Goffinet, M.C. and Koeller, W.D. (1995) Chemical nature of a semipermeable layer in seed coats of leek, onion (Liliaceae), tomato and pepper (Solanaceae). *Seed Science and Technology* 23, 135–145.

Bewley, J.D. (1997) Breaking down the walls – a role for endo-β-mannanase in release from seed dormancy? *Trends in Plant Science* 2, 464–469.

Bewley, J.D. and Black M. (1994) *Seeds: Physiology of Development and Germination*, 2nd edn. Plenum Press, New York.

Bewley, J.D., Burton, R.A., Morohashi, Y. and Fincher, G.B. (1997) Molecular cloning of a cDNA encoding a (1–4)-β-mannan endohydrolase from the seeds of germinated tomato (*Lycopersicon esculentum*). *Planta* 203, 454–459.

Black, M., Corbineau, F., Grzesik, M., Guy, P. and Côme, D. (1996) Carbohydrate metabolism in the developing and maturing wheat embryo in relation to its desiccation tolerance. *Journal of Experimental Botany* 47, 145–159.

Bray, C.M. (1995) Biochemical processes during the osmopriming of seeds. In: Kigel, J. and Galili, G. (eds) *Seed Development and Germination*. Marcel Dekker, New York, pp. 767–789.

Brenac, P., Horbowicz, M., Downer, S.M., Dickerman, A.M., Smith, M.E. and Obendorf, R.L. (1997) Raffinose accumulation related to desiccation tolerance during maize (*Zea mays* L.) seed development and maturation. *Journal of Plant Physiology* 150, 481–488.

Brummell, D.A., Harpster, M.H. and Dunsmuir, P. (1999) Differential expression of expansin gene family members during growth and ripening of tomato fruit. *Plant Molecular Biology* 39, 161–169.

Chen, F., and Bradford, K.J. (1998) Expansin genes are expressed in tomato seeds in association with germination. *Plant Biology '98*. American Society of Plant Physiologists, Rockville, Maryland, abstract no. 47.

Chien, C.-T., Bartel, P.L., Sternglanz, R. and Field, S. 1991. The two-hybrid system: a method to identify and clone genes for proteins that interact with a protein of interest. *Proceedings of the National Academy of Sciences, USA* 88, 9578–9582.

Cooley, M.B., Yang, H., Dahal, P., Mella, R.A., Downie, B., Haigh, A.M. and Bradford, K.J. (1999) Expression of vacuolar H^+-ATPase in response to gibberellin during tomato seed germination. *Plant Physiology* 121, 1339–1347.

Cosgrove, D.J. (1997) Creeping walls, softening fruit, and penetrating pollen tubes: the growing roles of expansins. *Proceedings of the National Academy of Sciences, USA* 94, 5504–5505.

Cosgrove, D.J. (1998) Cell wall loosening by expansins. *Plant Physiology* 118, 333–339.

Cosgrove, D.J., Scherban, T.Y. and Durachko, D.M. (1998) Highly specific and distinct expression patterns for two alpha-expansin genes in *Arabidopsis*. *Plant Biology '98*. American Society of Plant Physiologists, Rockville, Maryland, abstract no. 175.

Dahal, P. and Bradford, K.J. (1990) Effects of priming and endosperm integrity on seed germination rates of tomato genotypes. II. Germination at reduced water potential. *Journal of Experimental Botany* 41, 1441–1453.

Dahal, P., Downie, B., Yang, H., Mella, R.A. and Bradford, K.J. (1997a) An endosperm-specific mRNA induced by gibberellin during tomato seed germination. *Plant Physiology* (Suppl.) 114, 292.

Dahal, P., Nevins, D.J. and Bradford, K.J. (1997b) Relationship of endo-β-D-mannanase activity and cell wall hydrolysis in tomato endosperm to germination rates. *Plant Physiology* 113, 1243–1252.

Dahal, P., Downie, B., Mella, R.A., Yim, K.-O. and Bradford, K.J. (1998) A GA-stimulated transcript (*GAST*) is expressed during tomato seed germination. *Plant Biology '98*. American Society of Plant Physiologists, Rockville, Maryland, abstract no. 49.

Downie, B., Dirk, L.M.A., Hadfield, K.A., Wilkins, T.A., Bennett, A.B. and Bradford, K.J. (1998a) A gel diffusion assay for quantification of pectin methylesterase activity. *Analytical Biochemistry* 264, 149–157.

Downie, B., Gurusinghe, S., Dahal, P. and Bradford, K.J. (1998b) A galactinol synthase gene is up-regulated in tomato seeds when germination is prevented by various factors, but does not respond to ABA. *Plant Biology '98*. American Society of Plant Physiologists, Rockville, Maryland, abstract no. 61.

Downie, B., Gurusinghe, S.H. and Bradford, K.J. (1999) Internal anatomy of individual tomato seeds: relationship to abscisic acid and germination physiology. *Seed Science Research* 9, 117–128.

Fleming, A.J., McQueen-Mason, S., Mandel, T. and Kuhlemeier, C. (1997) Induction of leaf primordia by the cell wall protein expansin. *Science* 276, 1415–1418.

Fry, S.C., Smith, R.C., Renwick, K.F., Martin, D.J., Hodge, S.K. and Matthews, K.J. (1992) Xyloglucan endotransglycosylase, a new wall-loosening enzyme activity from plants. *Biochemical Journal* 282, 821–828.

Gomez-Cadenas, A., Verhey, S.D., Holappa, L.D., Shen, Q., Ho, T.-H.D. and Walker-Simmons, M.K. (1998) An abscisic acid-induced protein kinase, PKABA1, mediates abscisic acid-suppressed gene expression in barley aleurone layers. *Proceedings of the National Academy of Sciences, USA* 99, 1767–1772.

González-Carranza, Z.H., Lozoya-Gloria, E. and Roberts, J.A. (1998) Recent developments in abscission: shedding light on the shedding process. *Trends in Plant Science* 3, 10–14.

Groot, S.P.C. and Karssen, C.M. (1987) Gibberellins regulate seed germination in tomato by endosperm weakening: a study with gibberellin-deficient mutants. *Planta* 171, 525–531.

Groot, S.P.C. and Karssen, C.M. (1992) Dormancy and germination of abscisic acid-deficient tomato seeds. Studies with the *sitiens* mutant. *Plant Physiology* 99, 952–958.

Groot, S.P.C., Kieliszewska-Rokicka, B., Vermeer, E. and Karssen, C.M. (1988) Gibberellin-induced hydrolysis of endosperm cell walls in gibberellin-deficient tomato seeds prior to radicle protrusion. *Planta* 174, 500–504.

Hadfield, K.A. and Bennett, A.B. (1998) Polygalacturonases: many genes in search of a function. *Plant Physiology* 117, 337–343.

Haigh, A.H. (1988) Why do tomato seeds prime? Physiological investigations into the control of seed germination and priming. PhD dissertation. Maquarie University, Sydney, Australia.

Halford, N.G. and Hardie, D.G. (1998) SNF1-related protein kinases: global regulators of carbon metabolism in plants? *Plant Molecular Biology* 37, 735–748.

Halford, N.G., Purcell, P.C. and Hardie, D.G. (1999) Is hexokinase really a sugar sensor in plants? *Trends in Plant Science* 4, 117–120.

Hardie, D.G., Carling, D. and Carlson, M. (1998) The AMP-activated/SNF1 protein kinase subfamily: metabolic sensors of the eukaryotic cell? *Annual Review of Biochemistry* 67, 821–855.

Herzog, M., Dorne, A.-M. and Grellet, F. (1995) *GASA*, a gibberellin-regulated gene family in *Arabidopsis thaliana* related to the tomato *GAST1* gene. *Plant Molecular Biology* 27, 743–752.

Hilhorst, H.W.M. and Downie, B. (1995) Primary dormancy in tomato (*Lycopersicon esculentum* cv. Moneymaker): studies with the *sitiens* mutant. *Journal of Experimental Botany* 47, 89–97.

Hilhorst, H.W.M., Groot, S.P.C. and Bino, R.J. (1998) The tomato seed as a model system to study seed development and germination. *Acta Botanica Neerlandica* 47, 169–183.

Holappa, L.D. and Walker-Simmons, M.K. (1995) The wheat abscisic acid-responsive protein kinase mRNA, PKABA1, is up-regulated by dehydration, cold temperature, and osmotic stress. *Plant Physiology* 108, 1203–1210.

Horbowicz, M. and Obendorf, R.L. (1994) Seed desiccation tolerance and storability: dependence on flatulence-producing oligosaccharides and cyclitols – review and survey. *Seed Science Research* 4, 385–405.

Jiang, R. and Carlson, M. (1996) Glucose regulates protein interactions within the yeast SNF1 protein kinase complex. *Genes and Development* 10, 3105–3115.

Kelly, M.O., and Bradford, K.J. (1986) Insensitivity of the *diageotropica* tomato mutant to auxin. *Plant Physiology* 82, 713–717.

Kermode, A.R. (1995) Regulatory mechanisms in the transition from seed development to germination: interactions between the embryo and the seed environment. In: Kigel, J. and Galili, G. (eds) *Seed Development and Germination.* Marcel Dekker, New York, pp. 273–332.

Lakatos, L., Klein, M., Höfgen, R. and Bánfalvi, Z. (1999) Potato StubSNF1 interacts with StubGAL83: a plant protein kinase complex with yeast and mammalian counterparts. *Plant Journal* 17, 569–574.

Lashbrook, C.C., Tieman, D.M. and Klee, H.J. (1998) Differential regulation of the tomato ETR gene family throughout plant development. *Plant Journal* 15, 243–252.

Leubner-Metzger, G., Fründt, C., Vögeli-Lange, R. and Meins, F. Jr. (1995) Class I β-1,3-glucanases in the endosperm of tobacco during germination. *Plant Physiology* 109, 751–759.

Leubner-Metzger, G., Fründt, C. and Meins, F. Jr. (1996) Effects of gibberellins, darkness and osmotica on endosperm rupture and class I β-1,3-glucanase induction in tobacco seed germination. *Planta* 199, 282–288.

Leviatov, S., Shoseyev, O. and Wolf, S. (1995) Involvement of endomannanase in the control of tomato seed germination under low temperature conditions. *Annals of Botany* 76, 1–6.

Liang, P. and Pardee, A.B. (1992) Differential display of eukaryotic messenger RNA by means of the polymerase chain reaction. *Science* 257, 967–971.

Linthorst, H.J.M. (1991) Pathogenesis-related proteins of plants. *Critical Reviews in Plant Science* 10, 123–150.

Liu, J.-J., Odegard, W. and de Lumen, B.O. (1995) Galactinol synthase from kidney bean cotyledon and zucchini leaf. Purification and N-terminal sequences. *Plant Physiology* 109, 505–511.

Liu, J.-J., Krenz, D.C., Galvez, A.F. and de Lumen, B.O. (1998) Galactinol synthase (GS): increased enzyme activity and levels of mRNA due to cold and desiccation. *Plant Science* 134, 11–20.

Maclachlan, G. and Brady, C. (1994) Endo-1,4-β-glucanase, xyloglucanase, and xyloglucan endo-transglycosylase activities versus potential substrates in ripening tomatoes. *Plant Physiology* 105, 965–974.

Maeshima, M., Hara-Nishimura, I., Takeuchi, Y. and Nishimura, M. (1994) Accumulation of vacuolar H^+-pyrophosphatase and H^+-ATPase during reformation of the central vacuole in germinating pumpkin seeds. *Plant Physiology* 106, 61–69.

Mauch, F., Mauch-Mani, B. and Boller, T. (1988) Antifungal hydrolases in pea tissue: inhibition of fungal growth by combinations of chitinase and β-1,3-glucanase. *Plant Physiology* 88, 936–942.

McQueen-Mason, S. and Cosgrove, D.J. (1995) Expansin mode of action on cell walls: analysis of wall hydrolysis, stress relaxation, and binding. *Plant Physiology* 107, 87–100.

McQueen-Mason, S., Durachko, D.M. and Cosgrove, D.J. (1992) Two endogenous proteins that induce cell wall expansion in plants. *The Plant Cell* 4, 1425–1433.

Mella, R.A., Maldonaldo, S. and Sánchez, R.A. (1995) Phytochrome-induced structural changes and protein degradation prior to radicle protrusion in *Datura ferox* seeds. *Canadian Journal of Botany* 73, 1371–1378.

Ni, B.-R. and Bradford, K.J. (1993) Germination and dormancy of abscisic acid- and gibberellin-deficient mutant tomato (*Lycopersicon esculentum*) seeds. Sensitivity of germination to abscisic acid, gibberellin and water potential. *Plant Physiology* 101, 607–617.

Nishitani, K. and Tominaga, R. (1992) Endo-xyloglucan transferase, a novel class of glycosyltransferase that catalyzes transfer of a segment of one xyloglucan molecule to another xyloglucan molecule. *Journal of Biological Chemistry* 267, 21058–21064.

Nonogaki, H. and Morohashi, Y. (1996) An endo-β-mannanase develops exclusively in the micropylar endosperm of tomato seeds prior to radicle emergence. *Plant Physiology* 110, 555–559.

Nonogaki, H., Nomaguchi, M., Okumoto, N., Kaneko, Y., Matsushima, H. and Morohashi, Y. (1998) Temporal and spatial pattern of the biochemical activation of the endosperm during and following imbibition of tomato seeds. *Physiologia Plantarum* 102, 236–242.

Obendorf, R.L. (1997) Oligosaccharides and galactosyl cyclitols in seed desiccation tolerance. *Seed Science Research* 7, 63–74.

Osborne, D.J. and Jackson, M.B. (eds) (1989) *Cell Separation in Plants: Physiology, Biochemistry, and Molecular Biology*. Springer-Verlag, Berlin.

Peretto, R., Favaron, F., Bettini, V., De Lorenzo, G., Marini, S., Alghisi, P., Cervone, F. and Bonfante, P. (1992) Expression and localization of polygalacturonase during the outgrowth of lateral roots in *Allium porrum* L. *Planta* 188, 164–172.

Purcell, P.C., Smith, A.M. and Halford, N.G. (1998) Antisense expression of a sucrose non-fermenting-1-related protein kinase sequence in potato results in decreased expression of sucrose synthase in tubers and loss of sucrose-inducibility of sucrose synthase transcripts in leaves. *Plant Journal* 14, 195–202.

Reinhardt, D., Wittwer, F., Mandel, T. and Kuhlemeier, C. (1998) Localized upregulation of a new expansin gene predicts the site of leaf formation in the tomato meristem. *The Plant Cell* 10, 1427–1438.

Rose, J.K.C., Lee, H.H. and Bennett, A.B. (1997) Expression of a divergent expansin gene is fruit-specific and ripening-regulated. *Proceedings of the National Academy of Sciences, USA* 94, 5955–5960.

Sánchez, R.A. and de Miguel, L. (1997) Phytochrome promotion of mannan-degrading enzyme activities in the micropylar endosperm of *Datura ferox* seeds requires the presence of the embryo and gibberellin synthesis. *Seed Science Research* 7, 27–33.

Sánchez, R.A., de Miguel, L.C. and Mercuri, O. (1985) Phytochrome control of cellulase activity in *Datura ferox* seeds, its relationship with germination. *Journal of Experimental Botany* 37, 1574–1580.

Schopfer, P. and Plachy, C. (1985) Control of seed germination by abscisic acid. III. Effect on embryo growth potential (minimum turgor pressure) and growth coefficient (cell wall extensibility) in *Brassica napus* L. *Plant Physiology* 77, 676–686.

Schröder, R., Atkinson, R.G., Langenkämper, G. and Redgwell, R.J. (1998) Biochemical and molecular characterisation of xyloglucan endotransglycosylase from ripe kiwifruit. *Planta* 204, 242–251.

Shcherban, T.Y., Shi, J., Durachko, D.M., Guiltinan, M.J., McQueen-Mason, S., Shieh, M. and Cosgrove, D.J. (1995) Molecular cloning and sequence analysis of expansins – a highly conserved, multigene family of proteins that mediate cell wall extension in plants. *Proceedings of the National Academy of Sciences, USA* 92, 9245–9249.

Shi, L., Gast, R.T., Gopalraj, M. and Olszewski, N.E. (1992) Characterization of a shoot-specific, GA3- and ABA-regulated gene from tomato. *Plant Journal* 2, 153–159.

Sitrit, Y., Hadfield, K.A., Bennett, A.B., Bradford, K.J. and Downie, B. (1999) Expression of a polygalacturonase associated with tomato seed germination. *Plant Physiology* 121, 419–428.

Smith, P.T., Kuo, T.M. and Crawford, C.G. (1991) Purification and characterization of galactinol synthase from mature zucchini squash leaves. *Plant Physiology* 96, 693–698.

Stephenson, M.B. and Hawes, M.C. (1994) Correlation of pectin methylesterase activity in root caps of pea with root border cell separation. *Plant Physiology* 106, 739–745.

Still, D.W., Dahal, P. and Bradford, K.J. (1997) A single-seed assay for endo-β-mannanase activity from tomato endosperm and radicle tissues. *Plant Physiology* 113, 13–20.

Swanson, S.J. and Jones, R.L. (1996) Gibberellic acid induces vacuolar acidification in barley aleurone. *The Plant Cell* 8, 2211–2221.

Sze, H., Ward, J.M. and Lai, S. (1992) Vacuolar H^+-translocating ATPases from plants: structure, function, and isoforms. *Journal of Bioenergetics and Biomembranes* 24, 371–381.

Taylor, B.H. and Scheuring, C.F. (1994) A molecular marker for lateral root initiation: the *RSI-1* gene of tomato (*Lycopersicon esculentum* Mill.) is activated in early lateral root primordia. *Molecular and General Genetics* 243, 148–157.

Toorop, P.E. (1998) The role of endo-β-mannanase activity in tomato seed germination. PhD Dissertation, Agricultural University, Wageningen, The Netherlands.

Toorop, P.E., Bewley, J.D. and Hilhorst, H.W.M. (1996) Endo-β-mannanase isoforms are present in the endosperm and embryo of tomato seeds but are not essentially linked to the completion of germination. *Planta* 200, 153–158.

Vögeli-Lange, R., Fründt, C., Hart, C.M., Beffa, R., Nagy, F. and Meins, F. Jr (1994) Evidence for a role of β-1,3-glucanase in dicot seed germination. *Plant Journal* 5, 273–278.

Voigt, B. and Bewley, J.D. (1996) Developing tomato seeds when removed from the fruit produce multiple forms of germinative and post-germinative endo-β-mannanase. Responses to desiccation, abscisic acid and osmoticum. *Planta* 200, 71–77.

Welbaum, G.E., Bradford, K.J., Yim, K.-O., Booth, D.T. and Oluoch, M.O. (1998) Biophysical, physiological and biochemical processes regulating seed germination. *Seed Science Research* 8, 161–172.

Wu, C.-T., Bradford, K.J., Leubner-Metzger, G. and Meins, F.Jr. (1998) Class I beta-1,3-glucanase and chitinase are expressed in the endosperm caps of tomato seeds prior to radicle emergence. *Plant Biology '98.* American Society of Plant Physiologists, Rockville, Maryland, abstract no. 46.

Yang, H., Cooley, M.B., Dahal, P., Downie, B., Mella, R.A. and Bradford, K.J. (1996) Isolation and characterization of GA-induced germination-specific genes in tomato by differential display. *Plant Physiology* (Suppl.) 111, 53.

Yang, H., Dahal, P., Cheng, Z., Abe, H. and Bradford, K.J. (1997) Expression of a SNF4-like protein kinase-related gene in tomato seeds. *Plant Physiology* (Suppl.) 114, 270.

23 Characterization of Germination-related Genes in *Avena fatua* seeds

R. JOHNSON

Department of Biology, Colby College, Waterville, ME 04901, USA

A few genes are transiently expressed in non-dormant *Avena fatua* (wild oat) seeds during early imbibition before visible germination occurs, indicating that they may encode important regulatory proteins that are involved in the initiation of germination. Three of these genes, *AFN1*, *AFN2*, and *AFN5*, reach peak levels of expression at 6–12 h of imbibition and then decline to barely detectable levels by 24 h. These genes are now being further characterized to determine what roles they may play in controlling seed germination. *AFN5* mRNA was found to be localized in both the embryonic axis and the scutellum of imbibing seeds. In the embryonic axis, peak levels of this transcript were observed at 3–6 h after the start of imbibition, while in the scutellum, peak mRNA levels were observed at 6–12 h. The biological roles of the transcripts encoded by *AFN1*, *AFN2*, and *AFN5* are still unknown. A portion of the *AFN2* sequence has a moderate degree of similarity to sequences (encoding glutamate and aspartate rich regions) found in other genes including a *Schizosaccharomyces pombe* H^+-ATPase. *AFN5* encodes a protein whose C terminus is very similar to the C terminus of the *Arabidopsis thaliana* thylakoidal processing peptidase.

Introduction

Seed germination is a critically important juncture in the plant life cycle and the decision made by an imbibing seed to initiate germination can be considered to be a critical regulatory step in plant development. The ability of seeds to germinate readily when conditions are suitable for successful growth and the ability to avoid germination at inappropriate times, through the maintenance of dormancy, are thus essential to the survival of a plant species. Seed germination is presumably under the control of specific genes whose expression, or lack thereof, is required to allow the process to proceed. The molecular biology and biochemistry of this decision-making process are not only of interest from the point of view of understanding an important process in developmental biology, but they may also provide insights that would allow better control of germination and dormancy characteristics of economically important crop seeds and weed seeds.

Experiments have shown that the physiology of dormant and non-dormant seeds is very similar during the initial stages of imbibition, but that dormant seeds do not continue the processes necessary to complete germination (Hegarty, 1978; de Klerk, 1987; Bewley and Black, 1994). For example, both dormant and non-dormant *Avena fatua* (wild oat) seeds, upon hydration at suitable temperatures, imbibe water and begin to respire. After 10–12 h of imbibition, however, respiration in dormant embryos reaches a plateau while respiration continues to increase steadily in the non-dormant embryos, eventually leading to cell expansion and radicle emergence (Simmonds and Simpson, 1971).

The actual cellular and molecular mechanisms that lead to the initiation of germination are still not known. Several physical factors may be involved in the germination process, including a lowering of the solute potential of radicle cells which would result in increased water uptake and turgor pressure, cell wall loosening in the radicle cells which would facilitate cell expansion under these conditions, and a weakening of the tissues adjacent to the radicle tip which would allow radicle extension to occur (Bewley and Black, 1994; Welbaum *et al.*, 1998). Which factor is most limiting for germination differs among species, and the ability of a seed to germinate may depend upon complex interactions between different processes (Schopfer and Plachy, 1985; Still and Bradford, 1997). For any seed, however, no matter what the limiting factor(s) for germination-related growth may be, these factors must be under the correct regulatory control. When the seed is dormant, some mechanism exists to prevent the initiation of germination-related growth even under appropriate conditions of imbibition. After the necessary treatment to remove dormancy, such blocks are removed and germination is initiated upon imbibition.

Both abscisic acid (ABA) and gibberellic acid (GA) are known to play significant roles in the regulation of seed dormancy and germination. Imposition of dormancy during embryogenesis requires ABA (Skriver and Mundy, 1990), but the role of ABA in mature imbibing seeds is much less certain. A number of reports suggest that the role of ABA in germination is complex and that localization of ABA production as well as sensitivity to the hormone may prove to be equally as important to the regulation of germination as are hormone levels (LePage-Degivry and Garello, 1992; Bewley and Black, 1994; Bewley, 1997).

Applications of GA stimulate dormant seeds of many species to germinate (Simpson, 1990), and a number of mutants that lack the ability to produce GA are unable to germinate unless this hormone is supplied exogenously (Groot and Karssen, 1987; Dubreucq *et al.*, 1996). However, even though GA treatment is widely used in dormancy and germination studies, it has not been established that the sequence of physiological events taking place during GA-induced germination of dormant seeds is the same as that occurring during 'normal' germination of non-dormant seeds. In fact, there is evidence that both gene expression (Johnson *et al.*, 1995) and embryo growth patterns

(Myers *et al.*, 1997) differ significantly between 'normal' and 'GA-induced' germination.

Identification of Germination-related Genes

It is very likely that the process of seed germination is initially stimulated by the action of hormone(s) or other signalling molecules, which may then initiate an appropriate signalling cascade leading to radicle emergence. Such a signalling system is likely to involve changes in the expression of important regulatory genes whose products are required to stimulate or inhibit the initiation of germination. Efforts have been undertaken to identify genes involved in the regulation of seed germination through the creation and characterization of seed germination mutants (Dubreucq *et al.*, 1996; Leon-Kloosterziel *et al.*, 1996). However, it is likely that a number of germination-related genes with critical roles in initiating growth may not be identified through this type of screening. An alternative approach is to identify genes associated with the germination process on the basis of their expression patterns during seed imbibition. Of particular interest are those genes whose expression patterns are different between non-dormant seeds (which will germinate as a result of imbibition) and dormant seeds (which will not). Genes that are preferentially expressed in dormant seeds may be involved in the prevention of germination or in the protection of dormant seeds from damage from repeated imbibition and long-term storage. Genes that are preferentially expressed in non-dormant seeds may be involved in initiating germination or may be involved in the early metabolic processes required for germination (Johnson *et al.*, 1995).

It is now well established that differential gene expression does indeed occur during imbibition of dormant and non-dormant seeds of a number of species (Morris *et al.*, 1991; Anderberg and Walker-Simmons, 1992; Hong *et al.*, 1992; Aalen *et al.*, 1994) and several differentially expressed genes have been identified in *Avena fatua* (Johnson *et al.*, 1995, 1996; Li and Foley, 1995). In order to identify those genes that are actually involved in the stimulation (or inhibition) of germination, efforts are now being concentrated on genes that are transiently expressed during early imbibition before germination occurs and that are expressed preferentially in non-dormant (or dormant) seeds. Several such genes have been identified in *A. fatua*, and are undergoing characterization.

AFD1

The *A. fatua* gene *AFD1* encodes a mRNA that is present in the embryos of seeds at 3–24 h of imbibition. When whole embryos are assayed, *AFD1* mRNA is found to be somewhat more abundant in dormant seeds than in non-dormant seeds throughout this period. After 24 h, the time at which germination of the seeds is observed, the *AFD1* mRNA rapidly disappears (Johnson, *et al.*, 1995). A more precise analysis, utilizing isolated embryonic axes and scutella, reveals that *AFD1* is not expressed uniformly throughout the embryo. As shown in Fig. 23.1, *AFD1* mRNA is present in both the axes and scutella of both non-dormant and dormant seeds during early imbibition, exhibiting a similar time course of accumulation in the two different regions of the embryo. Interestingly however, the difference in *AFD1* mRNA abundance between non-dormant and dormant seeds is much more striking in the scutellum than it is in the axis. At the beginning of imbibition, *AFD1* mRNA is indeed more abundant in the axis of dormant seeds than in the axis of non-dormant seeds. This difference, however, largely disappears by 12 h. In contrast, the difference in expression between the scutellum of dormant and non-dormant seeds is maintained as the level of *AFD1* mRNA accumulation in the scutellum of non-dormant seeds remains significantly below that seen in the scutellum of dormant seeds.

Although a complete cDNA sequence for *AFD1* has been obtained, the function of this gene remains unknown as it has no significant similarity to any previously identified gene. The pattern of expression of *AFD1* is suggestive of a possible role in the prevention of germination, perhaps by preventing some critical process such as nutrient uptake through the scutellum.

AFN1, *AFN2* and *AFN5*

Several genes that are preferentially expressed in non-dormant seeds have also been identified in *A. fatua*. *AFN1*, *AFN2*, and *AFN5* mRNAs are all substantially more abundant in embryos derived from non-dormant seeds at 6–12 h of imbibition (Johnson *et al.*, 1995, 1996). All of these mRNAs decrease greatly in abundance well before 24 h, when radicle emergence is first observed. The

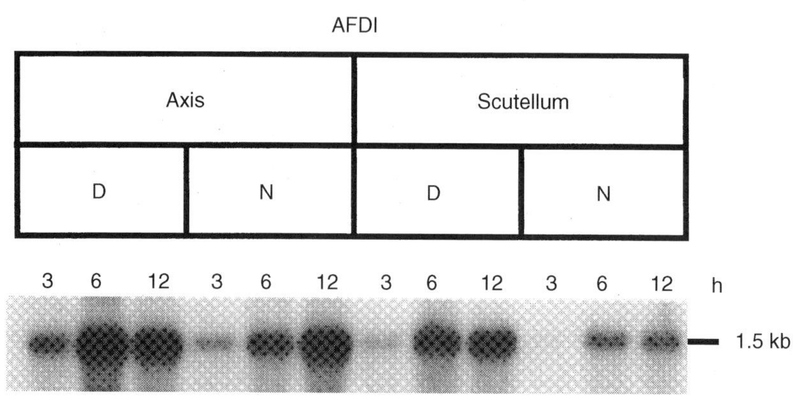

Fig. 23.1. Expression of *AFD1* in axis and scutellum tissues of imbibing *Avena fatua* seeds. After imbibition of dormant (D) and non-dormant (N) seeds for 3, 6, or 12 h in distilled water at 14°C, embryonic axes and scutella were dissected from the seeds and total RNA was isolated from these samples. RNA (5 μg per lane) was electrophoresed in an agarose gel containing formaldehyde, transferred to a nylon membrane, and hybridized with cDNA clone *AFD1*.

fact that these mRNAs are not very abundant and disappear well before germination is completed suggests that they are likely to be involved in 'switching' processes that stimulate germination rather than in the metabolic processes required to support seedling growth.

Besides investigating the temporal patterns of expression of these genes, additional work has been initiated to determine the localization of their expression. This should provide more information about the possible biological roles that could be played by the genes. As shown in Fig. 23.2, *AFN5* mRNA is present in both the embryonic axis and the scutellum of non-dormant seeds. The mRNA both appears and disappears more rapidly in the axis, while there is a lag period of 3–6 h before similar expression patterns are observed in the scutellum. The fact that the RNA can be detected in both axis and scutellum tissues before germination suggests that it might serve as a signal for initiation of growth-related events in the axis and for events related to beginning the uptake of endosperm nutrients through the scutellum.

The actual biological roles of the transcripts encoded by *AFN1*, *AFN2*, and *AFN5* are still unknown. A portion of the *AFN2* sequence has a moderate degree of similarity to sequences (encoding glutamate and aspartate rich regions) found in other genes including a *Schizosaccharomyces pombe* H^+-ATPase. *AFN5* encodes a protein whose C terminus is very similar to the C terminus of an *Arabidopsis thaliana* thylakoidal processing peptidase (TPP) (Chaal, 1998). The *Arabidopsis* TPP is not known to have any specific role during seed germination. A pea protein that is immunologically cross-reactive to the *Arabidopsis* TPP has been localized to the thylakoid membrane (Chaal, 1998), but no information is yet available about the timing or tissue specificity of either the *Arabidopsis* or pea TPP gene expression.

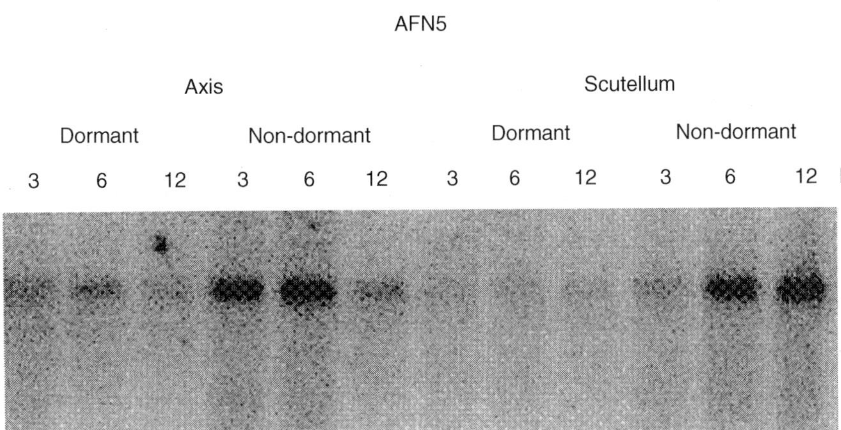

Fig. 23.2. Expression of *AFN5* in axis and scutellum tissues of imbibing *Avena fatua* seeds. After imbibition of seeds for 3, 6, or 12 h in distilled water at 14°C, embryonic axes and scutella were dissected from the seeds and total RNA was isolated from these samples. RNA (5 μg per lane) was electrophoresed in an agarose gel containing formaldehyde, transferred to a nylon membrane, and hybridized with cDNA clone *AFN5*. The RNA bands shown are approximately 2.3 kb.

Conclusions

A number of genes are transiently expressed in seeds during early imbibition before germination occurs. Some of these genes may prove to play important roles in stimulating the germination of non-dormant seeds or in preventing the germination of dormant seeds. Cloning of cDNAs for several germination-related genes has allowed the initial characterization of their expression patterns and sequences. Determination of the biological roles played by these genes will require identification of the activities of the proteins that they encode. Confirmation that these genes indeed play a critical role in stimulating or preventing germination will have to be established through the use of antisense genes or other knockout techniques that will allow observation of the phenotype that results from elimination of a particular gene's function.

References

Aalen, R.B., Opsahl-Ferstad, H.G., Linnestad, C. and Olsen, O.-A. (1994) Transcripts encoding an oleosin and dormancy related protein are present in both the aleurone layer and the embryo of developing barley (*Hordeum vulgare* L.) seeds. *Plant Journal* 5, 385–396.

Anderberg, R.J. and Walker-Simmons, M.K. (1992) Isolation of a wheat cDNA clone for an abscisic acid inducible transcript with homology to protein kinases. *Proceedings of the National Academy of Sciences, USA* 89, 10183–10187.

Bewley, J.D. (1997) Seed germination and dormancy. *Plant Cell* 9, 1055–1066.

Bewley, J.D. and Black, M. (1994) *Seeds: Physiology of Development and Germination.* Plenum Publishing Company, New York.

Chaal, B.K. (1998) Characterization of a cDNA encoding the thylakoidal processing peptidase from *Arabidopsis thaliana. Journal of Biological Chemistry* 273, 689–692.

de Klerk, G.J. (1987) Release of dormancy during after-ripening of *Agrostemma githago* seeds. *Physiologia Plantarum* 139, 209–217.

Dubreucq, B., Grappin, P. and Caboche, M. (1996) A new method for the identification and isolation of genes essential for *Arabidopsis thaliana* seed germination. *Molecular and General Genetics* 252, 42–50.

Groot, S.P.C. and Karssen, C.M. (1987) Gibberellins regulate seed germination in tomato by endosperm weakening; a study with gibberellin deficient mutants. *Planta* 171, 525–531.

Hegarty, T.W. (1978) The physiology of seed hydration and dehydration and the relation between water stress and the control of germination: a review. *Plant, Cell and Environment* 1, 101–119.

Hong, B., Bong, B. and Ho, T.D. (1992) Developmental and organ-specific expression of an ABA and stress induced protein in barley. *Plant Molecular Biology* 18, 663–674.

Johnson, R.R., Cranston, H.J., Chaverra, M.E. and Dyer, W.E. (1995) Characterization of cDNA clones for differentially expressed genes in embryos of dormant and nondormant *Avena fatua* L. caryopses. *Plant Molecular Biology* 28, 113–122.

Johnson, R.R., Cranston, H.J., Chaverra, M.E. and Dyer, W.E. (1996) Analysis of cDNA clones for differentially expressed genes in dormant and non-dormant wild oat

(*Avena fatua* L.) embryos. In: Lang, G.A. (ed.) *Plant Dormancy.* CAB International, Wallingford, pp. 293–300.

Leon-Kloosterziel, K.M., van de Bunt, G.A., Zeevaart, J.A.D. and Koornneef, M. (1996) Arabidopsis mutants with a reduced seed dormancy. *Plant Physiology* 110, 233–240.

LePage-Degivry, M.T. and Garello, G. (1992) In situ abscisic acid synthesis. A requirement for induction of embryo dormancy in *Heliantus annuus. Plant Physiology* 98, 1386–1390.

Li, B. and Foley, M.E. (1995) Cloning and characterization of differentially expressed genes in imbibed dormant and afterripened *Avena fatua* embryos. *Plant Molecular Biology* 29, 823–831.

Morris, C.F., Anderberg, R.J., Goldmark, P.J. and Walker-Simmons, M.K. (1991) Molecular cloning and expression of abscisic acid responsive genes in embryos of dormant wheat seeds. *Plant Physiology* 95, 814–821.

Myers, S. P., Foley, M.E. and Nichols, M.B. (1997) Developmental differences between afterripened and dormant excised *Avena fatua* embryos. *Annals of Botany* 79, 19–23.

Schopfer, P. and Plachy, C. (1985) Control of seed germination by abscisic acid: effect of embryo growth potential and growth coefficient in *Brassica napus* L. *Plant Physiology* 77, 676–686.

Simmonds, J.A. and Simpson, G.A. (1971) Increased participation of pentose phosphate pathway in response to afterripening and gibberellic acid treatment in caryopses of *Avena fatua. Canadian Journal of Botany* 49, 1833–1840.

Simpson, G.M. (1990) *Seed Dormancy in Grasses.* Cambridge University Press, Cambridge, UK.

Skriver, K. and Mundy, J. (1990) Gene expression in response to abscisic acid and osmotic stress. *Plant Cell* 2, 503–512.

Still, D.W. and Bradford, K.J. (1997) Endo-β-mannanase activity from individual tomato endosperm caps and radicle tips in relation to germination rates. *Plant Physiology* 113, 21–29.

Welbaum, G.E., Bradford, K.J., Yim, K., Booth, D.T. and Oluoch, M.O. (1998) Biophysical, physiological, and biochemical processes regulating seed germination. *Seed Science Research* 8, 161–172.

24 Cell Cycle Control during Maize Germination

J.M. VÁZQUEZ-RAMOS

Departamento de Bioquímica, Facultad de Química, Universidad Nacional Autónoma de México, Av. Universidad y Copilco, 04510 México D.F., México

Establishment of the cell cycle during maize germination has been followed using several cell cycle proteins as markers. Also, the effect of plant phytoregulators such as cytokinins and abscisic acid (ABA) has been studied. Dry seeds seem to contain basal levels of proteins such as replicative-type DNA polymerases, p34cdc2 kinase and proliferating-cell nuclear antigen (PCNA) as determined using homologous antibodies; putative cyclin B, cyclin D, CDK4 kinase, p53 and E2F proteins are also present as determined using heterologous antibodies. The behaviour of these proteins varies during germination: the amount and/or activity of DNA polymerases, proliferating-cell nuclear antigen and p34/cyclin B increases with germination, whereas the amount of p53 and cyclin D is reduced. Incubation in the presence of cytokinins enhances activity of DNA polymerases p34/cyclin B and, temporarily, that of CDK4/cyclin D. PCNA and cyclin D are found forming a kinase activity-containing complex during early germination and this complex seems to dissociate at the time at which cells start replicating DNA, an event that occurs faster if germination is accelerated by cytokinins. Abscisic acid appears to delay these events in accordance with its reported inhibitory effect on DNA metabolism and seed germination.

Introduction

Cell division and growth are processes that take place early during seed formation, leading to morphogenetic events in the embryo and endosperm. These and other structures mature as seeds develop while seed reserve formation and deposition proceed simultaneously. Several weeks elapse from the end of the proliferation stage during embryogenesis till seed maturation. Imbibition of water by mature seeds will trigger the germination process, during which the cells in the different maize tissues will be reactivated

 Seed Biology: Advances and Applications (eds M. Black, K.J. Bradford and J. Vázquez-Ramos)

thus re-starting metabolism, ultimately resulting in radicle protrusion and the end of the germination process. Respiration and carbohydrate metabolism are fundamental for germination, implying that organelles are also activated (Bewley and Black, 1994).

Synthesis of macromolecules, such as proteins and nucleic acids, can be observed within minutes after imbibition with an increasing rate as germination advances. Replicative-type DNA synthesis is a late event and it is not evident that a full cell cycle takes place before the radicle emerges. In fact, very little is known about the timing of initiation of the first cell cycle during germination and virtually nothing about the way it is controlled, especially if we recall the time elapsed since the last cell division took place during embryogenesis.

Our studies are devoted to understanding DNA metabolism and cell cycle events during maize germination, trying to establish a temporal course of events and, by modifying it through accelerating germination either by phytohormones such as cytokinins, or by pre-priming seeds, to understand their importance and control during germination.

DNA Metabolism during Maize Germination

Evidence of DNA synthesis can be obtained within the first hour of imbibition of maize embryonic axes and subsequently and its location can be both nuclear and organellar (Zlatanova *et al.*, 1987; Vázquez-Ramos, 1993). Nuclear replicative DNA synthesis starts after 12–15 h of germination, reaching a peak by 25 h. Mitotic figures in meristematic tissues can be observed after around 28 h of germination, with a peak by 36 h (Baiza *et al.*, 1989). DNA polymerase activity can be detected in protein extracts prepared from axes removed from dry seeds. This extract contains at least three different DNA polymerases named 1, 2 and 3 according to their elution profile through a DEAE-Sepharose column (Coello *et al.*, 1992). The activities of DNA polymerases 1 and 3 do not seem to vary much during the first 24 h of germination, whereas the activity of DNA polymerase 2 shows a sharp increase by 11 h of germination with a peak by 24 h. DNA polymerases 1 and 2 appear to be replicative-type enzymes while DNA polymerase 3 would seem to be of repair-type (García *et al.*, 1997). DNA polymerase 2 is a phosphoprotein that becomes phosphorylated at around the time it is activated (11 h germination) which also coincides with the onset time of the S phase (around 12 h). The increase in DNA polymerase 2 activity, it should be noted, is not due to an increase in the amount of the protein and may be related to its phosphorylated state (Coello and Vázquez-Ramos, 1995).

The characteristics shown by DNA polymerase 1 would define it as a δ-type replicative polymerase. DNA polymerase 2 has several characteristics similar to those shown by α-type enzymes, but most significantly, the presence of a DNA primase activity within the protein complex that conforms to DNA polymerase 2, even in the highest purified form (Kornberg and Baker, 1992; García *et al.*, 1997).

Stimulation of DNA Metabolism and the Cell Cycle during Germination by Cytokinins

Addition of cytokinins (benzyladenine, BA) to germinating maize embryonic axes accelerates cell cycle-related events. BA-treated axes start replicating its nuclear DNA by 4–6 h of germination (12–15 h in control cells) and by 15 h there is evidence of mitotic figures (24–28 h in control cells), with a peak by 24 h germination (Reyes *et al.*, 1991). Intense repair-like synthesis can be seen in the early hours and the activity of repair-related enzymes such as poly-ADP ribosyl polymerases is stimulated (Zúñiga-Aguilar *et al.*, 1995).

DNA polymerase activity is also enhanced by BA. The replicative-type DNA polymerases seem to be the targets since partially purified fractions of DNA polymerases 1 and 2 from BA-treated axes show much higher activity than the corresponding enzymes in control fractions. Table 24.1 shows that the activity of polymerase 1 in BA-treated axes is 2.9-fold higher at 6 h and 1.5-fold higher at 20 h, compared with the activity in controls. The activity of polymerase 2 in BA-treated axes is 4.8-fold higher at 6 h and 2.9-fold higher at 20 h. This higher polymerase activity at 6 h of germination correlates well with the previous finding that the S phase starts by 4–6 h in cells of BA-treated axes (Reyes *et al.*, 1991). At least for DNA polymerase 2, this enhanced activity is not due to an increase in the amount of the proteins that form part of this holoenzyme as demonstrated by Western blot analysis using homologous antibodies, indicating that a post-translational control may be involved in DNA polymerase stimulation. It was stated before that DNA polymerase 2 is a phosphoprotein; therefore, phosphorylation ahead of time could be part of the mechanism by which BA activates this enzyme.

It should be noted that the three DNA polymerases studied are present in dry seeds and that the amount of DNA polymerase 2 does not vary during the 0–48 h period of germination, even in BA-stimulated axes.

Perhaps not surprisingly, protein kinase activity is also enhanced by BA as early as 3 h of maize germination (Zúñiga-Aguilar *et al.*, 1995). In eukaryotes, signal transduction is dependent on cascades of protein phosphorylation and dephosphorylation and thus it is most interesting that the response of maize cells to BA involves the stimulation of this mechanism of protein modification.

Table 24.1. Stimulation of replicative-type DNA polymerases by BA during germination.

Imbibition time	DNA polymerase 1 (δ-type) (times over control)	DNA polymerase 2 (α-type) (times over control)
6 h germination	2.9-fold	4.8-fold
20 h germination	1.5-fold	2.9-fold

Cell Cycle Proteins during Maize Germination

The cell cycle is controlled, at least partially, by a series of Ca^{2+}-independent protein kinases that phosphorylate substrates that in turn regulate critical steps in the cycle. These kinases are themselves regulated by phosphorylation and by association with a group of unstable proteins called cyclins. Since cyclins indicate the stage of the cell cycle in which the kinases are acting, the associated kinases are called cyclin-dependent kinases or CDKs (Hunter, 1993). Only one such kinase is found in yeast cells, $p34^{cdc2}$, an essential protein required for the transitions G1/S and G2/M, bound to different cyclins (Morgan, 1997). Several CDKs are found in higher eukaryotes, associated with distinct cyclins, governing the different transitions in the cycle (Morgan, 1997). The homologue protein to yeast $p34^{cdc2}$, also called CDK1, functions mainly in the G2/M transition and was the first CDK-like protein to be found in plants (John *et al.*, 1989).

A Ca^{2+}-independent kinase activity is present during early maize germination and this activity increases notably when axes are imbibed in the presence of BA, showing a peak by 15 h of germination, a time when the first mitotic figures start appearing under these conditions (Herrera-Teigeiro *et al.*, 1999).

Since $p34^{cdc2}$-type kinases are responsible for the passage of cells from G2 to M, the presence and activity of one such kinase was searched for in maize embryo axes. Both the transcript for maize $p34^{cdc2}$ and the corresponding protein are present in cells from dry seeds and their amount does not vary substantively throughout the germination process (Herrera-Teigeiro *et al.*, 1999). $p34^{cdc2}$ forms a complex with cyclin B and evidence of cyclin B-like proteins in plants has been reported (Renaudin *et al.*, 1994). One such protein has been immunorecognized in extracts from dry maize axes and this protein shows minor variations as seeds germinate. Antibodies against human cyclin B immunoprecipitate maize $p34^{cdc2}$ and the reverse is also true, antibodies against $p34^{cdc2}$ immunoprecipitate the cyclin B-type protein. More relevant is that these immunoprecipitates can phosphorylate histone H1, the preferred substrate of this kinase complex (and of most cyclin-CDK complexes). Moreover, the activity of this complex is stimulated by BA, showing a peak by 15 h of germination, the time when mitotic figures start appearing under these conditions. Further evidence has shown that the treatment with BA stimulates the transfer of the cyclin-CDK complex to nuclei, its site of action (Morgan, 1997; Herrera-Teigeiro *et al.*, 1999). In mammals and yeast cells, the cyclin B-$p34^{cdc2}$ complex is regulated at three levels: synthesis-degradation of cyclin B, phosphorylation-dephosphorylation of $p34^{cdc2}$ and nuclear translocation of the complex, an event that takes place at the beginning of the M phase (Morgan, 1997). In maize cells, the transport of the kinase complex to nuclei appears as early as 15 h of germination under stimulated conditions and several hours later in control cells. Since both $p34^{cdc2}$ and the putative cyclin B seem to be stable proteins and are present at similar levels during the germination process, then cytokinins should stimulate mitosis during germination by activating and mobilizing the kinase complex thus driving the G2/M transition (Herrera-Teigeiro *et al.*, 1999). Evidence of cytokinins promoting plant $p34^{cdc2}$ protein kinase phosphorylation has also been provided (Zhang *et al.*, 1996).

The S phase initiation time in BA-stimulated maize axes cells is advanced several hours compared to the timing in control cells. Thus, the action of cytokinins must also be exerted at a very early stage, soon after imbibition of water, to promote cell cycle events.

Before entering S phase, cells must survey the environment to check that the appropriate conditions for proliferation are attainable. Cells in dry seeds should remain in a stationary state of the cycle, or G0 and they would shift into the G1, metabolic phase as imbibition proceeds. In germinating maize seeds, cells should traverse these G0/G1 phases and go into the S phase in approximately 10–12 h. It is noteworthy that the S phase starts after only 4 h in cells of BA-treated axes, indicating that the mechanism of cell cycle promotion by cytokinins involves a notable shortening of the G0/G1 stages (a similar timing is observed in germinating, osmoprimed maize seeds).

The G1 phase of the cell cycle in higher eukaryotes is a complex stage, controlled by a plethora of proteins that have fundamental roles in sensing the metabolic and nutritional status, chromosomal integrity, the population in tissues and thus the 'convenience' of proliferation. Having met these criteria, cells can establish a commitment to undergo a complete cell cycle, regulated by cyclin-CDK complexes.

Several cell cycle proteins participating in the G1 phase of the cycle have been followed during maize germination and also the influence of cytokinins in their behaviour has been studied. Heterologous (anti-human) antibodies against cyclin D, CDK4, p53 and E2F and homologous antibodies against maize proliferating-cell nuclear antigen (PCNA) have been used.

Cyclin D, together with its kinase partner CDK4, controls the entry of cells into a committed G1 phase, i.e. this complex is a sensor of the cellular metabolic state and under appropriate conditions will allow cells to undergo a complete cell cycle, after which cyclin D may be degraded (Won *et al.*, 1992; Morgan, 1997). The main task of this complex appears to be to phosphorylate and thus inactivate a protein, pRB (Ewen *et al.*, 1993), that represses the progression of the cycle by sequestering a transcription factor, E2F, necessary for the expression of genes whose protein products are essential for events at late G1 and S phases (Sherr, 1994); DNA metabolism enzymes and CDKs are some of these protein products (Fig. 24.1).

PCNA is an essential accessory protein for DNA polymerase δ (Tan *et al.*, 1986), the polymerase involved in processive DNA replication and thus it is an excellent S phase marker. However, in recent years evidence has accumulated placing PCNA as a participant in the G1 phase. Association between PCNA and cyclin D has been found and it is speculated that the main purpose of this association would be to keep PCNA away from the replication apparatus (Fig. 24.1), thus regulating the initiation of the S phase (Pagano *et al.*, 1994). The role of p53 protein is to constantly survey chromosomal integrity, being responsible for blocking the progression of the cell cycle given any chromosomal damage; therefore, p53 helps to prevent accumulation of chromosomal errors (Lane, 1992). In extreme cases, cells will undergo p53-mediated apoptosis (cell death).

All cell cycle-like proteins enlisted above are present in cells from dry maize embryo axes, as determined by Western blot; nonetheless, they show different behaviour during germination: the amount of the E2F-like protein remains constant during the period measured (0–24 h of germination). Cyclin D and p53-type proteins disappear as germination advances so that there is very little protein by 15 h (Fig. 24.2; Cruz-García *et al.*, 1998). Antibodies against recombinant maize PCNA (López *et al.*, 1995; I. Herrera, M.P. Sánchez and J.M. Vázquez-Ramos, unpublished) recognize a protein that gradually increases during germination, reaching a peak by 20 h (Fig. 24.2); finally, the putative CDK4 protein decreases to very low levels by 15 h.

Germination in the presence of BA does not change the behaviour of the E2F-like protein; however, the putative cyclin D and p53 proteins disappear much faster under the accelerated germination conditions (Cruz-García *et al.*, 1998). Proliferating-cell nuclear antigen protein shows a sharp increase by 3 h of germination, reaching a peak by 6 h, after which its amount remains constant (Fig. 24.2); the CDK4-type protein seems to be stabilized under these conditions. Imbibition of maize axes in the presence of a protein synthesis inhibitor, cycloheximide, causes a rapid loss of both PCNA and the putative cyclin D in the early hours of germination, suggesting that these proteins are

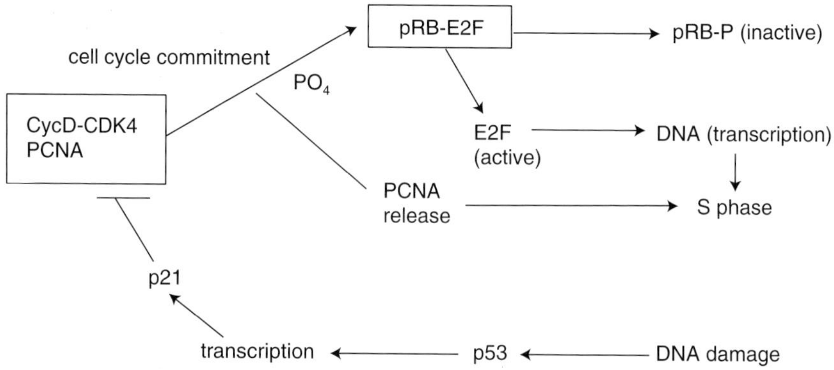

Fig. 24.1. Critical G1 regulatory steps.

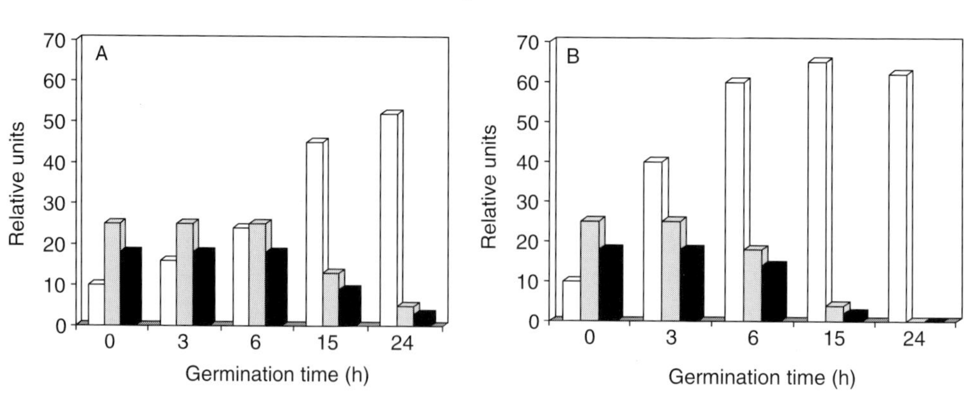

Fig. 24.2. Behaviour of cell cycle proteins (□, PCNA; ▩, cyclin D; ■, p53) during maize germination. A, control; B, benzyladenine treated seeds.

synthesized *de novo*. Degradation of the putative cyclin D and p53 proteins should perhaps be taken as a signal in the cell that a committed cell cycle has been initiated, promoted by the action of cyclin D and that cell cycle blockers such as p53 should be eliminated. On the contrary, PCNA amount should increase to match the requirements that the replicative machinery have for this protein factor.

Cyclin D and CDK4-type proteins can be found forming a complex. Antibodies against cyclin D co-immunoprecipitate the CDK4-type protein and the reverse is also true. In either case, histone H1 kinase activity can be measured in the precipitates. During germination, histone H1 kinase activity is higher at 6 h and then declines to undetectable levels, a behaviour reminiscent of that of the putative cyclin D protein, which disappears as germination advances. This process is accelerated if maize axes are imbibed in the presence of BA. The concomitant disappearance of both the cyclin D-like protein and the kinase activity would further reinforce the notion of CDK4 and cyclin D-type proteins forming a complex.

Moreover, maize PCNA and the putative cyclin D protein also form a complex, just as has been demonstrated for mammalian cells (Xiong *et al.*, 1992). These proteins can be found coeluting through a Superdex column, as determined by Western blot of proteins in fractions (Fig. 24.3). Furthermore, this association is not an artefact as antibodies against cyclin D coimmunoprecipitate PCNA and antibodies against PCNA coimmunoprecipitate the putative cyclin D. A search for the presence of the putative CDK4 in these pellets has shown that this kinase is also present within the complex. Thus, as in mammalian cells, the complex of maize putative cyclin D and CDK4 proteins contains PCNA. Incubation of maize embryo axes in the presence of BA promotes a

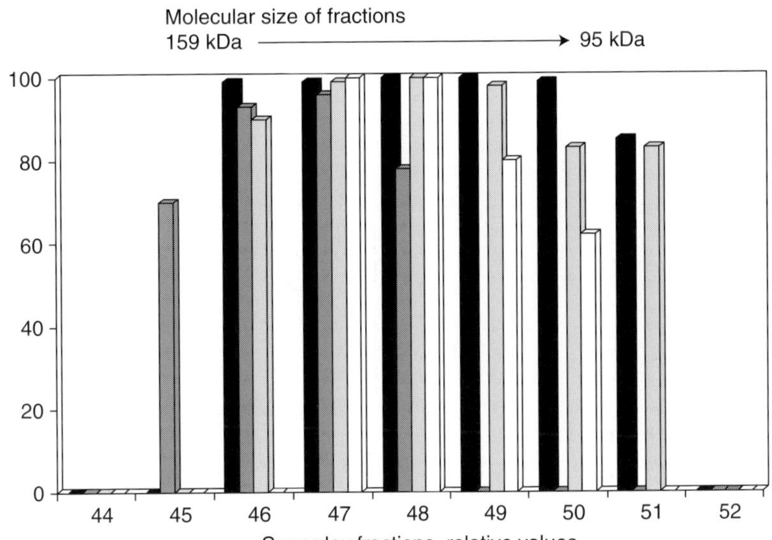

Fig. 24.3. Proliferating-cell nuclear antigen (■, 3 h; ▩, 6 h) and cyclin D (□, 3 h; □, 6 h) co-elution during germination.

rapid release of PCNA from this 'ternary' complex, an event that occurs at a time when DNA replication starts. Then, the suggested role of the cyclin-kinase complex as a PCNA chelator to control the time of initiation of the S phase may just be very likely.

In summary, all cell cycle proteins studied in this work are present in dry maize axes. However, their behaviour varies as germination advances and, in general, this variation is more pronounced if germination is accelerated. Apparently, post-translational protein modification may be a strong component of the regulatory events triggering the cell cycle during early germination, although transcriptional or translational regulatory events cannot be ruled out. Signal transduction may be the key event to control cell cycle and seed germination.

Acknowledgements

The author acknowledges support from DGAPA IN-209895 and CONACYT 3407P-N.

References

Baiza, A.M., Vázquez-Ramos, J.M. and Sánchez de Jiménez, E. (1989) DNA synthesis and cell division in embryonic maize tissues during germination. *Journal of Plant Physiology* 135, 416–421.

Bewley, J.D. and Black, M. (1994) *Seeds. Physiology of Development and Germination*, 2nd edn. Plenum Press, New York.

Coello, P. and Vázquez-Ramos, J.M. (1995) Maize DNA polymerase 2 is a phosphoprotein with increasing activity during germination. *European Journal of Biochemistry* 231, 99–103.

Coello, P., García, E., Rodríguez, R. and Vázquez-Ramos, J.M. (1992) A DNA polymerase from maize axes: its purification and possible role. *Plant Molecular Biology* 20, 1159–1168.

Cruz-García, F., Zúñiga-Aguilar, J.J. and Vázquez-Ramos, J.M. (1998) Effect of stimulating maize germination on cell cycle proteins. *Physiologia Plantarum* 102, 573–581.

Ewen, M.E., Sluss, H.K., Sherr, C.J., Matsushime, H., Kato, J.Y. and Livingston, D.M. (1993) Functional interactions of the retinoblastoma protein with the mammalian D-type cyclins. *Cell* 73, 487–497.

García, E., Orjuela, D., Camacho, Y., Zúñiga, J.J., Plasencia, J. and Vázquez-Ramos, J.M. (1997) Comparison among DNA polymerases 1, 2 and 3 from maize embryo axes. A DNA primase activity copurifies with DNA polymerase 2. *Plant Molecular Biology* 33, 445–455.

Herrera-Teigeiro, I., Jiménez-García, L.F. and Vázquez-Ramos, J.M. (1999) Benzyl adenine promotes early activation of $p34^{cdc2}$-like kinase(s) during maize germination. *Seed Science Research* 9, 55–62.

Hunter, T. (1993) Breaking the cycle. *Cell* 75, 839–841.

John, P.C.L., Sek, F.J. and Lee, M.G. (1989) A homolog of the cell cycle control protein $p34^{cdc2}$ participates in the division cycle of *Chlamydomonas*, and a similar protein is detectable in higher plants and remote taxa. *Plant Cell* 1, 1185–1193.

Kornberg, A. and Baker, T. (1992) *DNA Replication*, 2nd edn. W.H. Freeman and Co.

Lane, D.P. (1992) P53, guardian of the genome. *Nature* 358, 15–16.

López, I., Khan, S., Vázquez, J. and Hussey, P. (1995) Molecular cloning of a maize cDNA clone encoding a putative proliferating cell nuclear antigen. *Biochimica et Biophysica Acta* 1260, 119–121.

Morgan, D.O. (1997) Cyclin-dependent kinases: engines, clocks and microprocessors. *Annual Review of Cell Developmental Biology* 13, 261–291.

Pagano, M., Theodoras, A.M., Tam, S.W. and Draetta, G.F. (1994) Cyclin D1-mediated inhibition of repair and replicative DNA synthesis in human fibroblasts. *Genes and Development* 8, 1627–1639.

Renaudin, J.P., Colasanti, J., Rime, H., Yuan, Z. and Sundaresan, V. (1994) Cloning of four cyclins from maize indicates that higher plants have three structurally distinct groups of mitotic cyclins. *Proceedings of the National Academy of Sciences, USA* 91, 7375–7379.

Reyes-Jiménez, J., Jiménez-García, L.F., González, M.A. and Vázquez-Ramos, J.M. (1991) Benzyladenine stimulation of nuclear DNA synthesis and cell division in germinating maize. *Seed Science Research* 1, 113–117.

Sherr, C.J. (1994) G1 phase progression cycling on cue. *Cell* 79, 812–821.

Tan, C.K., Castillo, C., So, A.G. and Downey, K.M. (1986) An auxiliary protein for DNA polymerase δ from fetal calf thymus. *Journal of Biological Chemistry* 261, 12310–12316.

Vázquez-Ramos, J.M. (1993) Maize germination, DNA metabolism and cytokinins. In: Côme, D. and Corbineau, F. (eds) *Proceedings of the 4th Workshop on Seeds: Basic and Applied Aspects of Seed Biology*, Vol. 2. AFSIS, Paris, pp. 311–316.

Won, K.A., Xiong, Y., Beach, D. and Gilman, M.Z. (1992) Growth-regulated expression of D-type cyclin genes in human diploid fibroblasts. *Proceedings of the National Academy of Sciences, USA* 89, 9910–9914.

Xiong, Y., Zhang, H. and Beach, D. (1992) D-type cyclins associate with multiple protein kinases and the replication and repair factor PCNA. *Cell* 71, 505–514.

Zhang, K., Letham, D.S. and John, P.C.L. (1996) Cytokinin controls the cell cycle at mitosis by stimulating the tyrosine dephosphorylation and activation of $p34^{cdc2}$-like histone kinase. *Planta* 200, 2–12.

Zlatanova, J.S., Ivanov, P.S., Stoilov, L.M., Chimshirova, K.V. and Stanchev, B.S. (1987) DNA repair precedes replicative synthesis during early germination in maize. *Plant Molecular Biology* 10, 139–144.

Zúñiga-Aguilar, J.J., López, I., Gómez, A. and Vázquez-Ramos, J.M. (1995) Does benzyladenine stimulate DNA metabolism by modifying gene expression during maize germination? *Seed Science Research* 5, 219–226.

25 Recent Advances in ABA-regulated Gene Expression in Cereal Seeds: Evidence for Regulation by PKABA1 Protein Kinase

M.K. WALKER-SIMMONS

USDA Agriculture Research Service, Wheat Genetics, Quality, Physiology and Disease Research Unit, Washington State University, Pullman, WA 99164-6420, USA

The plant hormone, abscisic acid (ABA) has an essential role in the physiological processes that affect seed survival and reproduction. Many of these responses to ABA occur through ABA signalling processes that stimulate or suppress ABA-responsive gene expression. A protein kinase mRNA called PKABA1 (protein kinase – ABA-responsive), that is a potential intermediate in ABA signal transduction has been cloned from dormant wheat seeds. Constitutive PKABA1 suppresses gibberellin (GA)-regulated α-amylase and protease gene expression in barley aleurone. A brief overview of this protein kinase research and other recent advances in ABA signalling processes in cereals seeds is presented.

Introduction

Abscisic acid (ABA) is involved in many physiological processes in seeds. It is required for induction of dormancy and acquisition of desiccation tolerance during seed development. Applied ABA is an effective germination inhibitor and can prolong dormancy. Abscisic acid can suppress gibberellin (GA)-mediated genes encoding hydrolytic enzymes in germinating seeds, and in seedlings ABA can induce tolerance to dehydration and cold temperatures. The study of ABA-deficient and ABA-insensitive mutants has established the essential role of ABA in these physiological processes (reviewed in Hilhorst, 1995; Bonetta and McCourt, 1998). We are seeking to identify critical regulatory genes, particularly protein kinases, involved in ABA-mediated events in seeds. Recent progress is reviewed in this chapter.

CAB *International* 2000. *Seed Biology: Advances and Applications* (eds M. Black, K.J. Bradford and J. Vázquez-Ramos)

ABA Signalling Processes in Cereal Seeds

Most of the physiological processes involving ABA are mediated through ABA effects on gene expression. Many of the advances in our understanding of hormonal regulation of seed germination have come from studying the effects of ABA and GA on gene expression in the cereal aleurone layer. In this tissue ABA suppresses GA-inducible genes including those for α-amylases and proteases required for postgerminative growth. Abscisic acid also exhibits positive effects on gene expression in the aleurone cells. Abscisic acid increases the amount of LEA (late embryogenesis abundant) proteins and other proteins associated with desiccation tolerance and long-term seed survival.

Promoter elements and trans-acting factors that affect ABA-responsive gene expression in the cereal aleurone layer have been identified. Promoters of some ABA-responsive genes contain conserved ABA-responsive elements along with coupling elements that together provide the minimal promoter unit necessary for specific ABA-responsive gene expression (Shen and Ho, 1995; Shen *et al.*, 1996). Trans-acting factors affecting ABA-responsive gene expression in seeds include EmBP-1 (wheat embryo), VP1 (maize seed), and AB14 (*Arabidopsis* seed) (reviewed in Bonetta and McCourt, 1998; Grill and Himmelbach, 1998). Alteration of these trans-acting proteins can affect ABA responsiveness in seeds. For example, alteration of the amino acid sequence of AB14 at the carboxy-terminus results in loss of ABA sensitivity (Finkelstein *et al.*, 1998). A trans-acting factor that affects GA-mediated-amylase expression has also been identified. This is the MYB transcription factor that is expressed in response to GA (Gubler *et al.*, 1995). MYB transcriptionally activates expression of GA-regulated genes encoding hydrolytic enzymes including α-amylase in cereal aleurone tissue (Gubler *et al.*, 1998).

Other elements involved in ABA signalling have also been identified, but their connections with each other and relative position in the signalling pathway have not yet been determined. Phospholipase D has been identified as an intermediate in ABA signal transduction in barley aleurone cells (Ritchie and Gilroy, 1998a). Addition of ABA to this tissue causes an increase in phospholipase D activity. GA-responsive α-amylase gene expression is controlled by HvSPY, which has sequence homology to N-acetylglucosamine transferases (Robertson *et al.*, 1998). Mutation analysis of HvSPY indicates that HvSPY is a negative regulator of GA action. A calcium-dependent protein kinase identified in barley aleurone cells has been proposed to inhibit GA responses therein (Ritchie and Gilroy, 1998b).

The importance of protein kinases and reversible phosphorylation in ABA signalling is increasingly apparent (Verhey and Walker-Simmons, 1997; Grill and Himmelbach, 1998). This is not surprising since protein kinases that phosphorylate proteins and protein phosphatases that dephosphorylate proteins are components in most signalling pathways. Substantial evidence for the role of phosphorylation has come from the characterization of the ABA-insensitive (*abi*) *Arabidopsis* mutants. Seeds of these mutants can germinate in the presence of high ABA amounts and exhibit reduced dormancy (Karssen *et al.*, 1983). ABI1 is a protein phosphatase 2C, which when mutated has reduced protein phosphatase activity (Sheen, 1998). The *abi1* mutant is dominant. Possibly, the dominant effect occurs when the mutated protein phosphatase partially binds or interferes with a substrate, which results in blocking the signalling pathway (Grill and Himelbach, 1998).

PKABA1 – ABA-responsive Protein Kinase

A protein kinase with a potential role in ABA-signal transduction in cereal seeds is PKABA1 (protein kinase – ABA responsive). PKABA1 mRNA was cloned from a wheat embryo cDNA library prepared from dormant seed (Anderberg, 1992). PKABA1 was identified by screening with a degenerative DNA probe developed from conserved protein kinase catalytic sequences.

PKABA1 mRNA production is ABA-responsive (Anderberg and Walker-Simmons, 1992) and amounts of the mRNA increase in developing wheat seeds as ABA content rises. PKABA1 mRNA contents reach their highest level in mature seeds. The PKABA1 mRNA declines in germinating seeds. However, if germination is inhibited by applied ABA, PKABA1 mRNA contents remain high. Also, PKABA1 mRNA remains high in hydrated dormant seeds that are growth-arrested. PKABA1 mRNA is also induced in seedling crowns, shoots and roots by ABA increases resulting from cold temperature or dehydration (Holappa and Walker-Simmons, 1995).

Because PKABA1 is an ABA-responsive protein kinase, it is a potential candidate for an intermediate in the ABA signalling pathway. To learn if PKABA1 is involved in ABA signalling processes, we determined the effects of constitutive expression of PKABA1 in barley aleurone cells. PKABA1 effect was measured in the absence of the hormone ABA (Gomez-Cadenas *et al.*, 1999). The *PKABA1* coding sequence was linked to a constitutive promoter. This *PKABA1* construct was cobombarded into barley aleurone tissue with reporter constructs consisting of GA-responsive promoters linked to the marker gene, β-glucuronidase. GA-responsive promoters for both low- and high-pI α-amylase or a GA-responsive protease, EPB-1, were used. After bombardment of the constitutive *PKABA1* and a GA-responsive promoter construct, the aleurone were incubated in GA for 24 h, and then PKABA1 effects on GA-responsive gene expression were measured. Results showed the constitutive PKABA1 markedly suppressed GA-responsive gene expression in the absence of ABA. PKABA1 suppressed GA-responsive α-amylase and protease gene expression. Constitutive PKABA1 has the same magnitude of suppressive effect in aleurone cells as ABA.

Mutational analysis was used to determine if an active PKABA1 kinase is required for the suppressive effect on GA action in aleurone tissue (Gomez-Cadenas *et al.*, 1999). A null-mutant of PKABA1 was constructed by deleting the glycine-rich loop part of the ATP-binding site from the *PKABA1* coding sequence and then linking this mutated sequence to the constitutive promoter. Impairing ATP binding markedly reduces activity of protein kinases. When the null-mutant of PKABA1 was cobombarded with the GA-responsive α-amylase reporter construct, no significant suppressive activity on GA-responsive α-amylase gene expression was detected. Thus, results with the null-mutant indicate that PKABA1 kinase activity is required for the suppressive effects observed on GA-mediated gene expression in aleurone layers.

Effects of constitutive PKABA1 were also tested on ABA-responsive gene expression in barley aleurone cells. Constitutive PKABA1 expression was tested on the ABA-responsive promoter of the LEA gene, *HVA1*. Constitutive PKABA1 had only a small effect on ABA-responsive gene expression as measured with the *HVA1* promoter. Effects were small compared with much larger stimulatory effects of the hormone ABA.

Results from determining effects of constitutive PKABA1 in barley aleurone tissue indicate that the ABA signal pathway in aleurone cells may be branched. PKABA1 appears to be a major intermediate in the pathway leading to suppression of GA-responsive gene expression. However, PKABA1 has only a small effect on the pathway that induces the ABA-responsive *LEA* genes.

PKABA1 Sequence Analysis

Sequence analysis has revealed that the deduced amino acid sequence of PKABA1 contains all the conserved sequence features of a serine/threonine protein kinase. As shown in Fig. 25.1 there is an ATP-binding site near the amino terminus adjacent to the kinase catalytic domain. Beyond the catalytic domain there is an unknown region presumed to contain regulatory sequences. A notable feature is a stretch of acidic amino acids, mainly aspartic acid, near the carboxyl terminus. A number of plant protein kinases with sequence similarity to *PKABA1* including the acidic amino acid stretch have now been cloned. Other members of the *PKABA1* subfamily have been identified in wheat, soybean, rapeseed, tobacco, ice plant and *Arabidopsis* (reviewed by Holappa and Walker-Simmons, 1997). A wheat homologue, *TaPK3*, has 97% sequence similarity to *PKABA1*, but is differentially expressed (Holappa and Walker-Simmons, 1997). *TaPK3* is expressed in greening wheat seedlings and has not been detected in seeds. A rice endosperm PKABA1 homologue has been recently identified that is expressed in developing seed (Hotta *et al.*, 1998).

PKABA1 sequence comparison studies provide additional evidence supporting a functional role of PKABA1 in metabolic regulation, particularly in protecting cells during nutritional or environmental stress. Members of the *PKABA1* subfamily have the highest sequence similarity with the mammalian AMP-activated protein kinase and the yeast SNF1 protein kinase (Halford and

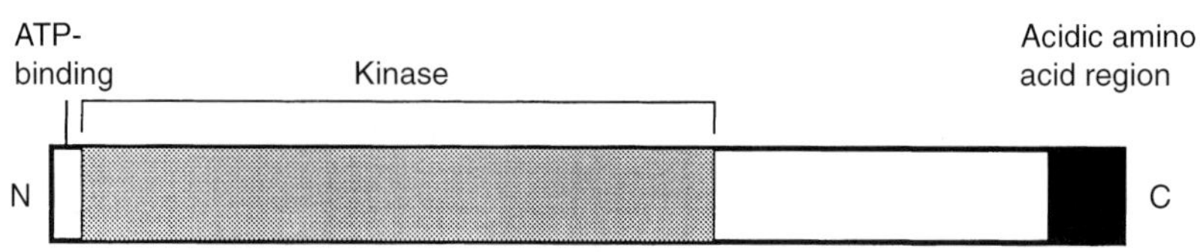

Fig. 25.1. Domain structure of PKABA1.

Hardie, 1998). Mammalian AMP-activated protein kinase is activated when cells become depleted for ATP (Hardie and Carling, 1997). This mammalian kinase suppresses biosynthetic processes requiring ATP such as fatty acid and sterol synthesis. Yeast SNF1 is activated by glucose starvation and is required for derepression of glucose-repressed genes. The sequence similarity of PKABA1 with these mammalian and yeast kinases that regulate metabolism during nutritional and environmental stress suggests a similar function for *PKABA1*. Like the mammalian AMP-activated protein kinase and the yeast SNF1, PKABA1 may also have a protective role when rapid growth is unfavourable in seeds.

Conclusions

Elements in the ABA signalling pathway are beginning to be identified in seeds. The importance of protein phosphorylation processes in ABA signalling has been established by genetic evidence indicating that protein phosphatases are involved in ABA responses and by the identification of ABA-responsive protein kinases. A novel protein kinase, *PKABA1*, that is ABA-responsive has been cloned from dormant seed embryos. Constitutive expression of *PKABA1* markedly suppresses GA induction of low- and high-pI α-amylase and protease genes in barley aleurone tissue. A null-mutant of *PKABA1* did not have a significant effect on α-amylase expression. These results suggest that PKABA1 is an intermediate in the ABA signalling pathway leading to suppression of GA-responsive gene expression in cereal aleurone cells. *PKABA1* has sequence similarity with mammalian and yeast protein kinases that suppress metabolic processes during nutrient and environmental stress. The sequence similarities suggest that PKABA1 has a similar protective role in seeds in suppressing GA-inducible genes encoding hydrolytic enzymes required for postgerminative growth.

References

Anderberg, R.J. and Walker-Simmons, M.K. (1992) Isolation of a wheat cDNA clone for an abscisic-acid-inducible transcript with homology to protein kinases. *Proceedings of the National Academy of Sciences, USA* 89, 10183–10187.

Bonetta, D. and McCourt, P. (1998) Genetic analysis of ABA signal transduction pathways. *Trends in Plant Science* 3, 231–235.

Finkelstein, R.R, Wang, M.L., Lynch, T.J., Rao, S. and Goodman, H.M. (1998) The *Arabidopsis* abscisic aicd response locus ABI4 encodes an APETALA 2 domain protein. *Plant Cell* 10, 1043–1054.

Gomez-Cadenas, A., Verhey, S.D., Holappa, L.D., Shen, Q. Ho, T.-H.D. and Walker-Simmons, M. (1999) An abscisic acid-induced protein kinase, PKABA1, mediates abscisic acid-suppressed gene expression in barley aleurone layers. *Proceedings of the National Academy of Sciences, USA* 96, 1767–1772.

Grill, E. and Himmelbach, A. (1998) ABA signal transduction. *Current Opinion in Plant Biology* 1, 412–418.

Gubler, F., Kalla, R., Roberts, J.K. and Jacobsen, J.V. (1995) Gibberellin-regulated expression of a *myb* gene in barley aleurone cells: evidence for Myb transactivation of a high-pI α-amylase gene. *Plant Cell* 7, 1879–1891.

Gubler, F., Raventos, D., Keys, M., Watts, R., Mundy, J. and Jacobsen, J.V. (1998) Target genes and regulatory domains of the GAMYB transcriptional activator in cereal aleurone. *Plant Journal* 17, 1–9.

Halford, N.G. and Hardie, D.G. (1998) SNF1-related protein kinases: global regulators of carbon metabolism in plants? *Plant Molecular Biology* 37, 735–748.

Hardie, D.G. and Carling, D. (1997) The AMP-activated protein kinase – fuel gauge of the mammalian cell? *European Journal of Biochemistry* 246, 259–273.

Hilhorst, H.W. (1995) A critical update on seed dormancy. I. Primary dormancy. *Seed Science Research* 5, 61–73.

Holappa, L. and Walker-Simmons, M.K. (1995) The wheat ABA-responsive protein kinase mRNA, PKABA1, is upregulated by dehydration, cold temperature and osmotic stress. *Plant Physiology* 108, 1203–1210.

Holappa, L.D. and Walker-Simmons, M.K. (1997) The wheat protein kinase gene, *TaPK3*, of the *PKABA1* subfamily is differentially regulated in greening wheat seedlings. *Plant Molecular Biology* 33, 935–941.

Hotta, H., Aoki, N., Matsuda, T. and Adachi, T. (1998) Molecular analysis of a novel protein kinase in maturing rice seed. *Gene* 213, 47–54.

Karssen, C., Swan, D.B.-v. d., Breekland, A. and Koorneef, M. (1983) Induction of dormancy during seed development by endogenous abscisic acid: studies on abscisic acid deficient genotypes of *Arabidopsis thaliana* (L.) Heynh. *Planta* 157, 158–165.

Ritchie, S. and Gilroy, S. (1998a) Abscisic acid signal transduction in the barley aleurone is mediated by phospholipase D activity. *Proceedings of the National Academy of Sciences, USA* 95, 2697–2702.

Ritchie, S. and Gilroy, S. (1998b) Calcium-dependent protein phosphorylation may mediate the gibberellic acid response in barley aleurone. *Plant Physiology* 116, 765–776.

Robertson, M., Swain, S., Chandler, P.M. and Olszewski, N.E. (1998) Identification of a negative regulator of gibberellin action, HvSPY, in barley. *Plant Cell* 10, 995–1008.

Sheen, J. (1998) Mutational analysis of protein phosphatase 2C involved in abscisic acid signal transduction in higher plants. *Proceedings of the National Academy of Sciences, USA* 95, 975–980.

Shen, Q. and Ho, T.-H.D, (1995) Functional dissection of an abscisic acid (ABA)-inducible gene reveals two independent ABA-responsive complexes each containing a G-box and a novel *cis*-acting element. *Plant Cell* 7, 295–307.

Shen, Q., Zhang, P. and Ho, T.D.-H. (1996) Modular nature of abscisic acid (ABA) response complexes: composite promoter units that are necessary and sufficient for ABA induction of gene expression in barley. *Plant Cell* 8, 1107–1119.

Verhey, S.D. and Walker-Simmons, M.K. (1998) Abscisic acid-mediated responses in seeds involving protein kinases and phosphatases. In: Ellis, R.H., Black, M., Murdoch, A.J. and Hong, T.D. (eds) *Basic and Applied Aspects of Seed Biology.* Kluwer Academic, Dordrecht, pp. 225–233.

26 Lettuce Endosperm Weakening: a Role for Endo-β-mannanase in Seed Germination at High Temperature*

D.J. Cantliffe, W.M. Nascimento, Y. Sung and D.J. Huber

University of Florida, IFAS, Horticultural Sciences Department, Gainesville, FL 32611-0690, USA

Under high temperatures, seed germination of most lettuce genotypes can be erratic or completely inhibited. Weakening of the endosperm layer is a prerequisite to radicle protrusion at high temperatures. Enzyme-mediated degradation of endosperm cell walls may be a crucial factor for lettuce germination at high temperatures. The galactomannan polysaccharides in lettuce endosperm cell wall are mobilized by endo-β-mannanase. The involvement of endo-β-mannanase during germination of lettuce seeds at high temperature (35°C) was investigated by gel-diffusion assay. Genotype, seed maturation temperature, and seed priming enhanced lettuce seed germination at high temperature. Less force to penetrate the endosperm was required by seeds of thermotolerant genotypes or primed seeds, and structural alterations of endosperm in the micropylar area were also observed before radicle protrusion. Higher endo-β-mannanase activity in lettuce endosperm before radicle protrusion was verified in these situations. Moreover, at high temperature, thermotolerant genotypes produced more ethylene, more endo-β-mannanase, and germinated better than thermosensitive genotypes. An inhibitor of ethylene action, silver thiosulphate, inhibited both mannanase activity and germination. A precursor of ethylene, ACC, induced endo-β-mannanase activity and germination of thermosensitive 'Dark Green Boston' at 35°C. These results suggest that increased endo-β-mannanase activity might contribute to lettuce endosperm weakening, especially at high temperatures.

* Florida Agricultural Experiment Station Journal Series No. R-06799.

Introduction

Lettuce seed germination is strongly temperature dependent. The optimum temperature for germination is around 20°C, and most lettuce genotypes fail to germinate at temperatures above 30°C. When high temperature conditions occur during seed imbibition, two different phenomena may occur: (i) thermoinhibition, a reversible condition, since germination will occur when the temperature decreases to a suitable level; and (ii) thermodormancy, where seeds do not germinate after the alleviation of high temperature.

Different strategies to alleviate the problems of thermoinhibition and/or thermodormancy have been used. Some thermotolerant genotypes have been developed, but environmental effects on expression of tolerance have been observed (Sung *et al.*, 1998a). Environmental factors during seed maturation can also influence the threshold temperature for seed germination. Seed priming has been successfully used to overcome the problem of high temperature inhibition of lettuce seed germination (Guedes *et al.*, 1979; Guedes and Cantliffe, 1980). Hormones, such as ethylene (Braun and Khan, 1976), have significantly improved lettuce seed germination at high temperatures. Nonetheless, the physiological and biochemical processes that control lettuce seed dormancy at high temperature are still not understood.

The lettuce seed embryo is completely enclosed within a two- to four-cell layer endosperm whose cell walls are comprised mainly of galactomannan polysaccharides. The endosperm may delay or prevent germination by acting as a physical barrier to radicle protrusion, especially under unfavourable conditions. Thus, weakening of the endosperm layer of lettuce seeds is a pre-requisite to radicle protrusion at high temperatures. Since the lettuce endosperm cell walls are composed largely of mannans, endo-β-mannanase could be a potential regulatory enzyme in endosperm weakening.

The objective of this research was to investigate the involvement of the hormone ethylene and the enzyme endo-β-mannanase in lettuce seed germination under high temperature conditions. Several approaches were taken, including: (i) studying the interaction between light and temperature on seed germination; (ii) examining genotypes that have exhibited different capacities to germinate at high temperature; (iii) determining how seed priming enhances lettuce germination at high temperature; and (iv) using inhibitors of ethylene action during seed germination.

Materials and Methods

Plant material

Five lettuce (*Lactuca sativa* L.) genotypes varying in levels of thermotolerance were used in this study: 'Dark Green Boston', 'Valmaine', 'Floricos 83', 'Everglades' and PI 251245. Thermotolerance was defined as the ability of seeds to germinate above 90% at temperatures up to 35°C in light (Guzman *et al.*, 1992; Sung, 1996). All seeds were produced in the same season and region of the San Joaquin Valley, California, in 1994. Seeds were stored at 10°C, 40% RH until used.

Seed maturation study

Lettuce plants of thermosensitive 'Dark Green Boston' and thermotolerant 'Everglades' were produced under greenhouse conditions until flowering and then transferred to growth chambers at 12 h photoperiod (day/night) and temperature regimes (day/night) of 20/10° or 30/20°C according to the methods used by Sung *et al.* (1998a). At maturity, seeds were harvested, threshed, and cleaned manually. Seeds were stored at 10°C and 45% RH until used. 'Dark Green Boston' and 'Everglades' were chosen in this study because of the genetic relationship between these two genotypes (Guzman *et al.*, 1992).

Seed puncture test

The puncture test was conducted using an Instron Universal Testing Machine as previously described (Sung *et al.*, 1998b).

Seed priming

Seeds were primed in 200 mm test tubes for 3 days at 15°C with constant light (~26 μmol m^{-2} s^{-1}) in an aerated solution of polyethylene glycol (osmotic potential of −1.2 MPa).

Seed germination

Four replications of 25 seeds were placed on two layers of 5 cm diameter blotter paper (Anchor Paper, Hudson, Wisconsin) moistened with 3 ml of distilled water. Distilled water was added as needed to keep the filter paper moist. Blotters were covered with 5.5 cm Petri dish lids and incubated at 20° or 35°C under constant light (~26 μmol m^{-2} s^{-1}) on a one-dimensional thermogradient bar (Type DB 5000, Van Dok & De Boer, B.V., The Netherlands). Seeds were also germinated in 10 mM of 1-aminocyclopropane-1-carboxylic acid (ACC), in 10 mM of aminoethoxyvinylglycine (AVG), or in 20 mM of silver thiosulphate (STS) solutions. Germinated seeds were counted daily.

Enzyme activity

A gel-diffusion assay (Downie *et al.*, 1994; Still and Bradford, 1997) was used to measure endo-β-mannanase (EC 3.2.1.78) during seed germination. Twenty-eight individual endosperms from the micropylar region and/or 14 pairs (radicle tip + the remaining endosperm, referred to as lateral endosperm) from lettuce seeds imbibed at different temperatures for different periods of time were used on each plate.

Ethylene determination

Three replications of 0.2 g of dry seeds were placed on two layers of 3.0 cm diameter germination paper (Anchor Paper, Hudson, Wisconsin) which were at the base of 38 ml volume vials sealed with rubber septa. The seeds in the vials were moistened with 3 ml of distilled water and then incubated under the same conditions as the standard germination procedures. One ml gas samples were withdrawn using a gas-tight hypodermic syringe. After sample withdrawal, the vials were flushed with air and sealed again for additional sampling. Ethylene was assayed using a gas chromatograph (Hewlett-Packard 5890 Series II) equipped with an alumina column and a flame ionization detector. The carrier gas was nitrogen and the column temperature was 100°C.

Results and Discussion

Five genotypes were germinated over a range of temperatures to determine their level of thermotolerance. At 27°C, germination of 'Dark Green Boston' was 74% in light, but did not germinate in the dark or at any temperature above 24°C in the dark (data not shown). Germination of 'Valmaine' (a thermosensitive genotype) was partially inhibited in light at 30°C and seeds did not germinate above 30°C. 'Floricos 83', 'Everglades', and PI 251245 germinated well at 30°C in light or dark. Above 30°C, germination of 'Everglades' declined in light and dark. For 'Floricos 83' and PI 251245, germination in light at 33°C was 83% and 94%, respectively, but fell to ≈40% and 50% in the dark. At 36°C, PI 251245 germination was 31% in light and 18% in dark. Thus, temperature had a profound influence on the ability of these genotypes to germinate. Light generally had a small influence. Some genotypes were clearly more thermotolerant than others, i.e. 'Dark Green Boston' and 'Valmaine' were not thermotolerant while 'Floricos 83', 'Everglades', and PI 251245 were thermotolerant.

When seeds of two genotypes representing each class of thermotolerance were matured under either low or high seed developmental temperature, subsequent germination at 35°C was affected (Table 26.1). Germination was markedly reduced for seeds matured at 20°/10°C compared with 30°/20°C. The degree of thermotolerance was improved in both thermotolerance classes when seeds were matured under high temperature.

The force required to penetrate lettuce seed endosperm tissue at 33°C after 6 h of imbibition was less in thermotolerant genotypes ('Floricos 83', 'Everglades', PI 251245) compared with thermosensitive genotypes ('Dark Green Boston' and 'Valmaine') (Table 26.2). When thermosensitive 'Dark Green Boston' seeds were primed, they germinated 100% at 35°C (Table 26.3).

Thermotolerant 'Everglades' germinated 100% at 35°C, regardless of whether or not they were primed.

Ikuma and Thimann (1963) proposed that the action of an enzyme produced by the lettuce embryo enabled the radicle tip to penetrate through the restricting tissues. Possible chemical weakening of the endosperm near the radicle end was suggested by Pavlista and Haber (1970). Since lettuce endosperm cell walls are composed largely of galactomannans (Halmer *et al.*, 1975), endo-β-mannanase might be an important enzyme in the cell wall degradation leading to endosperm weakening and subsequent radicle protrusion.

The time of appearance of endo-β-mannanase activity during germination has been debated. In some studies, mannanase activity detected in dry seeds or after the first hours of imbibition might have been due to activation of preexisting enzyme (Dutta *et al.*, 1994) or the retention of enzyme produced during seed development (Hilhorst and Downie, 1995). Thus, some researchers have speculated that growth conditions during seed development might affect endo-β-mannanase levels. In lettuce, early studies detected mannan hydrolysis only as a post-germinative event (after radicle protrusion) (Halmer *et al.*, 1975, 1976; Bewley and Halmer, 1980/81). For example, mannanase

Table 26.1. Germination percentage at 35°C of lettuce seeds matured at two different temperatures.

	Day/night developmental temperature (°C)		
Genotype	20/10	30/20	
Dark Green Boston	2	74	**
Everglades	50	98	**

**Significant F test at 1% level.

Table 26.2. Mean force required to penetrate lettuce seed endosperm tissue after 6 h imbibition at 33°C.

Genotype	Puncture force endosperm (N)
'Dark Green Boston'	0.104a*
'Valmaine'	0.118a
'Floricos 83'	0.093b
'Everglades'	0.088b
PI 251245	0.092b

*Means followed by the same letter are not statistically different at $P < 0.05$.

Table 26.3. Germination of primed lettuce seeds at 35°C.

Genotype	Non-primed	Primed
'Dark Green Boston'	4	100
'Everglades'	100	100

activity increased about 100-fold in all regions of the endosperm during 15 h following germination (Halmer *et al.*, 1978). Recently, Dutta *et al.* (1997) reported that a cell-wall-bound endo-β-mannanase is expressed in lettuce seed endosperm prior to radicle protrusion and is regulated by the same conditions that govern seed germination. These authors suggested that this enzyme is likely to be involved in the weakening of the endosperm cell walls.

Germination (Table 26.1) and endo-β-mannanase activity (Table 26.4) were affected by seed developmental temperature. When 'Dark Green Boston' seeds were matured at 20°/10°C and imbibed at 35°C for 24 h no endo-β-mannanase activity was detected and only 2% of the seeds germinated. When seeds of this thermosensitive genotype were developed under 30°/20°C, mannanase activity was observed at 24 h (1 h before radicle protrusion), and 74% of the seeds germinated at 35°C.

Seed priming also improved germination at 35°C of thermosensitive 'Dark Green Boston' (Table 26.3). Endo-β-mannanase activity was observed during priming and after reimbibition, but before radicle protrusion, when seeds of either 'Dark Green Boston' or 'Everglades' were germinated under 35°C temperatures (data not shown). Higher endo-β-mannanase activity was found in primed 'Dark Green Boston' compared with non-primed seeds. Enzyme activity in the micropylar area was high (about 50% of the total activity), considering the smaller mass of the micropylar tissue compared to the lateral part. Prior to radicle protrusion, endo-β-mannanase activity in the micropylar region was approximately three times higher in primed 'Dark Green Boston' seeds compared with nonprimed seeds germinated at 20 or 35°C. Endo-β-mannanase activity was observed after 2 h of imbibition in 'Dark Green Boston' primed seeds whereas in nonprimed seeds, enzyme activity was observed only after 24 h at 35°C.

Endo-β-mannanase was also observed in the lateral endosperm region before radicle protrusion, even at an early stage of incubation. In tomato, endo-β-mannanase is first produced in the micropylar area, and after radicle protrusion, in the entire endosperm tissue (Nonogaki and Morohashi, 1996; Toorop *et al.*, 1996; Voigt and Bewley, 1996). Nonogaki and Morohashi (1996) reported some differences in the products of galactomannan hydrolysis in tomato endosperm between the pre-germinative and post-germinative enzymes, indicating that the action pattern was different between the two types of enzymes. Bewley (1997) suggested that there are isoforms of the enzyme involved in tomato germination and other isoforms associated with post-germination cell wall mannan mobilization. This could also be true for lettuce.

Table 26.4. Endo-β-mannanase activity of 'Dark Green Boston' lettuce seeds subjected to various maturation temperatures and imbibed for 24 h at 35°C.

Developmental temperature	Enzyme activity (p mol min^{-1})
20/10	0
30/20	1.6

Table 26.5. Endo-β-mannanase activity and germination of 'Dark Green Boston' lettuce seed imbibed for 24 h at 35°C in ACC or STS.

Treatment	Mannanase activity (p mol min^{-1})	Germination (%)
Control	0.0 b*	33 b*
ACC	1.0 a	92 a
STS	0.0 b	0 c

*Means in each column followed by the same letter are not statistically different at $P < 0.05$.

Ethylene has also been reported to stimulate germination of lettuce at high temperature (Braun and Khan, 1976) and the synthesis of some enzymes (Cervantes *et al.*, 1994; Hasegawa *et al.*, 1995). Separation of cells due to the activation of cell wall dismantling enzymes, such as endo-β-1,4-glucanases, was reported recently as an ethylene effect (Casadoro *et al.*, 1998). Other cell wall-degrading enzymes show ethylene-dependency, such as endopolygalacturonase, some isoforms of α-galactosidase, β-arabinosidase, and galactanase (Pech *et al.*, 1998). Thus, it is reasonable to assume that ethylene might overcome the inhibitory effect of high temperature on lettuce seed germination by activating cell wall enzymes responsible for endosperm digestion. In the present work, thermotolerant genotypes produced more ethylene during seed germination at high temperature than the thermosensitive genotypes (data not shown). Conversely, seeds from thermosensitive 'Dark Green Boston' produced the least amount of ethylene at 35°C. These results agree with Prusinski and Khan (1990), who reported that the ability of lettuce genotypes to produce ethylene during high temperature stress corresponded with their ability to germinate. They suggested this as a criterion to select thermotolerant lettuce cultivars. The thermotolerant genotypes in the present work that produced more ethylene at high temperature also produced more endo-β-mannanase and germinated earlier. Further, when thermosensitive 'Dark Green Boston' was imbibed at 35°C, no endo-β-mannanase activity could be detected and few seeds germinated (Table 26.5). When ACC was supplied in the imbibition solution, endo-β-mannanase activity was detected before radicle protrusion and 92% of the seeds germinated. When STS, an inhibitor of ethylene action, was applied, no mannanase activity or germination was recorded. Thus, an association between ethylene evolution, endo-β-mannanase activity prior to radicle protrusion, and seed germination at high temperature was verified.

References

Bewley, J.D. (1997) Seed germination and dormancy. *The Plant Cell* 9, 1055–1066.

Bewley, J.D. and Halmer, P. (1980/81) Embryo-endosperm interactions in the hydrolysis of lettuce seed reserves. *Israel Journal of Botany* 29, 118–132.

Braun, J.W. and Khan, A.A. (1976) Alleviation of salinity and high temperature stress by plant regulators permeated into lettuce seeds via acetone. *Journal of the American Society for Horticultural Science* 101, 716–721.

Casadoro, G., Trainotti, L. and Tomasin, C.A. (1998) Expression of abcission-related endo-β-1,4-glucanases. *Biology and Biotechnology of the Plant Hormone Ethylene II* (Island of Santorini, Cyclades, Greece). Abstract no. 23, p. 39. Sept. 5–8, 1998.

Cervantes, E., Rodriguez, A. and Nicolas, G. (1994) Ethylene regulates the expression of a cysteine proteinase gene during germination of chick-pea *Cicer arietinum* L. *Plant Molecular Biology* 25, 207–215.

Downie, B., Hilhorst, H.W.M. and Bewley, J.D. (1994) A new assay for quantifying endo-β-D-mannanase activity using congo red dye. *Phytochemistry* 36, 829–835.

Dutta, S., Bradford, K.J. and Nevins, D.J. (1994) Cell-wall autohydrolysis in isolated endosperms of lettuce (*Lactuca sativa* L.). *Plant Physiology* 104, 623–628.

Dutta, S., Bradford, K.J. and Nevins, D.J. (1997) Endo-β-mannanase present in cell wall extracts of lettuce endosperm prior to radicle emergence. *Plant Physiology* 133, 155–161.

Guedes, A.C., Cantliffe, D.J., Shuler, K.D. and Munter, E. (1979) Overcoming thermodormancy in lettuce by seed priming. *Florida State Horticultural Society Proceedings* 92, 130–133.

Guedes, A.C. and Cantliffe, D.J. (1980) Germination of lettuce seeds at high temperature after seed priming. *Journal of the American Society for Horticultural Science* 105, 777–781.

Guzman, V.L., Nagata, R.T., Datnoff, L.E. and Raid, R.N. (1992) 'Florida 202' and 'Everglades': new butterhead lettuce cultivars adapted to Florida. *HortScience* 27, 852–853.

Halmer, P., Bewley, J.D. and Thorpe, T.A. (1975) Enzyme to break down lettuce endosperm cell wall during gibberellin-and-light-induced germination. *Nature* 258, 716–718.

Halmer, P., Bewley, J.D. and Thorpe, T.A. (1976) An enzyme to degrade lettuce endosperm cell walls. Appearance of a mannanase following phytochrome- and gibberellin-induced germination. *Planta* 130, 189–196.

Halmer, P., Bewley, J.D. and Thorpe, T.A. (1978) Degradation of the endosperm cell walls of *Lactuca sativa* L., cv. Grand Rapids. Timing of mobilization of soluble sugars, lipid and phytate. *Planta* 139, 1–8.

Hasegawa, R., Maruyama, A., Nakaya, M., Tsuda, S. and Esashi, Y. (1995) The presence of two types of β-cyanoalanine synthase in germinating seeds and their responses to ethylene. *Physiologia Plantarum* 93, 713–718.

Hilhorst, H.W.M. and Downie, B. (1995) Primary dormancy in tomato (*Lycopersicon esculentum* cv. Moneymaker): studies with the *sitiens* mutant. *Journal of Experimental Botany* 47, 89–97.

Ikuma, H. and Thimann, K.V. (1963) The role of seed-coats in germination of photosensitive lettuce seeds. *Plant Cell Physiology* 4, 169–185.

Nonogaki, H. and Morohashi, Y. (1996) An endo-β-mannanase develops exclusively in the micropylar endosperm of tomato seeds prior to radicle emergence. *Plant Physiology* 110, 555–559.

Pavlista, A.D. and Haber, A.H. (1970) Embryo expansion without protrusion in lettuce seeds. *Plant Physiology* 46, 636–637.

Pech, J.C., Ayub, R. and Latche, A. (1998) Ethylene-dependent and ethylene independent pathways in the melon. *Biology and Biotechnology of the Plant Hormone Ethylene II* (Island of Santorini, Cyclades, Greece). Abstract no. 26, p. 46. Sept. 5–8, 1998.

Prusinski, J. and Khan, A.A. (1990) Relationship of ethylene production to stress alleviation in seeds of lettuce cultivars. *Journal of the American Society for Horticultural Science* 115, 294–298.

Still, D.W. and Bradford, K.J. (1997) Endo-β-mannanase activity from individual tomato endosperm caps and radicle tips in relation to germination rates. *Plant Physiology* 113, 21–29.

Sung, Y. (1996) Identification and characterization of thermotolerance in lettuce seed germination. PhD dissertation, University of Florida, Gainesville.

Sung, Y., Cantliffe, D.J. and Nagata, R.T. (1998a) Seed developmental temperature regulation of thermotolerance in lettuce. *Journal of the American Society for Horticultural Science* 123, 700–705.

Sung, Y., Cantliffe, D.J. and Nagata, R.T. (1998b) Using a puncture test to identify the role of seed coverings on thermotolerant lettuce seed germination. *Journal of the American Society for Horticultural Science* 123, 1102–1106.

Toorop, P.E., Bewley, J.D. and Hilhorst, H.W.M. (1996) Endo-β-mannanase isoforms are present in the endosperm and embryo of tomato seeds, but are not essentially linked to the completion of germination. *Planta* 200, 153–158.

Voigt, B. and Bewley, J.D. (1996) Developing tomato seeds when removed from the fruit produce multiple forms of germinative and post-germinative endo-β-mannanase. Responses to desiccation, abscisic acid and osmoticum. *Planta* 200, 71–77.

27 Effect of Free Mg^{2+} on the Kinetics of the Plasma Membrane H^+-ATPase from Dry and Hydrated Maize Embryos

S. SÁNCHEZ-NIETO[1], R. RODRÍGUEZ-SOTRES[1], M. TUENA DE GÓMEZ-PUYOU[2], A. CARBALLO[3] AND M. GAVILANES-RUÍZ[1]

[1]*Departamento de Bioquímica, Facultad de Química, Conjunto E, Universidad Nacional Autónoma de México, Ciudad Universitaria, 04510 México D.F., México;* [2]*Departamento de Genética Molecular, Instituto de Fisiología Celular, Universidad Nacional Autónoma de México, Ciudad Universitaria, 04510 México D.F., México;* [3]*Centro de Genética, Colegio de Posgraduados, Montecillo, Estado de México, 56230 México*

The H^+-ATPase is a transmembrane protein which pumps H^+ from the cytosol to the apoplastic space at the expense of ATP hydrolysis. It is believed that this function is important to promote secondary transport of solutes through the plasma membrane, to facilitate cell elongation and to contribute to pH regulation. At the onset of germination, changes in the biochemical properties of the enzyme have been found, which could be related to differences in protein structure. In this work, the effect of free Mg^{2+} concentrations on the kinetics of the enzyme at different imbibition times was evaluated. It was found that 200 μM of free Mg^{2+} may inhibit the ATPase activity, but the extent of this inhibition differs for each time of imbibition tested and it is related to the free Mg^{2+} concentration and to the substrate concentration. Free Mg^{2+} also exerts an influence on the vanadate inhibition of the enzyme.

Introduction

The H^+-ATPase from plant plasma membrane is a protein of 100 kDa that traverses the membrane with 10–12 hydrophobic stretches and that contains several hydrophilic loops, the major one possessing the catalytic site facing the cytosol. The hydrophobic moiety presumably transports H^+ from the cell to the apoplastic space against its concentration gradient; therefore, energy must be spent at the expense of ATP hydrolysis, which is carried out by the hydrophilic sector of the enzyme (Serrano, 1990). This ATPase is the primary pump of higher abundance and three physiological roles have been identified for this enzyme: cell elongation (Hager *et al.*, 1991; Frias *et al.*, 1996), secondary transport (Bush *et al.*, 1996) and pH maintenance (Gout *et al.*, 1992). Important processes such as cell nutrition, stomatal movements and long distance transport of metabolites are supported by the function of this enzyme.

During seed germination some of these processes occur, i.e. cell elongation and nutrition take place in order to promote growth of the embryo cells (Mayer and Poljakoff-Mayber, 1989; Bewley and Black, 1994). We have found that the form of the H^+-ATPase present in the dry maize embryo is replaced by another form that is found at 5 h of imbibition (Sanchez-Nieto *et al.*, 1998). The data suggest that both enzyme forms may differ in structural characteristics such as phosphorylation or primary structure. In addition, the kinetic constants of these enzyme forms are strongly influenced by the membrane milieu (S. Sánchez-Nieto *et al.*, unpublished data). In this work, we investigated the influence of free Mg^{2+} on the kinetics of the enzyme from dry and hydrated embryos.

Materials and Methods

Biological material

Embryos were dissected from mature, dry maize seeds (*Zea mays* L. hybrid Montecillos A_60_2, Colegio de Posgraduados, Montecillo, Edo. de Mexico) using a scalpel and removing the endosperm.

Germination assays

These were performed as described elsewhere (Sánchez-Nieto *et al.*, 1998).

Purification of plasma membrane vesicles

Embryos imbibed for different times were frozen with liquid N_2 and homogenized as described by Sánchez-Nieto *et al.* (1997). The homogenate was centrifuged at 1000 *g* for 7 min. The supernatant was used to obtain an enriched plasma membrane vesicles fraction as described by Sánchez-Nieto *et al.* (1997). The enrichment of the plasma membrane preparation after two successive phase-partitioning steps was approximately 95%, evaluated as vanadate-sensitive ATP hydrolysis.

Determination of ATPase activity

ATP hydrolysis from plasma membrane was essentially measured as described in Sánchez-Nieto *et al.* (1998). Sodium orthovanadate was prepared according to Gallagher and Leonard (1982). ATPase activity was initiated by the addition of 3.3 μg of membrane protein and incubated for 3 h at 30°C. The reaction was stopped by adding 150 µl 24% SDS. Released Pi was determined with the method of González-Romo *et al.* (1992). All assays were done in six replicates for each treatment and all experiments were repeated with three different membrane preparations.

Kinetic measurements

Data of the ATPase activity in the presence of vanadate were made with the aid of non-linear regression algorithms implemented in the software package Origin (Microcal Software, Inc. Northampton, Massachusetts, USA) by fitting the experimental point to the following equation:

$$v = V_0\ I_{50}/(I_{50} + [VO_4^{3-}]) \tag{27.1}$$

where V_0 is the activity in the absence of inhibitor, v the observed activity and I_{50} the concentration of vanadate that causes 50% inhibition.

For determinations of the ATPase activity at various substrate concentrations, we added $MgCl_2$ and ATP/BTP, pH 7.0, in two different ways: one of these was to mix identical concentrations of both solutions to obtain the desired Mg:ATP complex, without regulating the concentration of free Mg^{2+}; the other was to add the $MgCl_2$ and ATP/BTP at the concentrations required to keep constant the free Mg^{2+} concentration. In both cases the concentration of the $MgHATP^-$ complex was varied from 0.05 to 8.11 mM. The concentrations of ATP and $MgCl_2$ necessary to obtain the desired concentration of the $MgHATP^-$ and free Mg^{2+} were calculated as in Rodríguez-Sotres and Muñoz-Clares (1990), using the stability constants reported by O'Sullivan and Smitters (1979).

The kinetic constants for the ATPase activity were obtained by fitting the experimental data to the following equations:

$$\text{Michaelis–Menten model: } v = V_M\ [MgHATP^-]/(K_m + [MgHATP^-]) \tag{27.2}$$

$$\text{Hill model: } v = V_m\ [MgHATP^-]^n/(S_{0.5}{}^n + [MgHATP^-]^n) \tag{27.3}$$

where v is the initial velocity, V_m is the maximum velocity, K_m is the Michaelis constant, $S_{0.5}$ is the half-saturating $MgHATP^-$ concentration, and n the Hill number.

Protein determination

The method of Peterson (1977) with bovine serum albumin as standard was used.

Results

Former results in our laboratory indicated that there is a relay of ATPase enzyme forms during the first 5 h of imbibition of maize embryos (Sanchez-Nieto *et al.*, 1998). Although the differences between the two forms of the enzyme seem to be of structural nature, it is still unclear whether the variations are due to amino acid sequence and/or to some covalent modification.

With the aim of establishing if this shift in enzyme form was related to changes in the kinetic constants throughout the first 5 h of imbibition, we determined the ATPase activity against the substrate ($MgHATP^-$) concentration in plasma membrane vesicles obtained from 0, 2 and 5 h imbibed embryos. The results are shown in Fig. 27.1 and the kinetic constants derived from these curves are presented in Table 27.1. It was found that in the three cases, 0, 2 and 5 h, the enzyme presented a Michaelian behaviour, with Hill coefficients very close to 1.0 and with a lower K_m for the enzyme from 0 h and a higher

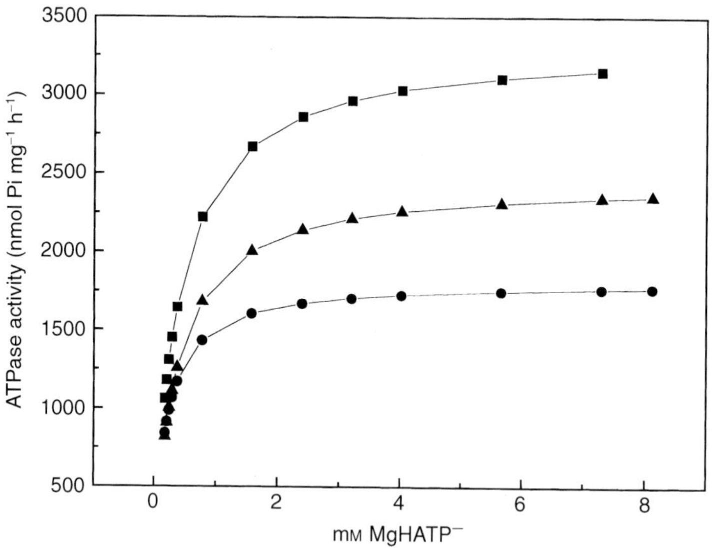

Fig. 27.1. H^+-ATPase activity from maize embryos imbibed at different times. ATP hydrolysis was measured in purified plasma membrane vesicles in a medium containing 250 mM sucrose, 20 mM MOPS/BTP pH 7.0, 7 μM CCCP, 2 mM NaN_3, 54 μM lysophosphatidylcholine and equimolar ratios of ATP/BTP pH 7.0 and $MgCl_2$. The reaction started with the addition of 3.3 μg of membrane protein and the assay was carried out at 30°C for 3 h. The reaction was terminated by the addition of SDS to a final concentration of 12%. The ATPase activity was measured as Pi release by the method of González-Romo *et al.* (1990). Imbibition times: 0 h (●), 2 h (■) and 5 h (▲).

Table 27.1. Kinetic parameters of the ATPase activity from plasma membranes of embryos at different times of imbibition.

Imbibition time (h)	K_m (mM)	V_{max} (nmol Pi mg^{-1} h^{-1})	n
0	0.207 ± 0.02	1337 ± 32.51	1.0149 ± 0.088
2	0.382 ± 0.07	2024 ± 37.5	1.0037 ± 0.003
5	0.359 ± 0.02	1433 ± 85.5	0.9213 ± 0.119

ATP hydrolysis was measured as described in Fig. 27.1. The kinetic constants were obtained by fitting the experimental points to the Hill model.

V_{max} for the 2 h enzyme. However, in these experiments the concentration of free Mg^{2+} was not maintained constant and fluctuated between 30 and 230 μM. Since high Mg^{2+} concentrations may result in inhibition of ATPase activity (Briskin and Poole, 1983), probably due to binding of Mg^{2+} to a specific site (Brooker and Slayman, 1983), we studied the possible differences in the plasma membrane ATPase activity from the 0, 2 and 5 h as regards its response to free Mg^{2+} concentrations, using two substrate concentrations: one in the K_m value (200 μM) and another one fourfold below the K_m (50 μM). The results in Fig. 27.2 show bell-shaped curves in the three different preparations. These

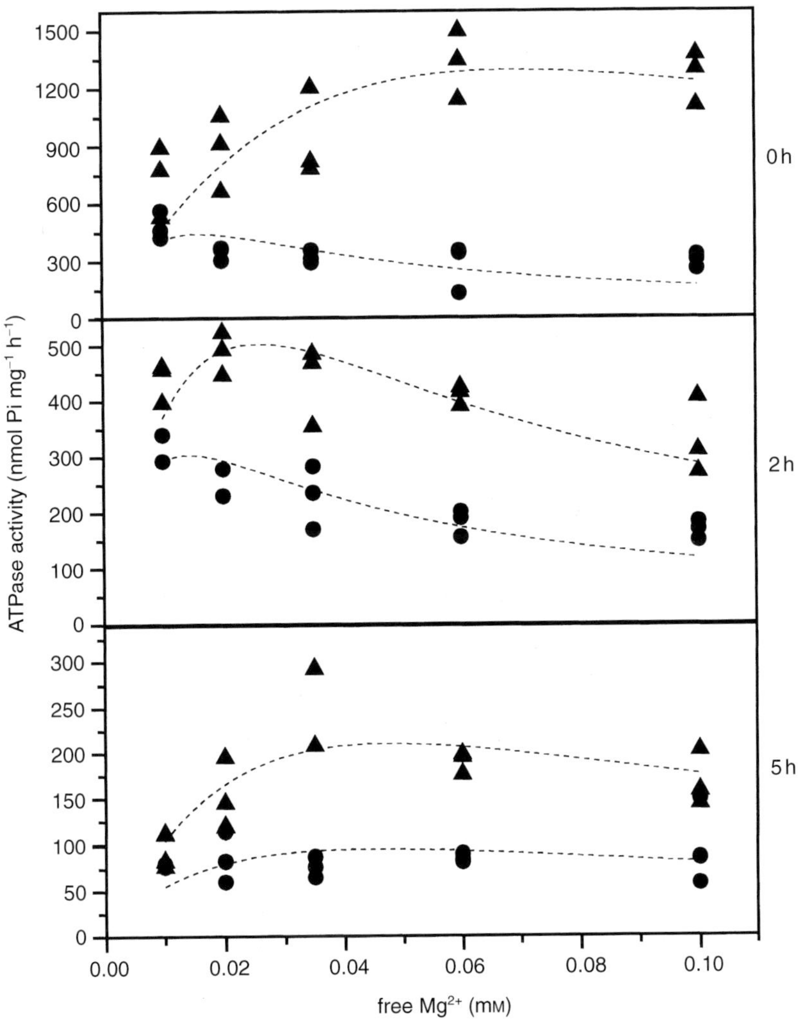

Fig. 27.2. Effect of free Mg^{2+} on the rate of ATP hydrolysis by the plasma membrane of embryos at different times of imbibition. The ATPase activity was measured as described in the legend of Fig. 27.1, but in this experiment concentrations of free Mg^{2+} were varied at two fixed $MgHATP^-$ concentrations of 50 (●) and 200 (▲) μM.

results are compatible with the random addition of free and complexed species to the active site. However, the curves were not quantitatively identical for the three imbibition times.

These findings also suggest that free Mg^{2+} could affect to different extents the ATPase activity of the three membrane preparations at almost all concentrations of $MgHATP^-$ used in the experiment of Fig. 27.1. To test this possibility, we obtained the substrate saturation curves of the ATPase activity at two free Mg^{2+} concentrations: 35 and 200 μM (Fig. 27.3). These results support the idea that high free Mg^{2+} concentration produces a significant inhibition of the activity from the plasma membrane H^+-ATPase in a substrate concentration range between 1.0 and 8.1 mM.

Free Mg^{2+} also interfered with vanadate inhibition. Figure 27.4 shows that the kinetic parameters of the H^+-ATPase from dry embryos in the absence and presence of different vanadate concentrations were also modified by the free Mg^{2+} concentration: 150 μM of free Mg^{2+} decreased the apparent K_m of the enzyme at the control and at the two vanadate concentrations tested. High free

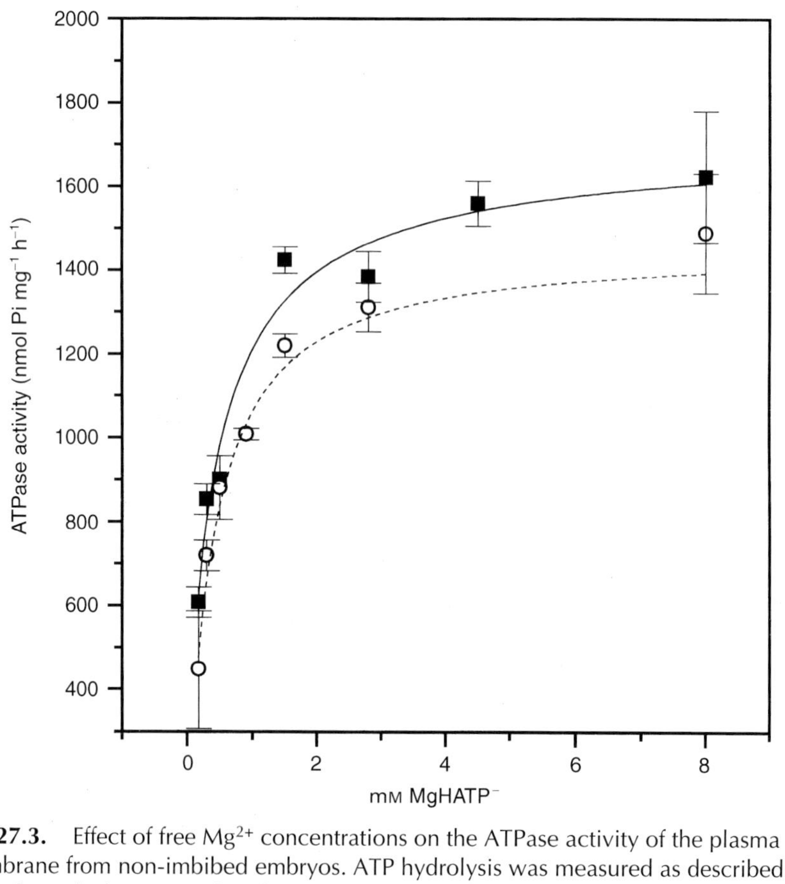

Fig. 27.3. Effect of free Mg^{2+} concentrations on the ATPase activity of the plasma membrane from non-imbibed embryos. ATP hydrolysis was measured as described in the legend of Fig. 27.1, but the experiment was carried out at two fixed free Mg^{2+} concentrations of 35 (■) and 200 (○) μM.

Mg^{2+} also modified V_{max} values. From these experiments, it was clear that the free Mg^{2+} concentration modified the extent of the inhibitory effect of vanadate. To confirm this possibility, the ATPase activity from 0, 2 and 5 h imbibed embryos was measured at three different substrate concentrations and varying vanadate concentration. The I_{50} values from the curves obtained are presented in Table 27.2. It is shown that the I_{50} values for vanadate decreased as substrate concentration increased. In addition, it can be noted that 0 and 2 h activities had marked less sensitivity to vanadate as compared with 5 h activity.

Discussion

We measured the kinetics of the H^+-ATPase of plasma membranes obtained from maize embryos imbibed for 0, 2 and 5 h, expecting some differences in the K_m and V_{max} values, according to the two enzyme forms previously

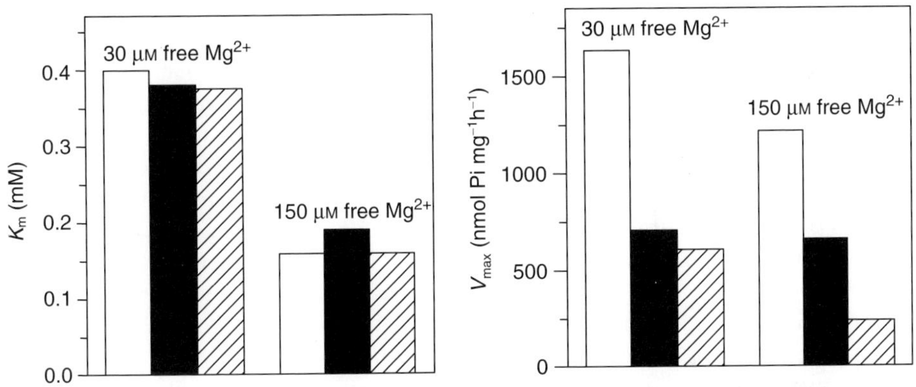

Fig. 27.4. Effect of vanadate and free Mg^{2+} on the kinetic parameters of the plasma membrane H^+-ATPase from non-imbibed embryos. The ATPase activity was measured as described under Materials and Methods, at two fixed free Mg^{2+} concentrations as indicated and the vanadate was varied from 40 to 300 μM (■, 0 μM; ■, 40 μM; ▨, 300 μM);. The kinetic parameters were obtained by fitting the experimental data to the Michaelis– Menten equation.

Table 27.2. Inhibition constants for Na_3VO_4 (I_{50}) of ATPase activity at different substrate concentrations ($MgHATP^-$).

Imbibition time (h)	0.24 mM $MgHATP^-$ (230 μM free Mg^{2+})	0.77 mM $MgHATP^-$ (70 μM free Mg^{2+})	8.11 mM $MgHATP^-$ (40 μM free Mg^{2+})
0	232.4 ± 56	311.9 ± 135	26.1 ± 10
2	102.4 ± 43	80.7 ± 76	22.6 ± 11
5	13,415 ± 1570	362.7 ± 315	131.7 ± 31

I_{50} values are expressed in μM Na_3VO_4.
Membrane protein (3.3 μg) from plasma membrane vesicles obtained from embryos imbibed for 0, 2 and 5 h was added to a reaction medium to measure ATP hydrolysis as described under Materials and Methods.

described (Sánchez-Nieto *et al.*, 1998). The Michaelian behaviour obtained agreed with the reports of several groups (Du Pont *et al.*, 1981; Perlin and Spanswick, 1981; Briskin and Poole, 1984). In fact, we found some slight differences in the kinetic constants. We considered the fact that the addition of equimolar concentrations of ATP and Mg^{2+} produces variable concentrations of free Mg^{2+} (Balke and Hodges, 1975; Briskin and Poole, 1983). In order to rule out a possible inhibitory effect of free Mg^{2+} that could be masking major kinetic differences among the 0, 2 and 5 h preparations, we studied the effect of free Mg^{2+} concentrations on the enzyme kinetics. Our results indicated that free Mg^{2+} could exert an inhibitory effect on the plasma membrane H^+-ATPase activity in our preparations. The extent of this inhibition depended on the substrate concentration and on the free Mg^{2+} concentration. This inhibitory effect varied for each imbibition time.

The effect at non-saturating substrate concentrations may suggest a Mg^{2+} competition with the substrate for the catalytic site, since the results are in agreement with a random binding of the substrate to the enzyme (Segel, 1993). In addition, the magnitude of the Mg^{2+} inhibition at high substrate concentrations in the three membrane preparations could be due to a difference in the affinity of a second Mg^{2+} binding site different from the catalytic site. Brooker and Slayman (1983) have suggested at least one Mg^{2+} binding site of this type in *Neurospora crassa*. In addition, it has been proposed that Mg^{2+} may help to stabilize the intermediate enzyme-ATP, optimizing the transfer of γ-ATP to the enzyme (Briskin and Poole, 1983). However, in our case, even when the K_m decreased 2.5-fold in the presence of 150 μM free Mg^{2+} as compared to the value at 30 μM of the cation, the V_{max} value did not increase.

The vanadate effect on the enzyme was also influenced by free Mg^{2+} in a manner suggesting a non-competitive mode of action of the inhibitor, as previously reported (Gallagher and Leonard, 1982; Gibrat *et al.*, 1988). The fact that vanadate sensitivity was lower in 0 and 2 h preparations as compared with the 5 h preparation, indicates that it was possible to detect important kinetic differences between both forms of enzyme, when the inhibitory effect of Mg^{2+} is absent.

Acknowledgements

We thank Consuelo Enriquez-Arredondo for excellent technical assistance. The technical help of Laurel Fabila-Ibarra is also recognized. This work was supported by Grants 4836-N9406 from CONACYT, México and IN206691 and IN204097 from DGAPA, University of México (UNAM).

References

Balke, E.N. and Hodges, T.K. (1975) Plasma membrane adenosine triphosphatase of oat roots. Activation and inhibition by Mg^{2+} and ATP. *Plant Physiology* 55, 83–86.

Bewley, J.D. and Black, M. (1994) *Seeds: Physiology of Development and Germination.* Plenum, New York, pp. 147–197.

Briskin, D.P. and Poole, R.J. (1983) Role of magnesium in the plasma membrane ATPase of red beet. *Plant Physiology* 71, 969–971.

Brooker, R.J. and Slayman, C.W. (1983) Effects of Mg^{2+} ions on the plasma membrane [H^+]-ATPase of *Neurospora crassa. Journal of Biological Chemistry* 22, 8833–8838.

Bush, D.R., Chou, T.J. and Chen, L. (1996) Molecular analysis of plant sugar and amino acid transporters. *Journal of Experimental Botany* 47, 1205–1210.

DuPont, F.M., Burke, L.L. and Spanswick, R.M. (1981) Characterization of a partially purified adenosine triphosphatase from corn root plasma membrane fraction. *Plant Physiology* 67, 59–63.

Frías, I., Caldeira, M.T., Pérez-Castiñeira, J.R., Navarro-Aviño, J.P., Culiañez-Maciá, F.A., Kuppinger, O., Stransky, H., Pagés, M., Hager, A. and Serrano, R. (1996) A major isoform of the maize plasma membrane H^+-ATPase: characterization and induction by auxin in coleoptiles. *Plant Cell* 8, 1533–1544.

Gallager, S.R. and Leonard, R.T. (1982) Effect of vanadate, molybdate and azide on membrane-associated ATPase and soluble phosphatase activities of corn roots. *Plant Physiology* 60, 1335–1340.

Gibrat, R., Grouzis, J.P., Rigaud, J., Galtier, N. and Grignon, C. (1988) Electrostatic analysis of effects of ion on the inhibition of corn root plasma membrane Mg^{2+}-ATPase by the bivalent orthovanadate. *Biochimica et Biophysica Acta* 979, 46–52.

González-Romo, P., Sánchez-Nieto, S. and Gavilanes-Ruíz, M. (1992) A modified colorimetric method for the determination of orthophosphate in the presence of high ATP concentration. *Analytical Biochemistry* 200, 235–238.

Gout, E., Bligny, R. and Douce, R. (1992) Regulation of intracellular pH values in higher plant cells. *Journal of Biological Chemistry* 267, 13903–13909.

Hager, A., Debus, G., Edel, H.G., Stransky, H. and Serrano, R. (1991) Auxin induces exocytosis and the rapid synthesis of a high turnover pool of plasma membrane H^+-ATPase. *Planta* 185, 527–537.

Mayer, A.M. and Poljakoff-Mayber, A. (1989) *The Germination of Seeds.* Pergamon Press, Oxford, pp. 111–173.

O'Sullivan, W.J. and Smithers, G.W. (1979) Stability constants for biologically important metal-ligand complexes. *Methods in Enzymology* 63, 294–336.

Perlin, D.S. and Spanswick, R.M. (1981) Characterization of ATPase activity associated with corn leaf plasma membranes. *Plant Physiology* 68, 521–526.

Peterson, G.L. (1977) A simplification of the protein assay method of Lowry *et al.* which is more generally applicable. *Analytical Biochemistry* 83, 346–356.

Rodríguez-Sotres, R. and Muñoz-Clares, R.A. (1990) Kinetic evidence of the existence of a regulatory phosphoenolpyruvate binding site in maize leaf phosphoenolpyruvate carboxylase. *Archives of Biochemistry and Biophysics* 276, 180–190.

Sánchez-Nieto, S., García-Rubio, O., Pacheco-Moisés, F., Carballo, A., Rodríguez-Sotres R. and Gavilanes-Ruíz, M. (1997) Purification of plasma membranes from dry maize embryos. *Physiologia Plantarum* 101, 157–164.

Sánchez-Nieto, S., Tuena de Gomez Puyou, M., Rodríguez-Sotres, R., Carballo, A. and Gavilanes-Ruíz, M. (1998) Comparison of plasma membrane H^+-ATPase activity in vesicles obtained from dry and hydrated maize embryos. *Biochimica et Biophysica Acta* 1414, 175–187.

Segel, I.H. (1993) *Enzyme Kinetics. Behavior and Analysis of Rapid Equilibrium and Steady State Enzyme Systems.* John Wiley & Sons, New York, pp. 346–464.

Serrano, R. (1990) Plasma membrane ATPase. In: Larsson, C. and Møller, I.M. (eds) *The Plant Plasma Membrane.* Springer Verlag, Berlin, pp. 127–122.

28 The Barley Scutellar Peptide Transporter: Relationship to Germination and Loss of Seed Viability

W.M. WATERWORTH, C.E. WEST, M.I. DAWS AND C.M. BRAY

School of Biological Sciences, University of Manchester, Oxford Road, Manchester M13 9PT, UK

Peptide transport activity arises via *de novo* protein synthesis in the scutellum of germinating barley grains immediately following imbibition and precedes the development of appreciable amino acid transport activity in the scutellum. The temporal and spatial expression pattern of the barley scutellar peptide transporter (*HvPTR1*) is indicative of a central role for *HvPTR1* in peptide transport during barley grain germination. Peptide transport activity also appears to be regulated at the translational or post-translational level by products arising from barley endosperm reserve mobilization including glucose and amino acids. The levels of peptide transport activity in the scutellum of imbibing barley grains decrease in parallel with loss of viability of a seed lot and are an early and sensitive indicator of the viability of barley grains.

Introduction

The onset of seed germination is associated with a rapid resumption of cellular RNA and protein synthesis whilst the replication of DNA is often delayed for several hours (Bray, 1979). This delay allows repair processes that are thought to be operative during the first few hours of germination to be completed prior to resumption of growth processes (Osborne, 1983). In barley grains, the scutellum, a modified cotyledon which functions in nutrient transport, accounts for almost all the protein synthesis proceeding in barley embryos after 4.5 h imbibition with much lower levels found in root and coleoptile structures (Stoddart *et al.*, 1973). During the hours following imbibition, the scutellum synthesizes a variety of specific carrier proteins which are subsequently localized to the plasma membrane of scutellar epithelium. These carrier proteins function to transport the degradation products arising from the mobilization of the cereal endosperm reserves, e.g. glucose, peptides, amino acids, across the scutellum from where they are translocated either with or without further metabolism into the embryo (Bewley and Black, 1994).

Associated with the germination of cereal grains, storage proteins in the endosperm are hydrolysed to form a reservoir of small peptides and amino acids which are then translocated across the scutellum to supply the embryo with organic nitrogen to support growth processes during germination and early seedling establishment. Both peptides and amino acid transporters have been demonstrated in the scutellum of several cereals during germination (Salmenkallio and Sopanen, 1989). Peptide transport in the scutellum of germinated barley (*Hordeum vulgare*) grain was first demonstrated by Higgins and Payne (1977) and has subsequently been well characterized in terms of kinetics, specificity and ontogeny (reviewed in Hardy and Payne, 1992). Peptides appear to play at least as important a role as amino acids in the transfer of stored nitrogen from the barley endosperm to the embryo following germination (Higgins and Payne, 1981). In the barley scutellum, peptide transport is strongly pH dependent, proton coupled, utilizes di- and tripeptides and possibly some tetra- and pentapeptides, and must, of necessity, be stereospecific.

Recently we have cloned the barley scutellar peptide transporter (*HvPTR1*) and characterized it by functional expression in *Xenopus* oocytes (West *et al.*, 1998). Cloning of the barley scutellar peptide transporter has facilitated a molecular characterization of the synthesis, regulation and role of the transporter following germination as well as increasing our knowledge of the basic properties of a plant plasma membrane protein. One aspect of an ongoing study which concerns investigations into the role of *HvPTR1* during germination and loss of seed viability is reported here.

Experimental procedures

Plant material and growth conditions

Barley grain (*H. vulgare* L. cv Maris Otter 1993 harvest) was germinated at 23°C in the dark on moist filter paper. Embryos and/or scutella were excised from the endosperm using a scalpel blade and treated as described for peptide transport assays with or without prior incubation with glucose or amino acids as appropriate.

Solute transport by barley scutella

Transport of Ala-[^{14}C]Phe and [^{14}C]alanine (Amersham International) was monitored by uptake of radiolabel into barley scutella tissue as described previously (Walker-Smith and Payne, 1984).

Isolation of total RNA and Northern analysis

RNA was isolated using a modified version of the method of Knight and Gray (1994) as described previously (West *et al.*, 1998).

RNA was separated by formaldehyde agarose gel electrophoresis and blotted onto nylon membrane (Nytran-N, Schleicher and Schuell, London, UK) using standard methods (Sambrook *et al.*, 1989). Hybridization conditions, autoradiography and image analysis were performed as described previously (West *et al.*, 1998).

Accelerated ageing of seeds

Seed moisture contents, expressed on a fresh weight basis, were determined gravimetrically on three samples of 100 seeds after drying for 1 h in an oven at 130°C, according to the ISTA recommendations for non-oily seeds. Seeds (150) were then sealed in glass ampoules (19 × 32 mm) and conditioned to 15.6% moisture content (fresh weight basis) by the addition of distilled water. Seeds were allowed to equilibrate to 15.6% moisture content at 4°C before being transferred to 43°C for periods of up to 7 days. After ageing, seeds were sown on filter paper moistened with water and their germination performance assessed at 23°C. Germination of seeds was recorded when radicles had emerged by at least 1 mm.

Substrate regulation of peptide uptake in isolated barley embryos

Barley embryos with scutellum attached were dissected from whole grains imbibed for 20 h and placed scutellum side down onto 1.2% (w/v) agar (BDH Ltd, Poole, Dorset, UK) buffered with 50 mM sodium phosphate-citrate buffer (pH 3.8) containing either no additives, 5 mM sorbitol, 5 mM total amino acid mixture or 5 mM glucose. Embryos/scutella were incubated in the dark at 23°C for 4 h and then assayed for peptide uptake.

Incorporation of radiolabelled amino acids into protein

The rate of protein synthesis in isolated barley embryos/scutella was determined by the incorporation of [U-^{14}C]leucine (2 mM, 2.5 μCi ml^{-1}; Amersham International) as described previously (Mans and Novelli, 1961).

Results

Peptide transport activity development during germination

The ability of isolated barley grain scutella to transport peptides and amino acids during germination was determined by measurement of Ala-[^{14}C]Phe and [^{14}C]Ala uptake respectively (Fig. 28.1). Peptide transport activity exhibits a rapid increase from 6 h imbibition up to a maximum at 18–24 h after which time there is relatively little change up to 3 days imbibition. A maximum rate of peptide transport of 130 nmol peptide g^{-1} FW min^{-1} is observed at this time. A 30% decline in peptide transport activity occurs between 3 and 5 days imbibition at which time peptide transport rates of 80 nmol peptide g^{-1} FW min^{-1} are measurable. This pattern of development of peptide transport activity contrasts with that of amino acid transport activity which develops in scutellar tissue later in germination. Although low but significant levels of amino acid transport activity (~40 nmol amino acid g^{-1} FW min^{-1}) can be detected up to 2 days imbibition, it is only over the subsequent 24 h (2–3 days imbibition) period that rapid increases in amino acid transport activity, up to 100 nmol g^{-1} FW min^{-1}, are observed. These amino acid transport activity levels are then maintained in scutellar tissue up to at least 5 days post imbibition when peptide transport activity levels are falling (Fig. 28.1). Inhibition of development of peptide transport activity in scutellar tissue during germination by the protein synthesis inhibitor cycloheximide and the RNA polymerase II inhibitor α-amantin (West *et al.*, 1998) are indicative of the *de novo* synthesis of the peptide transporter during barley grain germination.

Analysis of peptide transporter gene expression

The recent cloning of the barley scutellar peptide transporter *HvPTR1* (West *et al.*, 1998) has permitted an analysis of expression patterns of this transporter with respect to both tissue specificity and developmental regulation during germination. Northern analysis has demonstrated that *HvPTR1* expression is highly tissue specific. High levels of *HvPTR1* transcripts are only detected in imbibing scutellar tissue from viable barley grains whilst expression appears absent in roots or leaves of the mature barley plant and also in embryonic axis tissue (Fig. 28.2). It cannot be entirely discounted that there may be very low levels of expression of *HvPTR1* in these other tissues but they are not detectable by Northern analysis under the conditions employed in this study. The only other tissue in which *HvPTR1* expression has been detected is in the embryo, but not the endosperm, of developing barley grains where *HvPTR1* transcripts have been detected via a reverse transcription-polymerase chain reaction (RT-PCR) approach. Sequence analysis of this PCR cDNA product from developing grains indicates that it is identical to the *HvPTR1* gene product expressed in germinating barley scutella (W.M. Waterworth, unpublished data). *HvPTR1* transcript levels increase rapidly in scutella reaching maximum

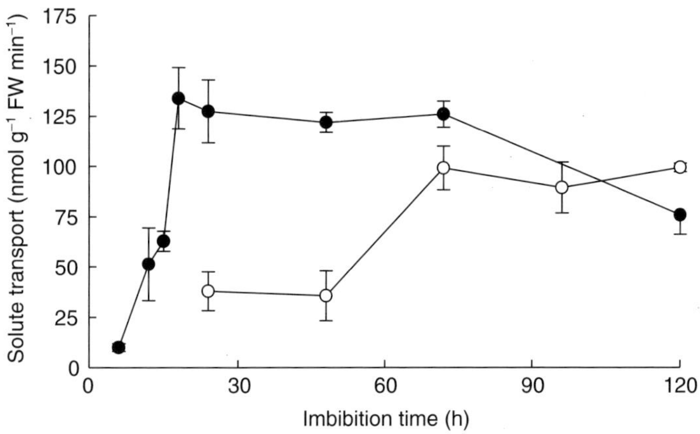

Fig. 28.1. Development of peptide and amino acid transport by isolated scutella during germination. Ala-[^{14}C]Phe transport (●) or [^{14}C]Ala transport (○) was assayed in scutella isolated from barley grains imbibed for the time indicated as described in Experimental Procedures. Values are the mean ± standard error of 2–4 independent assays.

Fig. 28.2. Northern analysis of different tissues. Lane 1: mature leaf; lane 2: 5 day shoot; lane 3: scutellum at 24 h imbibition; lane 4: mature root; lane 5: 5 day root; lane 6: whole embryo at 24 h imbibition; lane 7: embryonic axis at 24 h imbibition.

levels at 24 h imbibition and remaining high up to at least 4 days seedling growth (Fig. 28.3).

Substrate regulation of peptide transport in barley scutellum

When excised barley embryos are incubated on agar containing different supplements, a range of effects on the development of peptide transport activity is observed. The presence of amino acids in the agar at concentrations (5 mM) representative of those found in the endosperm of germinating barley grains (Higgins and Payne, 1981) results in a significant decrease (~60%) in peptide transport activity in scutellar tissue over a 4 h incubation period (Table 28.1). Glucose is the major breakdown product of starch in the endosperm of germinating barley and is the form in which most of the carbohydrate reserves of the endosperm are transported into the scutellum prior to conversion to sucrose and onward transport to the embryo during germination (Bewley and Black, 1994). Isolated barley embryos incubated on agar in the presence of 5 mM glucose show a 65% increase in peptide transport activity over a 4 h incubation period (Table 28.1). Ongoing studies indicate that *HvPTR1* transcript levels are unaffected during incubation with either amino acids or

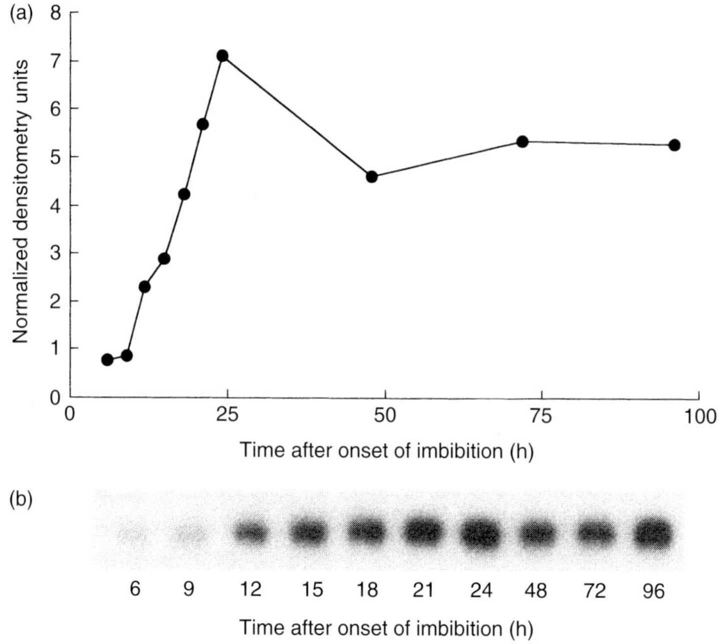

Fig. 28.3. Developmental regulation of *HvPTR1* expression during early germination. (a) Normalized densitometric analysis of hybridization signals. The amount of probe hybridizing to the blot was quantified using a PhosphoImager (Fuji) and normalized to take account of small differences in RNA loading using densitometry analysis of 28S RNA cDNA probe hybridizing to the same blot. (b) Northern analysis of *HvPTR1* expression during germination. Total RNA (20 μg) was isolated from barley scutellum at times indicated after the onset of imbibition. Northern analysis was performed to determine levels of *HvPTR1* transcript.

glucose (W.M. Waterworth, unpublished data) suggesting that regulation of peptide transporter activity is occurring at the translational or post-translational level. The presence of an inert osmolyte, 5 mM sorbitol, in the agar in control experiments demonstrates that these effects on peptide transport activity by the presence of glucose and amino acids cannot be attributable simply to an effect of changes in osmolarity of the medium surrounding the excised embryos during incubation on agar.

Peptide transport activity and seed viability

The level at which cereal embryos are able to support protein synthesis, particularly under germination conditions employing either an imposed osmotic or temperature stress, has been demonstrated to be indicative of the vigour or viability rating of a seed lot (Blowers *et al.*, 1980; Smith and Bray, 1984). Isolated non-viable barley embryos (with scutellum attached) are still capable of supporting a low level of protein synthesis which approaches ~40% the level found in viable embryos germinated for 3 days at 23°C (Fig. 28.4a).

Table 28.1. Effect of glucose, amino acids and sorbitol on peptide transport activity in imbibing barley scutella.

Supplement to agar	Peptide transport activity (% of control values)
None	100
5 mM glucose	165 ± 23.4
5 mM amino acids	42.5 ± 13
5 mM sorbitol	90.3 ± 13

Embryos were isolated from barley grain imbibed for 20 h and placed on agar with or without the addition of appropriate supplements for 4 h prior to assay of peptide transport activity. Peptide transport rates in scutella placed on agar containing no supplements were adjusted to 100% for each separate experiment. These peptide transport rates are equivalent to 192 ± 28 nmol g^{-1} FW min^{-1} (glucose experiment), 263 ± 19 nmol g^{-1} FW min^{-1} (amino acid experiment) and 156 ± 11 nmol g^{-1} FW min^{-1} (sorbitol experiment). Each data point represents the standard error of the mean of four replicate assays.

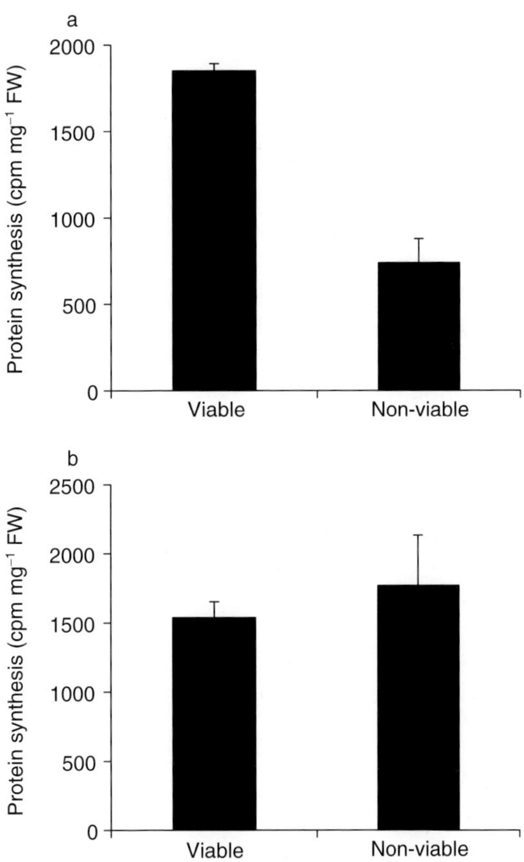

Fig. 28.4. Protein synthesis during imbibition in viable and non-viable seeds. (a) Protein synthesis in viable and non-viable whole barley embryos. (b) Protein synthesis in scutellar tissue from viable and non-viable barley.

Surprisingly, when scutellar tissue is isolated from viable or from non-viable embryos and rates of protein synthesis in these scutellar preparations are determined then a distinctly different result is obtained. Scutella from either viable or non-viable barley grains are equally capable of supporting protein synthesis at 3 days imbibition (Fig. 28.4b) and no difference in rates of protein synthesis related to viability scores were found in the barley seed lots employed in this study. Thus, rates of protein synthesis in barley scutellar tissue during germination are not indicative of the viability of a seed. It is probable that the significant levels of protein synthesis detectable in whole non-viable barley embryos (Fig. 28.4a) resides mainly in the scutellar tissue attached to these embryos.

Although the scutellum from 3 day germinated non-viable barley grain can support levels of protein synthesis comparable to those found in the scutellum of viable grain, peptide transport activity does not develop in the scutellum of the non-viable grain (Fig. 28.5). When viable barley grains are subjected to an accelerated ageing regime the level of peptide transport activity in the scutella also reflects the viability rating of the seed lot (Table 28.2). Thus, at least for barley grains, the level of peptide transporter activity in scutellar tissue appears to be a more reliable indicator of viability ratings than protein synthesis levels alone.

Discussion

The development of peptide transport activity in the scutellum of germinating barley grains precedes that of amino acid transport and rates of amino acid transport only reach those of peptide transport after 4 days imbibition (Fig. 28.1). Peptides produced as products of endosperm protein hydrolysis accumulate at millimolar concentrations in the endosperm of germinating barley grains during the first few days following imbibition (Higgins and Payne,

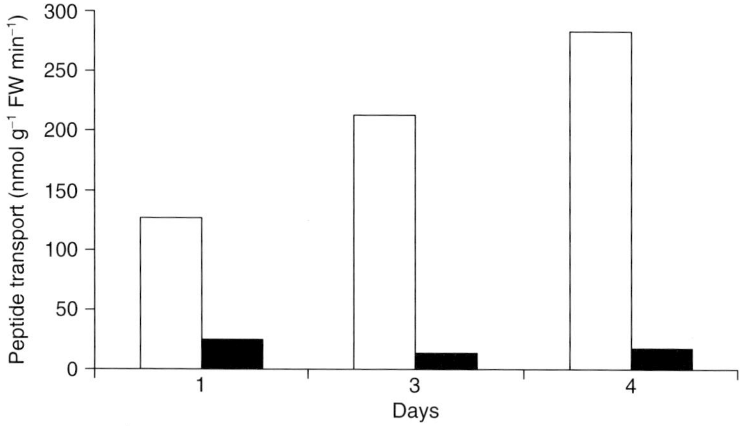

Fig. 28.5. Peptide transport activity in scutella of viable and non-viable grain over 1–4 day period. Scutella from viable barley grains (□); non-viable barley grains (■).

Table 28.2. The effect of accelerated ageing on viability and peptide uptake by barley scutella after 48 h imbibition.

Duration of accelerated ageing (days) at 43°C and 15.6% moisture content	Viability (%)	Peptide transport activity (nmol Ala[^{14}C]Phe transported g^{-1} FW min^{-1})
0	95.5 ± 1.5	95.5 ± 3.5
1	94 ± 4.0	86.5 ± 0.5
2	73 ± 3.0	67 ± 8.0
3	41 ± 3.0	38.5 ± 7.5
6	0	0

Values represent means of two independent experiments.

1981). These observations suggest that peptide transport by the scutellum may provide the embryo with a significant source of organic nitrogen especially during germination and the very early stages of seedling development. *HvPTR1*, the barley scutellar peptide transporter, codes for a 579 amino acid polypeptide with a theoretical molecular mass of 63 kDa (West *et al.*, 1998) and exhibits highest amino acid sequence homology (58% identity) with an *Arabidopsis* peptide transporter *AtPTR2-B* (Rentsch *et al.*, 1995; Song *et al.*, 1997). Sequence alignments have demonstrated that *HvPTR1* is a member of the proton-coupled oligopeptide transporter (PTR) family of peptide transporters (Paulsen and Skurray, 1994). Whilst hydropathy analysis (Kyte and Doolittle, 1982) of HvPTR1 reveals 12 putative transmembrane domains (West *et al.*, 1998), sequence analysis of HvPTR1 fails to identify any consensus target signals to the endomembrane system although HvPTR1 has been shown to be located in the scutellar plasma membrane (W.M. Waterworth, C.E. West and C.M. Bray, unpublished data). How HvPTR1 is targeted to the scutellar plasma membrane after synthesis remains to be determined.

HvPTR1 transcript levels remain high up to 4 days seedling growth but peptide transport activity levels remain unchanged over this period. Previous studies (Walker-Smith and Payne, 1985) demonstrated that the turnover of the peptide transporter in barley scutellar plasma membrane was low over this period and peptide transport at this time was unaffected by inhibition of protein synthesis. Collectively, these observations suggest that there may be some translational control of HvPTR1 synthesis at these later post imbibition stages.

Little is known concerning those factors controlling expression of peptide transport activity during germination at either the transcriptional or post-transcriptional level although evidence exists of control at both levels. Removal of the endosperm has no effect upon peptide transport activity development in barley during the first 12 h imbibition although under these conditions activity declines rapidly after 24 h imbibition in contrast to the situation in intact viable grain (Sopanen, 1979). The repression and induction of peptide transport activity in isolated scutellar tissue by amino acids and glucose respectively is indicative of a role for these metabolites either directly or indirectly in the control of peptide transport activity during germination of barley grains. However, it remains to be shown definitively whether the control is at the transcriptional, translational or post-translational level.

Loss of viability in barley grains is reflected in the absence of development of peptide transport activity in the scutellum after imbibition. Thus, scutellar peptide transporter activity levels during imbibition are an early and sensitive indicator of the viability rating of a seed lot. Surprisingly, scutellar tissue from imbibed grains of the non-viable seeds used in this study could still support protein synthesis at rates comparable to those found in the scutella from imbibed viable grains. Lack of induction of peptide transport activity observed in the scutella of non-viable grains may be the result of the absence of an appropriate signal, e.g. glucose and/or peptides, reaching the scutellum because of the failure to initiate hydrolysis of endosperm reserves in non-viable seeds. An alternative explanation is that the scutellar epithelial tissue of non-viable grain is unable to perceive or respond to appropriate signals and so fails to induce *HvPTR1* expression. However, more information is required concerning the physiological factors which control *HvPTR1* expression before any of these possibilities can be confirmed.

Peptide transport in plants is usually associated with those tissues in which rapid protein mobilization is taking place. This includes seed storage protein proteolysis during germination, organic nitrogen redistribution during leaf senescence and protein deposition during seed development. Only a few reports of peptide transport in different plant tissues exist although proton-coupled peptide transport has been characterized in the plasma membrane of broad bean (*Vicia faba*) mature leaf mesophyll cells (Jamai *et al.*, 1996). Additionally, two peptide transporters have been isolated from *Arabidopsis thaliana*. A root-specific peptide transporter *AtPTR2-A* (Steiner *et al.*, 1994) is expressed at low levels in root tissue whilst a second peptide transporter *AtPTR2-B* (Rentsch *et al.*, 1995; Song *et al.*, 1997) is expressed in most *Arabidopsis* tissues with particularly high levels in developing seed pods. Transgenic *Arabidopsis thaliana* plants with reduced levels of AtPTR2-B showed abnormal seed development with fewer, larger seeds per silique than wild type plants (Song *et al.*, 1997). This observation is consistent with a significant role for peptide transport in seed development. It is of interest to note that *HvPTR1* expression has also been detected in the embryo tissue of developing barley seeds but a role for *HvPTR1* and peptide transporter activity in grain filling in barley has yet to be established.

Relatively few genes that exhibit germination-specific expression have been identified to date and the mobilization of cereal endosperm reserves is usually considered to be a post-germination event (Bewley, 1997). Significant amounts of amino acids and peptides have been shown to be present in the endosperm of unimbibed cereal grains, but only the levels of peptide transport and not amino acid transport can account for the rates of nitrogen transfer from endosperm to embryo in barley grains during early germination (Higgins and Payne, 1981). The peptide transporter *HvPTR1* now joins one of only a small number of characterized genes whose expression is induced during germination and this early expression of *HvPTR1* in scutellar tissue is suggestive of a more central role for peptide transport in the onset of growth processes during germination *per se* than has previously been thought.

Acknowledgements

The authors thank Caroline Bowsher who kindly provided the barley 28S rRNA cDNA clone and Steve Brown (PBI, Cambridge) who provided the Maris Otter barley grain. This work was supported by a research grant from the Biotechnology and Biological Sciences Research Council.

References

Bewley, J.D. (1997) Seed germination and dormancy. *Plant Cell* 9, 1055–1066.

Bewley, J.D. and Black, M. (1994) *Seeds. Physiology of Development and Germination*, 2nd edn. Plenum Press, New York and London, 445 pp.

Blowers, L.E., Stormonth, D.A. and Bray, C.M. (1980) Nucleic acid and protein synthesis and loss of vigour in germinating wheat embryos. *Planta* 150, 19–25.

Bray, C.M. (1979) Nucleic acid and protein synthesis in the embryo of germinating cereals. In: Laidman, D.L. and Wyn Jones, R.G. (eds) *Recent Advances in the Biochemistry of Cereals.* London, Academic Press, pp. 147–173.

Hardy, D.J. and Payne, J.W. (1992) Amino acid and peptide transport in higher plants. *Plant Physiology (Life Science Advances)* 11, 59–73.

Higgins, C.F. and Payne, J.W. (1977) Characterisation of active dipeptide transport of germinating barley embryos: effects of pH and metabolic inhibitors. *Planta* 136, 71–76.

Higgins, C.F. and Payne, J.W. (1981) The peptide pools of germinating barley grains: relation to hydrolysis and transport of storage proteins. *Plant Physiology* 67, 785–792.

Jamai, A., Laloi, M., Bourbouloux, A., Valantin, M. and Delrot, S. (1996) Characterisation of leucine-leucine transport in leaf tissues. *Journal of Experimental Botany* 47, 1223–1227.

Knight, J.S. and Gray, J.C. (1994) Expression of genes encoding the tobacco chloroplast phosphate translocator is not light-regulated and is repressed by sucrose. *Molecular and General Genetics* 242, 586–594.

Kyte, J. and Doolittle, R.F. (1982) A simple method for displaying the hydropathic character of a protein. *Journal of Molecular Biology* 157, 105–132.

Mans, R.J. and Novelli, G.D. (1961) Measurement of the incorporation of radioactive amino acids into protein by a filter paper disc method. *Archives of Biochemistry and Biophysics* 94, 48–53.

Osborne, D.J. (1983) Biochemical control of systems operating in the early hours of germination. *Canadian Journal of Botany* 61, 3568–3577.

Paulsen, I.T. and Skurray, R.A. (1994) The POT family of transport proteins. *Trends in Biochemical Sciences* 18, 404.

Rentsch, D., Laloi, M., Rouhara, I., Schmeizer, E., Delrot, S. and Frommer, W.B. (1995) NTR1 encodes a high affinity oligopeptide transporter in *Arabidopsis. FEBS Letters* 370, 264–268.

Salmenkallio, M. and Sopanen, T. (1989) Amino acid and peptide uptake in the scutella of germinating grains of barley, wheat, rice and maize. *Plant Physiology* 89, 1285–1291.

Sambrook, J., Fritsch, E.F. and Maniatis, T. (1989) *Molecular Cloning: A Laboratory Manual*, 2nd edn. Cold Spring Harbor Laboratory Press, Cold Spring Harbor.

Smith, C.A.D. and Bray, C.M. (1984) Polyadenylated RNA levels and macromolecular synthesis during loss of seed vigour. *Plant Science Letters* 34, 335–343.

Song, W., Koh, S., Czako, M., Marton, L., Dreukard, E., Becker, J.M. and Stacey, G. (1997) Antisense expression of the peptide transport gene *AtPTR2-B* delays flowering and arrests seed development in transgenic *Arabidopsis* plants. *Plant Physiology* 114, 927–935.

Sopanen, T. (1979) Development of peptide transport activity in barley scutellum during germination. *Plant Physiology* 64, 570–574.

Steiner, H-Y., Song, W., Zhang, L., Naider, F., Becker, J.M. and Stacey, G. (1994) An *Arabidopsis* peptide transporter is a member of a new class of membrane transport proteins. *Plant Cell* 6, 1289–1299.

Stoddart, J.L., Thomas, H. and Robertson, A. (1973) Protein synthesis patterns in barley embryos during germination. *Planta* 112, 309–321.

Walker-Smith, D.J. and Payne, J.W. (1984) Characteristics of the protein carrier of the peptide-transport system in the scutellum of germinating barley embryos. *Planta* 162, 166–173.

Walker-Smith, D.J. and Payne, J.W. (1985) Synthesis of the peptide-transport carrier of the barley scutellum during the early stages of germination. *Planta* 164, 550–556.

West, C.E., Waterworth, W.M., Stephens, S.M., Smith, C.P. and Bray, C.M. (1998) Cloning and functional characterisation of a peptide transporter expressed in the scutellum of barley grain during the early stages of germination. *The Plant Journal* 15, 221–229.

29 Molecular Cloning and Possible Role of an Asparaginyl Endopeptidase (REP-2) from Germinated Rice Seeds

H. KATO AND T. MINAMIKAWA

Department of Biological Sciences, Tokyo Metropolitan University, Minami-ohsawa 1-1, Hachioji, Tokyo 192-0397, Japan

Two major cysteine endopeptidases, designated REP-1 and REP-2, were separated from extracts of germinated rice seeds by hydrophobic chromatography. REP-1 digested *in vitro* both the acidic and basic subunits of glutelin, the major storage protein of rice. REP-1 was synthesized as a precursor which has a long prosequence. Accumulation of the REP-1 protein in seeds was controlled by the plant hormones, gibberellic acid and abscisic acid. The accumulation of REP-1 mRNA is mainly regulated at the transcriptional level, but the endopeptidase activity in germinating rice seeds develops later than the appearance of REP-1 mRNA, indicating that the expression of REP-1 is regulated also by post-transcriptional events. In the present study, REP-2 was purified from extracts of 9-day germinated rice seeds. In the final step of the purification by gel filtration chromatography, the enzyme was separated into two forms, REP-2α (39 kDa) and REP-2β (40 kDa). The amino-terminal amino acid sequence of REP-2 showed strong similarity to that of asparaginyl endopeptidases (Asn-EPases). The enzyme was immunoreactive to an antiserum raised against legumain, an Asn-EPase from jack bean and had a substrate specificity toward Asn residues. These data indicate that REP-2 is a member of the Asn-EPases that have a strict substrate specificity toward the carboxyl side of Asn residues. A single-stranded cDNA was synthesized from a poly(A) RNA fraction of rice seedlings, and a cDNA product (0.48 kb) amplified by the reverse transcription-polymerase chain reaction (RT-PCR) was obtained using the amino-terminal amino acid sequence and the conserved sequences of known Asn-EPases as primers. The deduced amino acid sequence of this cDNA showed 72% identity to that of VmPE-1, the Asn-EPase from *Vigna mungo*. A possible role of REP-2 acting on the post-translational processing of REP-1 as well as on the restricted hydrolysis of seed storage proteins is discussed.

Separation of Two Major Endopeptidases, REP-1 and REP-2, from Germinated Rice Seeds

When extracts prepared from 9-day germinated rice seeds were chromatographed on a butyl-Cellulofine column, endopeptidase activity was separated into two major fractions, REP-1 and REP-2 (Fig. 29.1). Experiments with various protease inhibitors showed that both enzymes are cysteine endopeptidases. We further purified REP-1 to homogeneity and demonstrated that the enzyme is the major endopeptidase that digests rice storage glutelin (Kato and Minamikawa, 1996). The amino acid sequence deduced from REP-1 cDNA indicated that REP-1 is synthesized as a precursor, which has a long prosequence similar to other cysteine endopeptidases, SH-EP from *V. mungo* (Mitsuhashi and Minamikawa, 1989) and EP-B from barley (Koehler and Ho, 1990), both of which are responsible for the degradation of seed storage proteins. The accumulation of REP-1 mRNA and protein is regulated by both gibberellin and abscisic acid. Increase in the amount of REP-1 mRNA after onset of imbibition preceded that of the REP-1 activity, indicating that the expression of REP-1 is regulated not only at the transcriptional level but also at the post-transcriptional level (Shintani *et al.*, 1997). SH-EP is also known to be synthesized as a large precursor. The prosequence of the SH-EP precursor is cleaved by an Asn-EPase, named VmPE-1 (Okamoto and Minamikawa, 1995). Thus, we postulated that in germinating rice seeds, the prosequence of REP-1 precursor is processed to form the mature form by REP-2.

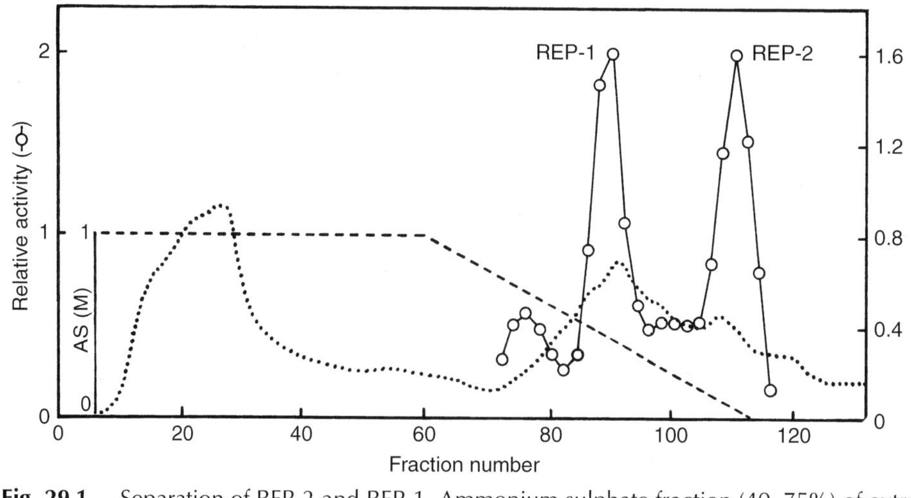

Fig. 29.1. Separation of REP-2 and REP-1. Ammonium sulphate fraction (40–75%) of extracts from 9-day rice seedlings was loaded onto a column of butyl-Cellulofine. Endopeptidase activity in fractions from the column was assayed using azoalbumin as a substrate.

Purification and Characterization of REP-2

Substrate specificities of REP-1 and REP-2 were examined using several synthetic substrates. REP-1 showed a substrate specificity toward Z-Phe-Arg-MCA, which was also efficiently cleaved by endopeptidases of the papain family. REP-2 efficiently cleaved Z-Ala-Ala-Asn-MCA, a specific substrate for Asn-EPases, but not Z-Phe-Arg-MCA (Table 29.1), indicating that REP-2 is a member of Asn-EPases. The REP-2 fraction used had aminopeptidase activity. This may be due to aminopeptidases coexisting in the fraction.

REP-2 was further purified by successive chromatography on QA-52 (Whatman) and Sephacryl S-200 (Pharmacia) columns. After the gel filtration chromatography, REP-2 was separated into two forms, REP-2α (39 kDa) and REP-2β (40 kDa). Both forms were immunoreactive to the antiserum against legumain. Ten amino acid residues of the amino-terminal sequences of both enzymes were identical. pH-dependence of both forms of REP-2 showed sharp pH profiles with optima at pH 6.0. Both forms were stable only around pH 6.0. Western blot analysis with the antiserum against legumain showed that both REP-2α and REP-2β were detected in dry seeds as well as in germinated seeds. Amounts of both forms decreased gradually during germination and they became undetectable in 15-day seedlings (Fig. 29.2).

Molecular Cloning of REP-2 cDNA and Deduced Amino Acid Sequence

An RT-PCR technique was employed to analyse the primary structure of REP-2. Three conserved sequences were selected from other reported Asn-EPases and used as primers 2, 3 and 4. The sequence of the amino-terminal region of REP-2 was used as primer 1. Primers 1 and 4 were external primers, and primers 2 and 3 were internal primers. As presented in Fig. 29.3A, mRNA was prepared from 4-day germinated rice seeds using an oligo (dT) cellulose column (Pharmacia). Single-stranded cDNA was synthesized by a first strand

Table 29.1. Substrate specificities of REP-2 and REP-1.

	Relative activity (%)	
Substrate	REP-2	REP-1
Z-Ala-Ala-Asn-MCA	100	45
Z-Phe-Arg-MCA	0	100
Z-Arg-Arg-MCA	4	0
Arg-MCA	110	0
Ala-MCA	4	0
Leu-MCA	56	0
Phe-MCA	32	0

REP-1 and REP-2 used were obtained from hydrophobic chromatography. Z, carbobenzoxy group; MCA, 4-methyl-coumaryl-7-amide.

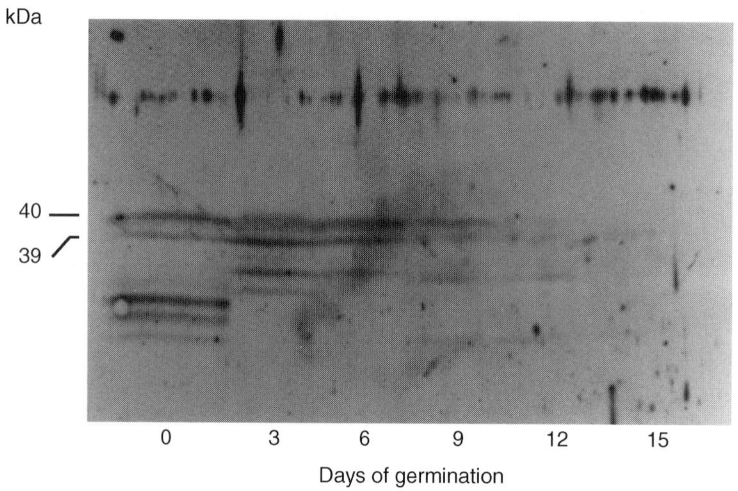

Fig. 29.2. Change with time in the amounts of 39 kDa REP-2α and of 40 kDa REP-2β. Seeds were germinated for 0–15 days at 27°C under darkness. The extract (10 μl each) of seedlings was analysed by immunoblotting using an antiserum against legumain after SDS-PAGE. The antiserum was a generous gift from Dr Y. Miura-Izu.

cDNA synthesis kit (Takara Shuzo) using the oligo (dT) primer. The first PCR was conducted using the external primers, and the primary product of the PCR was used as the template for the second PCR. The final product (0.48 kb), named pREP2ins, was amplified by the second PCR using primers 2 and 3 (Fig. 29.3B). pREPins was subcloned into the pUC118 cloning vector and its nucleotide sequence was determined by the dideoxyribonucleotide sequencing method. The deduced amino acid sequence of pREP2ins is highly homologous to known Asn-EPases: 75% with legumain (Takeda *et al.*, 1994), 76% with proteinase B (Becker *et al.*, 1995) and 72% with VmPE-1 (Okamoto and Minamikawa, 1999). These results indicate that pREP2ins is a partial cDNA for REP-2, a member of Asn-EPases.

Possible Roles of REP-2

Plant Asn-EPases are known to be involved in the processing of seed storage proteins (Hara-Nishimura *et al.*, 1991; Takeda *et al.*, 1994) and also in the processing of proteinases that are responsible for the degradation of storage proteins (Okamoto and Minamikawa, 1995). An Asn-EPase, named proteinase B, from germinated vetch seeds plays a role in the degradation of seed storage proteins during germination (Becker *et al.*, 1995). VmPE-1, an Asn-EPase from germinated *V. mungo* seeds, is suggested to be involved in the processing of SH-EP, the major proteinase that digests storage proteins during germination and early seedling growth, whereas another Asn-EPase from *V. mungo*, named VmPE-1A, is possibly involved in the restricted digestion of the storage proteins (Okamoto and Minamikawa, 1999) similarly to proteinase B.

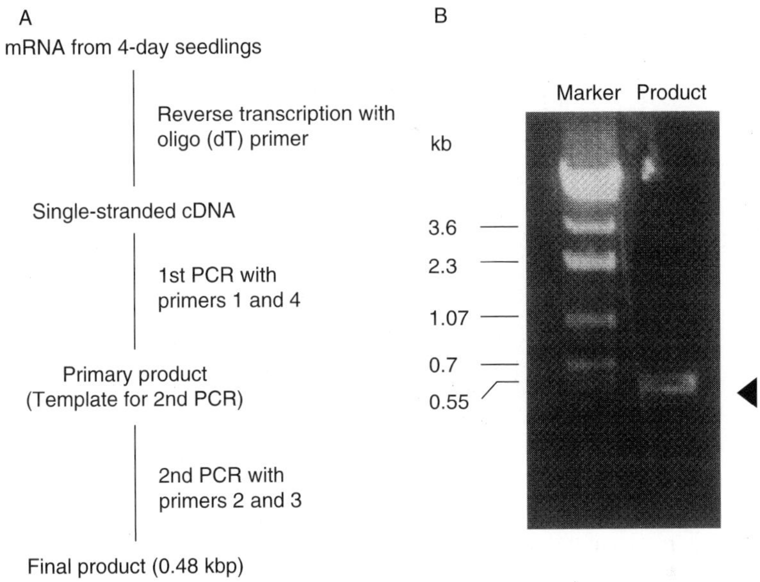

Fig. 29.3. The procedure of RT-PCR and analysis of the product. A. Single-stranded cDNA from 4-day seedlings was used as a template for the first PCR using primers 1 and 4. The inner region of the primary product was further amplified by the second PCR using primers 2 and 3. B. The final product of 0.48 kb was separated on a 1% agarose gel.

Rice glutelin, the major storage protein in rice, is synthesized as a large precursor consisting of acidic and basic subunits following a signal peptide, and the precursor is post-translationally processed into two subunits by being cleaved at the carboxy side of an Asn residue (Takaiwa *et al.*, 1986). Since two forms of REP-2 occur in both dry and germinated rice seeds, we postulate that the enzyme is involved both in the processing of glutelin precursors during seed maturation and in the restricted digestion of glutelin as well as the processing of REP-1 after imbibition of seeds. The relevance of the two forms of REP-2 to their roles remains to be elucidated.

Acknowledgement

This work was supported in part by a grant-in-aid (no. 09640776) from the Ministry of Education, Science, and Culture of Japan.

References

Becker, C., Shutov, A.D., Nong, Van H., Senyuk, V.I., Jung, R., Horstmann, C., Fischer, J., Nielsen, N.C. and Müntz, K. (1995) Purification, cDNA cloning and characterization of proteinase B, an asparagine-specific endopeptidase from germinating vetch (*Vicia sativa* L.) seeds. *European Journal of Biochemistry* 228, 456–462.

Hara-Nishimura, I., Inoue, K. and Nishimura, M. (1991) A unique vacuolar processing enzyme responsible for conversion of several proprotein precursors into the mature form. *FEBS Letters* 294, 89–93.

Kato, H. and Minamikawa, T. (1996) Identification and characterization of a rice cysteine endopeptidase that digests glutelin. *European Journal of Biochemistry* 239, 310–316.

Koehler, S.M. and Ho, T.-H.D. (1990) Hormonal regulation, processing, and secretion of cysteine proteinases in barley aleurone layers. *The Plant Cell* 2, 769–783.

Mitsuhashi, W. and Minamikawa, T. (1989) Synthesis and posttranslational activation of sulfhydryl-endopeptidase in cotyledons of germinating *Vigna mungo* seeds. *Plant Physiology* 89, 274–279.

Okamoto, T. and Minamikawa, T. (1995) Purification of a processing enzyme (VmPE-1) that is involved in post-translational processing of a plant cysteine endopeptidase (SH-EP). *European Journal of Biochemistry* 231, 300–305.

Okamoto, T. and Minamikawa, T. (1999) Molecular cloning and characterization of *Vigna mungo* processing enzyme 1 (VmPE-1), an asparaginyl endopeptidase possibly involved in post-translational processing of a vacuolar cysteine endopeptidase (SH-EP). *Plant Molecular Biology* 39, 63–73.

Shintani, A., Kato, H. and Minamikawa, T. (1997) Hormonal regulation of expression of two cysteine endopeptidase genes in rice seedlings. *Plant and Cell Physiology* 38, 1242–1248.

Takaiwa, F., Kikuchi, S. and Oono, K. (1986) The structure of rice storage protein glutelin precursor deduced from cDNA. *FEBS Letters* 206, 33–35.

Takeda, O., Miura, Y., Mitta, M., Matsushita, H., Kato, I., Abe, Y., Yokosawa, H. and Ishii, S. (1994) Isolation and analysis of cDNA encoding a precursor of *Canavalia ensiformis* asparaginyl endopeptidase (legumain). *Journal of Biochemistry* 116, 541–546.

30 Metabolism of Essential Oils during Inhibition of Wheat Seed Germination

N. DUDAI[1], O. LARKOV[1], A.M. MAYER[2], A. POLJAKOFF-MAYBER[2], E. PUTIEVSKY[1] AND H.R. LERNER[2]

[1]*Division of Aromatic Plants, ARO, Newe Yaar PO Box 1021, Ramat Yishai 30095, Israel;* [2]*Department of Botany, The Hebrew University of Jerusalem, Jerusalem 91904, Israel*

Screening of essential oils from aromatic plants showed that many contain germination inhibitors active at quite low concentrations (about 0.3 mM), if applied as oil in the gaseous phase. The amount of the inhibitory compound found per embryo was 1.3 nmol. Among the most active essential oils were those from *Cymbopogon citratus*, *Poaceae*, containing 80% citral and from *Micromeria fruticosa*, *Lamiaceae*, containing 70% pulegone. The essential oils inhibited germination, growth and development of wheat. Wheat seed treated with the pure constituents of the essential oils metabolized them. Citral (composed of geranial and neral) was converted to the corresponding oxidation and reduction derivatives neric and geranic acid, geraniol and nerol. The metabolic products were significantly less inhibitory than the parent compound. Other aldehydic compounds such as citronellal, vanillin and decanal were also metabolized, the metabolic products being less toxic than the parent substance. The ketonic monoterpene pulegone was metabolized to the less toxic menthofurane and isomenthone. The phenolic monoterpene carvacrol was metabolized to an as yet unidentified compound. It appears that metabolism and especially reduction of aldehydic monoterpenes to the corresponding alcohol is a mechanism for detoxification. We will discuss the possible ecological and applied significance of these findings.

Introduction

The effect of essential oils as inhibitors of germination has been described previously (Lerner and Evenari, 1961; Asplund, 1968; Reynolds, 1987; Friedman, 1995; Dudai *et al.*, 1999). These investigators studied the effect of monoterpenes when applied in the liquid phase. Among others, pulegone and citral have been shown to be active inhibitors under these conditions at concentrations of the order of 0.1 mM. Such studies are dependent on the solubility of the monoterpenes in water, which is not very great (Weidenhamer *et al.*, 1993). In most of the previous work little attention was paid to the precise chemical nature of the essential oils and frequently the inhibition of subsequent seedling development was stressed. The mechanism of the inhibition of germination by monoterpenes has not been studied in any detail previously, but the inhibitory effect of aldehydes and especially acetaldehyde is well known (Zhang *et al.*, 1997).

In our previous work we showed that the essential oils extracted from three species, *Cymbopogon citratus*, *Micromeria fruticosa* and *Origanum syriacum* were very active inhibitors of the germination of wheat seeds. When these oils were applied to wheat seeds in the gaseous phase at concentrations of 25–80 nl ml^{-1} they inhibited germination to 50% or more. The major components of the essential oils of these three species represent different monoterpenes, citral (aldehydic) in *Cymbopogon*, pulegone (ketonic) in *Micromeria* and carvacrol (phenolic) in *Origanum*. These major components are available in pure form and it therefore became possible to try and study the accumulation of these compounds in the seed and its major components, embryo and endosperm, to follow the kinetics of their accumulation and to investigate their metabolism. We report on these aspects of the effect of essential oils on seed germination.

Materials and Methods

The extraction of essential oils from the plants, the methods of germination of wheat seeds and the system for exposing seeds to the essential oils has been described by Dudai *et al.* (1999), the monoterpenes always being applied in the gaseous phase. In experiments in which metabolism was followed, the seeds were disinfected with 1% hypochlorite and then rinsed with sterile distilled water prior to germination. To study the effect of the essential oils on the seeds, they were either exposed to the essential oil, rinsed and then placed in distilled water, or the seeds were first placed in water for various periods of time and then exposed to the essential oils. In this way, the reversibility of inhibition and its kinetics could be established. When required, seeds were separated into embryo and endosperm by cutting out the endosperm using a sharp scalpel.

The monoterpenes in the seeds were examined using GC-MS. For analysis of the composition of the monoterpenes in the seeds or its parts, after supplying them with known compounds, these were extracted with shaking for 24 h using *tert*-butyl methyl ether. The extract was applied to a HP-5 capillary column (30 cm × 0.25 mm) and the column eluted with helium at a rate of 1 ml min^{-1}. The eluate was analysed by comparing the peaks with known standards and with library GC-MS data on the elution patterns of monoterpenes and their derivatives.

Results and Discussion

The inhibition of germination by some of the pure monoterpenes is shown in Fig. 30.1, which clearly shows the dependence of inhibition on concentration. From the data in Figs 30.2 and 30.3, using pulegone as the monoterpene, it can be seen that at a given monoterpene concentration, inhibition depends on the length of exposure of the seeds to the essential oil. A minimal period of time seems to exist before inhibition becomes evident, which suggests that a certain amount of monoterpene must accumulate in the embryo in order to inhibit

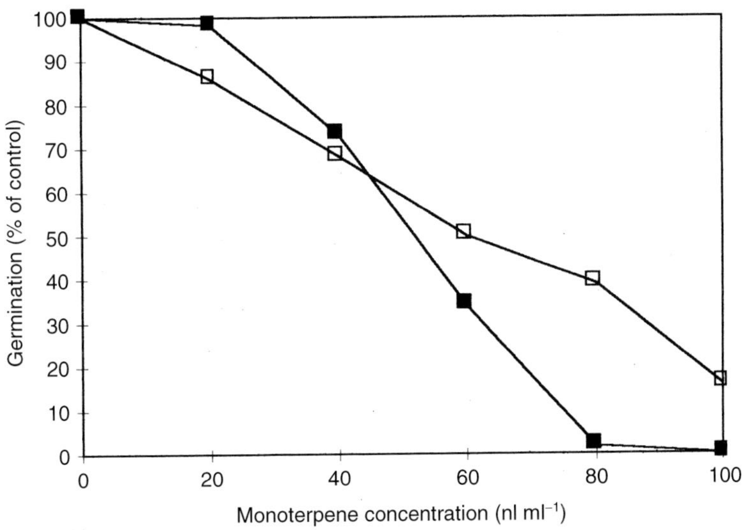

Fig. 30.1. Effect of two monoterpenes, citral (■) and carvacrol (□) on the germination of wheat.

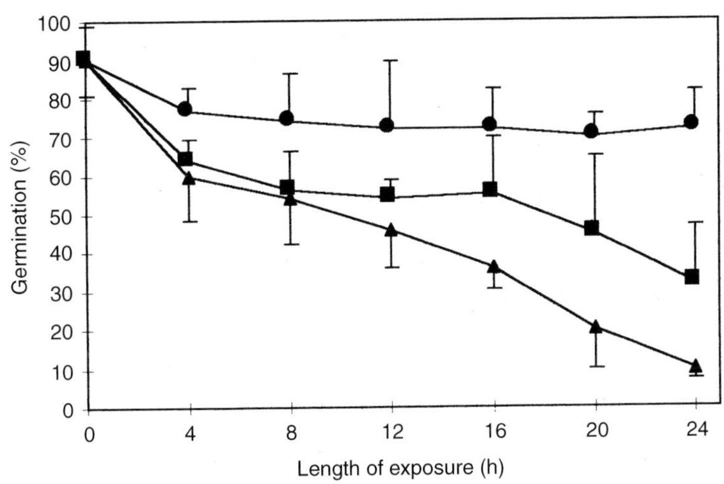

Fig. 30.2. Effect of duration of exposure to pulegone (at 40 (●), 80 (■) or 120 (▲) nl ml^{-1} in gaseous phase) on the germination of wheat.

germination. Inhibition is reversible until a critical time has been reached, after which removal of the monoterpene no longer results in resumption of germination. Even as little as exposure for 4 h at the beginning of imbibition was sufficient to cause some inhibition of germination. As expected, after 16 h imbibition followed by exposure to the essential oil, little or no inhibition occurred as the seeds had already started to germinate. Essentially similar results were obtained when the monoterpene was citral or carvacrol.

When seeds were exposed under aseptic conditions to citral, composed of a mixture of neral and geranial, it was possible to demonstrate the metabolism to mainly geraniol and nerol, the corresponding alcohols, and very small amounts of the corresponding acids, geranic acid and neric acid (Figs 30.4 and 30.5).

The amount of citral and its products in the embryo and the endosperm was extremely low amounting to about of 1.3 nmol per embryo. Similarly,

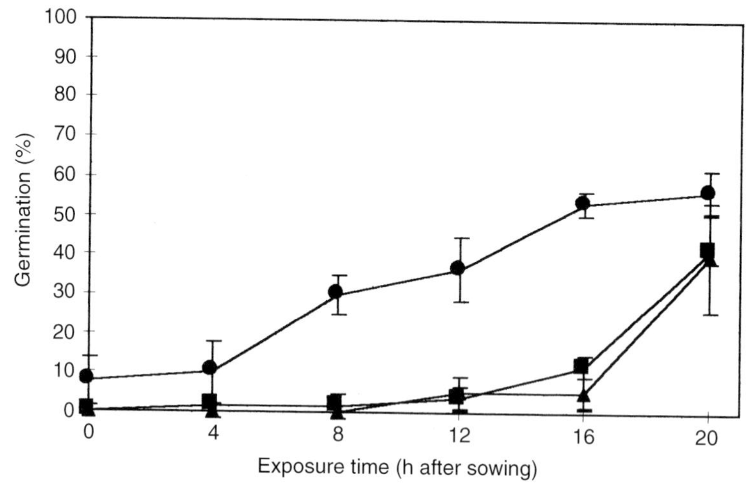

Fig. 30.3. Effect of time of exposure to pulegone (at 40 (●), 80 (■) or 120 (▲) nl ml^{-1} in gaseous phase) after beginning of sowing on germination of wheat.

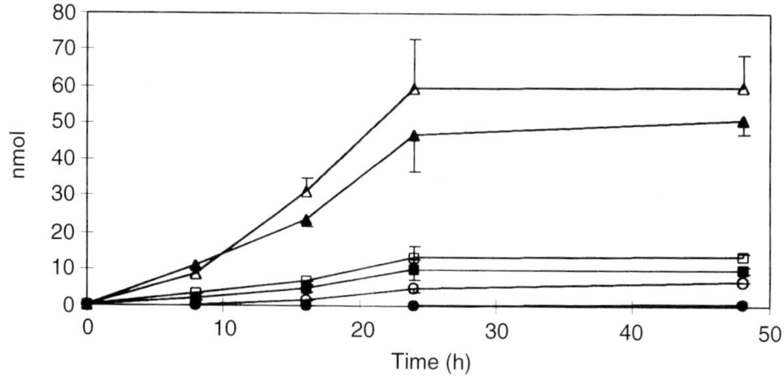

Fig. 30.4. Time course of metabolism of citral in embryos of wheat. ○, Geranic acid; □, geranial; △, geraniol; ●, neric acid; ■, neral; ▲, nerol.

other aldehydic compounds citronellal, vanillin and decanal were metabolized mainly to the corresponding alcohols, with small amounts of the corresponding acids being formed. The derivatives of all the metabolic products were less active as germination inhibitors than the parent compound. The ketonic monoterpene pulegone was also metabolized, in this case to the less toxic menthofurane and isomenthone. The phenolic monoterpene carvacrol was also metabolized, but the products have not yet been identified. When similar experiments were carried out without previously disinfecting the seeds, the metabolism was rather different indicating that microbial contaminants could effect the metabolism of the monoterpenes.

Using the GC-MS techniques we were able to follow quantitatively the amount of monoterpenes appearing in the embryo and endosperm, following exposure to the various monoterpenes. Citral was supplied to seeds and its accumulation followed (Table 30.1). It can be seen that the amount of citral per endosperm was about six times that in the embryo, and the amount in the endosperm increased slightly with time. However the fresh weight of the endosperm after 24 h imbibition was about 20 times that of the embryo. As a result the effective concentration in the embryo was significantly greater than in the endosperm. The amount present clearly was sufficient to inhibit germination.

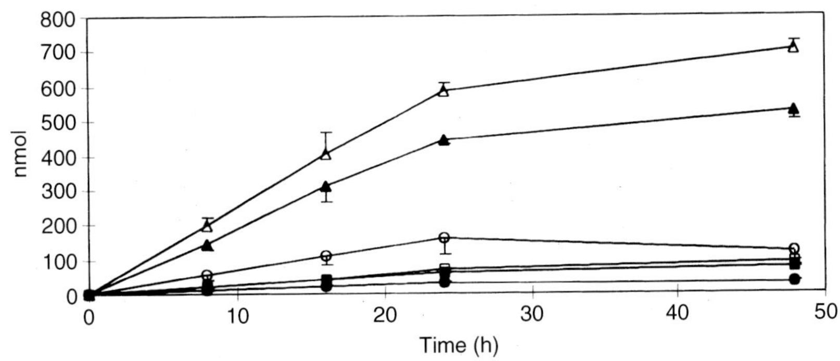

Fig. 30.5. Time course of metabolism of citral in endosperm of wheat. ○, Geranic acid; □, geranial; △, geraniol; ●, neric acid; ■, neral; ▲, nerol.

Table 30.1. Fresh weight and amount of citral** and its metabolic products (in nmol) in embryos and endosperm of wheat after 24 and 48 h of exposure.

Seed part	Time after sowing (h)	Weight (mg)	Geraniol* + nerol	Geranial* + neral	Geranic acid* + neric acid
Embryo	24	50 ± 4	106 ± 23	23 ± 7	5 ± 1
	48	54 ± 6	110 ± 3	23 ± 2	7 ± 1
Endosperm	24	1228 ± 22	1024 ± 34	128 ± 1	234 ± 51
	48	1272 ± 30	1224 ± 50	165 ± 8	147 ± 29

*Amount for 20 embryos or endosperm.
**160 nl ml^{-1} citral applied in gaseous phase.

The process by which aldehydic compounds are reduced to the corresponding alcohol could be due to the action of a non-specific alcohol dehydrogenase, acting as a detoxifying agent. This contrasts with the highly specific geraniol dehydrogenase reported to be present in the leaves of *Cymbopogon* (Sangwan *et al.*, 1993) which contain essential oils. Thus, in wheat seeds and other species not containing essential oils, detoxification could occur, until the detoxifying system is saturated. It might be suggested that very small amounts of essential oils would not inhibit germination until a threshhold value is reached.

The fact that the essential oils and the monoterpenes in them are volatile has implications for their possible mechanism of action under natural conditions. We were able to show that leaves of plants containing essential oils, when incorporated into the soil, were capable of inhibiting the germination of some species, e.g. wheat or tomato. The active compounds diffuse out and are capable of acting at a distance from their source. This is a mechanism by which plants containing volatile monoterpenes might affect competitors by allelopathic interactions (Einhellig, 1995). Such a mechanism might be utilized in order to exploit essential oils as agents for weed control.

However, it must be stressed that until now our efforts to define the mechanism of action of monoterpenes as germination inhibitors have not been successful, despite their very high activity at low concentrations.

References

Asplund, R.O. (1968) Monoterpenes: relationship between structure and inhibition of germination. *Phytochemistry* 7, 1995–1997.

Dudai, N., Poljakoff-Mayber, A., Mayer, A.M., Putievsky, E. and Lerner, H.R (1999) Essential oils as allelochemicals and their potential use as bioherbicides. *Journal of Chemical Ecology* 25, 1079–1089.

Einhellig, F.A. (1995) Mechanism of action of allelochemicals in allelopathy. In: Einhellig, F.A. (ed.) *Allelopathy: Organisms Processes and Applications.* American Chemical Society, Washington DC, pp. 96–116.

Friedman, J. (1995) Allelopathy, autotoxicity and germination. In: Kigl, J. and Galili, G. (eds) *Seed Development and Germination.* Marcel Dekker Inc., New York, pp. 629–644.

Lerner, H.R and Evenari, M. (1961) The nature of the germination inhibitor present in the leaves of *Eucalyptus rostrata. Physiologia Plantarum* 14, 221–229.

Reynolds, T. (1987) Comparative effect of alicyclic compounds and quinones on inhibition of lettuce fruit germination. *Annals of Botany* 60, 215–223.

Sangwan, R.J, Singh-Snagwan, N. and Luthra, R. (1993) Metabolism of acyclic monoterpenes: partial purification of geraniol dehydrogenase from lemongrass (*Cymbopogon flexuosus* Stapf.) leaves. *Journal of Plant Physiology* 142, 129–134.

Weidenhamer, J.D., Macias, F.A., Fisher, N.M. and Williamson, G.B. (1993) Just how insoluble are monoterpenes? *Journal of Chemical Ecology* 19, 1799–1803.

Zhang, M., Nagata, S., Miyazawa, K., Kikuchi, H. and Esashi, Y. (1997) A competitive enzyme linked immunosorbent assay to quantify acetaldehyde-protein adducts that accumulate in dry seeds during aging. *Plant Physiology* 113, 397–402.

V Dormancy

31 Genetic Model for Dormancy in Wild Oat

M.E. Foley

United States Department of Agriculture, Agricultural Research Service, Biosciences Research Laboratory, PO Box 5674, State University Station, Fargo, ND 58105-5674, USA

Seed dormancy is a key characteristic associated with the weedy nature of wild oat (*Avena fatua*). Wild oat seeds require afterripening under warm-dry conditions to exhibit a normal-rapid onset and rate of germination. Wild oat seeds in natural populations range from highly dormant to non-dormant. Genetic and environmental factors influence germinability of wild oat seeds. A highly dormant (M73) and a non-dormant (SH430) inbred line were cross-pollinated. F_2, backcross ($BC_{1ND}F_1$ and $BC_{1D}F_1$) and recombinant inbred populations were developed. Germinability of seeds in these populations was evaluated and data were used to propose a three-locus model to explain dormancy. Bulked segregant analysis and random amplified polymorphic DNA (RAPD) techniques were used to identify molecular markers linked to quantitative trait loci (QTL) that regulate germinability in wild oat.

Introduction

Untimely rain before harvest, but after maturation of some small-grain crops, leads to premature seed germination in the florets called pre-harvest sprouting. Resistance to pre-harvest sprouting in barley, oats, rice and wheat is correlated with the level of dormancy in the mature seeds. A low level of dormancy in cultivars grown in areas prone to pre-harvest sprouting is desirable to reduce its occurrence. Since resistance to pre-harvest sprouting and seed dormancy are associated, new insights may be gained by investigating dormancy of non-domesticated or weedy small-grain species.

Dormancy is of interest to weed scientists because it is a key characteristic associated with weeds in agroecosystems. Dormancy optimizes seed germination over time and dictates the need to apply weed control measures year after year. Simpson (1990) defines dormancy as the temporary failure of a viable seed to germinate, after a specific length of time, in a particular set of environmental conditions that later evoke germination when the restrictive state has

CAB *International* 2000. *Seed Biology: Advances and Applications*
(eds M. Black, K.J. Bradford and J. Vázquez-Ramos)

been terminated by either natural or artificial conditions. In context with our investigations, dormant seeds are characterized by a delayed onset and/or reduced rate of germination, relative to fully afterripened or non-dormant seeds. There are two common ways by which dormancy is imposed: coat-imposed dormancy enforced by restrictive seed coverings (hull, pericarp and testa), and embryo dormancy where control of dormancy resides within the embryo itself.

Afterripening is loss of the dormant state over some period of time through exposure of the seed to a set of environmental conditions after maturation and separation from the parent plant (Simpson, 1990; Li and Foley, 1997). Afterripening does not cause an abrupt change from a dormant to a germinable state. Rather, seeds in a population become more responsive to a range of conditions that promote germination, and less responsive to conditions that restrict germination. For example, as seeds in a population afterripen they can germinate over a wider range of temperatures than before afterripening. Barley, oats, rice, wheat and wild oat seeds normally require afterripening under warm-dry conditions (Simpson, 1990; Strand, 1991).

Wild oat is a serious weed worldwide and is used as a model system to investigate the fundamental basis for dormancy in weed seeds (Simpson, 1990). Dormancy in wild oat can result from both coat-imposed and embryo dormancy (Simpson, 1990; Foley, 1992). The long-term goal of my research is to elucidate the physiological basis for dormancy and afterripening-induced breaking of dormancy. Toward this goal, we are investigating the role of genetic and environmental factors on germinability of wild oat seeds (Fennimore, 1997; Foley and Fennimore, 1998).

Materials and Methods

Based on the cross-pollination of an inbred dormant and non-dormant line (M73 × SH430), we have developed several populations segregating for dormancy (F_2, $BC_{1ND}F_1$, $BC_{1D}F_1$), including a population of 127 F_2-derived recombinant inbred lines (RI) produced by single-seed descent. Reciprocal crosses were not made because previous research determined that maternal effects were not significant (Jana *et al.*, 1979). The germination phenotypes were classified based on germination at 15°C using a modification of the procedures outlined by Jana *et al.* (1979) and Paterson *et al.* (1989). Progeny testing was conducted by randomly selecting ten seeds from each F_3 and $BC_{1D}F_2$ family tested.

Tissue samples were harvested from leaves of 3- to 6-week-old plants for DNA marker analysis. DNA from eight true-breeding non-dormant and eight true-breeding dormant F_2 individuals were pooled to form non-dormant and dormant bulks, respectively for bulked segregant analysis (Michelmore *et al.*, 1991; Fennimore, 1997). Polymerase chain reactions (PCR) were conducted according to Williams *et al.* (1990). The bulks were screened with about 800 decamer random primers (Operon kits A-AN)[1]. Regression analysis was conducted to identify significant associations between single RAPD markers and days to germination. Multiple regression models were tested that included significant RAPD markers and interactions. SAS/CLUSTER[1] procedures were used to classify families as progenies of non-dormant, intermediate and dormant individuals.

Results and Discussion

All SH430 caryopses germinated in less than 10 days, while 98% of the M73 caryopses remained ungerminated for 44 days and required gibberellic acid to induce germination. F_1 caryopses completed germination in less than 25 days, and were classified as non-dormant. About 85% of the F_2 caryopses germinated within 14 days of imbibition. At the end of 44 days, about 5% of F_2 caryopses remained ungerminated. Nearly all of the $BC_{1ND}F_1$ caryopses germinated within 10 days and were classified as non-dormant. Sixty five percent of the $BC_{1D}F_1$ caryopses germinated during 44 days of imbibition. Further investigations with several generations revealed that mean germination after 10 days of imbibition was greater at 15 than at 20°C. Thus, non-dormancy is conditioned by temperature. A rudimentary model was developed using histograms of relative frequency versus days to germination, trial and error, and iteration until statistical tests (chi-square) suggested that the model fit the phenotypic data. The expected values for one three-locus model fit the observed values (Table 31.1). The dominant alleles G_1 and G_2 at the G_1 and G_2 loci and the recessive allele d promote early *G*ermination. The expression of D for late germination (*D*ormancy) is dependent upon the alleles present at G_1 and G_2 (and perhaps temperature). According to the proposed model, an embryo is non-dormant if it has at least two copies of dominant G_1 or G_2 alleles regardless of the genotype at the D locus (Table 31.2). A $G_1g_1G_2g_2$ individual will be non-dormant if the genotype at the D locus is Dd or dd. If only one allele of either G_1 or G_2 or neither is present and the genotype is dd, then the phenotype will also be non-dormant. The $G_1g_1G_2g_2DD$, $G_1g_1g_2g_2DD$, $G_1g_1g_2g_2Dd$, and $g_1g_1G_2g_2Dd$ individuals will have a phenotype intermediate between non-dormant and dormant. If the genotype is $g_1g_1G_2g_2DD$ or $g_1g_1g_2g_2D\_$ then the phenotype will be dormant. We hypothesize the genotype of SH430 to be $G_1G_1G_2G_2dd$ and that of M73 to be $g_1g_1g_2g_2DD$. Our genetic model is governed by the parents used and environmental conditions during seed development and germination. Determining a model for dormancy is subjective in that quantitative differences in a polygenic character at the phenotypic level are used to predict unknown genetic properties of the populations. Thus, like previous models (Jana *et al.*, 1979), our model is tentative and may be disproved or revised based on future data.

Random primers OPF-17, OPT-04, and OPX-06 were found to amplify polymorphic loci between the dormant and non-dormant bulks (Table 31.3).

[1] Mention of trademark or proprietary product does not constitute a guarantee or warranty of the product by the United States Department of Agriculture and does not imply its approval to the exclusion of other products that may also be suitable.

Table 31.1. Three-gene model: genotypes and their frequencies.

Non-dormant	Frequency	Intermediate	Frequency	Dormant	Frequency
$G_1g_1G_2g_2d\_$	3/16	$G_1g_1G_2g_2DD$	1/16	$g_1g_1G_2g_2DD$	1/32
$G_1G_1G_2G_2\_\ \_$	1/16	$G_1g_1g_2g_2D\_$	3/32	$g_1g_1g_2g_2D\_$	3/64
$G_1G_1G_2g_2\_\ \_$	1/8	$g_1g_1G_2g_2Dd$	1/16		
$G_1g_1G_2G_2\_\ \_$	1/8				
$G_1G_1g_2g_2\_\ \_$	1/16				
$g_1g_1G_2G_2\_\ \_$	1/16				
$G_1g_1g_2g_2dd$	1/32				
$g_1g_1G_2g_2dd$	1/32				
$g_1g_1g_2g_2dd$	1/64				

The blank spaces indicate either the dominant or recessive form of the allele could be present and the phenotype would remain the same.

Table 31.2. Model analysis: families partitioned into groups by cluster analysis of family variances. The number of observed families were compared to the expected number predicted by the genetic model.

	Observed no.			Expected no.				
Population	ND	IN	D	ND	IN	D	χ^2	*P* value
F_3	71	21	5	68	21	8	1.0	0.6065
$BC_{1D}F_2$	14	35	32	10	40	30	2.3	0.3119
F_7 RI lines	112	–	14	110	–	16	0.2	0.6315

ND, non-dormant; IN, intermediate; D, dormant.

Table 31.3. RAPD polymorphisms among the dormant and non-dormant parent, and dormant and non-dormant F_2 bulks.

Primer	Primer sequence 5′ to 3′	Size (bp)†	M73	SH430	D bulk	ND bulk
OPF-17	AACCCGGGAA	1591	+	–	+	–
OPT-04	CACAGAGGGA	1373	–	+	–	+
OPX-06	ACGCCAGAGG	1545	–	+	–	+

†Approximate size of polymorphic locus in base pairs based on migration distance relative to DNA size markers.
D, dormant; ND, non-dormant.

Regression analysis revealed significant association of markers OPT-04 and OPX-06 and early germination (non-dormancy). Chi-square analysis for marker segregation in 127 RI lines revealed that markers OPF-17/OPT-04 were linked in repulsion. Marker pairs OPT-04/OPX-06 and OPF-17/OPX-06 segregated independently. A multiple regression model that included markers OPF-17, OPT-04, OPX-06 and OPF-17 * OPT-04 interaction explained about 21% of the phenotypic variation in 97 random F_2 individuals.

Conclusions

We have proposed a genetic model to explain dormancy in wild oat and identified three molecular markers for dormancy QTL. Further research is needed to identify the chromosomal position of these molecular markers, identify additional markers and markers more tightly linked to dormancy QTL in wild oat. Molecular markers will facilitate improvements in our model and genotypic classification of RI lines. Investigation of allelic interactions, epistatic interactions, and genotype by environmental interactions will be critical to understanding the fundamental basis for dormancy and resistance to pre-harvest sprouting in domesticated and non-domesticated small grain species.

References

Fennimore, S.A. (1997) Genetic analysis of seed dormancy in wild oat (*Avena fatua*). PhD dissertation, Purdue University.

Foley, M.E. (1992) Effect of soluble sugars and gibberellic acid in breaking of dormancy of excised wild oat (*Avena fatua*) embryos. *Weed Science* 40, 208–214.

Foley, M.E. and Fennimore, S.A. (1998) Genetic basis for seed dormancy. *Seed Science Research* 8, 173–182.

Jana, S., Acharya, N. and Naylor, J.M. (1979) Dormancy studies in seeds of *Avena fatua*. 10. On the inheritance of germination behavior. *Canadian Journal of Botany* 57, 1663–1667.

Li, B. and Foley, M.E. (1997) Genetic and molecular control of seed dormancy. *Trends in Plant Science* 2, 384–389.

Michelmore, R.W., Paran, I. and Kesseli, R.V. (1991) Identification of markers linked to disease-resistance genes by bulked segregant analysis: a rapid method to detect markers in specific genomic regions by using segregating populations. *Proceedings of the National Academy of Sciences, USA* 88, 9828–9832.

Paterson, A.H., Sorrells, M.E. and Obendorf, R.L. (1989) Methods of evaluation for pre-harvest sprouting resistance in wheat breeding programs. *Canadian Journal of Plant Science* 69, 681–689.

Simpson, G.M. (1990) *Seed Dormancy in Grasses*. Cambridge University Press, Cambridge, UK, 297 pp.

Strand, E. (1991) Studies on seed dormancy in small grain species. III. Oats. *Norwegian Journal of Agricultural Sciences* 5, 51–59.

Williams, J.G.K., Kubelik, A.R., Livak, J., Rafalski, J.A. and Tingey, S.V. (1990) DNA polymorphisms amplified by arbitrary primers are useful as genetic markers. *Nucleic Acids Research* 18, 6531–6535.

32 Characterization and Expression of Two Protein Kinase Genes and an *EIN3*-like Gene, which are Regulated by ABA and GA_3 in Dormant *Fagus sylvatica* Seeds

O. Lorenzo, D. Rodríguez, C. Nicolás and G. Nicolás

Departamento de Fisiología Vegetal, Universidad de Salamanca, 37007 Salamanca, Spain

Fagus sylvatica **seeds (beechnuts) exhibit primary dormancy that may be released by stratification at 4°C. Application of abscisic acid (ABA) induces and maintains seed dormancy while gibberellic acid (GA_3) and ethylene break dormancy and allow seed germination to occur. From a cDNA library constructed from mRNAs of *Fagus sylvatica* dormant seeds and by means of differential screening or RT-PCR approach, three cDNA clones named FsPK1, FsPK2 and FsEINL1 have been isolated. Nucleotide sequences and their deduced amino acid sequences have shown that the FsPK1 clone has high homology with protein kinases and contains the putative consensus sequences of serine/threonine as well as tyrosine protein kinases. FsPK2 and FsEINL1 show homology with *CTR1* and *EIN3* genes respectively, both factors involved in the ethylene response pathway in *Arabidopsis*. These cDNA clones are differentially expressed in dormant seeds in response to the addition of ABA or GA_3 and our results suggest that their corresponding proteins may be involved in the mechanisms that regulate the maintenance or breaking of seed dormancy in *F. sylvatica*.**

Introduction

Seed dormancy is an adaptative mechanism to promote plant survival by distributing germination in both time and space. In many seeds it can be overcome by chilling, light, plant hormones, temperature and osmotic shock (Schneider and Gifford, 1994). Research in this field has focused on the physiological differences between dormant seeds and non-dormant seeds and on the role that the plant hormones, abscisic acid (ABA), gibberellic acid

(GA_3) and ethylene play in the induction, maintenance and breaking of seed dormancy.

Fagus sylvatica seeds exhibit endogenous dormancy that is eliminated by cold treatment at 4°C over a period longer than 8 weeks: application of ABA prevents the effects of cold on the breaking of seed dormancy (Nicolás *et al.*, 1996). Likewise, application of gibberellic acid (GA_3) and ethephon proved to be efficient in releasing beechnuts from dormancy and in substituting for cold treatment (Falleri *et al.*, 1997). Furthermore, GA_3 antagonizes the effects of ABA (Nicolás *et al.*, 1996, 1997). Responses to these hormones in dormant tissues include the induction of specific changes in gene expression although neither the function of these genes in adaptation to environmental changes nor the steps in the transduction pathway are known. Protein kinases are important in eukaryotic signal transduction pathways (Botella *et al.*, 1996) and there is increasing evidence that protein kinases have a role in ABA and ethylene-mediated responses (Kieber *et al.*, 1993; Koontz and Choi, 1993; Leung *et al.*, 1994; Holappa and Walker-Simmons, 1995; Walker-Simmons, 1998).

Our research focus is to study the role of ABA in the induction and maintenance of dormancy and the role of GA_3 and ethylene in the breaking of dormancy in *F. sylvatica* seeds, as well as in the expression of specific genes that could be involved in this developmental process.

Materials and Methods

Plant material and germination conditions

Seeds of *F. sylvatica* were obtained from the Danish State Forestry Improvement Station. Seeds were dried to a moisture content of 21% and stored in sealed jars. Their viability was approximately 97% (checked by standard procedures recommended by the suppliers). Seed germination was carried out as described in Nicolás *et al.* (1996). The imbibition media used were sterile water plus 100 μM ABA, 100 μM GA_3, 10 μM Paclobutrazol, 100 μM ABA + 100 μM GA_3, 1 mM $CaCl_2$, 100 μM ABA + 1 mM $CaCl_2$, 100 μM ABA + 2 mM EGTA or 100 μM ABA + 200 μM TMB-8. For all the experiments the pericarp was manually removed before imbibition and seeds were maintained at 4°C for 1–6 weeks.

RNA extraction

Total RNA was extracted using Qiagen pack-500 cartridge (Qiagen Inc., California, USA), following the manufacturer's protocol. Poly (A^+) RNA was purified from total RNA by affinity chromatography in oligo(dT)-cellulose columns using the mRNA Purification Kit (Pharmacia Biotech).

Screening of a cDNA library

By differential screening, two partial clones of about 0.65 and 1.8 kbp, called FsPK1 and FsEINL1 respectively, were isolated from a cDNA library constructed using as a template poly (A^+) RNA from seeds imbibed in 100 μM ABA for 2 weeks (Nicolás *et al.*, 1997).

Cloning FsPK1 by 5′ rapid amplification of cDNA ends (RACE)

From a partial FsPK1 cDNA clone obtained by differential screening, a full length FsPK1 cDNA was produced by using 5′ RACE to extend sequences. The 5′ RACE reactions were carried out by using the Rapid Amplification of cDNAs Ends system, version 2.0 (Gibco-BRL). First-strand cDNA was synthesized from 1 μg of poly (A^+) RNA from ABA-treated seeds using primer SP1-PK (as gene-specific primer one), 5′-CATCCCACATGAAAGCAAATGTACAAG-3′, and tailing with poly(G) following the supplier's procedure. The first round PCR amplification was carried out with primer SP2-PK (as gene-specific primer two), 5′-CTCGAGCTGGTGAACTACAAAAGCCAGGTC-3′ and AAP (5′ RACE Abridged Anchor Primer, 5′-GGCCACGCGTCGACTAGTACGGGIIGGGII-GGGIIG-3′, supplier-provided) in PCR mix. The PCR protocol was: one cycle of 94°C (5 min); followed by 35 cycles of 94°C (1 min), 55°C (1 min), 72°C (2 min), and one cycle of 72°C (10 min). The first-round amplification product was diluted 1:100 and then used as a template for a second-round of amplification with SP2-PK and AUAP (Abridged Universal Amplification Primer, 5′-GGCCACGCGTCGACTAGTAC-3′, supplier-provided). The reaction conditions and protocol for second-round PCR were the same as for the first. The 1500 bp PCR product obtained was subcloned into pCR 2.1 vector (Original TA Cloning kit, Invitrogen) and sequenced.

Cloning of a partial cDNA encoding FsPK2 by RT-PCR

A fragment of 400 pb was amplified by RT-PCR of 1 μg of poly (A^+) RNA from ABA-treated seeds using degenerated oligonucleotides corresponding to two subdomains conserved among the serine/threonine kinases (Hanks *et al.*, 1988). The forward primer consisted of a 17-mer of the sequence 5′-GA(T,C)CT(G,T)AA(A,G)CCNGA(A,G)AA-3′ encoding the DLKPEN amino acid sequence (domain VI) and the reverse primer was a 26-mer of the sequence 5′-TC(A,G)GG(A,G)GC(A,G)TAGTACTC(A,T)GG(A,G)GTNCC-3′ corresponding to the GTPEYIAPE sequence (subdomain VIII). The amplified fragment (FsPK2) was cloned into the pCR 2.1 vector (Original TA Cloning kit, Invitrogen) and sequenced. This revealed that the reverse primer alone amplified a serine/threonine kinase encoding subdomains IV to VIII.

DNA sequencing

Plasmid DNA templates were isolated by the Wizard *Plus* Minipreps DNA Purification System (Promega). Determination of the nucleotide sequence of the cDNA clone was performed by the method of Sanger *et al.* (1977). Both the DNA and deduced protein sequences were compared to other sequences in the EMBL GenBank and SwissProt databases, respectively, using the FASTA algorithm (Pearson and Lipman, 1988).

Northern analysis

Northern analysis was performed as previously described by Nicolás *et al.* (1997). Blotted membranes were exposed to X-Omat films (Kodak) and the autoradiographs analysed on a Bioimage 60S Image Analyzer (Millipore, Visage 4.6K Software).

Results and Discussion

Cloning and characterization of FsPK1

Using 5′ RACE we reconstructed a partial cDNA clone isolated from a cDNA library using the differential screening described in the Materials and Methods section. FsPK1 is a full-length cDNA clone of 1535 bp long which contains an open reading frame of 1440 nucleotides (with the ATG start codon and lacking the 5′end as determined by comparison with GmPK6 and by estimation of its mRNA size by Northern blot). This open reading frame encodes a 480 amino acid protein of 54.8 kDa. The amino acid deduced sequence of the FsPK1 cDNA is shown in Fig. 32.1.

A search of the sequence databases revealed that FsPK1 has a 72% amino acid identity with a novel protein kinase from soyabean (Feng *et al.*, 1993) (Fig. 32.1) and also shares a high degree of homology within the 11 subdomains (I–XI) in the catalytic domains of protein kinases. These results strongly suggest that FsPK1 is indeed a protein kinase. The sequences in subdomain VIb (HRDLKSEN) and VIII (GTYRWMAPE), which are believed to confer hydroxyamino acid substrate specificity, suggest that FsPK1 belongs to a serine/threonine protein kinase family. As is typical for protein kinases, less sequence identity was found in the regions spanning the catalytic subdomains. Currently, CDPKs (calcium-dependent protein kinases) are the most prevalent serine/threonine PKs found in higher plants (Sheen, 1996). FsPK1 does not seem to be a CDPK because of the lack at the C-terminus of the calmodulin-like domain conserved in all the plant CDPKs so far cloned (Stone and Walker, 1995).

Surprisingly, FsPK1 also contains a single tryptophan at position 412 within catalytic subdomain XI that is present in all tyrosine kinases (Hanks

```
FsPK1    1  MGEEGNSWIRRTKFSHTVCHRLDASRLASFPIIIRPERNSGLKSRPGATS
GmPK6    1  MGEDGNSWIRRTNFSHTVCHRLDPARLGSIPISVQSEQ----KSRP--SS
            ***.******** ********** .** * ** .. *.    ****  .*

FsPK1   51  SSEKFGTIYNKQRSLSPLPESRLSDVFKEARSETKRFMTPLPRRREQAKG
GmPK6   45  KAQRHPMTYK-QRSLSPLPETYLSEAFREARLEQKRFSTPNPRR-E--KR
            ...     *  *********. **. *.*** * *** ** *** *  *

FsPK1  101  VMGKFVHKDFHEVKASDSKSLANSSALRHLSSMKVNDKSKYRKESSWAKY
GmPK6   91  IMGKLLNKDSRETKESSSKSPS--------RSPNRQVKSKNRKDSAWTKL
            .*** ..** .* * * *** .         *   . *** **.*.*.*

FsPK1  151  FDHGGGRVNAVETSDEWTVDLSKLFLGLKFAHGAHSRLYHGIYNDEPVAV
GmPK6  133  LDNGGGKITAVETAEEWNVDMSQLFFGLKFAHGAHSRLYHGVYKDEAVAV
             *.***...****..**.**.*.** ***************.* ** ***
                I                                            II
FsPK1  201  KIIRVPEDDENGALGARLEKQFNREVTLLSRLHFHNIIKFVAACRKPPVY
GmPK6  183  KIIMVPEDDGNGALASRLEKQFIREVTLLSRLHHQNVIKFSAACRKPPVY
            *** ***** **** .****** ********** .*.*** *********
                                  III           IV
FsPK1  251  CVVTEYLSEGSLRAYLHKLERKSLPLQKLIAFALDIARGMEYIHSQGVIH
GmPK6  233  CIITEYLAEGSLRAYLHKLEHQTISLQKLIAFALDIARGMEYIHSQGVIH
            *..****.************....**************************
                      V                      VIa
FsPK1  301  RDLKPENVLIDQEFHLKIADFGIACEEAYCDSLADDPGTYRWMAPEMIKH
GmPK6  283  RDLKPENILINEDNHLKIADFGIACEEASCDLLADDPGTYRWMAPEMIKR
            *******.** .. ************** ** *****************.
               VIb                VII                  VIII
FsPK1  351  KSYGRKVDVYSFGLILWEMVAGTIPYEDMNPVQAAFAVVNKNLRPVIPRY
GmPK6  333  KSYGKKVDVYSFGLILWEMLTGTIPYEDMNPIQAAFAVVNKNSRPIIPSN
            ****.**************..**********.********** **.**
                    IX                      X
FsPK1  401  CPPAMRALIEQC[W]SLQSEKRPEFWQVVKVLEQFESSLARDGTLNLVQSLT
GmPK6  383  CPPAMRALIEQC[W]SLQPDKRPEFWQVVKILEQFESSLASDGTLSLVPN-P
            **************** .**********.********* **** **
                          XI
FsPK1  451  CQDHKKGLLHWIHKLGPVHPN-GSMPKPKLTZ
GmPK6  432  CWDHKKGLLHWIQKLGPLHQNSGPVPKPKFTZ
            * **********.****.* * * .**** **
```

Fig. 32.1. Comparison of the amino acid sequence deduced from the FsPK1 clone with a novel soyabean protein kinase (GmPK6) using the Clustal V method. Subdomains designated by Hanks *et al.* (1988) are shown below the aligned sequences. The tryptophan conserved in protein tyrosine kinases is boxed. The putative regulatory domain of both proteins is underlined.

et al., 1988) and absent in most serine/threonine protein kinases so far cloned. As far as we know, just a few protein kinases present characteristics of both serine/threonine and tyrosine protein kinases and in agreement with Feng *et al.* (1993) this raises the possibility that FsPK1 is a functional mosaic of both kinases.

Expression of FsPK1

Using this clone as a probe, we determined its expression under two conditions which previously we had found either to break dormancy (stratification) or to counteract the cold treatment (ABA) in *F. sylvatica* (Nicolás *et al.*, 1996). In Fig. 32.2 the Northern blot shows that in the presence of 100 μM ABA, which counteracts the cold effect in breaking dormancy, the accumulation of FsPK1 transcripts increases four to five times compared to stratified seeds, in which dormancy is released. The basal level of expression of FsPK1 at time 0 (dry dormant seeds) is very low, similar to that found after 6 weeks of stratification. Similar ABA responsiveness has been found in a protein kinase from wheat (Anderberg and Walker-Simmons, 1992). The correlation between the presence of FsPK1 in ABA-treated seeds and its disappearance in stratified seeds, suggest that FsPK1 is related to dormancy.

It is very interesting to note that supplying external calcium also stimulates FsPK1 expression (Fig. 32.2), which increases FsPK1 mRNA levels 20 times over the water controls when ABA and calcium are added together. However, there are no additional effects on dormancy when the seeds are incubated in both ABA and calcium (data not shown). The addition of two calcium chelators, EGTA (exogenous chelator) and TMB-8 (endogenous chelator) together with ABA (Fig. 32.3) reduces the expression of FsPK1 to basal levels, suggesting that calcium is needed for the ABA effect. It has been reported that increased expression of mechanical strain-induced genes occurs, in response to elevated amounts of extracellular calcium in *Arabidopsis* (Braam and Davis, 1990), and also in barley aleurone an ABA-induced RAB gene expression was influenced by external calcium concentration (Van der Meulen *et al.*, 1996). Botella *et al.* (1996) also reported calcium-dependent expression of a CDPK from mungbean. Finally, it has also been suggested that ABA may exert some of its physiological effects through this ion (De Silva *et al.*, 1985; Napier *et al.*, 1989; Colorado *et al.*, 1994; Nicolás *et al.*, 1996). Our results seem to indicate a synergistic effect of calcium by enhancing the ABA effects and also show, for

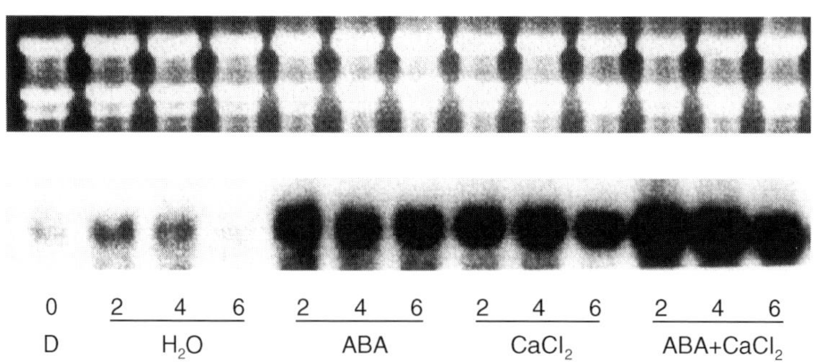

Fig. 32.2. Northern blot analysis of total RNA isolated from *Fagus sylvatica* dry dormant seeds (D) or dormant seeds sown at 4°C for 2, 4 and 6 weeks in water, 100 μM ABA, 1 mM $CaCl_2$ and 100 μM ABA + 1 mM $CaCl_2$. Ten micrograms of total RNA was used per lane and hybridized with a probe from the FsPK1 clone. Top panel: stained gel showing rRNAs.

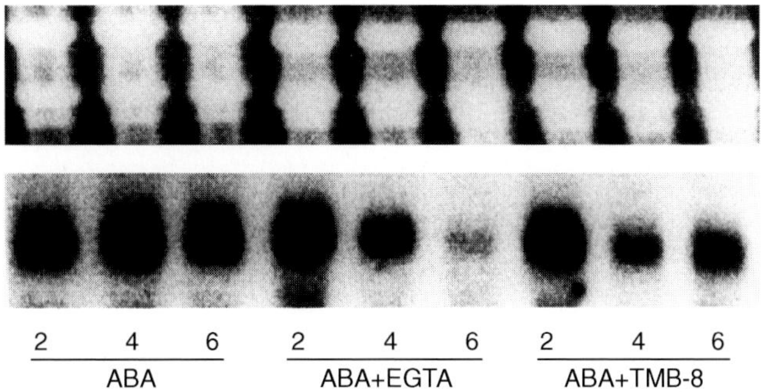

Fig. 32.3. Northern blot analysis of total RNA isolated from *Fagus sylvatica* dormant seeds sown at 4°C during 2, 4 and 6 weeks in water, 100 μM ABA, 100 μM ABA + 2 mM EGTA and 100 μM ABA + 0.2 mM TMB-8. Ten micrograms of total RNA was used per lane and hybridized with a probe from the FsPK1 clone. Top panel: stained gel showing rRNAs.

the first time, an ABA and calcium-dependent expression of a plant protein kinase gene, in hydrated dormant, growth-arrested seeds.

Cloning, comparison and expression of FsPK2

We obtained cDNA sequences encoding protein kinases from *F. sylvatica*, using degenerated oligonucleotides corresponding to two subdomains (VI and VIII) conserved among the serine/threonine kinases (Hanks *et al.*, 1988). By cloning and sequencing one of these PCR products, we have identified a fragment of 400 bp which shares a homology of > 35% with the CTR1 (for Constitutive Triple Response) protein kinase from *Arabidopsis* and also with MAP kinases such as MEK1 from *Oryza sativa*, in five (IV–VIII) of the eleven subdomains present in their catalytic domain (Fig. 32.4).

Since the FsPK2 clone was isolated using poly (A^+) RNA from ABA-treated seeds, and the fragment was not present in poly (A^+) RNA from GA_3 and ethephon-treated seeds, we investigated its relationship with dormancy by using it as a probe in Northern analysis. The level of FsPK2 transcript was higher in ABA and $CaCl_2$ treatments while GA_3 addition greatly reduced transcript accumulation. A high level of expression was maintained when ABA and $CaCl_2$ were added together (Fig. 32.5).

The *CTR1* gene encodes a putative serine/threonine protein kinase that is closely related to the Raf protein kinase family, which is involved in modulating cellular responses to extracellular signals (Kieber *et al.*, 1993). Also, loss of function of the *CTR1* gene causes constitutive activation of the ethylene response pathway.

In the present work we describe for the first time an ABA-responsive *CTR1*-like gene which acts as a negative regulator of a multistep signal transduction pathway by phosphorylation of other proteins in the cascade, being the final result the induction or maintenance of dormancy in *F. sylvatica* seeds.

```
                                   IV
FsPK2    Y S G V P L L K S L K H S N I V K F Y N S W I D D K N K T V N I I T E L F T S G   40
OsMEK1   Y C E I H L L K T L K H R N I M K F Y T S W V D V S R R N I N F I T E M F T S G  113
AtCTR1   L R E V A I M K R L R H P N I V L F M G A - - V T Q P P N L S I V T E Y L S R G  632

           V                                                      VIa
FsPK2    N L R Q Y C K K H K K V D M K A L K G W A R - - - - - Q I L T G L N Y L H S H S   75
OsMEK1   T L R Q Y R Q K H M R V N I W A V K H W C R - - - - - Q I L S G L L Y L H S H D  148
AtCTR1   S L - - Y R L L H K S G A R E Q L D E R R R L S M A Y D V A K G M N Y L H N R N  670

                    VIb                                  VII
FsPK2    P P I I H R D L K C D N I F I N G N Q G E V K I G D L G L A T V M E Q A - - N A  113
OsMEK1   P P I I H R D L K C D N I F V N G N Q G E V K I G D L G L A A I L R K S - - H A  186
AtCTR1   P P I V H R D L K S P N L L V D - K K Y T V K V C D F G L S R L K A S T F L S S  709

                                VIII
FsPK2    K S V I G T P E Y Y A P                                                         125
OsMEK1   V H C V G T P E F M A P                                                         198
AtCTR1   K S A A G T P E W M A P                                                         721
```

Fig. 32.4. Sequence comparison of FsPK2 kinase domains (IV–VIII) to other protein kinase sequences from *Oryza sativa* (OsMEK1) and *Arabidopsis thaliana* (AtCTR1).

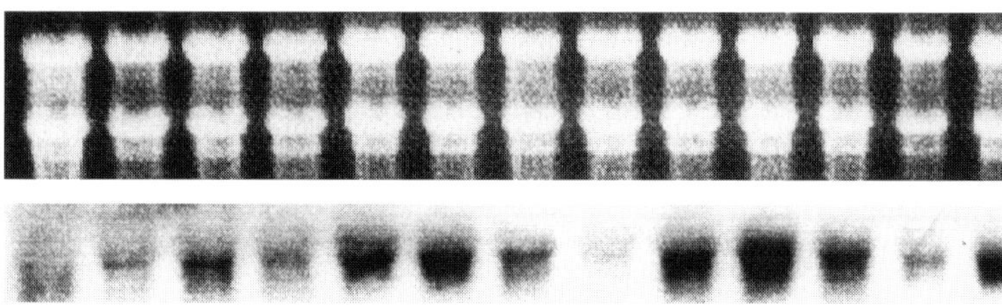

Fig. 32.5. Northern blot analysis of total RNA isolated from *Fagus sylvatica* dry dormant seeds (D) and dormant seeds sown at 4°C from 2 to 6 weeks in water, 100 µM ABA, 1 mM $CaCl_2$, 100 µM ABA + 1 mM $CaCl_2$, 100 µM GA_3 or 100 µM ABA + 100 µM GA_3. Ten micrograms of total RNA were used per lane and hybridized with a probe from the FsPK2 clone. Top panel: stained gel showing rRNAs.

Characterization and expression of FsEINL1

FsEINL1 was the second clone isolated from the cDNA library as previously described. The sequenced insert comprises a cDNA of 1670 bp which contains a partial open reading frame of 1140 nucleotides (lacks the ATG start codon and the 5′end and Northern analysis showed a transcript of 2600 pb). This open reading frame encodes a polypeptide of 380 amino acid residues, with a calculated molecular mass of 42.6 kDa.

A sequence similarity ranging from 45 to 50% was found by comparing the FsEINL1 predicted amino acid sequence with the ethylene-insensitive clones from *Arabidopsis thaliana* (Chao *et al.*, 1997) (Fig. 32.6). The FsEINL1 sequence starts at nucleotide 736 of the AtEIL3 (for Ethylene-Insensitive-Like) sequence, and the deduced amino acid sequence has a 45% amino acid

```
FsEINL1      1   VKLGLPHGQRPPYKKPHDLKKMWKVGVLTAVIKHMSPDIAKIRRHVRQSK
AtEIL3     212   VKLGLPKSQSPPYRKPHDLKKMWKVGVLTAVINHMLPDIAKIKRHVRQSK
                 ******. * ***.****************** ** ******.*******
                                     III                   IV

FsEINL1     51   CLQDKMTAKESAIWLGVLSREEALIRQPSSDNGTSGVTDMPRGGRDENKR
AtEIL3     262   CLQDKMTAKESAIWLAVLNQEESLIQQPSSDNGNSNVTETHRRGNNADRR
                 *************** ** .**.**.*******.*.**.  * *   ..*
                                                         IVa

FsEINL1    101   -AAVSSDSDYDVDGVDDGVSSVPSTDERRNQPMDVEPSDNLQNNTPVQDK
AtEIL3     312   KPVVNSDSDYDVDGTEEASGSVSSKDSRRNQIQKEQP---TAISHSVRDQ
                    * ********* ..   ** * * ****    .*         *.*.

FsEINL1    150   APGEKQPKRKRARVRSNCADQIPAPSHNEPLNVEPIITLPDVNHTDV-QV
AtEIL3     359   DKAEKHRRRKRPRIRSGTVNRQ---EEEQPE-AQQRNILPDMNHVDAPLL
                    **. .*** *.**..  .       .*   .    ***.** *   .
                      V

FsEINL1    199   GFQIHGDQQETGKIAALRLREKDFDVQPQLPVSEFN-HFSALPADNVIST
AtEIL3     405   EYNINGTHQED-DVVDPNIALGPEDNGLELVVPEFNNNYTYLPLVN--EQ
                  ..*.* .**   .    .     *   .* * *** ... **  *

FsEINL1    248   QSMYVDGRPLLYPVVQNTEMHHGDNYNFYNPSMEYGLTHDRQQSLIVMNE
AtEIL3     405   TMMPVDERPMLYGPNPNQELQFGSGYNFYNPSAVF--VHN-QEDDILHTQ
                   * ** **.**    * *.. * .*******  .   *  *.  *. ..

FsEINL1    298   PQIRPEEVGLHVPTLHGSSTELYVKDPFNNEQHRPVDSQFGSAIDSLSLD
AtEIL3     499   IEMNTQAPPHNSGFEEAPGGVLQPLGLLGNE-----DGVTGSELPQYQSG
                  ..  .    .          *      .**     *   ** .

FsEINL1    348   CGGLNSPFHYDIDGLSSLEDFLHDEDLIQYFGA
AtEIL3     544   ILSPLTDLDFDYGGFG---------DDFSWFGA
                      .   .*  *           *    ***
```

Fig. 32.6. Alignment of FsEINL1 and EIL3 from *Arabidopsis thaliana* amino acid sequences generated using the Clustal V method, showing their basic domains (III–V).

identity over a 380 amino acid sequence overlap. Other ethylene-insensitive proteins showed a high degree of sequence similarity to FsEINL1 within the basic domains (III–V), being nearly 50% for most of them. As is typical for EIN clones, less sequence identity was found in the regions spanning the basic domains and specially in the C-terminal region.

Northern blot analysis showed the correlation between the presence of FsEINL1 in GA_3-treated seeds and its disappearance in dormant seeds and in the presence of ABA, which suggests that FsEINL1 is related to the breaking of dormancy by gibberellins (Fig. 32.7). From the first week of treatment with GA_3, a high level of expression was maintained. In the presence of Paclobutrazol, a well known GA biosynthesis inhibitor, FsEINL1 showed no expression over 2, 4 and 6 weeks. Surprisingly, no expression was found following stratification. Probably, more than one mechanism is operating in breaking dormancy in *F. sylvatica* seeds.

EIN3 (for Ethylene-Insensitive) or *EIL3* genes act downstream of the histidine kinase ethylene receptor ETR1 and the Raf-like kinase CTR1. The *EIN3* gene encodes a novel nuclear-localized protein that shares similarity in sequence, structural features and genetic function with EIL3 protein (Chao *et al.*, 1997). In this work we also describe for the first time a GA-responsive *EIL3* gene that is activated by GA_3, and triggers the latest step in the ethylene signalling pathway breaking dormancy and inducing germination in *F. sylvatica* seeds. In support of our hypothesis is the fact that ethephon is able to break dormancy in *F. sylvatica* seeds (unpublished results). In Fig. 32.8 we summarize the possible role of ethylene, ABA and GA_3 in the dormancy of *F. sylvatica* seeds. Ethylene binds to ETR1 and represses CTR1 activity (Chao *et al.*, 1997) and negatively regulates the activity of downstream components in the pathway. The final result is the breakage of dormancy. ABA, by activating CTR1 gene, and GA, by activating EIL or EIN genes induce or break dormancy respectively.

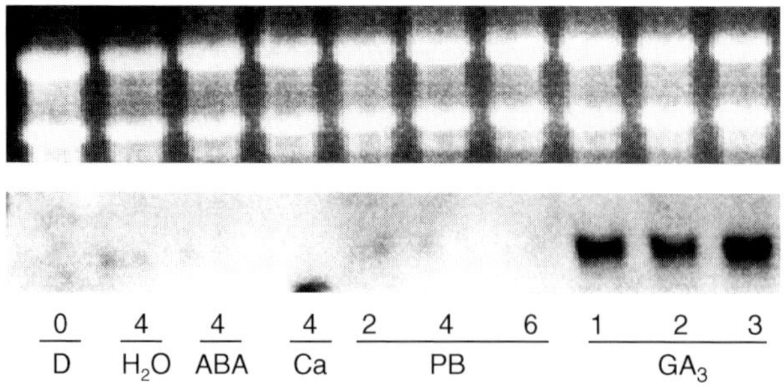

Fig. 32.7. Northern blot analysis of total RNA isolated from *Fagus sylvatica* dry dormant seeds (D) or dormant seeds sown at 4°C from 1 to 6 weeks in water, 100 μM ABA, 1 mM $CaCl_2$, 10 μM Paclobutrazol and 100 μM GA_3. Ten micrograms of total RNA were used per lane and hybridized with a probe from the FsEINL1 clone. Top panel: stained gel showing rRNAs.

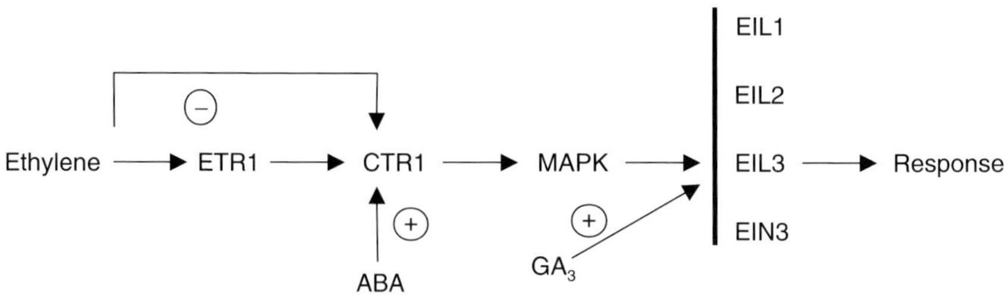

Fig. 32.8. Hypothetical action of ABA and GA_3 on the ethylene response pathway in *Fagus sylvatica* seed dormancy.

Acknowledgements

This work was supported by grants PB96-1313 from the Dirección General de Investigación Científica y Técnica (Spain) and AIR2-CT93-1667 from the European Union.

References

Anderberg, R.J. and Walker-Simmons, M.K. (1992) Isolation of a wheat cDNA clone for an abscisic acid inducible transcript with homology to protein kinases. *Proceedings of the National Academy of Sciences, USA* 89, 10183–10187.

Botella, J.R., Arteca, J.M., Somodevilla, M. and Arteca, R.N. (1996) Calcium-dependent protein kinase gene expression in response to physical and chemical stimuli in mungbean (*Vigna radiata*). *Plant Molecular Biology* 30, 1129–1137.

Braam, J. and Davis, R.W. (1990) Rain, wind and touch induced expression of calmodulin and calmodulin-related genes in *Arabidopsis*. *Cell* 60, 357–364.

Chao, Q., Rothenberg, M., Solano, R., Roman, G., Terzaghi, W. and Ecker, J.R. (1997) Activation of the ethylene gas response pathway in *Arabidopsis*, by the nuclear protein ethylene-insensitive3 and related proteins. *Cell* 89, 1133–1144.

Colorado, P., Rodríguez, A., Nicolás, G. and Rodríguez, D. (1994) Abscisic acid and stress regulate gene expression during germination of chickpea seeds. Possible role of calcium. *Physiologia Plantarum* 91, 461–467.

De Silva, D.L.R., Hetherington, A.M. and Mansfield, T.A. (1985) Synergism between calcium ions and abscisic acid in preventing stomatal opening. *New Phytologist* 100, 473–482.

Falleri, E., Muller, C. and Laroppe, E. (1997) Effect of ethephon on dormancy breaking in beechnuts. In: Ellis, R.H., Black, M., Murdoch, A.J. and Hong, T.D. (eds) *Basic and Applied Aspects of Seed Biology*. Kluwer Academic Publishers, Dordrecht, pp. 303–309.

Feng, X.H., Zhao, Y., Bottino, P.J. and Kung, S.D. (1993) Cloning and characterization of a novel member of protein kinase family from soyabean. *Biochimica et Biophysica Acta* 1172, 200–204.

Hanks, S.K., Quinn, A.N. and Hunter, T. (1988) The protein kinase family: conserved features and deduced phylogeny of the catalytic domains. *Science* 241, 42–52.

Holappa, L.D. and Walker-Simmons, M.K. (1995) The wheat abscisic acid-responsive protein kinase mRNA, PKABA1, is up regulated by dehydration, cold temperature, and osmotic stress. *Plant Physiology* 108, 1203–1210.

Kieber, J.J., Rothenberg, M., Roman, G., Feldmann, K.A. and Ecker, J.R. (1993) CTR1, a negative regulator of the ethylene response pathway in *Arabidopsis*, encodes a member of the Raf family of protein kinases. *Cell* 72, 427–441.

Koontz, D.A. and Choi, J.H. (1993) Protein phosphorylation in carrot somatic embryos in response to abscisic acid. *Plant Physiology and Biochemistry* 31, 95–102.

Leung, J., Bouvier-Durand, M., Morris, P.C., Guerrier, D., Chefdor, E. and Giraudat, J. (1994) *Arabidopsis* ABA response gene ABI1: feature of a calcium-modulated protein phosphatase. *Science* 264, 1448–1452.

Napier, J.A., Chapman, J.M. and Black, M. (1989) Calcium dependent induction of novel proteins by abscisic acid in wheat aleurone tissue of different development stages. *Planta* 179, 156–164.

Nicolás, C., Nicolás, G. and Rodríguez, D. (1996) Antagonistic effects of abscisic acid and gibberellic acid on the breaking of dormancy of *Fagus sylvatica* seeds. *Physiologia Plantarum* 96, 244–250.

Nicolás, C., Rodríguez, D., Poulsen, F., Eriksen, E.N. and Nicolás, G. (1997) The expression of an abscisic acid-responsive glycine-rich protein coincides with the level of seed dormancy in *Fagus sylvatica. Plant and Cell Physiology* 38, 1303–1310.

Pearson, W.R. and Lipman, D.J. (1988) Improved tools for biological sequence analysis. *Proceedings of the National Academy of Sciences, USA* 85, 2444–2448.

Sanger, F., Nicklen, S. and Coulson, A.R. (1977) DNA sequencing with chain terminating inhibitors. *Proceedings of the National Academy of Sciences, USA* 74, 5463–5467.

Schneider, W.L. and Gifford, D.J. (1994) Loblolly pine seed dormancy I. The relationship between protein synthesis and loss of dormancy. *Physiologia Plantarum* 90, 246–252.

Sheen, J. (1996) Ca^{2+} dependent protein kinases and stress signal transduction in plants. *Science* 294, 1900–1902.

Stone, J.M. and Walker, J.C. (1995) Plant protein kinase families and signal transduction. *Plant Physiology* 108, 451–457.

Van der Meulen, R.M., Visser, K. and Wang, M. (1996) Effects of modulation of calcium levels and calcium fluxes on ABA-induced gene expression in barley aleurone. *Plant Science* 117, 75–82.

Walker-Simmons, M.K. (1998) Protein kinases in seeds. *Seed Science Research* 8, 193–200.

33 Effects of Fusicoccin and Gibberellic Acid on Germination and α-Amylase Expression in Barley Grains

R.M. VAN DER MEULEN, J. VREDENBREGT-HEISTEK, M.P.M. CASPERS AND M. WANG

Center for Phytotechnology RUL/TNO, TNO Plant Biotechnology Department, Wassenaarseweg 64, 2333 AL Leiden, The Netherlands

Both gibberellic acid (GA_3) and fusicoccin (FC) were able to break dormancy of barley grains and to stimulate the germination rate of embryos isolated from such grains: they showed an additive effect on germination. Induced α-amylase mRNA expression is about ten times less sensitive to GA_3 in aleurone layers isolated from dormant grains than in those isolated from non-dormant grains. Embryos from dormant grains were about 100 times less sensitive to GA_3 than were 'non-dormant' aleurone cells. No GA-induced α-amylase mRNA could be detected in embryos isolated from non-dormant grains. Fusicoccin had no effect on GA_3-induced α-amylase mRNA expression in aleurone tissue isolated from non-dormant grains but was able to enhance the GA_3-induced responses in aleurone tissue isolated from dormant grains. In embryos from dormant grains, synergistic effects of FC and GA_3 were observed on induction of α-amylase mRNA expression. It was also found that FC was able to induce an acidification of extracellular pH (pH_e). The FC-induced enhancement of GA action is likely to be due to a decrease in pH_e.

Introduction

Germination is the starting point of the higher plant life cycle and is controlled by both internal and external factors. It is well known that gibberellin (GA) is able to break dormancy and induce germination, and that acidification of external pH is able to enhance GA-induced biological responses such as α-amylase production (Sinjorgo *et al.*, 1993). Fusicoccin (FC), a toxin produced by the fungus *Fusicoccum amygdali*, is able to induce a wide spectrum of physiological responses in plants (Marrè, 1979; De Boer, 1997; Wang *et al.*, 1998). Fusicoccin is able to break dormancy of intact barley grains and to stimulate the germination rate of embryos isolated from dormant grains (Lado *et al.*, 1974; Wang *et al.*, 1998). Gibberellic acid has a similar effect on breaking of barley grain dormancy (Wang *et al.*, 1998). In addition, several reports have demonstrated the occurrence of specific membrane-bound FC receptors, which are functionally connected to the plasma membrane H^+-ATPase (Marra *et al.*, 1992).

We are interested in processes involved in dormancy/germination regulation, the effect of FC in the dormant barley grains and the interaction between FC and classic plant growth regulators such as GA. In the current study, we investigated the effects of FC and GA on embryo germination and α-amylase mRNA expression in both embryo and aleurone of the barley grain.

Materials and Methods

Barley grains and germination test

Hordeum distichum L. cv. Triumph dormant and non-dormant grains were grown in a phytotron under the conditions described previously (Schuurink *et al.*, 1992). Germination tests were carried out as described by Wang *et al.* (1998). Mean values ± SD are presented unless stated otherwise.

pH measurements and Northern analysis

pH was measured with a Beckman (Fullerton USA) Φ 100 ISFET pH meter. In each Northern analysis experiment, ten dormant or non-dormant aleurone layers or embryos were incubated in 1 ml or 300 μl buffer in a 24-well plate at 20°C in the dark for 24 h before samples were collected. RNA isolation and Northern blotting (Wang *et al.*, 1992) was done with 0.8 kb SacI total α-amylase probe (Rogers, 1985) or a low pI amylase 5′ utr probe (0.26 kb) PCR fragment made on plasmid pAmySPGlu (Phillipson, 1993) using primers: 5′-CCATGGAGGCTGTGCC, 5′CCATGGCTCGAGCTCAA. The amount of ^{32}P-labelled probe hybridizing to the low pI α-amylase mRNA was semi-quantitatively determined by measuring the absorbance on autoradiographs with an UL-troscan KL densitometer (LKB).

Results and Discussion

The effects of FC and GA_3 on germination of embryos isolated from dormant grains was studied. Figure 33.1 shows that 16 h of incubation in GA_3 was able to stimulate the germination of embryos isolated from dormant grains. The half-maximum stimulation required about 5×10^{-7} M GA_3. Fusicoccin (10^{-5} M) was also able to stimulate germination and enhanced the stimulatory effects of GA_3. In the presence of FC, half-maximal stimulation of germination required only about 2×10^{-8} M GA_3 (Fig. 33.1).

We tested GA-induced α-amylase expression in both isolated embryos and aleurone layers. Figure 33.2 shows that after 24 h of incubation, 5×10^{-9} M GA_3

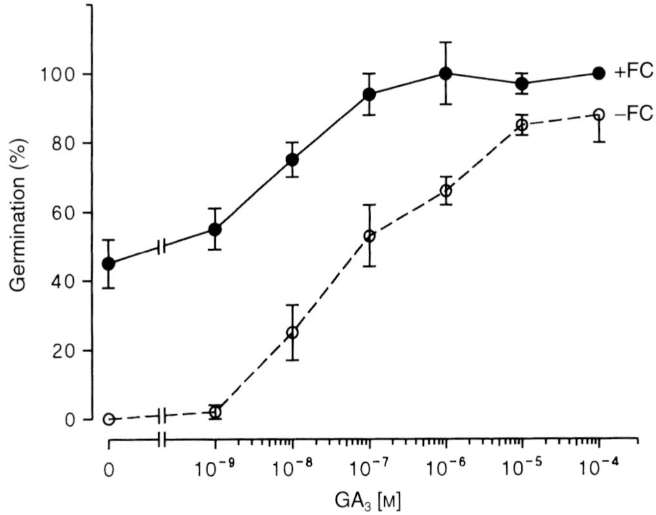

Fig. 33.1. Effect of fusicoccin (FC) and gibberellic acid (GA_3) on germination of embryos isolated from dormant grains: ten embryos isolated from dormant barley grains (in duplicate) were incubated in 300 μl water containing various concentrations of GA_3 in the presence (●) or absence (○) of FC (10^{-5} M) for 16 h in the dark. Embryos were considered germinated if the radicles were ≤ 1 mm. The means ± SD of five independent experiments are presented.

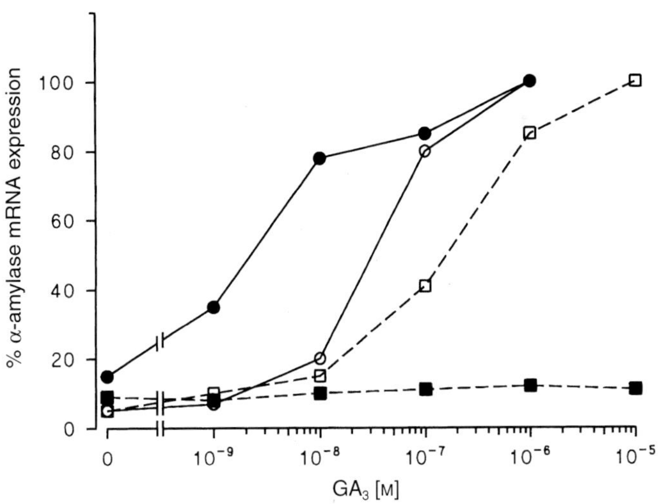

Fig. 33.2. Effect of gibberellic acid (GA_3) on α-amylase mRNA expression in isolated aleurone layers (●, ○) and embryos (■, □). A series of concentrations ranging from 10^{-9} to 10^{-5} M was used. In each experiment, dormant (open symbols) or non-dormant (closed symbols) aleurone layers or embryos were used. The α-amylase mRNA expression at 10^{-6} M GA_3 for aleurone layers and at 10^{-5} M GA_3 for dormant embryos was set as 100%. One representative experiment is presented from at least three independent experiments.

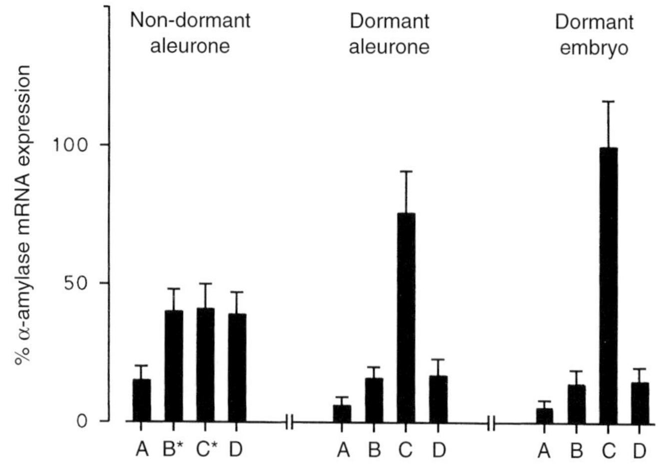

Fig. 33.3. Effect of fusicoccin (FC) on gibberellin (GA)-induced α-amylase mRNA expression in isolated aleurone layers and embryos. In each Northern analysis, samples were incubated with or without A: control (buffer), B*: 10^{-9} M GA_3, B: 10^{-8} M GA_3, C*: GA_3 (10^{-9} M) + FC, C: GA_3 (10^{-8} M) + FC or D: 10^{-5} M FC. The α-amylase mRNA expression at 10^{-6} M GA_3 was set as 100%. One representative experiment is presented from at least three independent experiments.

was able to give a half-maximum induction of α-amylase mRNA expression in aleurone layers isolated from non-dormant grains (Fig. 33.2). However, a ten times higher concentration of GA_3 was required for such an induction of α-amylase mRNA expression in aleurone from dormant grains (Fig. 33.2). In embryos isolated from non-dormant grains, GA_3 even at a concentration of 10^{-5} M was not able to induce α-amylase mRNA expression (Fig. 33.2). In embryos isolated from dormant grains, 5×10^{-7} M GA_3 was able to give a half-maximum induction of α-amylase mRNA expression (Fig. 33.2).

Figure 33.3 shows that addition of FC (10^{-5} M) had no significant effect on GA-induced α-amylase mRNA expression in aleurone tissue isolated from non-dormant grains (using 10^{-9} M GA_3), but had a synergistic effect on this response in aleurone layers isolated from dormant grains (using 10^{-8} M GA_3). 10^{-8} M GA_3 was not able to induce α-amylase mRNA expression in embryos isolated from dormant grains. However, in combination with 10^{-5} M FC, 10^{-8} M GA_3 was able to give the same level of α-amylase mRNA expression as compared with 10^{-6} M GA_3 alone (Figs 33.3 and 33.2).

A well known feature of FC is that it stimulates the activity of plasma membrane H^+-ATPases in plant cells (De Boer, 1997). The consequence of activation of plasma membrane H^+-ATPases can be an acidification of the external medium. Since GA-induced α-amylase production can be enhanced by acidification of the external medium (Sinjorgo *et al.*, 1993), we examined the effect of GA_3 and FC on the pH of the incubation medium in which aleurone layers or embryos from barley grains were incubated. As a control, ten isolated embryos or aleurone layers were incubated in 300 μl of water containing the same amount of carrier (0.07% ethanol) present in the GA_3 and FC solutions.

GA_3 was able to break dormancy in intact dormant barley grains (Wang *et al.*, 1998) and to stimulate the germination of embryos isolated from dormant grains (Fig. 33.1 and Wang *et al.*, 1998). After 24 h incubation, GA_3 showed no stimulation of acidification of the medium pH in which embryos or aleurone were incubated as compared with the water control (Table 33.1). Fusicoccin had the same effect as GA on breaking of dormancy and stimulation of germination (Fig. 33.1 and Wang *et al.*, 1998), but FC was also able to induce a strong acidification of the medium pH in which embryos or aleurone were incubated (Table 33.1).

The activation of plasma membrane H^+-ATPase by FC may cause the lowering of pH_e (Table 33.1). Sinjorgo *et al.* (1993) showed that the effect of GA_3 on α-amylase production in barley aleurone is enhanced at low pH_e. Fusicoccin was able to enhance GA-induced α-amylase mRNA expression of dormant aleurone and embryos (Fig. 33.3). Therefore, it is possible that the acidification of pH_e caused by FC is the reason for the enhancement of GA-induced α-amylase mRNA expression in aleurone. In addition, FC significantly reduced the media pH_e for embryos isolated from dormant grains (Table 33.1), while a decrease in pH_e had no significant effect on germination of embryos (data not shown). It is possible that the effects of FC on embryo germination and aleurone α-amylase production are via two different pathways. Our further investigation will focus on this aspect.

Acknowledgements

We would like to thank B. Van Duijn for stimulating discussion and suggestions. This work was partially supported by European Community program no. PL962275 and by the Dutch Technology Foundation project no. 805.22.765.

Table 33.1. Effect of gibberellic acid (GA_3) and fusicoccin (FC) on pH_e of embryos and aleurone isolated from dormant and non-dormant barley grains.

Samples	pH_e in D aleurone means ± SD	pH_e in ND aleurone means ± SD	pH_e in D embryo means ± SD
Control	5.90 ± 0.08	5.84 ± 0.03	5.23 ± 0.52
GA_3	5.37 ± 0.05	5.60 ± 0.06	5.43 ± 0.62
FC	4.72 ± 0.04	4.91 ± 0.11	4.56 ± 0.37
GA_3 + FC	4.78 ± 0.03	4.85 ± 0.04	4.44 ± 0.27

Ten embryos or aleurone layers (in duplicate) isolated from dormant grains were incubated in water (control) or solutions containing GA: GA_3 (10^{-9} M) for non-dormant aleurone and GA_3 (10^{-8} M) for dormant aleurone and embryos; FC: 10^{-5} M; or a combination of GA_3 and FC. After 24 h of incubation at 20°C in the dark, pH in the incubation media was measured. The means ± SD from three independent experiments are presented. D, dormant; ND, non-dormant.

References

De Boer, B. (1997) Fusicoccin – a key to multiple 14-3-3 locks. *Trends in Plant Science* 4, 60–66.

Lado, P., Rasi-Caldogno, F. and Colombo, R. (1974) Promoting effect of fusicoccin on seed germination. *Physiologia Plantarum* 31, 149–152.

Marra, M., Ballio, A., Fullone, M.R. and Aducci, P. (1992) Some properties of a functional reconstituted plasmalemma H^+-ATPase activated by fusicoccin. *Plant Physiology* 98, 1029–1034.

Marrè, E. (1979) Fusicoccin: a tool in plant physiology. *Annual Review of Plant Physiology* 30, 273–288.

Phillipson, B.A. (1993) Expression of a hybrid (1-3, 1-4)-β-glucanase in barley protoplasts. *Plant Science* 91, 195–206.

Rogers, J.C. (1985) Two barley alpha-amylase gene families are regulated differently in aleurone cells. *Journal of Biological Chemistry* 260, 3731–3738.

Schuurink, R.C., Van Beckum, J.M.M. and Heidekamp, F. (1992) Modulation of grain dormancy in barley by variation of plant growth conditions. *Hereditas* 117, 137–143.

Sinjorgo, K.M., De Vries, M.A., Heistek, J.C., Van Zeijl, M.J., Van der Veen, S.W. and Douma, A.C. (1993) The effect of external pH on the gibberellic acid response of barley aleurone. *Journal of Plant Physiology* 142, 506–509.

Wang, M., Van Duijn, B., Van der Meulen, R.M. and Heidekamp, F. (1992) Effect of abscisic acid analogues on intracellular calcium level and gene expression in barley aleurone protoplasts. In: Karssen, C.M. *et al.* (eds) *Plant Growth Substances.* Kluwer Academic Press, Dordrecht, The Netherlands, pp. 635–642.

Wang, M., Van der Meulen, R.M., Visser, K., Van Schaik, H.P., Van Duijn, B. and De Boer, A.H. (1998) Effects of dormancy-breaking chemicals on ABA levels in barley grain embryos. *Seed Science Research* 8, 129–137.

34 Smoke and Germination of Arable and Rangeland Weeds

S.W. Adkins, P.J. Davidson, L. Matthew, S.C. Navie, D.A. Wills, I.N. Taylor and S.M. Bellairs

School of Land and Food, The University of Queensland, St Lucia Campus, Brisbane, Qld 4072, Australia

Plant-derived smoke has been shown to stimulate the germination of a number of native Australian species; however, little is known of the effect of smoke on the germination of introduced weed species. A series of laboratory experiments was conducted to determine the effect of smoke on the germination of two arable grass weeds, wild oats (*Avena sterilis* ssp. *ludoviciana* L.) and paradoxa grass (*Phalaris paradoxa* L.), and a rangeland weed, parthenium (*Parthenium hysterophorus* L.). Both arable weeds were stimulated by a range of smoked water concentrations (5–100%) while the germination of the non-dormant seed of *P. hysterophorus* was reversibly depressed by moderate concentrations of smoked water (5–10%). In a further experiment with soil seed bank samples collected from five arable and one rangeland site near Brisbane, the soil was treated with aerosol smoke produced from burning native vegetation. The germination of three introduced weed species, molasses grass (*Melinus minutiflora* Beauv.), green panic (*Panicum maximum* Jacq.) and common verbena (*Verbena officinalis* L.) were stimulated by smoke. No other weed or native species was stimulated by the smoke treatment. These observations are discussed in the light that smoke may play an important ecological role in the management and control of introduced weeds in native and arable communities in Australia.

Introduction

Compounds produced by the combustion of plant matter, found in smoke and charred wood from fires, have been found to stimulate germination in a variety of species (Keeley *et al.*, 1985; Keeley and Pizzorno, 1986; Van de Venter and Esterhuizen, 1988; De Lange and Boucher, 1990; Baxter *et al.*, 1994; Dixon *et al.*, 1995; Enright *et al.*, 1997; Davidson and Adkins, 1997). To date, germination enhancement has been shown in more than 170 species from 37 families (Roche *et al.*, 1997). Most studies have focused upon species native to fire-prone areas, although some examination of the effect of smoke on vegetable seed has shown that germination can also be stimulated in lettuce (*Lactuca sativa* L.; Drewes *et al.*, 1995) and celery (*Apium graveolens* L.; Thomas and Van Staden, 1995).

The compound(s) in smoke that affect germination are as yet unidentified (Brown and Van Staden, 1997) but probably act internally in the seed (Keeley and Keeley, 1987). Keeley and Pizzorno (1986) stated that the effector is produced by a variety of wood types, it is water soluble and it is derived from the hemicellulose fraction of wood. Van Staden *et al.* (1995) reported attempts to purify the biologically active fraction of smoke. Twelve compounds were identified, not all of which were present in all smoke extracts examined.

Smoke is effective on species from a wide range of families, which vary in ecology, reproductive strategy, seed size and morphology (Dixon and Roche, 1995). For this reason, the commercial development of smoke treatments for the stimulation of seed germination is considered a viable option. Products currently in the market include 'Seed Starter, Australian Smoky Water', 'Regen 2000' and 'Kirstenbosch Instant Smoke Plus' seed primer. Brown and Van Staden (1997) found that both of these products significantly stimulated seed germination of a number of selected species.

The use of smoked water has applications in the rehabilitation of disturbed areas, horticultural industries, ecological management, and crop production. Such applications are primarily for species which are difficult to germinate under normal circumstances. Use of smoke or smoked water in bushland rehabilitation in Australia is thought to be very promising with native species selectively stimulated over introduced weedy species (O'Neill, 1997). However, very little examination of the effect of smoke on germination of weed species has been carried out. One feature of many annual weeds is their ability to persist in the soil seed bank for several years due to dormancy. Should smoke prove to be an effective agent in breaking annual-weed seed dormancy, then this will have implications for bushland rehabilitation, and may give rise to a treatment that can deplete arable weed seed banks.

The aims of the present study are firstly, to investigate the effect of smoked water upon the germination of two arable monocot weeds with dormant seed (*Avena sterilis* spp. *ludoviciana* L., wild oats, and *Phalaris paradoxa* L., paradoxa grass) and one rangeland dicot weed with non-dormant seed (*Parthenium hysterophorus* L., parthenium weed) under laboratory conditions; and secondly, to investigate the effect of a smoke application on the seedling emergence from soil seed banks containing native and introduced weed species.

Materials and Methods

Seed germination studies

Seeds were collected from wild oat (*Avena sterilis* ssp. *ludoviciana* L.) plants growing in northeast New South Wales, Australia. The primary seed was separated from the secondary seed and each lot stored in paper envelopes at room temperature until required for experimentation. Studies were undertaken on one seed lot that had been stored for 4 weeks (hereafter referred to as freshly harvested) while a second study used seed that was 14 months old (partly afterripened). All seed lots were stored with hulls in place but were germinated as either intact seeds or as naked caryopses (with hulls removed by hand). Mature seeds were also collected from paradoxa grass (*Phalaris paradoxa* L.) plants growing in southeast Queensland and stored dry at −18°C for 2 weeks. The seed was removed from the outer glumes leaving the pales intact and then subjected to the germination studies. Parthenium weed (*Parthenium hysterophorus* L.) achenes were collected at maturity from plants growing in the rangelands in central Queensland. The achenes were stored dry for 6 months at room temperature before being subjected to germination studies.

For the germination studies involving *A. sterilis* ssp. *ludoviciana*, four replicates of 20 seeds were removed from storage and placed in 9 cm Petri dishes lined with two layers of 9 cm Whatman No. 2 filter papers moistened with 7 ml of distilled water (control) or smoked water solution. All dishes were incubated in a 12 h day, 12 h night photoperiod at a constant temperature of 20°C in an atmosphere saturated with water vapour. For the germination study of *P. paradoxa*, four replicates of 30 seeds were placed in 9 cm Petri dishes lined with a single Advantec 424 filter paper moistened with 5 ml of distilled water (control) or smoked water solution. All dishes were incubated at a 12 h day, 12 h night photoperiod, as above, but a 20/9°C thermoperiod was used. In the germination study of *P. hysterophorus*, the conditions were similar to the first experiment, but four replicates of 25 seeds were used and a 25/20°C thermoperiod was introduced in addition to the 12 h photoperiod. Germination (protrusion of coleorhiza through testa and pericarp for the grass species and emergence of root through fruit layers for parthenium weed) was recorded periodically.

The first two experiments, involving *A. sterilis* ssp. *ludoviciana* and *P. paradoxa*, were run for 28 and 29 days respectively. The experiment involving *P. hysterophorus* was run for 17 days, then the seed were removed from the smoked water solutions and placed in distilled water for a further 8 days.

All smoked water solutions were diluted from a stock solution of 'Seed Starter, Australian Smoky Water®' obtained from Kings Park and Botanical Gardens, Perth, Australia. Different smoked water concentrations were used in the three experiments (*A. sterilis* ssp. *ludoviciana*: 0, 1, 2, 5, 10, 20 and 50%; *P. paradoxa*: 0, 5, 20, 50 and 100%; *P. hysterophorus*: 0, 0.01, 0.05, 0.1, 0.5, 1, 5 and 10%). The pH of the solutions ranged from 3.2 (100% solution) to 4.2 (1% solution) with the exception of the distilled water control which was 5.5. All experiments reported here were repeated at least once with similar results.

Soil seed bank studies

In February 1998, soil samples were collected from five fallow field or pasture locations in south-east Queensland and one site in the Raven Street Reserve, a bushland reserve in suburban Brisbane. From each site, six soil samples were taken. Each soil sample came from a 1 m^2 area which was at least 10 m from the next sample site. Following surface litter removal, the soil was sampled to a depth of 2–5 cm by inserting a PVC core (10 cm diameter) into the soil. The core was then closed by covering the bottom with a small trowel. Using this technique 5 l of soil was collected from each sample area giving a total of 30 l from each site.

The 5 l soil samples were then individually retained in clear plastic bags and the seed in the soil assessed for germination 2 days after collection. Each sample was sieved (5 mm) to remove stones and large pieces of organic matter. From each of the 5 l replicate soil samples two 1.4 l subsamples were taken, one for each treatment (smoke or no smoke). Two open plastic trays (34 × 28 cm) containing 7.5 cm of a steam-sterilized potting mix were placed one on top of another (two tray depth). Onto the top tray, a 1.5 cm layer of soil was placed from each of the two subsampled replicates per site. The smoke treatment was applied to half of these trays by placing them in a tent into which cool smoke was piped from an incinerator (Roche *et al.*, 1997). Trays were smoked for 60 min, after which they were removed from the smoking tent and placed in a glasshouse. Each tray was then moistened with tap water delivered through an atomizer to ensure that the smoke settled into the soil. These trays were not watered again for 24 h to prevent leaching of smoke. After this time all trays received a daily watering to field capacity. The control trays were similarly treated and all trays were examined weekly for emerging seedlings over the next 6 weeks. Seedlings were identified as soon as possible, however some seedlings, especially grasses and sedges, could only be identified upon flowering, which took over 8 months after emergence. Once identified, seedlings were removed from the trays.

Statistical analysis

The data from the experiment involving *A. sterilis* ssp. *ludoviciana* were analysed using a one-way analysis of variance (ANOVA) on the final germination percentage recorded after 28 days for each of the eight separate seed types with the eight smoked water concentrations as treatments. In the *P. paradoxa* experiment, a two-way ANOVA was performed on a larger data set from which the relevant information was extracted. Statistical analysis was conducted on the final germination recorded after 29 days. Finally, for the experiment involving *P. hysterophorus*, a one-way ANOVA was conducted on the germination achieved after 17 and 25 days.

The data obtained from the soil seed bank study were analysed by examining for each site the number of seed germinating from each species 2 and 6 weeks after treatment. The initial analysis was carried out for each species from each site using a three-way ANOVA with the three factors being treatment, site, and sample nested within site. Species that occurred at only one site had the main effects assessed using a two-way ANOVA (treatment and sample). To undertake a study of the rate of germination, the proportion of the final 6-week germination that had occurred by 2 weeks was determined for each species. These data were then analysed only for species occurring in high numbers in all six replicates at a site.

Results

Germination studies

Smoked water solutions did not affect the final germination of primary after-ripened seed of *A. sterilis* ssp. *ludoviciana* (both intact florets and caryopses), the germination of all treatments and the controls being relatively high (68–100%; Fig. 34.1). However, all smoked water treatments (1–50% solutions) significantly ($P < 0.05$) stimulated the final germination of secondary after-ripened seed (both intact florets and caryopses; Fig. 34.1). Smoke stimulation was even more apparent in the freshly harvested seed. The caryopses of primary and secondary seeds were significantly stimulated by all concentrations of smoke water used (Fig. 34.2). By 28 days, all smoked water treatments had induced at least 80% germination while the germination in the controls was still less than 10%. Intact primary florets were also stimulated by all smoked water concentrations, although the higher concentrations (5–50%) were more effective than the lower concentrations (1–2%; Fig. 34.2). Finally, in the study of intact secondary florets of freshly harvested seed, only the highest smoked water concentration (50%) significantly stimulated germination over that of the control (Fig. 34.2).

Hence, similar concentrations of smoked water elicited a diminishing germination response in primary caryopses > secondary caryopses > primary hulled seed > secondary hulled seed. The same treatments had less effect on freshly harvested seed than they did on partly afterripened seed.

In the *P. paradoxa* experiment, all smoked water concentrations (5–100%) used induced a significant germination response in comparison to that seen in the control (Fig. 34.3). The increase in germination was proportional to the smoked water concentration applied, however, a decline from the peak response (50% smoked water) was noticed with the application of the 100% solution of smoked water.

In the *P. hysterophorus* germination experiment the 5 and 10% smoked water solutions significantly inhibited germination after 17 days (Fig. 34.4). Germination in all other smoked water treatments (0.01–1%) and the controls was very high (> 90%) indicating little or no dormancy in these seed lots. When the seed was removed from the smoked water solutions and placed in distilled water, there was a rapid recovery (2 days) in the germination of all treatments, particularly the 5 and 10% smoked water treatments, up to maximum germination levels. Hence, there were no treatment differences ($P < 0.05$) between the final mean percentage germination values at day 25 (Fig. 34.4).

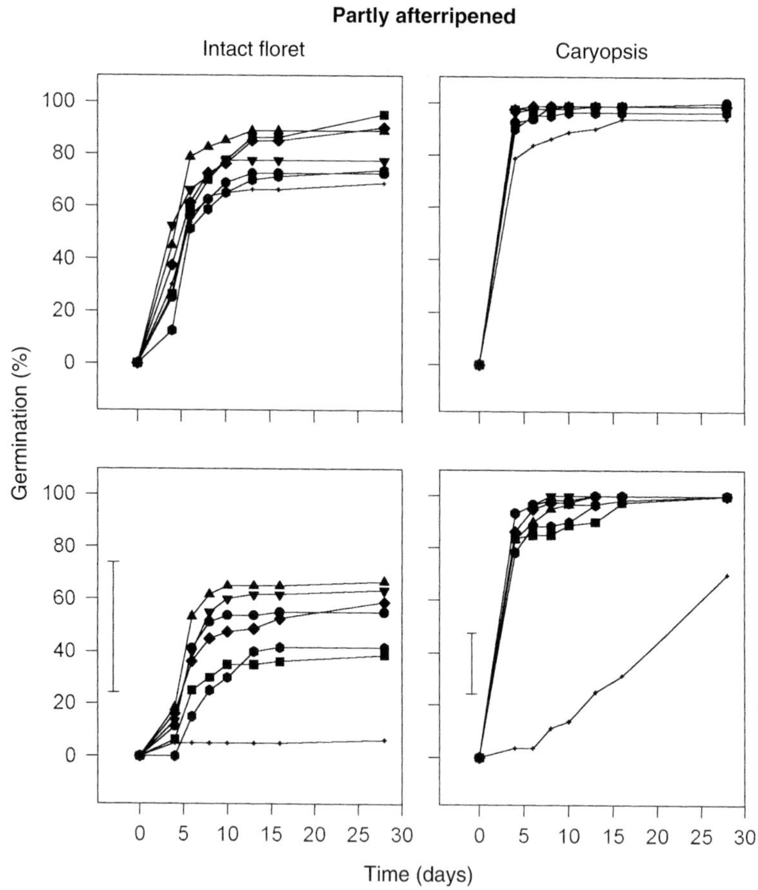

Fig. 34.1. The percentage germination of *Avena sterilis* spp. *ludoviciana* seed treated with one of seven smoked water treatments (+ 0%, ■ 1%, ▲ 2%, ♦ 5%, ▼ 10%, ● 20%, ⬢ 50%). The seeds were partly afterripened and were treated as intact florets (with pales in place) or as caryopses (with pales removed). The seeds were either primary from the proximal position in the spikelet or the smaller secondary seed.

Seed bank studies

From the six sites sampled, there were 3841 seeds germinating of 24 dicot and 17 monocot species. Most germination occurred from the samples from sites E and F while the application of smoke significantly increased total germination at sites C and F (Fig. 34.5). Six dicot species and eight monocot species had more than ten seedlings emerge per site for a given treatment. Of these, only three species showed a significant emergence response to smoke stimulation 6 weeks after the initial treatment, namely *Melinus minutiflora* Beauv. (molasses grass), *Panicum maximum* Jacq. (green panic) and *Verbena officinalis* L. (common verbena), all of which are weeds introduced to Australia (Fig. 34.6). For a further two species a significantly greater rate of germination was

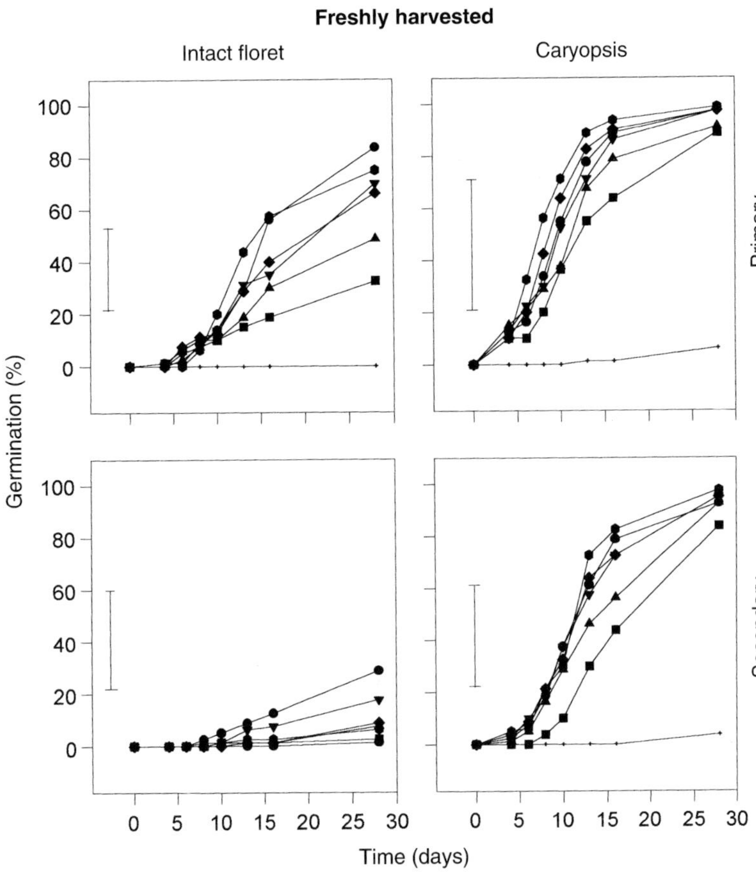

Fig. 34.2. The percentage germination of *Avena sterilis* spp. *ludoviciana* seeds treated with one of seven smoked water treatments (+ 0%, ■ 1%, ▲ 2%, ♦ 5%, ▼ 10%, ● 20%, ⬣ 50%). The seeds were freshly harvested and were treated as intact florets (with pales in place) or as caryopses (with pales removed). The seeds used were either primary from the proximal position in the spikelet or the smaller secondary seed.

achieved in smoked trays ($P < 0.05$), namely *Cyperus gracilis* R. Br. (slender sedge) and *C. polystachus* Rottb. (bunchy sedge); both are Australian natives (Fig. 34.7).

Discussion

Smoked water treatments and germination

Smoked water solutions significantly enhanced the germination of most *A. sterilis* ssp. *ludoviciana* seed lots. One seed lot (partly afterripened primary seed) that was not stimulated by smoked water exhibited little or no dormancy and germinated whether or not smoked water was applied. The only seed lot

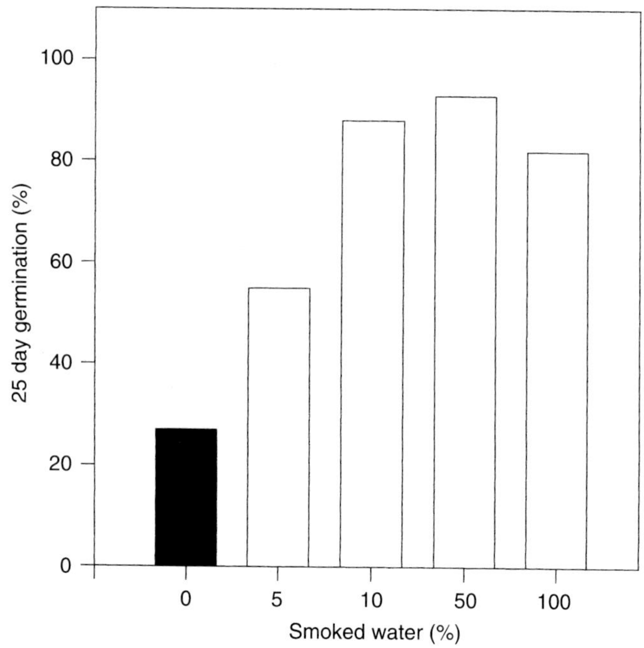

Fig. 34.3. The percentage germination of *Phalaris paradoxa* seed treated with one of five smoked water treatments. The seeds were freshly harvested and were treated as intact florets with pales in place.

that resisted germination in the presence of smoked water was the one exhibiting the highest degree of dormancy (i.e. freshly harvested secondary florets). Smoked water solutions were also very effective in stimulating the germination of dormant *P. paradoxa* seed. The highest concentration of smoked water tested (100%) did not elicit as good a response as the more moderate concentrations (20–50%), indicating that very high concentrations of smoked water may have an inhibiting effect on germination. These two weed species are thought to have originally come from fire-prone Mediterranean climates. Hence, it is not unexpected that their germination is stimulated by smoke.

A study on the less dormant rangeland weed *P. hysterophorus* indicated this species to be particularly sensitive to the inhibitory effects of smoked water, with germination being suppressed by low concentrations (5–10%). Unlike the above two grasses, this species is not thought to have originated from a fire-prone habitat. Thus it is possible that such species will be more sensitive to the effects of high concentrations of smoke. Past studies have shown that many native Australian grasses are stimulated by smoke (i.e. *Heteropogon contortus* (L.) Beauv. ex Roemer & Schultes: Campbell, 1995; *Themeda triandra* Forssk.: Baxter *et al.* 1994; *Triodia longiceps* N.Burb., Davidson and Adkins, 1997). Therefore, smoke may inhibit the germination of *P. hysterophorus* but stimulate the germination of native pasture species. Thus, smoked water may be useful in the restoration of native pastures invaded by this introduced weed.

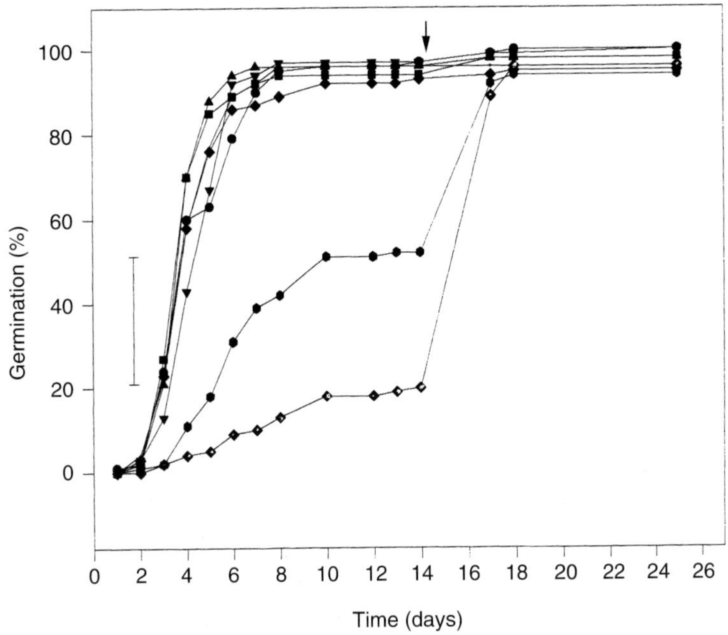

Fig. 34.4. The percentage germination of *Parthenium hysterophorus* achenes treated with one of eight smoked water treatments (+ 0%, ■ 0.01%, ▲ 0.05%, ♦ 0.1%, ▼ 0.5%, ● 1%, ● 5%, ◆ 10%). The achenes were freshly harvested and had been treated as intact achenes with all accessory structures in place. After 14 days (arrow), seeds from the 5% and 10% smoked water treatments were transferred to distilled water.

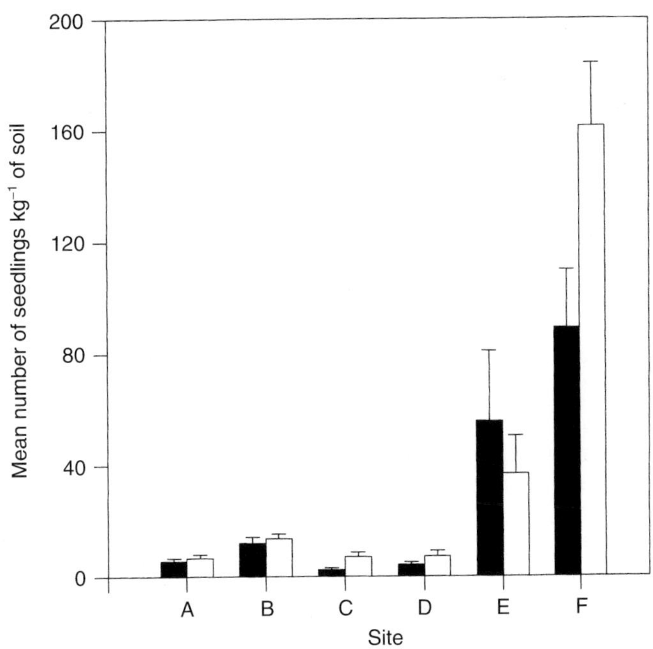

Fig. 34.5. The effect of a smoke treatment (□) on the emergence of seedlings from soil taken from six different sites when compared to a control (■).

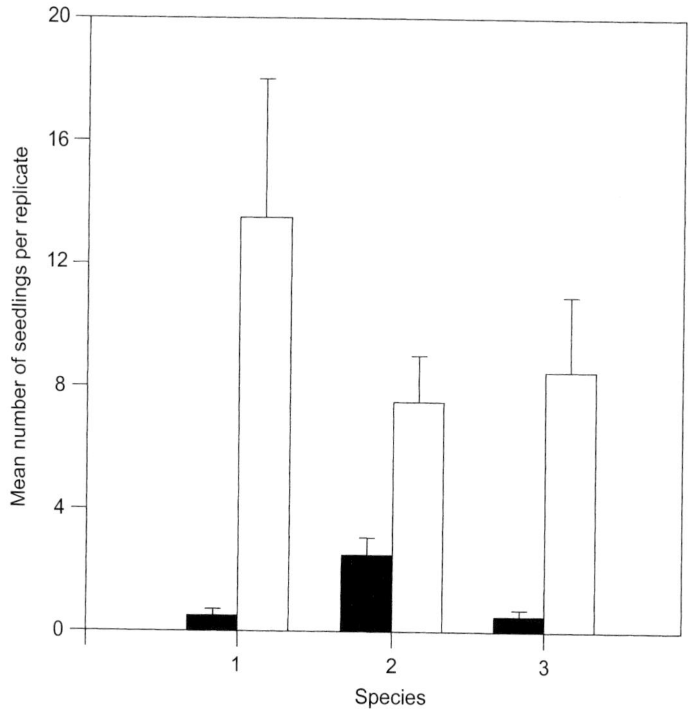

Fig. 34.6. The mean germination of *Melinus minutiflora* (1), *Panicum maximum* (2), and *Verbena officinalis* (3), species exhibiting a significantly greater emergence response to the smoke treatment (□) than in the control (■).

Smoke treatment and soil seed banks

There was considerable variation in the total number of seedlings that emerged from the six sites and this may have been due to effective weed control measures in the five arable sites. The final bushland site had a considerably larger seed bank present. In the two sites where smoke significantly stimulated more seedlings to emerge, three of the 14 common species found were significantly stimulated by smoke. Each of the smoke stimulated species (namely *M. minutiflora, P. maximum, V. officinalis*) is a major bushland weed and all are also considered minor arable weeds. This observation that three introduced weeds are stimulated by smoke where bushland natives are not is important for future bushland revegetation approaches.

Smoke had a significant effect in stimulating the germination of *Melinus minutiflora* L. (Fig. 34.6), which is a native of tropical African pastures (Swarbrick and Skarratt, 1994), and so may, in its native range, be subjected to fires, leading to the evolution of a smoke-sensitive germination trigger. The second species stimulated by smoke was *Panicum maximum* Jacq., which is also a native to African pastures (Swarbrick and Skarratt, 1994). Both are bushland weeds in Australia. The origin of the third smoke-stimulated species,

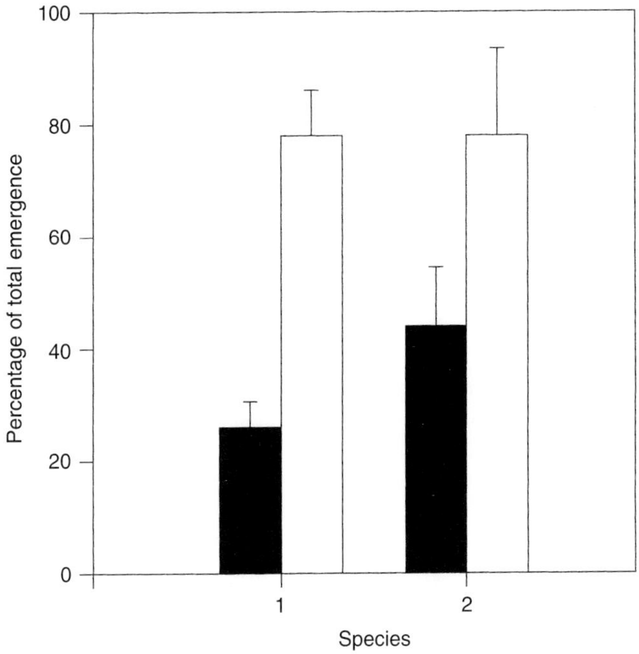

Fig. 34.7. Effect of smoke (□) or no smoke (■) on the rate of emergence of *Cyperus gracilis* (1) and *Cyperus polystachus* (2) seedlings, expressed as the percentage of total germination at 6 weeks that had occurred within 2 weeks of the application of the smoke treatment.

Verbena officinalis L. is disputed. It is recorded as a native in central America (Kleinschmidt and Johnson, 1987) while another report records it as European in origin (Swarbrick and Skarratt, 1994). Therefore it is difficult to determine whether it is a native of fire-prone environments. In Australia, it is a weed in both cultivated and bushland areas. As this weed is a close relative of a major Australian bushland weed, lantana (*Lantana camara* L.), the response of both these species to smoke deserves further investigation.

Two species were found to have their rate of germination increased by smoke but not their final germination percentage. These two species *Cyperus polystachus* and *C. gracilis* are Australian natives and such a treatment may enhance their establishment relative to other native and weed species.

Smoke treatment and ecosystem management

It is the view of several scientists that smoke treatments are useful in relation to bushland regeneration, as smoke seems selectively to promote native Australian species and not introduced weeds (O'Neill, 1997). The findings of this present investigation show that this is not always the case and some caution needs to be employed. While the treatment would still be of major benefit in relatively intact bushland, in disturbed weedy sites (such as urban nature reserves) it may lead to promotion of some weed species.

The finding that certain weed species can be stimulated by smoke, either in the laboratory or in the soil seed bank, is of great interest to those who are trying to purge the soil seed bank of dormant weed seeds. Using smoke to break seed dormancy in arable weeds can potentially provide major benefits to farmers, as it could be used in farm management to reduce the weed burden on crops and decrease the need for herbicides or physical weed control.

Acknowledgements

We wish to acknowledge the contributions of Andrea Adkins for undertaking the statistical analyses, the presentation of data and the typing of the manuscript, and Winston Bean for help in the glasshouse studies.

References

Baxter, B.J.M., Van Staden. J., Granger, J.E., and Brown, N.A.C. (1994) Plant-derived smoke and smoke extracts stimulate seed germination of the fire-climax grass *Themeda triandra*. *Environmental and Experimental Botany* 34, 217–223.

Brown, N.A.C. and Van Staden, J. (1997) Smoke as a germination cue: a review. *Plant Growth Regulation* 22, 115–124.

Campbell, S.D. (1995) Plant mechanisms that influence the balance between *Heteropogon contortus* and *Aristida ramosa* in spring burnt pastures. PhD thesis, The University of Queensland, Brisbane, Australia.

Davidson, P.J. and Adkins, S.W. (1997) Germination of *Triodia* grass seed by plant derived smoke. In: *Proceedings of the Australian Rangeland Conference*, Gatton Campus, University of Queensland, 1–4 December, 1997. Australian Rangeland Society, pp. 29–30.

De Lange, J.H. and Boucher, C. (1990) Autecological studies on *Audouinia capitata* (Bruniaceae). VIII. Role of fire in regeneration. *South African Journal of Botany* 58, 700–703.

Dixon, K.W. and Roche, S. (1995) The role of combustion products (smoke) in stimulating *ex-situ* and *in-situ* germination of Western Australian plants. *Proceedings of the International Plant Propagation Society* 45, 53–56.

Dixon, K.W., Roche, S. and Pate, J.S. (1995) The promotive effect of smoke derived from burnt native vegetation on seed germination of Western Australian plants. *Oecologia* 101, 185–192.

Drewes, F.E., Smith, M.T. and Van Staden, J. (1995) The effect of a plant-derived smoke extract on the germination of light sensitive lettuce seed. *Plant Growth Regulation* 16, 205–209.

Enright, N.J., Goldblum, D., Ata, P. and Ashton, D.H. (1997) The independent effects of heat, smoke and ash on emergence of seedlings from the soil seed bank of a healthy *Eucalyptus* woodland in Grampians (Gariwerd) National Park, western Victoria. *Australian Journal of Ecology* 22, 81–88.

Keeley, J.E. and Keeley, S.C. (1987) Role of fire in the germination of chaparral herbs and suffrutescents. *Madrono* 34, 240–249.

Keeley, J.E. and Pizzorno, M. (1986) Charred wood stimulation of two fire-following herbs of the California chaparral and the role of hemicellulose. *American Journal of Botany* 73, 1289–1297.

Keeley, J.E., Morton, B.A., Pedrosa, A. and Trotter, P. (1985) Role of allelopathy, heat and charred wood in the germination of chaparral herbs and suffrutescents. *Journal of Ecology* 73, 445–458.

Kleinschmidt, H.E. and Johnson, R.W. (1987) *Weeds of Queensland.* Queensland Department of Primary Industries, Brisbane, Australia, 369 pp.

O'Neill, G. (1997) Chemical Kiss of Life. *Sunday Herald Sun*, 8/6/97.

Roche, S., Dixon, K.W. and Pate, J.S. (1997) Seed aging and smoke: partner cues in the amelioration of seed dormancy in selected Australian native species. *Australian Journal of Botany* 45, 783–815.

Swarbrick, J.T. and Skarratt, D.B. (1994) *The Bushweed 2 Data Base of Environment Weeds in Australia.* The University of Queensland, Brisbane, Australia.

Thomas, T.H. and Van Staden, J. (1995) Dormancy break of celery (*Apium graveolens* L.) seeds by plant-derived smoke extract. *Plant Growth Regulation* 17, 195–198.

Van de Venter, H.A. and Esterhuizen, A.D. (1988) The effect of factors associated with fire on germination of *Erica sessiflora* and *E. hebecalyx* (Ericaceae). *South African Journal of Botany* 54, 301–304.

Van Staden, J., Drewes, F.E. and Brown, N.A.C. (1995) Some chromatographic characteristics of germination stimulants in plant-derived smoke extracts. *Plant Growth Regulation* 17, 241–249.

VI Ecology

35 Intermittent Germination: Causes and Ecological Implications

P.B. Cavers, M.M. Qaderi, R. Manku and M.P. Downs

Department of Plant Sciences, University of Western Ontario, London, Ontario, N6A 5B7, Canada

Seeds of many wild species, especially weeds, exhibit intermittent germination, a pattern in which the seeds produced by a single population in one season will germinate and emerge at irregular intervals over a period of months or years. Even the seeds from a single parent plant can show intermittent germination. An intriguing feature of some species with intermittent germination is that all or most of their seeds in the seed bank have been classed as non-dormant. Despite this, seeds from a population of such species vary in germination response. Factors acting during seed development (temperature, moisture, burial, damage from microorganisms) can influence subsequent germination performance. Developmental (e.g. age of the inflorescence) and positional (within or between inflorescences on the parent plant) effects can also be important. In the seed bank, after dispersal, factors such as depth in the soil, burial under leaf litter, and the orientation on the soil surface can affect the germination responses of individual seeds. A clearer understanding of the combined effects of these factors will enable agriculturalists and others to manipulate the overall germination patterns of seed populations and especially to decrease intermittency.

Introduction

The first detailed description of intermittent germination was provided by Sir Edward Salisbury in England (Salisbury, 1961). He defined intermittent germination as occurring 'with intervals of days, or even weeks, between the quite irregular appearance of plumules'. On the basis of this definition a more proper term should be 'intermittent emergence' or 'intermittent seedling emergence'. In his book entitled *Weeds and Aliens* Salisbury (1961) described intermittent germination for the seeds of many weed species including *Silene alba* (= *S. latifolia*), *Geranium dissectum*, *Trifolium micranthum* and *Veronica persica.* Two later publications by Salisbury (1963, 1965) provided detailed

©CAB *International* 2000. *Seed Biology: Advances and Applications* (eds M. Black, K.J. Bradford and J. Vázquez-Ramos)

studies of intermittent germination in *Capsella simplex* and *Plantago major*. In both of these papers he included graphs of intermittent germination over a period of eight months for seeds collected from a single parent plant.

The simplest pattern of intermittent germination has the first seedlings appearing shortly after they are dispersed from the parent plant, then new groups of seedlings appearing irregularly for weeks, months or years, whenever environmental conditions permit germination. Salisbury (1961) pointed out that patterns differ among species. For example, seeds of *Veronica persica* and four other species did not start to emerge until the spring following their production, but then appeared at intervals for more than two years. Leguminous species produce both 'soft' and 'hard' seeds. Soft seeds 'have readily permeable seed coats and germinate quickly', whereas hard seeds 'exhibit delayed (and intermittent) germination'. Since different strains of a clover species have different proportions of soft and hard seeds, Salisbury speculated that there would be varietal differences in germination patterns. Other species have more than one type of seed or fruit produced on the same plant. Salisbury (1961) noted that the germination patterns differ among these types. He cited *Atriplex patula*, where the large brown seeds usually germinate soon after dispersal but the smaller black seeds start to germinate later and then emerge more or less continuously for a period of weeks.

Of what value is intermittent germination? In his 1963 publication Salisbury reported that intermittent germination is 'prevalent . . . as a feature of our annual garden weeds'. He claimed that human activities, involving the destruction of larger and obvious seedling cohorts, have probably caused 'artificial selection . . . that has accentuated this feature'. There is no doubt that a gardener or a commercial grower finds it much easier to destroy a single cohort of seedlings appearing at one time than successive cohorts, each appearing very shortly after the previous one has been eliminated. It is not surprising that Salisbury's intermittent germination ideas were formulated in England, since he would have been aware of the many papers of Brenchley and Warington (e.g. 1930, 1933) who recorded intermittent emergence patterns for buried seed populations from soil cores. For some species in their experiments, new seedlings arose each year for ten years, despite no new additions of seeds over that period.

In spite of the obvious importance of intermittent germination patterns, there has been comparatively little study of them. One reason for this is that many germination experiments have been too brief to reveal intermittent patterns. Often, they have been terminated after 25 days or less. However, germination from some samples can spread over a much longer period. For example, data from a recent Petri dish study of two populations of Scotch thistle (*Onopordum acanthium*) by M. Qaderi (Fig. 35.1) shows that one population (Gibbons Park) had virtually complete germination within 11 days, but the second population (Environmental Sciences Western) had a more intermittent germination pattern that extended over almost 50 days. Further tests revealed that most of the remaining cypselas (seeds) at 50 days were viable, indicating that the intermittent pattern would have been extended if the experiment had been prolonged. A second reason is that comparatively few

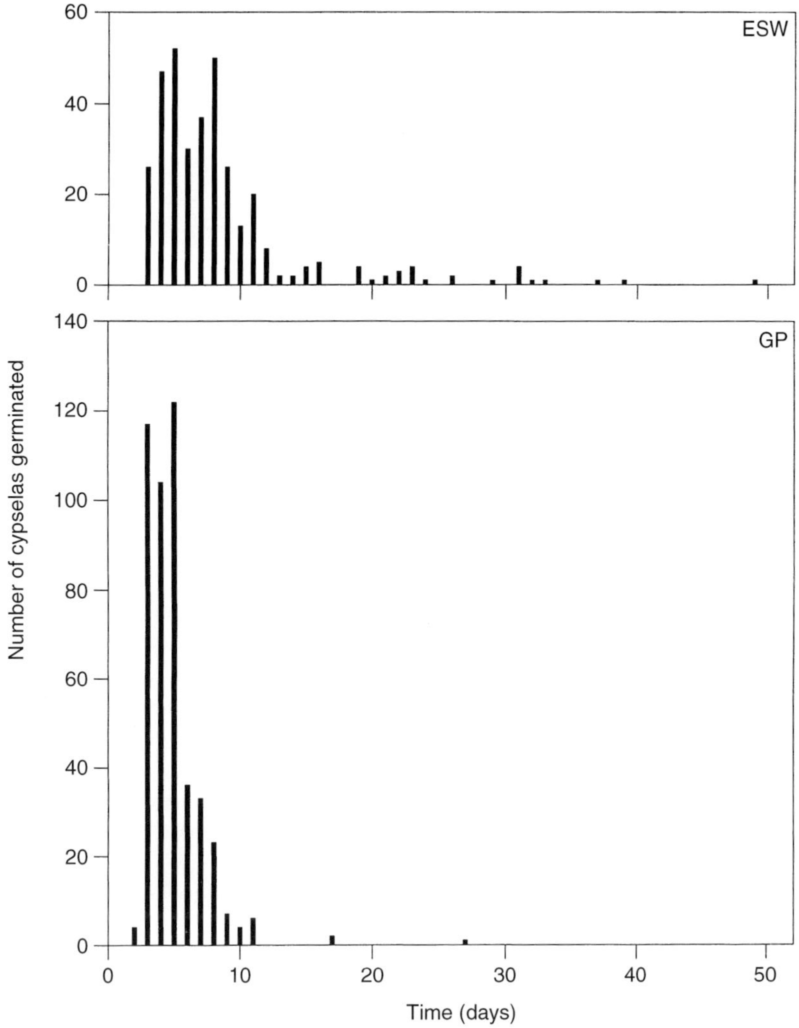

Fig. 35.1. Germination patterns of cypselas of two Scotch thistle (*Onopordum acanthium*) populations, Environmental Sciences Western (ESW) and Gibbons Park (GP), collected in September 1996 in London, Ontario and incubated at 25/10°C.

scientists have studied seedling emergence patterns from known seed samples in field tests. Salisbury (1961, 1963, 1965) was one of the few.

Factors Affecting Germination Patterns

What influences germination and seedling emergence patterns in the field? It is logical to predict that both genetic and environmental factors are involved.

Genetic factors

Baskin and Baskin (1998) cite many studies to show that different genotypes within a species can have different requirements for germination and/or different rates of germination under the same environmental conditions. There are also numerous studies that reveal large differences in germination among viable seeds from different parent plants in the same habitat (e.g. Cavers and Harper, 1966; Cavers, 1974; Sidhu and Cavers, 1977; Perez-Garcia, 1993; Qaderi, 1998). If a population consists of a number of different genotypes and if these genotypes vary in rate of germination, then intermittent germination can be predicted for that population.

Environmental factors

During seed maturation

There are many factors operating during the process of seed development that can influence germination patterns.

BURIAL OF DEVELOPING SEEDS. In black medick (*Medicago lupulina*), inflorescences are borne beneath the long trailing branches. If the plant grows as a colonizer on bare soil, developing seeds can be covered or partially covered by soil splashed on them during heavy rains. Such seeds become viviparous, germinating on the plant and forming small colonies of seedlings (Sidhu and Cavers, 1977). Emergence is intermittent over the growing season simply because new flowers are formed every 2–3 days under good growing conditions. In contrast, black medick plants growing in dense vegetation, such as in a lawn, will ripen hard dormant seeds that can live for years in a seed bank (Sidhu, 1971). Preliminary investigations with different durations of burial during seed ripening and different moisture levels in the soil suggest that seeds can be formed that are not viviparous but do have weaker dormancy than seeds ripening in lawns. In a habitat with a mixture of bare and vegetated areas, it is probable that the seeds of black medick would germinate intermittently, with seeds formed on bare soil germinating first and those formed in dense vegetation remaining dormant for months or years before germinating.

MICROBIAL ATTACKS ON FLOWERING PLANTS. Plants of fall panicum (*Panicum dichotomiflorum*) that were infested with the smut *Ustilago destruens* produced very few seeds compared with uninfected plants. However, the seeds that were produced on infected plants had weaker dormancy and tended to emerge earlier in the spring (Govinthasamy and Cavers, 1995). This would be advantageous in crops such as corn, where the early-emerging seedlings are strongly competitive with the crop. A mixed population of infected and uninfected plants produces a population of seeds that will show greater intermittency in emergence than seeds from a completely uninfected population of fall panicum.

TEMPERATURES DURING SEED MATURATION. More than 35 years ago Grant Lipp and Ballard (1963) allowed plants propagated vegetatively from a single clone of *Anagallis arvensis* to mature under different temperature regimes in greenhouses. Seeds produced under the warmest regime (30°C day–25°C night) had no dormancy, those matured at 25°C day–20°C night had moderate dormancy, whereas those matured at 20°C day–15°C night had strong dormancy that persisted through a full year of storage.

J. Barton and P.B. Cavers (unpublished) took freshly ripened seeds of dandelion (*Taraxacum officinale*) from large plants in the field. Seeds from each capitulum were put to germinate immediately in Petri dishes at 25°C day, 10°C night. The warmer the temperatures during the period from anthesis to seed maturity, the faster was the germination.

In several field and greenhouse experiments with cypselas (seeds) of *Onopordum acanthium*, Qaderi (1998 and unpublished) has found that cypselas from a single plant can be ripened in different capitula over a period as long as two to three months. In all of his experiments cypselas ripened under warmer temperatures germinated faster and to a higher percentage than cypselas ripened under cooler temperatures.

The results obtained in each of the above experiments would lead to greater intermittency than would be obtained from samples of seeds that have the same germination response, regardless of the temperatures during seed maturation.

During dispersal

Before seeds arrive in the microsite from which they will germinate, they can experience conditions that can affect their dormancy. For example, plants of *Polygonum lapathifolium* and *P. persicaria* are often moved in rivers and streams over winter or they may simply be in wet areas of fields that are inundated for a few weeks in the spring. In such positions they will receive the natural stratification needed to break their dormancy. The longer the stratification period, the faster and more complete will be the germination of such seeds (Staniforth and Cavers, 1979). After dispersal, a population composed of seeds that have received a variety of stratification experiences will show intermittent germination (R.J. Staniforth, Manitoba, 1999, personal communication).

The overwintering environment

The pivotal role of the overwintering environment on future seedling emergence patterns is well illustrated by an experiment conducted with *Rumex crispus* (curled dock). Although the seeds of this species are usually ripe by late August, they are shed gradually during the autumn. In many years in Ontario, Canada, hundreds of fruits remain on parent plants throughout the winter. In this experiment (P.B. Cavers, unpublished), the fruits (achenes), with perianth segments attached were taken from individual parent plants in late August 1966. The fruits from each of seven plants were divided into three equal lots and each lot was sewn into a flat 30 × 30 cm bag constructed from fibreglass window screening. One of these bags per plant was buried 20 cm deep, one was fastened to the soil surface and one was kept 20 cm above the soil in an arable field in London, Ontario. On May 12 of the following year three lots of 200 fruits each from each fibreglass bag were spread on top of a 5 cm deep layer of sterilized clay-loam soil in 30 × 15 cm wooden flats (total of nine flats per parent plant). The flats were kept in a greenhouse with no artificial lighting and with only enough heat in the winter to maintain temperatures at 5°C or above. The flats were watered regularly to keep them moist. Seedlings of *R. crispus* were counted and removed as they appeared from May 1967 to August 1968 (when some overzealous maintenance workers inadvertently destroyed the test flats). The pattern of emergence for achenes from one of the parent plants is portrayed in Fig. 35.2.

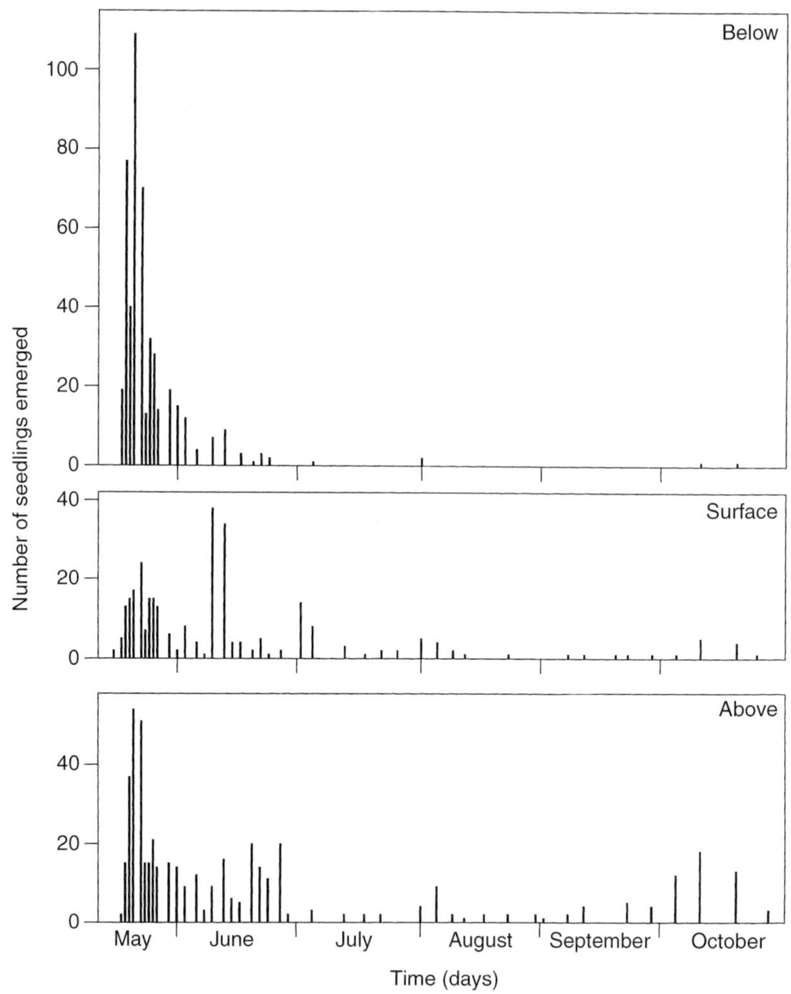

Fig. 35.2. Germination patterns of *Rumex crispus* seeds taken from one plant, then overwintered 20 cm below, on, and 20 cm above the soil surface outside, then put to germinate on sterilized soil in a greenhouse.

Although the total number of seedlings emerged was almost the same for the above- and below-ground storage treatments, the patterns of emergence were very different. Of all the seedlings that arose from the below-ground treatment, 95.9% appeared within one month of retrieval from the field. In contrast, only 62.8% of the seedlings that arose from the above-ground treatment appeared in the first month. The remaining seedlings appeared intermittently over the next 14 months. Total seedling emergence from the surface treatment (51.8%) was much less than from the other two treatments but the number arising after the first month (94) was much greater than the 20 seedlings from the below-ground treatment that appeared after the first month. Later-appearing seedlings were just as healthy as the first seedlings to arise. Results for the samples from the other six parent plants of *R. crispus* were similar to the patterns shown here. This experiment demonstrates that the conditions of overwinter storage of seeds can have dramatic effects on subsequent patterns of seedling emergence.

The microenvironment of the seed in the soil

IMPORTANCE OF BURIAL. Often, differences in patterns of seedling emergence can be seen by comparing the performance of seeds left on the surface with that of seeds that are buried. M. Downs compared emergence from seeds of *Cirsium vulgare* left on the surface of a moist soil in a greenhouse with that from seeds buried 1 cm deep (Fig. 35.3A, B). Although the final percent emergence was not different, the buried seeds emerged more slowly.

ORIENTATION OF SEEDS. Even the orientation of seeds on the surface of the soil can affect the germination pattern. Sheldon (1974) obtained greatly different patterns for seeds of *Taraxacum officinale* and *Sonchus oleraceus* sown in a variety of orientations on the surface.

Manku (1998) took a sample of achenes of *C. vulgare* and placed them in different positions on the surface of moist sand in a greenhouse. This sample of achenes (from London, Ontario) had no innate dormancy and all viable achenes germinated within 10 days when tested at 25°C day, 10°C night, in Petri dishes. The four achene positions used were: (i) flattened side down on the surface of the substrate (surface); (ii) with the edges of the sides slightly set into the surface of the substrate (surface-side); (iii) half buried with the elaiosome (micropyle) end in the substrate (radicle-up); and (iv) half buried with the radicle end (hilum) in the substrate (radicle-down). In Fig. 35.4 the results of daily counts of germination for achenes from the centres of the capitula are shown. In each treatment 400 achenes were used.

Even with these readily germinable achenes, germination patterns were completely different for the different orientation treatments (Fig. 35.4). Virtually all achenes in the radicle-down treatment germinated within 3 days. Surface-side achenes were slightly slower to germinate but in the radicle-up and surface treatments germination patterns were attenuated and clearly intermittent. In both of these latter treatments up to 25% of the seeds remained viable and ungerminated after the 30 day test period had elapsed, whereas no viable seeds were left after 30 days of the radicle-down treatment. During

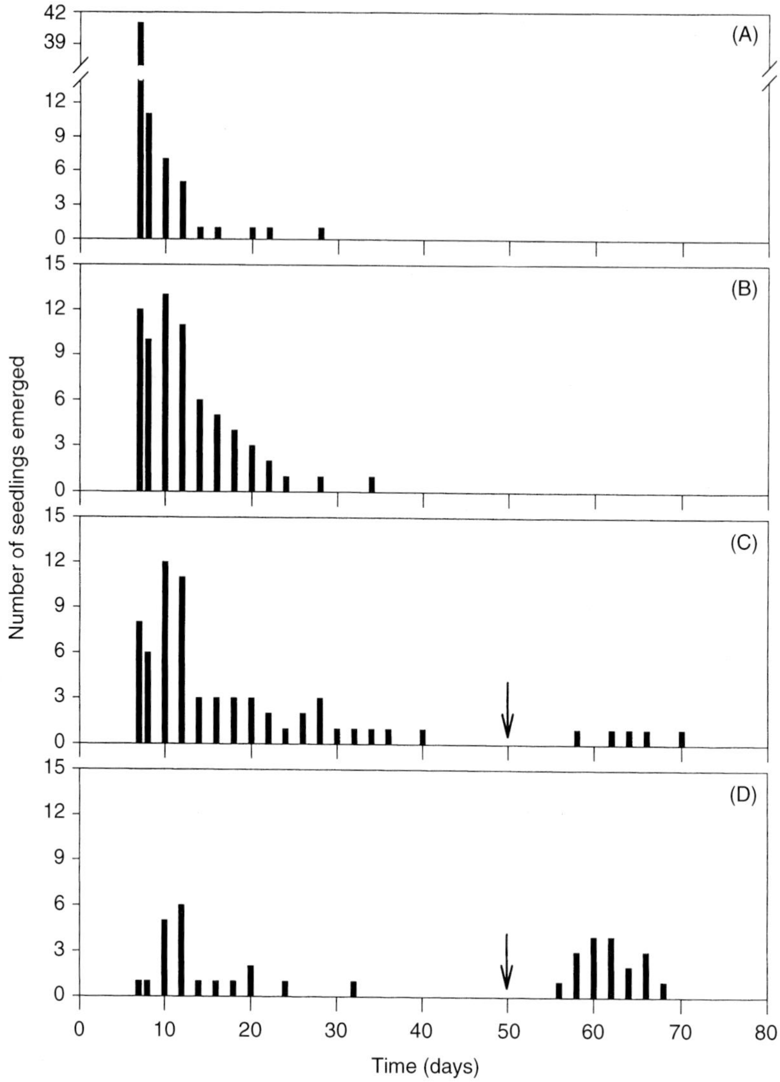

Fig. 35.3. Percent emergence of *Cirsium vulgare* in a greenhouse from achenes sown: (A) on the soil surface and (B) 1 cm below the surface. Treatments (C) and (D) were sown 1 cm below the surface then covered with two and five layers, respectively, of *Acer saccharum* leaves. The leaves were removed at times indicated by the arrows.

dispersal, an achene that is still attached to its pappus usually lands in the radicle-down position. In contrast, an achene that has become detached from its pappus can land in a variety of positions on the soil.

Manku (1998) also compared germination from flat achenes from the centre of a capitulum with that from curved achenes from the outer edges (periphery) of a capitulum. For some orientations there was little difference in the germination patterns, but for other orientations the central achenes

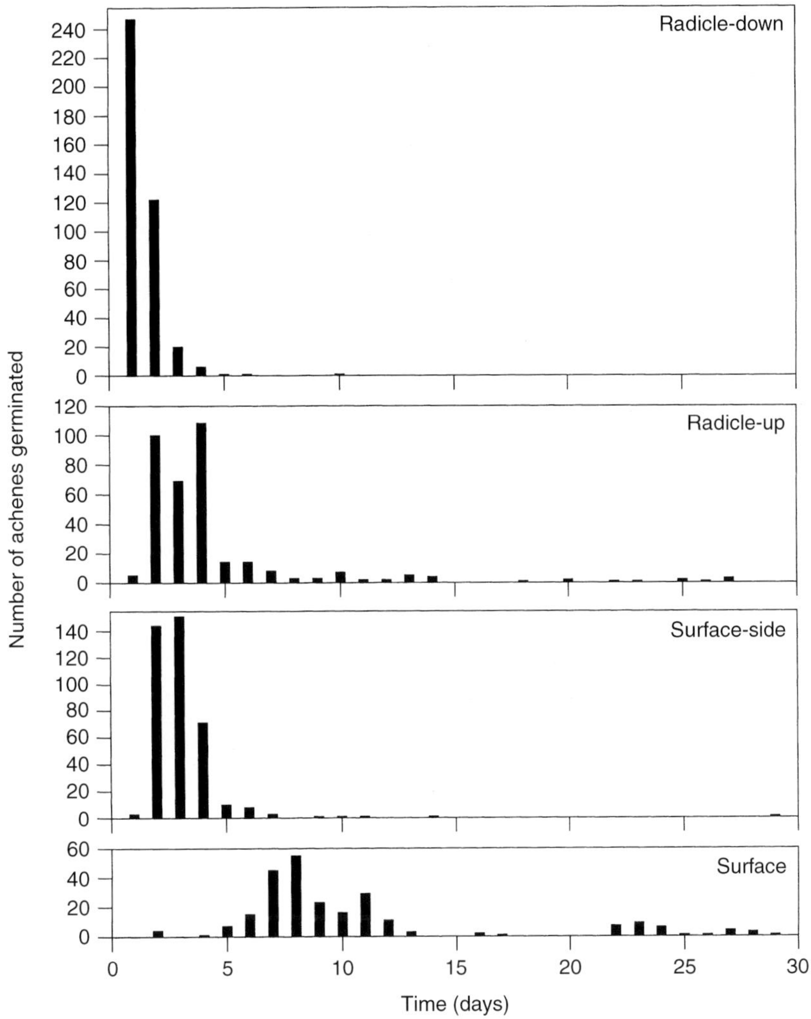

Fig. 35.4. Number of achenes of *Cirsium vulgare* that germinated per day from different orientations on moist sand.

germinated rapidly whereas peripheral achenes had intermittent germination throughout the 30 day test period (Fig. 35.5).

THE ROLE OF LITTER. After all of the factors discussed so far have operated, dead leaves falling on recently dispersed seeds can further extend the intermittent germination pattern. Downs (1998) found that leaf litter deposited on buried achenes of *C. vulgare* in the early autumn led to higher percent germination the following spring, in comparison to germination from a treatment with no litter cover. In a greenhouse experiment (Fig. 35.3), seedling emergence was delayed under two layers of sugar maple leaves and reduced (and delayed) under five layers of leaves. After a period of 10 days with no

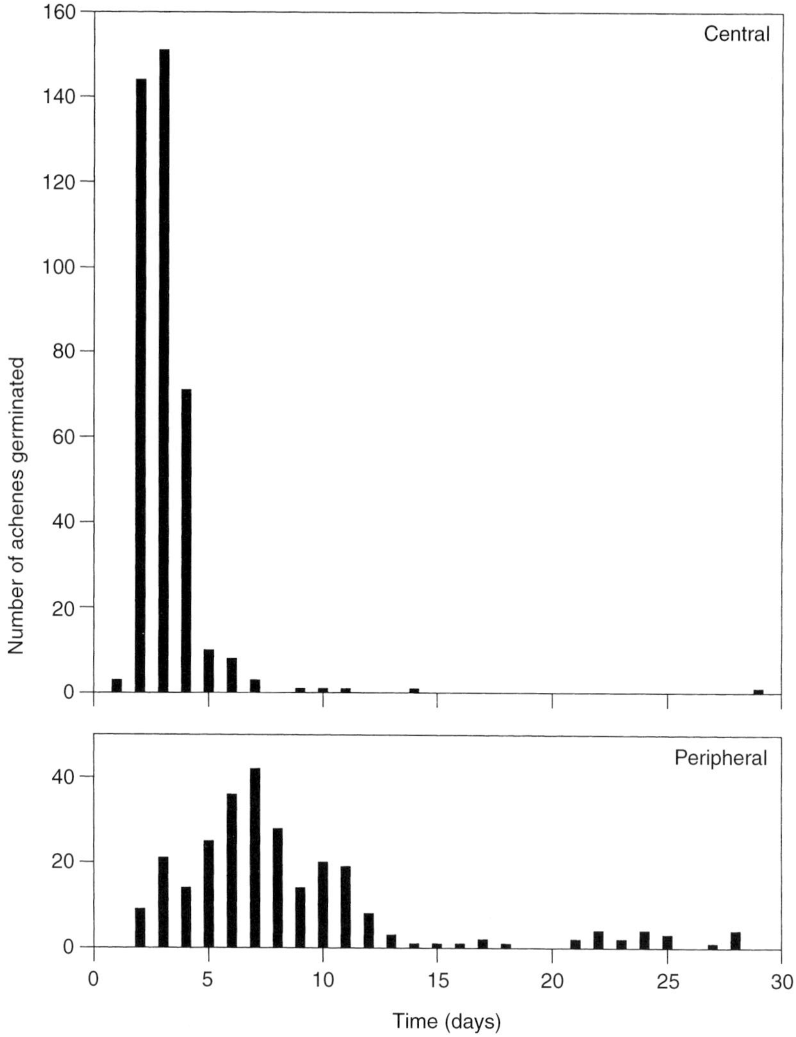

Fig. 35.5. Number of achenes of *Cirsium vulgare* from central and peripheral positions on the capitulum that germinated in the surface-side orientation on moist sand.

emergence in any treatment, the litter was removed (Fig. 53.3C, D). A flush of seedlings appeared shortly afterwards in these treatments but there was no further emergence from the treatments without litter.

Summary and Conclusions

These experiments pinpoint a number of means by which germination patterns of a variety of plant species can be made more intermittent. For those species that produce several cohorts of seeds during a growing season, the

Table 35.1. Seedling emergence from cypselas of *Onopordum acanthium* (ESW population) maturing and collected at four different dates in 1996 (total of 500 cypselas per collection).

Collection	Autumn 1996	Spring 1997	Summer 1997	Spring 1998	Summer 1998	Total to date
August 5 1996	1	169	1	86	0	257
August 20 1996	1	244	3	46	3	297
September 4 1996	69	115	9	25	0	218
September 19 1996	0	26	0	138	2	166

factors may have different effects on different cohorts. For example, on *Onopordum acanthium* plants seeds (cypselas) ripen from late July until November in southern Ontario. Cypselas collected at four different times had different emergence patterns over the next two years (Table 35.1). All cohorts contribute to the overall intermittent emergence pattern but it is apparent that there will be more cypselas from the September 19 cohort germinating in the future than from the other cohorts.

An important feature of these studies is that two species with clearly intermittent emergence patterns have been described as having little or no innate dormancy (*R. crispus* by Baskin and Baskin, 1985, and *C. vulgare* by Klinkhamer and de Jong, 1993). It is apparent that environmental factors can impose intermittent emergence patterns on seed samples that would otherwise have rapid and complete germination soon after dispersal. The greatest benefit of intermittent germination is that it provides a population with insurance against disaster (Salisbury, 1961). No single event that kills all seedlings will destroy the population because new seedlings will soon arise from the seed bank. On the other hand, a population of seeds that germinates rapidly and completely may be better able to exploit favourable conditions (Salisbury, 1961).

Acknowledgements

This work was funded through an Operating Grant from NSERC of Canada to P.B. Cavers. We thank Dr Sheila Macfie for helpful comments.

References

Baskin, C.C. and Baskin J.M. (1998) *Seeds. Ecology, Biogeography, and Evolution of Dormancy and Germination.* Academic Press, San Diego, 666 pp.

Baskin, J.M. and Baskin, C.C. (1985) Does seed dormancy play a role in the germination ecology of *Rumex crispus*? *Weed Science* 33, 340–343.

Brenchley, W.E. and Warington, K. (1930) The weed seed population of arable soil. I. Numerical estimation of viable seeds and observations on their natural dormancy. *Journal of Ecology* 18, 235–272.

Brenchley, W.E. and Warington, K. (1933) The weed seed population of arable soil. II. Influence of crop, soil and methods of cultivation upon the relative abundance of viable seeds. *Journal of Ecology* 21, 103–127.

Cavers, P.B. (1974) Germination polymorphism in *Rumex crispus*. The effects of different storage conditions on germination responses of seeds collected from individual plants. *Canadian Journal of Botany* 52, 575–583.

Cavers, P.B. and Harper, J.L. (1966) Germination polymorphism in *Rumex crispus* and *Rumex obtusifolius*. *Journal of Ecology* 54, 367–382.

Downs, M.P. (1998) Effects of leaf litter on seedling emergence of bull thistle, *Cirsium vulgare* (Savi) Ten. MSc thesis, The University of Western Ontario, London, Canada.

Govinthasamy, T. and Cavers, P.B. (1995) The effects of smut (*Ustilago destruens*) on seed production, dormancy, and viability in fall panicum (*Panicum dichotomiflorum*). *Canadian Journal of Botany* 73, 1628–1634.

Grant Lipp, A.E. and Ballard, L.A.T. (1963) Germination patterns shown by the light-sensitive seed of *Anagallis arvensis*. *Australian Journal of Biological Sciences* 16, 572–584.

Klinkhamer, P.G.L. and de Jong, T.J. (1993) Biological flora of the British Isles. *Cirsium vulgare* (Savi) Ten. *Journal of Ecology* 81, 177–191.

Manku, R. (1998) Achene variation in bull thistle, *Cirsium vulgare* (Savi) Ten. MSc thesis, The University of Western Ontario, London, Canada.

Perez-Garcia, F. (1993) Effect of the origin of the cypsela on germination of *Onopordum acanthium* L. (Asteraceae). *Seed Science and Technology* 21, 187–195.

Qaderi, M.M. (1998) Intraspecific variation in germination of Scotch thistle (*Onopordum acanthium* L.) cypselas. MSc thesis, The University of Western Ontario, London, Canada.

Salisbury, E.J. (1961) *Weeds and Aliens*. New Naturalist Series, Collins, London, 384 pp.

Salisbury, E.J. (1963) Intermittent germination of *Capsella*. *Nature* 199, 1303–1304.

Salisbury, E.J. (1965) Germination experiments with seeds of a segregate of *Plantago major* and their bearing on germination studies. *Annals of Botany, New Series* 29, 513–521.

Sheldon, J.C. (1974) The behaviour of seeds in soil. III. The influence of seed morphology and the behaviour of seedlings on the establishment of plants from surface-lying seeds. *Journal of Ecology* 62, 47–66.

Sidhu, S.S. (1971) Some aspects of the ecology of black medick (*Medicago lupulina* L.). PhD thesis, The University of Western Ontario, London, Canada.

Sidhu, S.S. and Cavers, P.B. (1977) Maturity-dormancy relationships in attached and detached seeds of *Medicago lupulina* L. (black medick). *Botanical Gazette* 138, 174–182.

Staniforth, R.J. and Cavers, P.B. (1979) Field and laboratory germination responses of achenes of *Polygonum lapathifolium, P. pensylvanicum and P. persicaria*. *Canadian Journal of Botany* 57, 877–885.

36 Seed Ecology at the Northern Limit of the Tropical Rain Forest in America

C. VÁZQUEZ-YANES,[†] A. OROZCO-SEGOVIA, M.E. SÁNCHEZ-CORONADO, M. ROJAS-ARÉCHIGA AND A.I. BATIS

Instituto de Ecología, Universidad Nacional Autónoma de México, Apartado Postal 70-275, Ciudad Universitaria, 04510 México D.F., México

Many years of seed research performed at the 'Los Tuxtlas' tropical rain forest reserve in México have produced a wealth of information unavailable for other tropical rain forests of the world. Information is presented on seed size, weight and moisture content, soil seed banks, dormancy mechanisms, seed germination behaviour, seed longevity and storage behaviour. Some research approaches that would provide valuable information for the understanding of seed ecology of the forest are suggested.

Introduction

During 1966 the National Autonomous University of Mexico obtained from the Federal Government a piece of land with a surface of 700 hectares covered with pristine tropical rain forest at the northern limit of its range in the American Continent. This forest is located in the volcanic coastal mountain range close to the Gulf of Mexico and is known by the local name as 'Los Tuxtlas' region in the Federal State of Veracruz, México. The Tropical Biology Station Los Tuxtlas was founded in this land, which is located at 18°35′ N, 96°06′ W with an altitude of around 100 to 400 m above sea level. The climate is warm-wet, with maximal and minimal temperature of 29 and 17°C respectively (average 25°C). Annual rainfall averages about 4500 mm. It may rain every month of the year but three different climate periods can be identified: a warm, summer, heavy rainy season (June–October), a cool, autumn and winter, moderate rainy season (November–February) and a spring, semi-dry season (March–May). The predominant vegetation is a highly diverse, tall evergreen tropical rain forest that has been the object of numerous studies by local and visiting researchers. The vegetation at the station contains a vascular flora of

[†]Deceased.

940 known species and shares many plant genera and species with forests located in southern Mexico, Central and South America (Ibarra-Manríquez and Sinaca-Colín, 1995, 1996a,b). The forest of Los Tuxtlas represents a precious biodiversity asset which also contains abundant valuable and potentially valuable germplasm for current and future human needs (Ibarra-Manriquez *et al.*, 1997).

The existence of the Los Tuxtlas Station has provided us with the opportunity to perform many long-term experiments on the physiological ecology of seeds of pioneer forest trees and shrubs and to some extent on seeds of tree species in the mature forest. Several reviews of the work done on seed ecophysiology and dormancy mechanisms complemented with information obtained by groups of researchers working in other tropical rain forests of the world have been published previously (Vázquez-Yanes and Orozco-Segovia, 1984, 1990, 1993, 1994, 1995).

In the following we synthesize the recent available information on tropical rain forest seed ecology from the research at Los Tuxtlas forest.

Tree Phenology and Seed Dispersal

As in other tropical rain forests, plants producing flowers and fruits can be found in the forest all the year round. During the dry season the majority of plants producing dry wind-dispersed seeds shed them. On the other hand, species with other dispersal syndromes like zoochorous and barochorous may shed the seed at any season, depending on the phenology of each taxa. However, during the year there are two relatively well defined peaks of heavy seed rain in the forest: in the early dry season and in the late rainy season (Ibarra-Manríquez and Oyama, 1992; Guevara and Laborde, 1993; Martínez-Ramos and Soto-Castro, 1993). Each species has one or more dispersal mechanisms, predominately dissemination by animals, mainly bats and birds but also other mammals, reptiles and insects. Pioneer trees such as *Cecropia obtusifolia* Bertol. and strangler figs often attract a wide variety of dispersers due to their abundant and nutritious fruit production (Vázquez-Yanes *et al.*, 1976; Estrada *et al.*, 1984).

An analysis of the relationship between seed weight and period of fruit production of 139 species of trees indicated that during the dry periods seeds tend to be smaller than those produced during the rainy period (Fig. 36.1).

Seed Size and Moisture Content

The seed production at Los Tuxtlas shows a wide range of variability in seed size, weight, and number per fruit. The smallest seed weights until now are in the range of *Myriocarpa longipes* Liebm (0.0009 g) and the heaviest is *Pouteria sapota* (Jack.) Moore & Stearn (22.5 g), representing about six orders of magnitude of difference. Pioneer plants that invade gaps and clearings

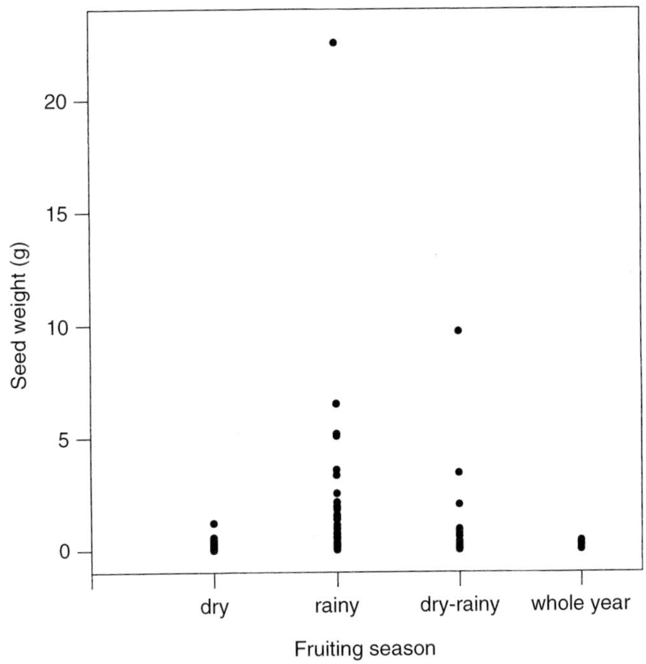

Fig. 36.1. Seed weight of 139 tree species from the tropical rain forest at Los Tuxtlas, Veracruz, México, with different fruiting periods. Data from Ibarra-Manríquez and Oyama (1992).

produce consistently smaller seeds than do mature forest trees (Ibarra-Manríquez and Oyama, 1992).

An evaluation of the moisture content (dry weight basis) of the seeds of 24 common tree species of the forest at the moment when they were being dispersed was performed. Results showed a normal distribution within a range of moisture contents of 8.28 and 72.74%. Hard-coated legumes and pioneers produce the driest seeds, while moistures above 25% come from soft-coated seeds of mature forest plants, which represent the majority of the seed species found at the site.

There is not any specific relationship between the seed size and moisture content (Fig. 36.2). Therefore, other physical characteristics apart from water content must contribute to differences in weight.

Soil Seed Bank

Pioneer studies on soil seed bank composition of tropical rain forests were made in Malaysia, Nigeria, Australia, and Tropical America (see Garwood, 1989). It is well known that in this type of plant community the most abundant

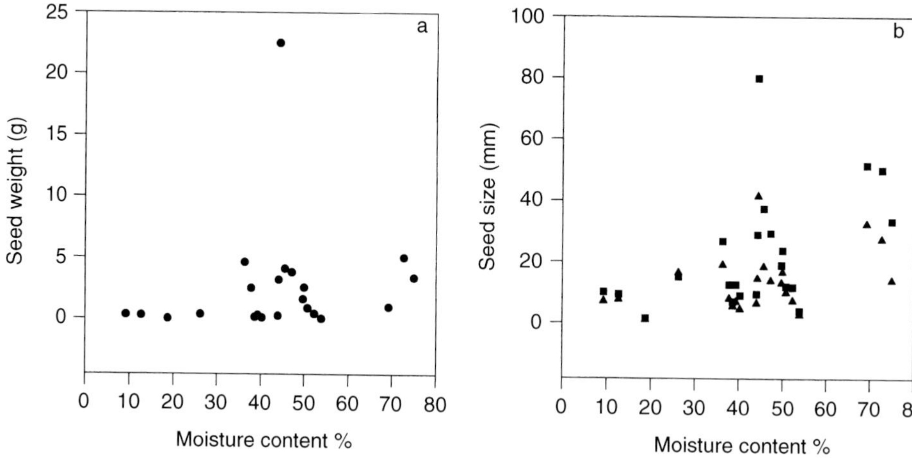

Fig. 36.2. Seed weight (a) and seed size (b) in relation to seed moisture content of 25 pioneer and forest trees from Los Tuxtlas. In (b), seed length (■) and seed width (▲). Data from Puchet (1986), Puchet and Vázquez-Yanes (1987), Vázquez-Yanes and Orozco-Segovia (1990), and Rodríguez-Hernández *et al.* (1999).

seeds found in the soil correspond to early colonizing or pioneer plants of early successional stages of the forest. Because tropical rain forests maintain almost continuously favourable conditions for seed germination, it is assumed that the presence of a persistent seed bank in these soils is associated with either: (i) abundant seed deposition facilitated by a long lasting or continuous fruiting season in some species combined with efficient animal or wind dispersal (Alvarez-Buylla and Martínez-Ramos, 1990); or (ii) a delay in germination in the soil produced by an enforced dormancy mechanism (*sensu* Murdoch and Ellis, 1992). The relative importance of seed deposition versus seed dormancy for seed bank persistence may be estimated by obtaining data on the amount and seasonality of seed deposition and on the physiology of dormancy and longevity in the soil seed bank (Vázquez-Yanes and Orozco-Segovia, 1994).

At Los Tuxtlas, a study of the seed bank of forest soils was made by transporting soil samples from two mature forest localities to an open place, a forest gap and a place beneath the canopy (Salmerón, 1984). Seeds of many species of pioneer plants germinated under the stimulus of the direct sunlight in the open situation and the forest gap but very few did so under the canopy (Fig. 36.3). The trees that predominated were *Heliocarpus appendiculatus* Tucz. and *C. obtusifolia*. The abundance of the first species was probably because the experiment was performed soon after end of the dispersal season of that abundant species. On the other hand, different species of the genus *Cecropia* are known to be among the most frequent components of the soil seed banks in several tropical American rain forests (Valio and Joly, 1979; Holthuijzen and Boerboom, 1982; Brokaw, 1998). Its dormancy mechanism is well known (Vázquez-Yanes and Smith, 1982).

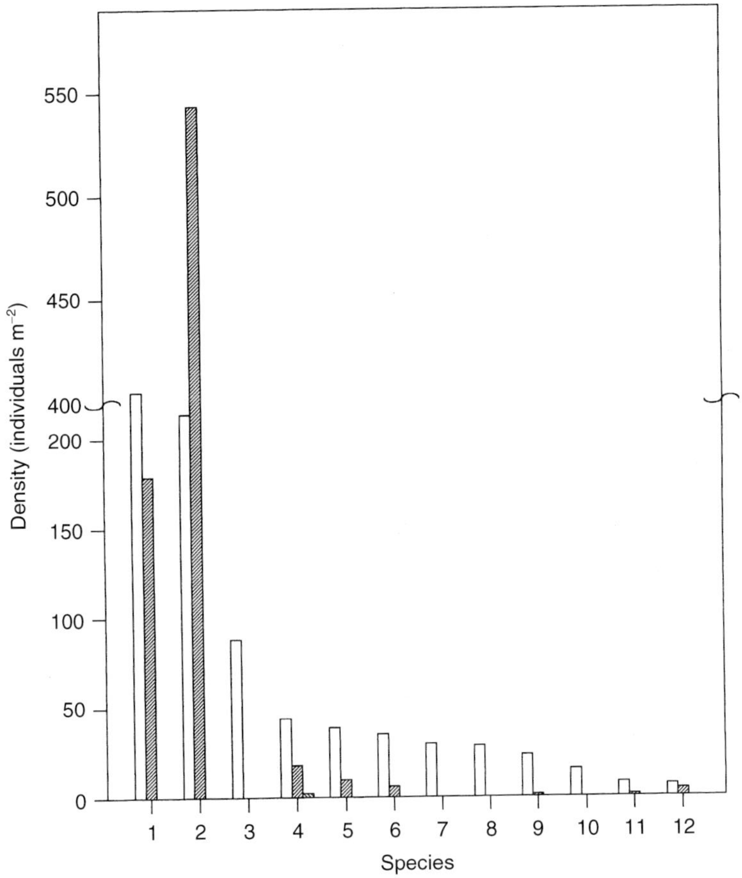

Fig. 36.3. Density of seedlings emerged in tropical rain forest soil samples when they were placed in an open place (■), a forest gap (▨), and in the forest (▧). 1, *Heliocarpus donnell-smithii* Rose; 2, *Cecropia obtusifolia* Bertol, F.L.; 3, *Clidemia* sp.; 4, *Piper hispidum* Sw.; 5, *P. auritum* Kunth; 6, *Urera caracasana* (Jacq.); 7, *Euphorbia* sp.; 8, *Phyllantus* sp.; 9, *Solanum diphyllum* L.; 10, *Acalypha* sp.; 11, *Iresine celosia* L.; 12, *Phytolacca rivinoides* Kunth & Bouché. Data from Salmerón (1984).

Dormancy Mechanisms

Enforced dormancy mechanisms of pioneer and mature forest woody species of Los Tuxtlas producing light-sensitive seeds have been the object of several publications for the species *C. obtusifolia* (Vázquez-Yanes and Smith, 1982), *Piper* spp. (Orozco-Segovia and Vázquez-Yanes, 1989), *Urera caracasana* (Jack.) Griseb (Orozco-Segovia *et al.*, 1987), *Ficus* spp. (Vázquez-Yanes *et al.*, 1996) and wild *Carica papaya* L. (Paz and Vázquez-Yanes, 1998). Many of the light-sensitive seeds may remain dormant for prolonged times at low red:far red ratios either beneath a dense plant canopy or beneath the leaf litter of the forest (Vázquez-Yanes *et al.*, 1990; Vázquez-Yanes and Orozco Segovia, 1992). However, some species may germinate at low red:far red ratios and therefore the ecological significance of photoblastism in these cases remains to be investigated further. The behaviour of photoblastic seed species and genera with respect to different light conditions is summarized (Table 36.1).

Dormancy overcome by the breaking of the seed coat or by exposure to daily alternations of temperature has also been investigated in *Ochroma lagopus* Swartz (Vázquez-Yanes, 1974) and *Heliocarpus donnell-smithii* Rose (Vázquez-Yanes and Orozco-Segovia, 1982a).

The great majority of the mature forest woody and herbaceous plants produce seeds that are light insensitive, soft-coated and capable of germinating rapidly at constant temperature. However, a significant number show a lag time between sowing and germination that often is longer than 4 weeks. This period can be considered to be the consequence of an endogenous dormancy (Baskin and Baskin, 1998). Some of them take longer times to complete germination, showing delayed or sporadic germination fluxes (Fig. 36.4). Previous partial drying may also alter the germination pattern by extending the germination period (Fig. 36.5). A similar effect of dehydration exists also among some palms (Dickie *et al.*, 1993). On the other hand, pioneer plants have a short lag time and fast germination when the environmental factors that trigger germination are present (Fig. 36.4).

Longevity in the Field

As the great majority of the mature rain forest species do not accumulate seeds in the soil seed bank, their longevity in natural conditions must be very short.

Table 36.1. Germination responses to different light conditions of photoblastic species and variability within genera.

Species	Darkness	White light	Far red	Understorey
Carica papaya	None	High	None	Very low
Cecropia obtusifolia	None	High	None	Very low
Ficus insipida	None	High	Medium	Low
F. petenensis	None	High	High	High
F. yoponensis	None	High	High	High
Piper aequale	None	High	High	High
P. amalago	High	High	–	High
P. auritum	None	High	None	Very low
P. hispidum	None	High	High	High
P. lapathifolium	High	High	–	High
P. sanctum	None	High	–	Very low
P. umbellatum	None	High	None	None
P. yzabalanum	None	High	–	None
Urera caracasana	None	High	None	Very low
Verbesina greenmanii	None	High	None	Low

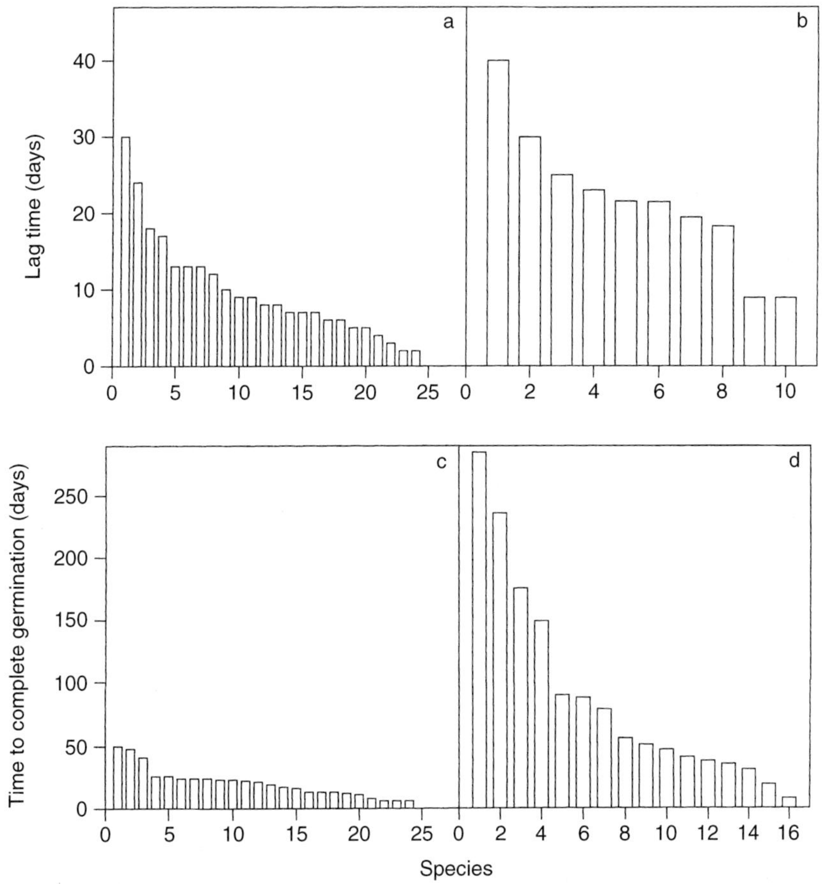

Fig. 36.4. Lag time and time to complete germination of 24 pioneer species (a, c) and more than ten forest tree species (b, d). Data from Vázquez-Yanes (1979), Vázquez-Yanes and Orozco-Segovia (1982b), Puchet (1986), Orozco-Segovia *et al.* (1987); Careaga-Olvera (1989), and Rodríguez-Hernández *et al.* (1999).

However, pioneer species often found as components of the soil seed bank appear to behave differently. Experiments conducted at the Station which have given data on seed longevity in soil storage conditions utilize the prolonged burial of nylon net bags containing seeds of pioneer plants in the forest soil. Many of the seeds survived in that condition for more than a year (Pérez-Nasser and Vázquez-Yanes, 1986).

The storage of pioneer plant seeds requiring light for germination which were kept for years fully hydrated in complete darkness at room temperature in Petri dishes demonstrated that these seeds can survive for very long times in these conditions, indicating the kind of storage longevity of moist seeds that may exist in the soil seed bank (Orozco-Segovia and Vázquez-Yanes, 1990; Vázquez-Yanes and Orozco-Segovia, 1996).

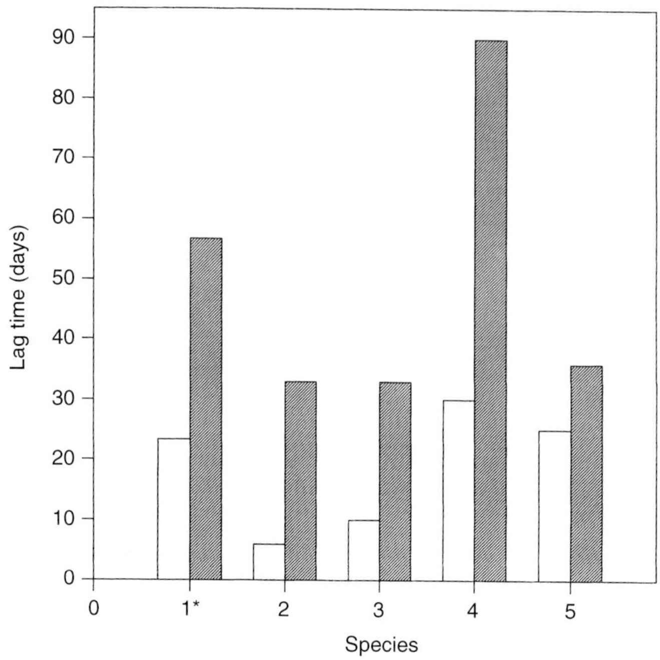

Fig. 36.5. Lag time for germination of five tree species from the rain forest at Los Tuxtlas under the following treatments: seeds not previously dehydrated and germinated at 25°C (□); seeds previously dehydrated and germinated at 25°C or germinated in conditions that could favour seed dehydration (fluctuating temperatures) (▨). Germination capacity was reduced drastically in the second treatment except in the marked species (*). 1, *Nectandra ambigens* (S.F. Blake); 2, *Dialium guianense* (Aubl.) Sandwith; 3, *Licaria velutina* (van der Werff); 4, *Rheedia edulis* (Seem.) Tr. & Pl.; 5, *Chamaedorea alternans* H. Wendl. Data from Puchet (1986) and Rodríguez-Hernández *et al.* (1999).

Longevity in Storage Conditions

Tropical American tree species seeds with orthodox storage behaviour often offered for storage and trade by seed suppliers in Latin America (Ettori *et al.*, 1988) exist at Los Tuxtlas, such as *Ceiba pentandra* (L.) Gaertn., *Cedrela* spp., *O. lagopus* and *Trema micrantha* (L.). Other pioneer trees, *Trichospermum mexicanum* (DC.) Baill. and *H. donnell-smithii*, and several legumes producing orthodox, hard-coated seeds like those of the genera *Enterolobium*, *Erythrina*, and *Dialium* are also frequent (Vázquez-Yanes, 1974, 1975, 1981; Vázquez-Yanes and Pérez-García, 1976, 1977; Vázquez-Yanes and Orozco-Segovia, 1982a). But we suspect the great majority of the less well known species forming the canopy and subcanopy of the forest should produce seed with recalcitrant storage behavior.

Several of the pioneer trees growing in the area common in early successional vegetation and large recent gaps produce seeds that can be stored for a long time in dry conditions at room temperature without losing their viability. Among them are plants of the genera *Cecropia, Piper, Urera,* and *Myriocarpa.* However, more research is needed to determine if these species produce orthodox or intermediate seed. In fact, a frequent pioneer tree in the area is the wild form of *Carica papaya,* which is known to produce seed with intermediate storage behaviour (Paz and Vázquez-Yanes, 1998).

Existing information on storage behaviour of many tropical forest trees of the world has been published and a CD-ROM database with the information was produced by the International Plant Genetic Resources Institute in Rome (Hong *et al.*, 1996). Data on the species found at Los Tuxtlas are shown (Table 36.2, pp. 384–385).

Conclusions

Of the hundreds of seed species existing at Los Tuxtlas we have partial information on relatively few of them even after many years of work at the Station. A great deal of effort is still required to understand the basis of seed behaviour, mainly among mature forest woody plants, understorey weeds, climbers and epiphytes. Our knowledge of seeds in general will be greatly increased when we develop more research in the tropical forests of the world.

References

Alvarez-Buylla, E.R. and Martínez-Ramos, M. (1990) Seed bank versus seed rain in the regeneration of a tropical pioneer tree. *Oecologia* 84, 314–325.

Baskin, C.C. and Baskin, J.M. (1998) *Seeds – Ecology, Biogeography, and Evolution of Dormancy and Germination.* Academic Press, San Diego, 665 pp.

Brokaw, N.V.L. (1998) *Cecropia schreberiana* in the Luquillo Mountains of Puerto Rico. *Botanical Review* 64, 91–120.

Careaga-Olvera, S.A. (1989) Efecto de la variación en el tamaño de las semillas sobre el desempeño de plántulas de especies tropicales. BSc thesis, Universidad Nacional Autónoma de México, México, D.F.

Dickie, J.B., Balick, M.J. and Linington, I.M. (1993) Studies on the practicality of *ex situ* preservation of palm seeds. *Principes* 37, 94–98.

Estrada, A., Coates-Estrada, R. and Vázquez-Yanes, C. (1984) Observations on fruiting and dispersers of *Cecropia obtusifolia* at Los Tuxtlas, México. *Biotropica* 16, 315–318.

Ettori, L.C., Baitello, J.B. and Figliolia, M.B. (1988) *Index Seminum.* Instituto Florestal Sao Paulo, Brazil, 15 pp.

Garwood, N.C. (1989) Tropical soil seed banks: a review. In: Leck, A.M., Parker, V.T. and Simpson, R.L. (eds) *Ecology of Soil Seed Banks.* Academic Press, San Diego, pp. 149–209.

Guevara, S. and Laborde, J. (1993) Monitoring seed dispersal at isolated standing trees in tropical pastures: consequences for local species availability. *Vegetatio* 107/108, 319–338.

Holthuijzen, A.M.A. and Boerboom, J.H.A. (1982) The *Cecropia* seed bank in the Surinam lowland rain forest. *Biotropica* 14, 62–68.

Table 36.2. Seed type (storage behaviour) of woody and herbaceous dicots found at Los Tuxtlas Biological Station in Veracruz, México (Hong *et al.*, 1996; Ibarra-Manríquez and Sinaca-Colín, 1995, 1996a,b).

Species of trees and shrubs*	Family	Seed type	Herbaceous species	Family	Seed type
Acacia cornigera (L.) Willd.	*Mimosaceae*	O	*Achyranthes aspera* L.	*Amaranthaceae*	O
Calatola laevigata Standl.	*Icacinaceae*	R?	*Aeschynomene americana* L.	*Fabaceae*	O
Calophyllum brasiliense Cambess.	*Clusiaceae*	R?	*Ageratum houstonianum* Mill.	*Asteraceae*	O
Carica papaya L.	*Caricaceae*	I	*Amaranthus spinosus* L.	*Amaranthaceae*	O
Cecropia obtusifolia Bertol.	*Cecropiaceae*	O?	*A. viridis* L.	*Amaranthaceae*	O
Cedrela odorata L.	*Meliaceae*	O	*Asclepias curassavica* L.	*Asclepiadaceae*	O?
Ceiba pentandra (L.) Gaertn.	*Bombacaceae*	O?	*Capsicum annuumm* L.	*Solanaceae*	O
Cojoba arborea (L.) Britt. *et* Rose	*Mimosaceae*	R	*Castilleja arvensis* Schltdl. *et* Cham.	*Scrophulariaceae*	O
Cordia alliodora (Ruiz *et* Pav.) Oken	*Boraginaceae*	O	*Crotalaria incana* L.	*Fabaceae*	O
Couepia polyandra (Kunth) Rose	*Chrysobalanaceae*	R	*Eclipta prostrata* (L.) L.	*Asteraceae*	O
Dialium guianense (Aubl.) Sandwith.	*Caesalpiniaceae*	O	*Erechtites hieracifolia* (L.) Raf. *ex* DC.	*Asteraceae*	O
Gliricidia sepium (Jacq.) Steud.	*Fabaceae*	O	*Galinsoga parviflora* Cav.	*Asteraceae*	O
Heliocarpus donnell-smithii Rose	*Tiliaceae*	O?	*Hyptis capitata* Jacq.	*Lamiaceae*	O
Indigofera suffruticosa Mill.	*Fabaceae*	O	*Lantana camara* L.	*Verbenaceae*	O?
Inga paterno Harms	*Mimosaceae*	R?	*Lobelia cardinalis* L.	*Campanulaceae*	O
Inga punctata Willd.	*Mimosaceae*	R	*Ludwigia leptocarpa* (Nutt.) Hara	*Onagraceae*	O
Malpighia glabra L.	*Malpighiaceae*	R?	*L. octavalvis* (Jacq.) Raven	*Onagraceae*	O
Myriocarpa longipes Liebm.*	*Urticaceae*	O	*Mimosa pudica* L.	*Mimosaceae*	O
Nectandra ambigens (S.F. Blake) C.K. Allen	*Lauraceae*	R	*Mirabilis jalapa* L.	*Nyctaginaceae*	O
			Mitracarpus villosus Cham. *et* Schltdl.	*Rubiaceae*	O

Species	Family	Storage	Species	Family	Storage
Ochroma pyramidale (Cav. *ex* Lam.) Urb.	*Bombacaceae*	O	*Physalis pubescens* L.	*Solanaceae*	O
Pimenta dioica (L.) Merr.	*Myrtaceae*	R?	*Senna obtusifolia* (L.) Irwin *et* Barneby	*Caesalpiniaceae*	O
Piper aequale Vahl	*Piperaceae*	O	*S. occidentalis* (L.) Irwin *et* Barneby	*Caesalpiniaceae*	O
Piper auritum Kunth*	*Piperaceae*	O?	*Sida rhombifolia* L.	*Malvaceae*	O
Piper hispidum Sw.*	*Piperaceae*	O	*Tridax procumbens* L.	*Asteraceae*	O
Piper umbellatum L.*	*Piperaceae*	O	*Urera lobata* L.	*Malvaceae*	O
Piper yzabalanum C. DC. *ex* Donn. Sm.*	*Piperaceae*	O?			
Platymiscium pinnatum (Jacq.) Dugand Steud.	*Fabaceae*	O?			
Poulsenia armata (Miq.) Standl.	*Moraceae*	O?			
Pouteria campechiana (Kunth) Baehni	*Sapotaceae*	R			
Sapindus saponaria L.	*Sapindaceae*	R?			
Senna multijuga (Rich.) Irwin *et* Barneby	*Caesalpiniaceae*	O			
Stemmadenia donnell-smithii (Rose) Woodson	*Apocynaceae*	O?			
Tabebuia guayacan (Seem.) Hemsl.	*Bignoniaceae*	O?			
Trema micrantha (L.) Blume	*Ulmaceae*	O?			
Trichilia havanensis Jacq.	*Meliaceae*	R?			
Urera caracasana (Jacq.) Griseb.*	*Urticaceae*	O			
Vochysia guatemalensis Donn. Sm.	*Vochysiaceae*	O?			

O, Orthodox; R, Recalcitrant; I, Intermediate.

Hong, T.D., Linnington, S. and Ellis, R.H. (1996) *Seed Storage Behaviour: a Compendium.* Handbooks for Genebanks No. 4, International Plant Genetic Resources Institute, Rome.

Ibarra-Manríquez, G. and Oyama, K. (1992) Ecological correlates of reproductive traits of Mexican rain forest trees. *American Journal of Botany* 79, 383–394.

Ibarra-Manríquez, G. and Sinaca-Colín, S. (1995) Lista florística comentada de la Estación de Biología Tropical 'Los Tuxtlas', Veracruz, México. *Revista de Biología Tropical* 43, 75–115.

Ibarra-Manríquez, G. and Sinaca-Colín, S. (1996a) Estación de Biología Tropical 'Los Tuxtlas' Veracruz, México: Lista florística comentada (Mimosaceae a Verbenaceae). *Revista de Biología Tropical* 44, 41–60.

Ibarra-Manríquez, G. and Sinaca-Colín, S. (1996b) Lista comentada de plantas de la Estación de Biología Tropical 'Los Tuxtlas', Veracruz, México: (Violaceae-Zingiberaceae). *Revista de Biología Tropical* 44, 427–447.

Ibarra-Manríquez, G., Ricker, M., Angeles, G., Sinaca-Colín, S. and Sinaca-Colín, M.A. (1997) Useful plants of the Los Tuxtlas rain forest (Veracruz, Mexico): considerations of their market potential. *Economic Botany* 51, 362–376.

Martínez-Ramos, M. and Soto-Castro, A. (1993) Seed rain and advanced regeneration in a tropical rain forest. *Vegetatio* 107/108, 299–318.

Murdoch, A.J. and Ellis, R.H. (1992) Longevity, viability and dormancy. In: Fenner, M. (ed.) *Seeds, the Ecology of Regeneration in Plant Communities.* CAB International, Wallingford, pp. 193–229.

Orozco-Segovia, A. and Vázquez-Yanes, C. (1989) Light effect on seed germination in *Piper* L. *Acta Oecologica, Oecologia Plantarum* 10, 128–141.

Orozco-Segovia, A. and Vázquez-Yanes, C. (1990) Effect of moisture on seed longevity in seeds of some tropical rain forest species. *Biotropica* 22, 215–216.

Orozco-Segovia, A., Vázquez-Yanes, C., Coates-Estrada, R and Pérez-Nasser, N. (1987) Ecophysiological characteristics of the seed of the tropical forest pioneer *Urera caracasana* (Urticaceae). *Tree Physiology* 3, 375–386.

Paz, L., and Vázquez-Yanes, C. (1998) Comparative ecophysiology of seed germination between wild and cultivated *Carica papaya. Tree Physiology* 18, 277–280.

Pérez- Nasser, N. and Vázquez-Yanes, C. (1986) Seeds from some tropical rain forest trees and shrubs of Veracruz, Mexico. *Malaysian Forester* 49, 352–356.

Puchet, C. (1986) Ecofisiología de la germinación de semillas de algunos árboles de la vegetación madura de la selva de 'Los Tuxtlas', Veracruz, México. BSc thesis, Universidad Nacional Autónoma de México, México, D.F.

Puchet, C. and Vázquez-Yanes, C. (1987) Heteromorfismo críptico en semillas recalcitrantes de tres especies arbóreas de la selva tropical húmeda de Veracruz, México. *Phytologia* 62, 100–106.

Rodríguez-Hernández, M.C., Orozco-Segovia A., Sánchez-Coronado, M.E. and Vázquez-Yanes, C. (1999) Patterns of seed germination of six mature Neotropical rain forest species in response to different degrees of dehydration. *Tree Physiology* (in press).

Salmerón Estrada, R. (1984) Germinación de semillas acumuladas en el suelo de una selva húmeda tropical 'Los Tuxtlas', Veracruz, México. BSc thesis, Universidad Nacional Autónoma de México, México, D.F.

Valio, I.F.M., and Joly, C.A. (1979) Light sensitivity of the seeds on the distribution of *Cecropia glaziovi* Snethlage (Moraceae). *Zeitschrift für Pflanzenphysiologie* 91, 371–376.

Vázquez-Yanes, C. (1974) Studies on the germination of seeds of *Ochroma lagopus* Sw. *Turrialba* 24, 176–179.

Vázquez-Yanes, C. (1975) The use of a thermogradient bar in the study of seed germination in *Ochroma lagopus* Sw. *Turrialba* 25, 328–330.

Vázquez-Yanes, C. (1979) Estudios sobre ecofisiología de la germinación en una zona cálido-húmeda de México. In: Gómez-Pompa, A., del Amo Rodríguez, S., Vázquez-Yanes, C. and Butanda-Cervera, A. (eds) *Investigaciones sobre la Regeneración de Selvas Altas en Veracruz, México.* Compañía Editorial Continental, México, D.F., pp. 279–387.

Vázquez-Yanes, C. (1981) Germinación de dos especies de tiliaceas arbóreas de la vegetación secundaria tropical: *Belotia campbellii* y *Heliocarpus donnell-smithii. Turrialba* 31, 81–83.

Vázquez-Yanes, C. and Orozco-Segovia, A. (1982a) Seed germination of a tropical rain forest pioneer tree *Heliocarpus donnell-smithii* in response to diurnal fluctuation of temperature. *Physiologia Plantarum* 56, 295–298.

Vázquez-Yanes, C. and Orozco-Segovia, A. (1982b) Longevidad, latencia y germinación de las semillas de *Verbesina grenmanii*: efecto de la calidad de la luz. *Turrialba* 31, 81–83.

Vázquez-Yanes, C. and Orozco-Segovia, A. (1984) Ecophysiology of seed germination in the tropical humid forests of the world: a review. In: Medina, E., Mooney, H.A. and Vázquez-Yanes, C. (eds) *Physiological Ecology of Plants of the Wet Tropics.* Task for Vegetation Science 12, Dr W. Junk Publishers, The Hague, The Netherlands, pp. 37–50.

Vázquez-Yanes, C. and Orozco-Segovia, A. (1990) Seed dormancy in a tropical rain forest. In Bawa, K.S. and Hadley, M. (eds) *Reproductive Biology of Tropical Plants. Man and the Biosphere Series.* UNESCO, Paris and Parthenon Publishing, Carnforth, pp. 247–259.

Vázquez-Yanes, C. and Orozco-Segovia, A. (1992) Effect of litter from a tropical rainforest on tree seed germination and establishment under controlled conditions. *Tree Physiology* 11, 391–400.

Vázquez-Yanes, C. and Orozco-Segovia, A. (1993) Patterns of seed longevity and germination in the tropical rainforest. *Annual Review of Ecology and Systematics* 24, 69–87.

Vázquez-Yanes, C. and Orozco-Segovia, A. (1994) Signals for seeds to sense and respond to gaps. In: Caldwell, M. and Pearcy, R. (eds) *Exploitation of Environmental Heterogeneity by Plants: Ecophysiological Processes Above and Below Ground.* Academic Press, New York, pp. 261–318.

Vázquez-Yanes, C. and Orozco-Segovia, A. (1995) Physiological ecology of seed dormancy and longevity in the tropical rain forest. In: Chazdon, R., Mulkey, S. and Smith, A. (eds) *Physiological Ecology of Tropical Forests.* Chapman and Hall, London, pp. 167–184.

Vázquez-Yanes, C. and Orozco-Segovia, A. (1996) Comparative longevity of seeds of five tropical rain forest woody species under different moisture conditions. *Canadian Journal of Botany* 74, 1635–1639.

Vázquez-Yanes, C. and Pérez-García, B. (1976) Notas sobre la morfología y la anatomía de la testa de las semillas de *Ochroma lagopus* Sw. *Turrialba* 26, 303–311.

Vázquez-Yanes, C. and Pérez-García, B. (1977) Notas sobre la morfología, la anatomía y la fisiología de la testa de las semillas de *Enterolobium cyclocarpum. Turrialba* 27, 427–429.

Vázquez-Yanes, C. and Smith, H. (1982) Phytochrome control of seed germination in two tropical rain forest pioneer trees: *Cecropia obtusifolia* and *Piper auritum* and its ecological significance. *New Phytologist* 92, 477–485.

Vázquez-Yanes, C., Orozco-Segovia, A., Francois, A. and Trejo, L. (1976) Some observations on seed dispersal by bats in a tropical humid region. *Biotropica* 7, 73–76.

Vázquez-Yanes, C., Orozco-Segovia, A., Rincón, E., Sánchez-Coronado, M.E., Huante, P., Barradas, V. and Toledo, J.R. (1990) Light beneath the litter in a tropical forest: effect on seed germination. *Ecology* 71, 1952–1958.

Vázquez-Yanes, C., Rojas-Aréchiga, M., Sánchez-Coronado, M.E. and Orozco-Segovia, A. (1996) Comparison of light-regulated seed germination in *Ficus* spp. and *Cecropia obtusifolia*: ecological implications. *Tree Physiology* 16, 871–875.

37 Genotypic and Phenotypic Germination Survival Strategies of Ecotypes and Annual Plant Species in the Negev Desert of Israel

Y. GUTTERMAN

Wyler Department of Drylands Agriculture, Jacob Blaustsein Institute for Desert Research and Department of Life Sciences, Ben-Gurion University of the Negev, Sede Boker Campus, 84990, Israel

Genotypic and phenotypic influences were found in local ecotypes of *Hordeum spontaneum* (wild barley) occurring in the Negev and other parts of Israel. These have varied responses to: (i) post-maturation temperatures and duration to allow afterripening; (ii) amounts of irrigation for germination in winter or summer conditions; and (iii) differences in seedling 'point of no return' and duration of drought tolerance according to ecotype and root length when dehydrated, as in some other annual plant species of the Negev Desert. Seed germination of *Plantago coronopus* of the Negev was regulated by local desert soils collected from ten natural locations along the annual average rainfall gradient, from 50 to 325 mm, which affects soil carbon and crust flora and thickness. The afterripening and optimal temperatures for *P. coronopus* seed germination are affected by post-maturation temperatures, duration, as well as temperatures and hours of light during wetting. Annual plants in the Negev were found to have different sets of survival strategies of seed dispersal and germination. The genotypic differences and phenotypic plasticity, affected by environmental factors during seed maturation include seed dispersal 'escape' or 'protection' strategies, preventing massive seed collection by seed eaters. The 'opportunistic' or 'cautious' germination strategies reduce the risk to species survival, as only a small part of the large seed bank is 'ready to germinate' after a rain event. The delayed point of no return and seedling drought tolerance may also increase species survival under unpredictable distribution of rain.

Introduction

Hordeum spontaneum *local ecotypes*

Hordeum spontaneum C. Koch (*Poaceae*) (wild barley) is distributed in the Irano-Turanian and east Mediterranean phytogeographic region, in a wide range of areas with different rainfall, elevation, temperatures and soils. Populations of *H. spontaneum* also occur in wadis in the Negev Desert (Table 37.1) (Nevo *et al.*, 1984; Zohary and Hopf, 1988; Feinbrun-Dothan and Danin, 1991).

Local environmental factors influence the appearance of local ecotypes of *H. spontaneum*, which are more adapted to a particular location. This is mainly because of the clastogamy and the atelechoric seed dispersal mechanisms whereby the spikelets containing one caryopsis are dispersed among the remains of the dry mother plant. These two strategies almost prevent gene flow between individual plants, and differences in the morphology and physiology among the local ecotypes and genetic lines are clear (Nevo *et al.*, 1984).

Local ecotypes of *H. spontaneum* from populations on north- and south-facing slopes of wadi Beit-Oren on Mount Carmel were compared with populations near Sede Boker. The germinabilities of the ecotypes from the south-facing slopes of Mount Carmel were nearer to those of the Sede Boker ecotype, despite the distance of about 200 km, compared with the germination of the ecotypes from north-facing slopes of Mount Carmel, at a distance of only about 200 m (Gutterman and Nevo, 1994). Local ecotypes with differing ecophysiological features can be found in different locations in Israel (Table 37.1) as well as in the same areas. Different ecotypes even exist along one hill slope, in one population, depending on the soil type and soil depth. These ecotypes may differ in their primary dormancy (afterripening), salt tolerance, percentages of germination under differing minimum amounts of water in the field, stages of germinating seedlings at which the point of no return is reached, as well as other physiological and morphological features. These differences were even found in *H. spontaneum* ecotypes (Table 37.1) collected

Table 37.1. Climatic data and soil type of the locations of populations from which *Hordeum spontaneum* caryopses were initially collected (adapted from Gutterman and Gozlan, 1998).

Location	Sede Boker	Tabigha	Neve Yaar	Hermon
Altitude (m)	450	0	100	1530
Mean annual temperatures (°C)	19	24	20	11
Mean annual rainfall (mm)	91	436	600	1600
Mean number of rainy days	15	45	50	70
Soil type	Loess	Terra rossa basalt	Dark rendzina	Terra rossa
Water holding capacity at 105°C (%)	2.4	–	7.3	5.1
Salt content (%)	0.180	–	0.076	0.064
Habitat	Wadi	Slope	Slope	Slope

and grown for three years and three generations in the same plot at Sede Boker with additional irrigation. The caryopses of the different genetic lines, which had originated from the different locations in Israel, were collected from the third generation grown at Sede Boker. They were tested for their dry afterripening (Fig. 37.1) and germination in salt solutions (Fig. 37.2), as well as in field conditions at Sede Boker in winter (Fig. 37.3A) and summer (Fig. 37.3B). Germinating seedlings were also tested for their 'point of no return', which is when rehydrated seedlings are no longer able to recover and develop into normal plants after the period of drought (Fig. 37.4) (Gutterman and Gozlan, 1998; Gozlan and Gutterman, 1999).

Afterripening Differences among *Hordeum spontaneum* Ecotypes

Differences in temperature requirements for afterripening were found in caryopses of *Hordeum spontaneum* (wild barley) ecotypes originating from different areas of Israel where there is also no rainfall during the hot summer. However, genetic differences in afterripening were found according to the elevation, location, soil type and soil depth from which the ecotypes were collected or from which these ecotypes had originated. Dormancy was reduced in the ecotype from Sede Boker after 70 days dry storage at 35 to 40°C but in the ecotype from Mount Hermon after 70 days of dry storage at 5 to 20°C (Table 37.1; Fig. 37.1) (Gutterman and Gozlan, 1998; Gozlan and Gutterman, 1999).

Afterripening has also been found in some other annual plant species occurring in the Negev. The caryopses of *Schismus arabicus* Nees (*Poaceae*) also require about 70 days at 40°C before they are capable of germinating. Such a mechanism prevents germination of the seeds after a late rain following

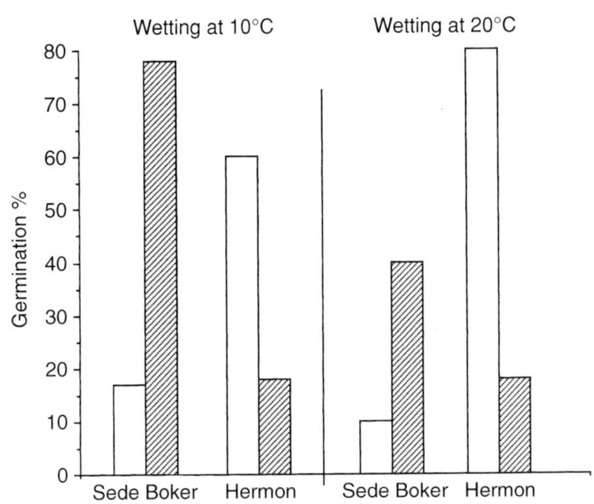

Fig. 37.1. Germination of Sede Boker and Mount Hermon *Hordeum spontaneum* ecotypes after 70 days of dry storage at 35–40°C (▨) or 5–20°C (□) and wetting at 20°C or 10°C (Gutterman and Gozlan, 1998).

their maturation at the beginning of the long, hot and dry summer (Gutterman, 1996a). Afterripening occurs in other *Poaceae* found in the Negev, including *Stipa capensis* Thunb., *Ammachloa palaestina* Boiss. and *Cutandia memphitica* (Sprengel) K. Richter, which are now under investigation, as well as in *Plantago coronopus* L. subsp. *commutata* (Guss.) Pilger (*Plantaginaceae*) (Evenari *et al.*, 1982).

Germination in Salt (NaCl) Concentrations

Differences were found in the germinability of *H. spontaneum* ecotypes from Tabigha basalt and Tabigha terra rosa after 15 days of germination in various salt solutions at 20°C (Fig. 37.2) (Gozlan and Gutterman, 1999).

Amounts of Water and Germination of *H. spontaneum* Ecotypes

Amounts of rain in mild winter temperatures

The different local ecotypes of *H. spontaneum* originating from Sede Boker, Neve Yaar deep or shallow soils, and Tabigha basalt soil, differ in their germination after differing minimum amounts of water (10 or 25 mm) by additional irrigation in winter. After only 10 mm of water, when the soil water content was about 3%, 40% of the caryopses originating from Sede Boker germinated but those from Tabigha basalt did not germinate. After 25 mm of irrigation, the ecotypes originating from the desert at Sede Boker, or the Mediterranean shallow soil at Neve Yaar germinated to above 90% but those from Neve Yaar deep soil and Tabigha basalt germinated to only 60% (Fig. 37.3A) (Gutterman and Gozlan, 1998).

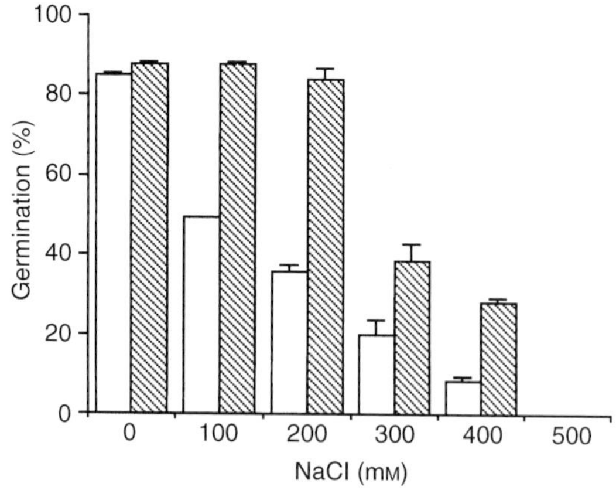

Fig. 37.2. Germination (% ± SE) of *Hordeum spontaneum* ecotypes from Tabigha basalt (□) and Tabigha terra rosa (▧) types in NaCl concentrations 15 days after wetting at 20°C (adapted from Gozlan and Gutterman, 1999).

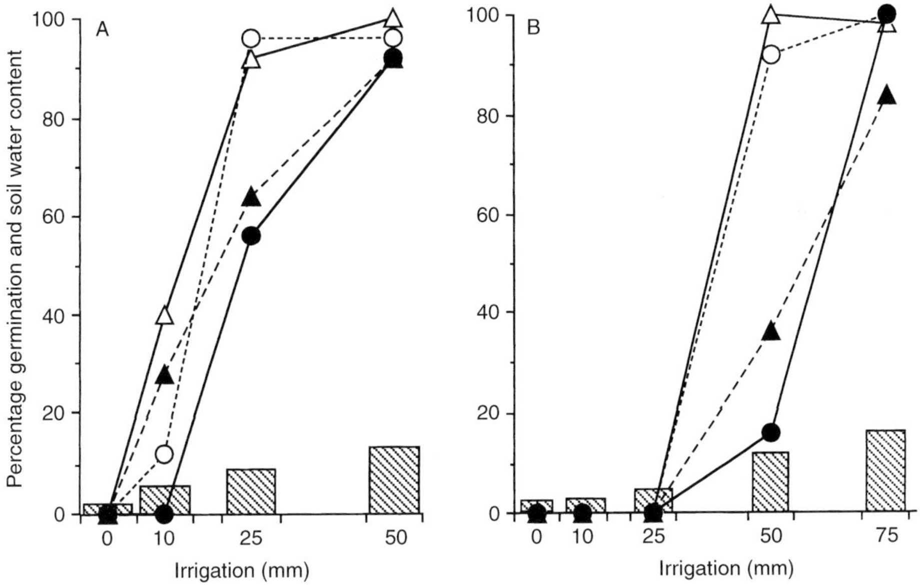

Fig. 37.3. Percentage of germination (3 days after wetting) and soil water content (▧) in field experiment of four *Hordeum spontaneum* ecotypes (△, Sede Boker; ○, Neve Yaar shallow soil; ●, Neve Yaar deep soil; ▲, Tabigha basalt) in (A) winter (11 March, 1997) and (B) summer (25 May, 1997) conditions with different amounts of irrigation (adapted from Gutterman and Gozlan, 1998).

The central part of the Negev receives unpredictable, small amounts of rain in winter averaging about 100 mm per year, and the summers are long, hot and dry (Evenari *et al.*, 1982; Gutterman, 1993). Seeds have very few chances during the season with sufficient rain to germinate, and most years this occurs between one and three times per year (Gutterman, 1993).

The lower the daily evaporation rates from a free water space the lower the amount of rain required to provoke germination of Negev annuals. The minimum requirement is about 10 to 15 mm of rain (Loria and Noy-Meir, 1979/80; Gutterman, 1993).

Germination in natural summer conditions and amounts of irrigation

Hordeum spontaneum seedlings may emerge during summer, when soil surface temperatures are high (about 53°C), daily RH is very low (about 27%), and evaporation rates from free water space is very high (about 300 mm per month), only after receiving high amounts of artificial irrigation equivalent to 50 to 75 mm of rain (Gutterman and Gozlan, 1998). Also in the summer experiment, the caryopses from the desert at Sede Boker or from Mediterranean shallow soil at Neve Yaar germinated to higher percentages (90 and 95% respectively) after irrigation of 50 mm, in comparison to Neve Yaar deep soil (16%) or Tabigha basalt soil (38%) (Fig. 37.3B).

Annual plant species occurring in areas receiving only winter rains may also germinate in summer but after much higher amounts of rain, even as much as ten times the requirement in winter. This has been found in *Schismus arabicus* (Gutterman and Evenari, 1994).

Drought Tolerance of *H. spontaneum* Seedlings

Among some plant species, the young germinating seedlings may have the ability to survive, even if there is a period of a few weeks of extreme drought between the rain that prompted germination and the following rain or rains. Differences in this ability were found in seedlings of *H. spontaneum* ecotypes with roots of 1, 5, 20 or even 40–50 mm long and a developed coleoptile (Fig. 37.4). When rehydrated, the seedlings produce adventitious roots and develop into normal plants even when the seedlings were in such dry conditions that their weight had reduced to below the weight of their caryopses

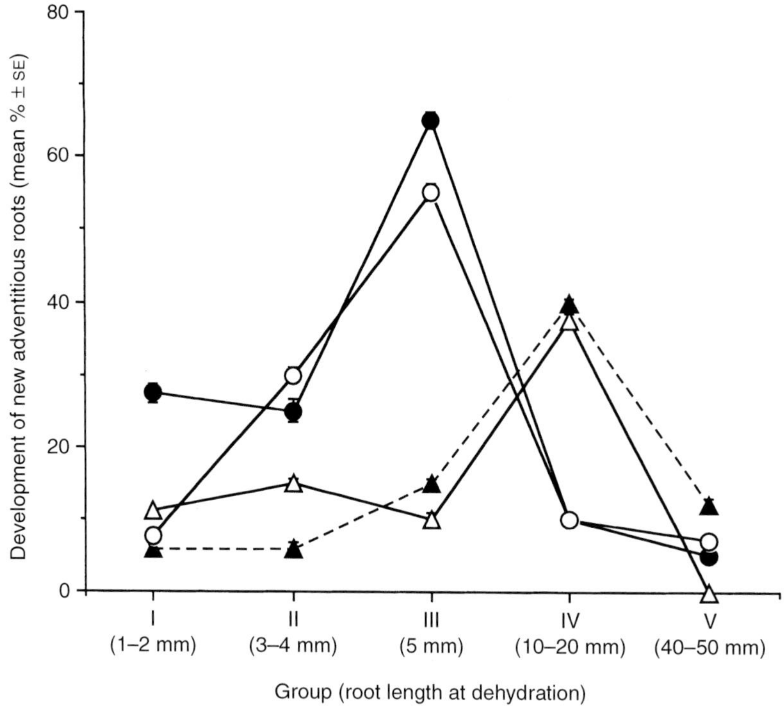

Fig. 37.4. Development of new adventitious roots on rehydrated, previously-dried germinated seeds of four *Hordeum spontaneum* ecotypes (△, Sede Boker; ▲, Tabigha basalt; ●, Neve Yaar deep soil; ○, Hermon). Seeds were germinated then given extreme dry storage for one week. They were grouped according to the length of the longest adventitious root produced on germination, rehydrated, and the average number of new seeds with adventitious roots produced by the seedlings was determined (adapted from Gutterman and Gozlan, 1998).

before the first wetting (Gutterman and Gozlan, 1998). This phenomenon has also been found in *Anastatica hierochuntica* L. (*Brassicaceae*) (Friedman *et al.*, 1981). The delay of the 'point of no return' may increase the chance of seedling survival when there is a period of drought shortly after the seedlings have emerged.

Negev Soils Affecting Seed Germination in *Plantago coronopus*

Seeds of *Plantago coronopus* that were germinated in optimal conditions of soil water content, temperature and light on soils collected from their natural habitats germinated to 30% or less in comparison to 60% in the filter paper control. In addition to many other germination-regulating mechanisms of seeds of annual plant species, the soil as a substrate also may regulate the germination of seeds that are adhered to the soil surface. *P. coronopus* seeds that were germinated on loess soils collected from different locations along the average rainfall gradient in the Negev Desert, from 50 in the centre to 325 mm rain per year in the northern part, differed in their germination percentages. The higher the average rainfall, the thicker the soil crust, from 1 to 15 mm, and the stronger the inhibition of germination. Along this rainfall gradient the amount of carbon in the soil samples was higher in samples collected from areas receiving the higher rainfall. The relationship between carbon and germination is not yet understood but there is a clear correlation between the amounts of carbon in the soil crust and the inhibitory effect along the rainfall gradient (Shem-Tov *et al.*, 1999) (Fig. 37.5).

Heteroblasty Affected by Seed Maturation Conditions

Under unpredictable amounts and distribution of rain per season in deserts, annual plants have developed different ecophysiological strategies of seed germination according to the amount of rain, temperatures and relative humidity during the period of wetting.

Only a small portion of the seed bank is 'ready to germinate' after a particular rainfall. This heteroblasty may vary in different plant species according to the environmental conditions during seed maturation as well as the position of the seeds in the capsule, on the inflorescence or on the plant.

Daylength and heteroblasty

The daylength during seed maturation was found in some of the annuals to have an influence on the plasticity of seed germination. Flowers that appear on different dates on one mother plant, or even along one branch, produce seeds under different daylengths with different germinability (Gutterman, 1993, 1994a, 1996b, 1997a).

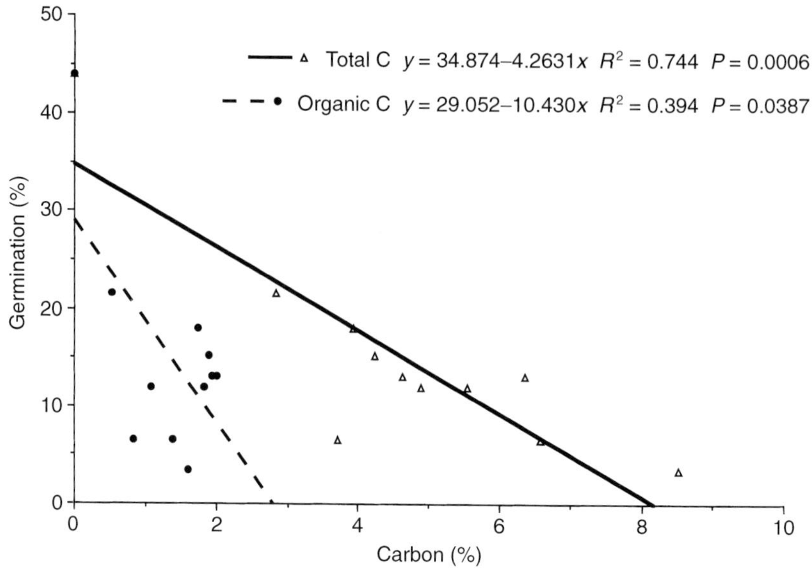

Fig. 37.5. Linear regression between the percentage of germination and carbon content in the soil. *Plantago coronopus* seeds were placed in Petri dishes on soil crusts collected in the Negev along the rainfall gradient from 50 to 325 mm, or on filter paper as a control. Germination was checked 3 days after wetting with distilled water at 25°C in light. The total carbon content (in %) of the soil samples includes organic carbon and inorganic carbonates (Shem-Tov *et al.*, 1999).

Position and age affecting seed weight, colour and germinability

In *Spergularia diandra* (Guss.) Heldr. et Sart. (*Caryophyllaceae*), the seeds that mature in first flowers are black, the heaviest and have the highest germinability. Seeds that mature in later flowers are brown and those from the last flowers on senescent plants are yellow. The yellow seeds are the lightest and have the highest dormancy percentage (Gutterman, 1994a,b). Many other position effects on seed germinability occur in other Negev annuals (Evenari *et al.*, 1982; Gutterman, 1993).

Seed Dispersal Strategies Avoiding Massive Seed Consumption in Plants Occurring in the Negev

Different annual plant species occurring in the Negev Desert have various dispersal strategies which may reduce massive seed consumption by the many seed eaters occurring in the desert. Two main, and extreme, strategies will be discussed: (i) the escape strategy; and (ii) the protection strategy.

Escape strategies of dispersal of tiny seeds by wind and opportunistic strategies of seed germination

Plant species such as *Schismus arabicus* and *Spergularia diandra* disperse by wind very large numbers of their tiny, dust-like seeds at the beginning of the summer (Loria and Noy-Meir, 1979/80; Gutterman, 1993, 1994a,b, 1996c, 1997a). Seeds settle in cracks in the soil crust or into small depressions, etc., where they are covered by dust shortly after dispersal; in this way, they may escape from massive consumption.

This dispersal strategy, together with their opportunistic strategy of seed germination, enables these plant species to appear in very great numbers even in years with less than the average annual rainfall (Gutterman, 1994a,b; 1996b, 1997a,c).

Synaptospermic seed protection and cautious germination strategies

In synaptospermic plants (Zohary, 1962) the seeds remain in groups, protected from the time of maturation in the lignified inflorescences on the dead mother plant for many years. This may prevent or reduce collection by seed eaters. In *Emex spinosa* (L.) Campd. (*Polygonaceae*) and *Gymnarrhena micrantha* Desf. (*Asteraceae*), the achenes or propagules remain in the subterranean lignified structures of the dead mother plant from the time of maturation until they germinate *in situ* (Koller and Roth, 1964; Evenari *et al.*, 1982; Gutterman, 1993).

Blepharis spp. (*Acanthaceae*) has a more efficient dispersal strategy of seeds by rain. All the seeds are protected on the dry mother plants and the few seeds that are dispersed from their hydrochastic inflorescences after a relatively heavy rainfall germinate within one to a few hours after dispersal. They germinate in a wide range of temperatures, in light or dark (Gutterman, 1993), during rain, when seed eaters are not active (Whitford, 1978; Hord, 1986). This plant produces relatively large, well protected seeds in relatively small numbers (Gutterman, 1993, 1994b). *Anastatica hierochuntica* seeds, as in *Blepharis* spp., have a high chance to germinate and develop into plants because of the short time (5 to 6 h) during a rainfall from seed dispersal to germination (Gutterman, 1993). In *Carrichtera annua* (L.) DC. (*Brassicaceae*) and *Plantago coronopus*, the time between seed dispersal by rain and seed germination is longer – a few days. The adherence of these mucilaginous seeds to the soil surface by the mucilage may reduce seed collection by ants (Gutterman and Shem-Tov, 1997).

In conclusion, the genotypic differences increase the fitness of the plant species to its habitat. The phenotypic plasticity reduces the risk to survival under unpredictable amounts of rain, whereby only some of the seeds germinate after one rainfall.

References

Evenari, M., Shanan, L. and Tadmor, N. (1982) *The Negev. The Challenge of a Desert*, 2nd edn. Harvard University Press, Cambridge, Massachusetts, 437 pp.

Feinbrun-Dothan, N. and Danin, A. (1991) *Analytical Flora of Eretz-Israel*. Cana, Jerusalem, 1040 pp.

Friedman, J., Stein, Z. and Rushkin. E. (1981) Drought tolerance of germinating seeds and young seedlings of *Anastatica hieronchuntica* L. *Oecologia* 51, 400–403.

Gozlan, S. and Gutterman Y. (1999) Dry storage temperatures, duration, and salt concentrations, affect germination of local and edaphic geno-ecotypes of *Hordeum spontaneum* (Poaceae) in Israel. *Biological Journal of the Linnean Society* 67, 163–180.

Gutterman, Y. (1993) *Seed Germination in Desert Plants. Adaptations of Desert Organisms*. Springer, Berlin, Heidelberg, New York, 253 pp.

Gutterman, Y. (1994a) In Memoriam – Michael Evenari and his desert. Seed dispersal and germination strategies of *Spergularia diandra* compared with some other desert annual plants inhabiting the Negev Desert of Israel. *Israel Journal of Plant Sciences* 42, 261–274.

Gutterman, Y. (1994b) Strategies of seed dispersal and germination in plants inhabiting deserts. *Botanical Review* 60, 373–425.

Gutterman, Y. (1996a) Temperatures during storage, light and wetting affecting caryopses germinability of *Schismus arabicus*, a common desert annual grass. *Journal of Arid Environments* 33, 73–85.

Gutterman, Y. (1996b) Effect of daylength during plant development and caryopsis maturation on flowering and germination, in addition to temperature during dry storage and light during wetting, of *Schismus arabicus* inhabiting the Negev Desert. *Journal of Arid Environments* 33, 439–448.

Gutterman, Y. (1997a) Effect of daylength on flowering and seed morphology of *Spergularia diandra* occurring in the Negev Desert, Israel. *Journal of Arid Environments* 37, 611–622.

Gutterman, Y. (1997b) Genotypic, phenotypic and opportunistic germination strategies of some common desert annuals compared with plants with other seed dispersal and germination strategies. In: Ellis, R.H., Black, M., Murdoch, A.J. and Hong, T.O. (eds) *Basic and Applied Aspects of Seed Biology*. Proceedings of 5th Workshop on Seeds, Reading, UK, 1995. Kluwer Academic Publishers, Dordrecht, pp. 611–622.

Gutterman, Y. and Evenari, M. (1994) The influences of amounts and distribution of irrigation during the hot and dry season on emergence and survival of some desert winter annual plants in the Negev Desert. *Israel Journal of Plant Sciences* 42, 1–14.

Gutterman, Y. and Gozlan, S. (1998) Amounts of winter or summer rain triggering germination and 'the point of no return' of seedling desiccation tolerance, of *Hordeum spontaneum* local ecotypes in Israel. *Plant and Soil* 204, 223–234.

Gutterman, Y. and Nevo, E. (1994) Temperatures and ecological-genetic differentiation affecting the germination of *Hordeum spontaneum* caryopses harvested from three populations: the Negev Desert and opposing slopes on Mediterranean Mount Carmel. *Israel Journal of Plant Sciences* 42, 183–195.

Gutterman, Y. and Shem-Tov, S. (1997) The efficiency of the strategy of mucilaginous seeds of some common annuals of the Negev adhering to the soil crust to delay collection by ants. *Israel Journal of Plant Sciences* 45, 317–327.

Hord, O. (1986) Collection of Food and Separation of Resources by Three Species of Harvesting Ants in Avdat. MSc thesis. Hebrew University of Jerusalem (in Hebrew with English summary).

Koller, D. and Roth, N. (1964) Studies on the ecological and physiological significance of amphicary in *Gymnarrhena micrantha* (Compositae). *American Journal of Botany* 51, 26–35.

Loria, M. and Noy-Meir, I. (1979/80) Dynamics of some annual populations in a desert loess plain. *Israel Journal of Botany* 28, 211–225.

Nevo, E., Beiles, A., Gutterman, Y., Storch, N. and Kaplan, D. (1984) Genetic resources of wild cereals in Israel and the vicinity: II Phenotypic variation within and between populations of wild barley, *Hordeum spontaneum*. *Euphytica* 33, 737–756.

Shem-Tov, S., Zaady, E., Groffman, P.M. and Gutterman, Y. (1999) Soil carbon content along a rainfall gradient and inhibition of germination: a potential mechanism for regulating distribution of *Plantago coronopus*. *Soil Biology and Biochemistry* 31, 1209–1217.

Whitford, W.G. (1978) Structure and seasonal activity of Chihuahuan desert ant communities. *Insectes Sociaux* 25, 79–88.

Zohary, D. and Hopf. M. (1988) *Domestication of Plants in the Old World*. Clarendon Press, Oxford, 249 pp.

Zohary, M. (1962) *Plant Life of Palestine*. Ronald, New York, 262 pp.

38 Hydrothermal Time as a Tool in Comparative Germination Studies

P.S. ALLEN[1], S.E. MEYER[2] AND M.A. KHAN[3]

[1]*Department of Agronomy and Horticulture, Brigham Young University, 275 Widstoe Building, Provo, UT 84602, USA;* [2]*USDA Forest Service Shrub Sciences Laboratory, Provo, Utah, USA;* [3]*Department of Botany, University of Karachi, Karachi, Pakistan*

Hydrothermal time (HTT) analysis is an effective method for quantifying germination response to temperature and water potential. Time course curves under any temperature and water potential conditions can be generated from knowledge of the HTT parameters mean base water potential, standard deviation of base water potentials, base temperature, and hydrothermal time requirement. Here we use these parameters as indices for making comparisons among 24 species including shrubs, grasses and perennial herbs from the deserts and semi-deserts of Asia and North America. Halophytes (salt-tolerant species) are characterized by low base water potential values but high hydrothermal time constants. Psammophytes (species that inhabit high-sand soils) are just the opposite, and bodenvags (generalist species with no special soil requirements) display a wide range in all parameter values. Variation in the distribution of base water potentials strongly influences uniformity of germination. The results illustrate that germination rate in water or at reduced water potentials is closely associated with HTT parameters. These findings also have important ecological relevance, in that they help explain differences in germination patterns associated with contrasting habitats.

Introduction

Temperature and water potential conditions strongly influence seed germination. Hydrothermal time (equation 38.1) describes progress toward seed germination under various combinations of incubation water potential and temperature:

$$\theta_{HT} = (\Psi - \Psi_b(g))(T - T_b)t_g \qquad (38.1)$$

where θ_{HT} is the hydrothermal time required for germination (e.g. MPa-degree-days), Ψ and T are the water potential and temperature of the incubation

©CAB *International* 2000. *Seed Biology: Advances and Applications* (eds M. Black, K.J. Bradford and J. Vázquez-Ramos)

medium, T_b is the base temperature at or below which germination will not occur, $\Psi_b(g)$ is the base water potential at or below which germination of the g fraction will not occur, and t_g is the time requirement for germination of fraction g.

Rearranging equation 1 and using a value of $g = 50$ illustrates that mean germination rate ($1/t_{50}$) is directly proportional to the differences between ($\Psi - \Psi_b(50)$) and ($T - T_b$), but is inversely proportional to θ_{HT}:

$$\frac{1}{t_{50}} = \frac{(\Psi - \Psi_b(50))(T - T_b)}{\theta_{HT}} \qquad (38.2)$$

Increases in θ_{HT}, T_b, and $\Psi_b(50)$ all reduce germination rate, while increases in Ψ (i.e. a less negative number) and T increase germination rate. Distribution of Ψ_b (g) is assumed to be approximately normal, with a mean base water potential ($\Psi_b(50)$) below which half of the seeds will not germinate and a standard deviation $\sigma_{\Psi b}$. This normal distribution of $\Psi_b(g)$ generates the characteristic inverse normal distribution of $t(g)$'s observed in germination time courses.

Hydrothermal time was first used to describe the germination rate for sugarbeet (Gummerson, 1986), but has since been applied to tomato (Dahal and Bradford, 1994; Cheng and Bradford, 1999). We have recently used hydrothermal time as a tool in predicting dormancy loss that occurs during dry afterripening, under both laboratory and field conditions (Christensen *et al.*, 1996; Allen and Meyer, 1998; Bauer *et al.*, 1998). Changes in germination time course curves as seeds afterripen are associated with a progressive decrease in $\Psi_b(50)$.

While hydrothermal time and the related hydrotime modelling approaches have received increased attention during the past decade (Bradford 1990, 1995; Finch-Savage *et al.*, 1998), relatively few species have been subjected to HTT analysis. In order to examine the suggestion that hydrothermal time theory has ecological significance as well as an underlying physiological basis, and is not just another way of empirically fitting germination time course data, we here present results of hydrothermal time analysis for 24 different species representing a range of growth habits and ecological habitats. This study had three specific objectives: (i) to evaluate hydrothermal time application across multiple species; (ii) to use hydrothermal time parameters to make comparisons among species; and (iii) to determine whether species with similar germination rates in water have similar hydrothermal time parameters.

Methods

The 24 species included in this study were grouped into three ecological classifications (Table 38.1). Halophytes are salt-tolerant species and were included because they germinate at low water potential (Ψ). Psammophytes are species that inhabit high-sand soils. Such soils are characterized by poor water retention and seeds of psammophytes frequently encounter rapid soil drying (i.e. widely fluctuating Ψ). Bodenvags represent generalist species with no specific soil type requirements.

Table 38.1. Descriptive information for 24 species included in this study.

Species	Family	Habit	Habitat	Distribution
Halophytes				
Arthrocnemum indicum	Chenopod	Succulent	Salt marsh	Old World tropics
Suaeda fruticosa	Chenopod	Succulent	Salt desert	Old World
Salicornia utahensis	Chenopod	Succulent	Salt desert	Temperate western US
Triglochin maritima	Arrowgrass	Perennial	Salt marsh	Temperate western US
Atriplex triangularis	Chenopod	Perennial	Salt marsh	North America
Polygonum aviculare	Buckwheat	Annual	Wide range	Cosmopolitan
Psammophytes				
Asclepias tuberosa	Milkweed	Perennial	Sandhills	Temperate western US
Artemisia cana	Aster	Shrub	Sandhills	Temperate western US
Eriogonum alatum	Buckwheat	Perennial	Sand desert	Temperate western US
Heterotheca villosa	Aster	Perennial	Sand desert	Temperate western US
Arabis pulchra	Crucifer	Perennial	Sand desert	Southwestern US
Stipa arida	Grass	Perennial	Sand desert	Southwestern US
Hymenoxys scaposus	Aster	Perennial	Sand desert	Western US
Bodenvags				
Kochia prostrata	Chenopod	Shrub	Cold desert	Central Asia
Poa secunda	Grass	Perennial	Cold desert	Western US
Carrichtera annua	Crucifer	Annual	Warm desert	Mediterranean
Elymus elymoides	Grass	Perennial	Cold desert	Western US
Asclepias asperula	Milkweed	Perennial	Cold desert	Western US
Ceratoides lanata	Chenopod	Shrub	Cold desert	Western US
Bromus tectorum	Grass	Annual	Cold desert	Cosmopolitan
Brachypodium distachyon	Grass	Annual	Warm desert	Mediterranean
Bromus fasciculatus	Grass	Annual	Warm desert	Mediterranean
Stipa capensis	Grass	Perennial	Warm desert	Mediterranean
Ephedra nevadensis	Ephedra	Shrub	Cold desert	Western US

All seeds included were non-dormant at the time studies were conducted. Seeds of several species required dry afterripening to relieve primary dormancy. Germination data for halophytes are from previously published work (Khan and Ungar, 1984, 1997, 2000a,b; Khan and Weber, 1986; Khan and Gul, 1998). For these species, germination at reduced water potentials was achieved by imbibing seeds in liquid contact with water or solutions of sodium chloride as described in the original publications. Germination experiments for all other species involved imbibing seeds on germination blotters saturated with water or polyethylene glycol solutions at a range of Ψ values (Christensen *et al.*, 1996; Bauer *et al.*, 1998).

Germination time course data for each species were analysed by repeated probit regression to calculate values for θ_{HT}, $\Psi_b(50)$, T_b and $\sigma_{\Psi b}(50)$ (the standard deviation of mean base water potentials, which is important in applying hydrothermal time to seed populations). This approach is described in detail by Christensen *et al.* (1996) and Bauer *et al.* (1998), based on earlier work by Ellis *et al.* (1986), Gummerson (1986) and Bradford (1990, 1995). The only modification from our earlier procedure was that T_b was also allowed to vary until the best model fit (highest R^2) was obtained, as outlined in Dahal and Bradford (1994).

Results and Discussion

Species varied considerably in T_b (Table 38.2), reflecting the range of ecological habitats represented in this study (Table 38.1). As expected, tropical and sub-tropical species had high T_b values; most temperate species had T_b values near 0°C. Values for θ_{HT}, $\Psi_b(50)$, $\sigma_{\Psi b}$ and mean germination rate also showed wide variation, sometimes with greater than a 10-fold increase from the lowest to the highest value obtained. In spite of these widely differing parameter values, the hydrothermal time model generally fit germination time course data quite well. When all probit-transformed germination data for a given species were regressed on $\Psi_b(g)$, only four species had R^2 values less than 0.7. Otherwise, R^2 values were similar to those reported for tomato (e.g. Cheng and Bradford, 1999) and sugarbeet (Gummerson, 1986). For those species with low R^2 values, the model fit was improved if a separate hydrothermal time equation was fitted for each incubation temperature (data not shown), as was observed with tomato (Dahal *et al.*, 1993).

Several generalizations regarding hydrothermal time parameters can be made for each ecological class (Table 38.2, Fig. 38.1). Values for $\Psi_b(50)$ and θ_{HT} were negatively correlated, largely due to the extreme values for halophytes (Fig. 38.1A). Halophytes were characterized by low $\Psi_b(50)$ values, with the extreme observation of −5.9 MPa for *Arthrocnemum indicum*. Halophytes also tended to have very large θ_{HT} and high $\sigma_{\Psi b}$ values, which resulted in a greater spread of germination times in water or at reduced water potentials (Bradford, 1990). Psammophytes were characterized by high $\Psi_b(50)$ and low θ_{HT} values and intermediate to low $\sigma_{\Psi b}$ values. This is of considerable ecological significance, as seeds of these species have a rapid germination rate in water but are strongly inhibited from germinating at reduced water potentials. This reduces risk of germinating into the rapidly drying seedzone environments characteristic of sandy soils. Bodenvags displayed a range of $\Psi_b(50)$ values, generally intermediate between the halophytes and psammophytes. The lowest $\Psi_b(50)$ for any bodenvag was −1.85 MPa (*Kochia prostrata*), as compared to −1.18 MPa, the minimum value for a psammophyte (*Asclepias tuberosa*). Values of θ_{HT} for bodenvags were distributed nearly across the entire range of values for all species. Data for two crop species, sugarbeet (Gummerson, 1986) and tomato (Cheng and Bradford, 1999), are included in Fig. 38.1 for comparison. Sugarbeet (diamond on regression line in Fig. 38.1A) had a $\Psi_b(50)$ of nearly −2 MPa, reflecting the halophytic origin of this species. In contrast, tomato (diamond near origin) has both a low $\Psi_b(50)$ and a low θ_{HT}. Note that both species have similar mean germination rates (Fig. 38.1B).

There was essentially no relationship between $\Psi_b(50)$ and mean germination rate (Fig. 38.1B). A low $\Psi_b(50)$, which would result in a greater

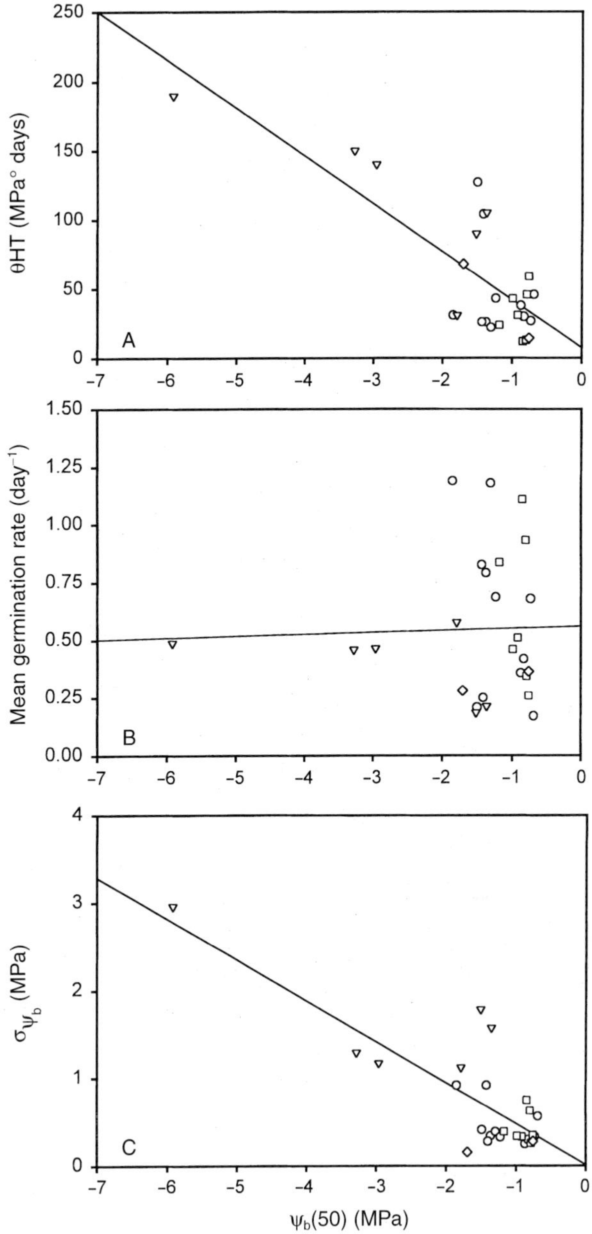

Fig. 38.1. Relationships between hydrothermal time parameters. (A) Relationship between mean base water potential ($\Psi_b(50)$) and hydrothermal time requirement (θ_{HT}). (B) Relationship between mean base water potential ($\Psi_b(50)$) and mean germination rate ($1/t_{50}$). (C) Relationship between mean base water potential ($\Psi_b(50)$) and standard deviation of base water potentials ($\sigma_{\Psi b}$). Data for crop species include previously published values for sugarbeet (Gummerson, 1986) and tomato (Cheng and Bradford, 1999). ○, Bodenvags; ▽, halophytes; □, psammophytes; ◇, crops.

Table 38.2. Hydrothermal time parameters and mean germination rates for 24 species representing halophytes (salt tolerant), psammophytes (high-sand soils), and bodenvags (no special soil requirements).

	T_b	θHT	$\sigma_{\psi b}$	R^2	$\Psi_b(50)$					Mean germination rate				
					10	15	20	25	30	10	15	20	25	30
Halophytes														
A. indicum	9	190	2.96	0.850	–	−5.92	−5.12	–	−4.42	–	0.19	0.30	–	0.49
S. fruticosa	9	150	1.29	0.774	–	−3.28	−3.28	–	−3.28	–	0.13	0.24	–	0.46
S. utahensis	8	140	1.17	0.756	–	−2.96	−2.96	–	−2.96	–	0.14	0.25	–	0.46
T. maritima	9	90	1.78	0.770	–	−1.51	−1.51	–	–	–	0.10	0.18	–	–
A. triangularis	3	105	1.57	0.688	−1.36	−1.21	–	−1.01	–	0.09	0.14	–	0.21	–
P. aviculare	0	31	1.12	0.745	−1.79	−0.94	−0.24	–	–	0.58	0.45	0.16	–	–
Psammophytes														
A. tuberosa	8	24	0.39	0.544	–	−0.75	–	−1.18	–	–	0.22	–	0.84	–
A. cana	0	43	0.34	0.708	−0.86	–	−0.99	–	–	0.20	–	0.46	–	–
E. alatum	3	31	0.33	0.764	−0.62	−0.92	−0.82	−0.72	–	0.14	0.36	0.45	0.51	–
H. villosa	1	12	0.75	0.746	−0.75	−0.85	−0.70	−0.50	–	0.56	0.99	1.11	1.00	–
A. pulchra	0	13	0.63	0.432	–	−0.80	–	−0.48	–	–	0.93	–	0.93	–
S. arida	0	46	0.26	0.883	−0.79	–	−0.79	–	–	0.17	–	0.34	–	–
H. scaposus	0	59	0.35	0.823	−0.51	–	−0.76	–	–	0.09	–	0.26	–	–

Bodenvags														
K. prostrata	0	31	0.92	0.856	−1.50	–	−1.85	–	–	0.48	–	1.19	–	–
P. secunda	0	127	0.41	0.773	−1.49	–	−1.31	–	–	0.12	–	0.21	–	–
C. annua	0	26	0.92	0.975	–	−1.43	–	+0.97	–	–	0.83	–	nil	–
E. elymoides	0	104	0.28	0.952	–	–	−1.41	–	−0.88	–	–	0.25	–	0.20
A. asperula	10	26	0.34	0.480	nil	−0.95	−1.17	−1.37	–	nil	0.18	0.45	0.79	–
C. lanata	0	22	0.39	0.900	−1.30	–	−1.30	–	–	0.59	–	1.18	–	–
B. tectorum	0	43	0.33	0.829	–	−1.23	–	−1.18	–	–	0.43	–	0.69	–
B. distachyon	0	38	0.25	0.897	–	−0.87	–	−0.54	–	–	0.34	–	0.36	–
B. fasciculatus	0	30	0.30	0.815	–	−0.83	–	−0.50	–	–	0.42	–	0.42	–
S. capensis	0	27	0.33	0.831	–	−0.73	–	−0.73	–	–	0.41	–	0.68	–
E. nevadensis	0	46	0.56	0.773	−0.69	–	−0.39	–	–	0.15	–	0.17	–	–

Model parameters were derived from repeated regression analyses using estimated base water potential as the independent variable and the probit-transformed germination fraction as the dependent variable.

accumulation rate for hydrothermal time, was typically offset by a high θ_{HT}. For example, the four halophytes with the lowest base water potentials had similar mean germination rates (approximately 0.5). Psammophytes tended to have closely similar base water potentials (Table 38.2); the variation in mean germination rates among these species was largely a function of variation in θ_{HT}. Bodenvags with rapid germination rates typically had a low $\Psi_b(50)$ or a low θ_{HT}, but usually not both.

A weak relationship existed between $\Psi_b(50)$ and $\sigma_{\Psi b}$, primarily due to the large $\sigma_{\Psi b}$ values of halophytes (Fig. 38.1C). Psammophytes and bodenvags had low to intermediate $\sigma_{\Psi b}$. While $\sigma_{\Psi b}$ does not affect mean germination rate, an increase in $\sigma_{\Psi b}$ leads to an increased germination rate for all fractions faster than the mean and a decreased germination rate for all fractions slower than the mean.

A distinct advantage of hydrothermal time analysis for seeds incubated across a range of T and Ψ conditions is that variation or similarity in germination rate can be ascribed to specific underlying factors. For example, a fast germination rate in water may be due either to a low hydrothermal time requirement or a low base water potential. By incubating seeds at various water potentials, the relative importance of $\Psi_b(50)$ and θ_{HT} can be evaluated.

The number of incubation temperatures included in this study ranged from two to four for a particular species. Variation in $\Psi_b(50)$ associated with incubation temperature produced distinct patterns (Table 38.2). First, for some species there was no difference in $\Psi_b(50)$ across the range of incubation temperatures tested (e.g. *Suaeda fruticosa* and *Salicornia pacifica*). For seeds of most species, $\Psi_b(50)$ increased with higher incubation temperature (e.g. *Arthrocnemum indicum* and *Bromus* spp.). In still other cases a higher incubation temperature corresponded to a lower $\Psi_b(50)$ (e.g. *Hymenoxys scaposus* and *Kochia prostrata*). We believe there is an underlying explanation for these patterns that would be more evident if species had been incubated across the entire range of germination-permissive temperatures. Our evidence suggests that there is a range of temperatures over which $\Psi_b(50)$ remains constant. Above that range, it increases linearly with temperature until the maximum (no germination). As incubation temperature approaches T_b, $\Psi_b(50)$ also appears to increase. This is most evident for species that were incubated at four temperatures (e.g. *Eriogonum alatum* and *Heterotheca villosa*), but is also supported by other studies in which application of hydrothermal time resulted in a poor fit at incubation temperatures near T_b unless $\Psi_b(50)$ was allowed to increase (S.E. Meyer and P.S. Allen, unpublished).

Temperature sensitivity varied considerably among species. Strongly temperature-dependent $\Psi_b(50)$ values may or may not result in a similar degree of variation in germination rate. For example, *Elymus elymoides* had a $\Psi_b(50)$ that progressively increased with incubation temperature (Table 38.2 and S.E. Meyer and P.S. Allen, unpublished). This offsets the increased hydrothermal time accumulation at increased incubation temperature and results in a germination rate that is nearly constant at incubation temperatures from 10 to 30°C. In contrast, the increasing $\Psi_b(50)$ with increasing temperature in *Arthrocnemum indicum* seeds is not great enough to offset the increased germination rate at higher incubation temperatures. With seeds of *Polygonum aviculare*, the increase in $\Psi_b(50)$ associated with increased incubation temperature is so large that mean germination rate decreases with increasing temperature.

Many simplifying assumptions associated with hydrothermal time theory were applied herein (e.g. constant θ_{HT}, T_b, and $\sigma_{\Psi b}$ across combinations of incubation Ψ and T), assumptions which occasionally have been shown to reduce reliability of model predictions (see Bradford, 1995). In spite of these limitations, hydrothermal time appears to provide a modelling approach that can be used to explain germination data across numerous experimental conditions and species.

Acknowledgements

This project was funded in part by Grant No. CSRS-93-3932 from the USDA Cooperative State Research Service and by a grant from the International Arid Lands Consortium. Thanks to Alisa Paulsen and Kerry Mitchell for able technical assistance, and to Drs Jaime Kigel and Avi Perevolotsky for seed collection.

References

Allen, P.S. and Meyer, S.E. (1998) Ecological aspects of seed dormancy loss. *Seed Science Research* 8, 183–192.

Bauer, M.C., Meyer, S.E. and Allen, P.S. (1998) A simulation model to predict seed dormancy loss in the field for *Bromus tectorum* L. *Journal of Experimental Botany* 49, 1235–1244.

Bradford, K.J. (1990) A water relations analysis of seed germination rates. *Plant Physiology* 94, 840–849.

Bradford, K.J. (1995) Water relations in seed germination. In: Kigel, J. and Galili, G. (eds) *Seed Development and Germination.* New York, Marcel Decker, Inc., pp. 351–396.

Cheng, Z. and Bradford, K.J. (1999) Hydrothermal time analysis of tomato seed germination responses to priming treatments. *Journal of Experimental Botany* 50, 89–99.

Christensen, M., Meyer, S.E. and Allen, P.S. (1996) A hydrothermal time model of seed after-ripening in *Bromus tectorum* L. *Seed Science Research* 6, 155–163.

Dahal, P. and Bradford, K.J. (1994) Hydrothermal time analysis of tomato seed germination at suboptimal temperature and reduced water potential. *Seed Science Research* 4, 71–80.

Dahal, P., Bradford, K.J. and Haigh, A.M. (1993) The concept of hydrothermal time in seed germination and priming. In: Côme, D. and Corbineau, F. (eds) *Proceedings, Fourth Internation Workshop on Seeds: Basic and Applied Aspects of Seed Biology.* Paris, ASFIS, pp. 1009–1014.

Ellis, R.H., Covell, S., Roberts, E.H. and Summerfield, R.J. (1986) The influence of temperature on seed germination rate in grain legumes. II. Intraspecific variation in chickpea at constant temperatures. *Journal of Experimental Botany* 37, 1503–1515.

Finch-Savage, W.E., Steckel, J.R.A. and Phelps, K. (1998) Germination and post-germination growth to carrot seedling emergence: predictive threshold models and sources of variation between sowing occasions. *New Phytologist* 139, 505–516.

Gummerson, R.J. (1986) The effect of constant temperatures and osmotic potentials on the germination of sugar beet. *Journal of Experimental Botany* 37, 729–741.

Khan, M.A. and Gul, B. (1998) High salt tolerance in the germinating dimorphic seeds of *Arthrocnemum indicum. International Journal of Plant Science* 159, 826–832.

Khan, M.A. and Unger, I.A. (1984) The effect of salinity and temperature on the germination of polymorphic seeds and growth of *Atriplex triangularis* Willd. *American Journal of Botany* 71, 481–489.

Khan, M.A. and Unger, I.A. (1997) Alleviation of seed dormancy in the desert forb *Zygophyllum simplex* L. from Pakistan. *Annals of Botany* 80, 395–400.

Khan, M.A. and Unger, I.A. (2000a) Effects of salinity on the germination of *Triglochin maritima* under various temperatures. *Seed Science and Technology* (in press).

Khan, M.A. and Unger, I.A. (2000b) Germination of the salt tolerant shrub *Suaeda fruticosa*: salinity and temperature responses. *Seed Science and Technology* (in press).

Khan, M.A. and Weber, D.J. (1986) Factors influencing seed germination in *Salicornia pacifica* var. *utahensis. American Journal of Botany* 73, 1163–1167.

39 Emergent Weedy Foxtail (*Setaria* spp.) Seed Germinability Behaviour

J. Dekker

Weed Biology Laboratory, Agronomy Department, Iowa State University, Ames, IA 50010, USA

The weedy foxtails (*Setaria* spp.) are an important group of colonizing plants whose biogeographical distribution is a function of many contributing sources. Weedy foxtail seed behaviour, the physical environment in which this behaviour occurs, and the seed morphology through which transduction of environmental signals to the embryo occurs are reviewed. Seed behaviour is characterized by seed development on the parent plant, afterripening in the soil seed bank, and seed germination and seedling establishment. The soil habitat can stimulate foxtail seed behaviour through the environmental signals of water, temperature and oxygen. The transduction of these signals are communicated through seed structures to the embryo. Notably, water and gas entry into the seed symplast is restricted to the placental pore region at the base of the seed. Water entry via the placental pore is never impeded when unfrozen, but the entry of gas occurs only when it is dissolved in imbibition water. The amount of dissolved gas that enters the seed is a function of temperature, solubility and availability from the soil atmosphere. The possible unifying role that soil oxygen–water availability might play in describing both seed behaviour and biogeographical distribution is discussed.

Introduction

Setaria Beauv. is a genus of about 125 species that includes food crops and a number of important weeds. Some of the weedy *Setaria* species are *S. viridis* (L.) Beauv. (green foxtail), *S. glauca* (Weigel) Hubb. (yellow foxtail), *S. faberii* Herrm. (giant foxtail), *S. verticillata* (L.) Beauv., and *S. geniculata* (Lamarck) Beauv. (knotroot foxtail) (Rominger, 1962). *Setaria viridis* sub-species *italica* (L.) Beauv. (foxtail millet), is an important world grain crop. It has been speculated that Africa is the original home of the genus because 74 out of 125 species occur on that continent (Stapf and Hubbard, 1930). Before being introduced to other continents, green foxtail's natural range was probably Eurasia

(Li *et al.*, 1942, 1945). It has been argued that many temperate weedy foxtails evolved from green foxtail-like ancestors (Li *et al.*, 1942, 1945; Rominger, 1962; Willweber-Kishimoto, 1962; Williams and Schreiber, 1976; Prasado Rao *et al.*, 1987).

Today green foxtail is primarily a temperate species but it is widely distributed between 45°S and 55°N latitudes (Holm *et al.*, 1977, 1997). It is one of the most widely distributed weedy foxtail species both globally and in the United States (Wang *et al.*, 1995a,b). It ranges from North America, through Central America to parts of South America; from Europe to northern Africa, and from east Asia to south Asia and Australia (Hafliger and Scholz, 1980). It is found in every state in the continental United States and every province in Canada (Lorenzi and Jeffery, 1987).

Foxtails are of considerable agronomic importance. Their associations with agriculture, as both crops and weeds, can be traced back thousands of years to ancient civilizations (Gao and Chen, 1988). Foxtail millet is one of the oldest cultivated cereals of China, dating back about 6000 years to the earliest agricultural settlements of the Yang-shao culture phase (Cheng, 1973). *Setaria geniculata* was an important wild cereal crop in Mexico before agriculture arose (Dewet and Harlan, 1975). *Setaria glauca* originated in Europe and was used centuries ago for flour and groats, especially from the 9th to 13th centuries of the Middle Ages when wheat was scarce (Dembinska, 1976). Today, these species are widely grown as crops in Africa, China, India and scattered areas throughout Eurasia (Kawase and Sakamoto, 1984; Gao and Chen, 1988). *Setaria viridis*, *S. glauca* and *S. faberii* are listed as major weeds worldwide and comprise the second most important weed group in the United States (Holm *et al.*, 1977, 1997). Foxtail seeds also serve as an important food source for wildlife (Martin *et al.*, 1961). Since their introduction to North America, foxtails have expanded in terms of range, population density, and the appearance of new morphological variants. Crop yield losses and herbicide expenditures make control of the foxtails a significant problem in crop production.

Substantial biological diversity, wide geographic distribution, and strong competitiveness in disturbed habitats all mark foxtails as highly successful weeds. Seed dormancy and the ability to survive for long periods in the soil is one of the most important traits possessed by the foxtails leading to their success and wide biogeographic distribution. More than one weedy foxtail species often coexist in the same agricultural field, exploiting slightly different niches. Significant heterogeneity in germinability among seeds shed by a single plant allow these species to emerge at appropriate times in the growing season and infest large areas of disturbed habitats (Dekker *et al.*, 1996). Weedy foxtail species are most frequently found in disturbed habitats, especially agroecosystems with annual tillage, planting and harvesting.

Foxtail biogeographical distribution is often associated with humid, oxygen-rich soils with distinctive, predictable seasonal and diurnal water and temperature fluctuations. These seasonal temperature–water cycles are correlated with the cyclical germination behaviour of these colonizing species. Several features characterize the cyclical nature of the seed bank environment that the weedy foxtails have adapted to (Silvertown, 1984; Forcella *et al.*, 1992,

1997; Dekker, 1997, 1998). These foxtail seed banks usually have an extended period of cool temperatures. Adequate moisture may be present during those periods, but often the water is frozen in winter, and unavailable. During this cool period, the day–night temperature fluctuations are low. Warming temperatures with adequate moisture follows the cool period. This warming period (spring) is associated with fluctuating diurnal temperatures. The peak foxtail germination period follows soon after this thawing and time of maximum day–night temperature differences. In late spring and early summer, soil temperature increases and less diurnal fluctuations occur. Foxtail seedling emergence continues through this period, but at much reduced numbers. In late summer and autumn foxtail seed is shed, when both daylength and temperature are decreasing. In late autumn day–night temperature differences increase, often with adequate moisture present. An increase in seedling emergence can occur in this autumn period, especially with *S. viridis.*

The purpose of this paper is to review the relevant literature about *Setaria* spp. seed behaviour, the physical environment in which this behaviour occurs, and the seed morphology through which transduction of environmental signals to the embryo occurs. From this observational foundation I speculate on the possible unifying role that soil oxygen–water availability might play in describing both seed behaviour and biogeographical distribution.

Foxtail Seed Behaviour

Seed development

The life cycle of foxtail seed begins with seed development, i.e. development of both seed envelopes, the endosperm and the embryo. Three separate nuclear genomes interact and produce the tissues that compose the foxtail seed. Parental tissues (2N) include the seed hull (glumes, hull (palea, lemmal)), many of the crushed layers forming the caryopsis coat, and vascular remnants and residual tracheary elements at the basal abscission area (Rost, 1973, 1975). The zygotic tissues include those of the endosperm (3N; aleurone, aleurone transfer cells) and the embryo (2N). The seed develops in approximately 11 days from anthesis until abscission (Dekker *et al.*, 1996) followed by dispersal of the seed from the parent panicle.

Embryogenesis

Giant foxtail embryos become capable of independent germination at about day 6 of embryogenesis (Dekker *et al.*, 1996). At that time, embryo germination is very high (i.e. 95%), and occurs in both parts of the axes (coleoptile, coleorhiza). By day 8, embryo germination decreases, becomes more variable, and axis-specific germination first appears (germination of only one part of the axis). This embryo dormancy induction period occurs from day 8 through to anthesis, when embryo germination is very low (i.e. 10%). The variability in embryo and caryopsis germination among individual seeds shed by a panicle increases from when dormancy is induced until after abscission.

Seed envelope genesis

The germination of giant foxtail embryos is modified by the presence of several seed envelopes, notably the hull (glumes, lemma, palea) and the caryopsis coat (Rost, 1973, 1975). Induction of dormancy in caryopses and seeds coincides with the sealing of the caryopsis by the caryopsis coat, and the maturation of the hull. The variability in germination among individual caryopses and seeds increases with time after dormancy induction. These events may describe a 'phenocritical' period in seed genesis when dormancy is induced (Christianson and Warnick, 1984).

The seed hull structures are present at anthesis, but change during seed ontogeny from open, soft tissues to a hard, enclosing structure at maturity (Dekker *et al.*, 1996). The caryopsis coat becomes fully developed at approximately day 4, and appears shiny, oily, and grey with dark spots. Sometime after day 4 it becomes gas impermeable, except at the proximal (basal) end of the caryopsis (the placental pad and pore region). Entry of water into the caryopsis occurs only through the placental pore after day 4 and is never restricted through this region of the seed from day 4 until germination occurs (Rost, 1971b, 1973, 1975).

During the ontogeny of the caryopsis, a single placental vascular bundle from the parent panicle supplies nutrients and water to the developing ovule (Rost, 1971b). The vascular bundle enters the ovule in a proximal position on the bottom surface of the developing caryopsis (Rost, 1971b). Later, when the caryopsis matures, a thick, dark, oval pad remains in the position occupied by the placental bundle. In longitudinal section this appears as an elongated multi-layered placental pad. The caryopsis matures and expands, filling the hull width by day 6, and its length by the time of abscission. The placental pad of the caryopsis shows evidence of reddish coloration at approximately day 9 which is probably a morphological indicator of physiological maturity of the seed (Dekker *et al.*, 1996).

Giant foxtail seeds are shed from the parent plant almost entirely dormant. Within those seeds, considerable heterogeneity in germinability exists. This diversity of germinability phenotypes of the seeds shed from the same parent plant then interacts with the soil seed bank environment to determine when those seeds germinate.

Seed afterripening in the soil

Once foxtail seeds enter the soil, they are influenced by the environmental conditions in the seed banks. Several important phenomena in seeds have been observed that provide clues as to the afterripening processes preceding seed germination and seedling establishment. Clues as to the nature of foxtail seed dormancy are provided by their response to stratification, high temperatures and seed damage.

Stratification

Dormant foxtail seed can be induced to germinate after a period (e.g. 1–12 weeks) of cool temperatures (e.g. 3–6°C), dark, and adequate moisture (Anderson, 1968). Foxtail seeds readily absorb water in all parts of the seed through the placental pore. The need for aeration directly adjacent to the seeds is not necessary for this afterripening to occur: submersion of the seeds in cool water produces similar effects as aerated stratification.

High temperature

Foxtail seeds subjected to high temperatures can respond in contradictory ways. In some instances, germination is increased by exposure to high temperatures (Taylorson and Brown, 1977). This increase in seed germination may be caused by cracks and damage to the seed envelopes due to high temperature without acclimation preceding the exposure. In other instances, the high soil temperatures of summer induce secondary dormancy in foxtail seed banks (data not presented).

Seed damage

Damage to the foxtail seed coat, such as puncturing the seed hull and caryopsis coat, often increases germination (Stanway, 1971; data not presented). Removal of the hull, and separation of the embryo from the caryopsis, both increase germination of the embryo (Dekker *et al.*, 1996). Removal or puncturing the caryopsis coat of caryopses with hulls removed also increased embryo germination.

Seed germination and seedling establishment

Seasonal germination

Foxtail seed germination exhibits an annual cycle of activity. The seeds begin to germinate in the spring, usually following an extended cool period. Peak periods of annual germination occur at the beginning of this soil warming period (e.g. in Iowa late April–early May). Most foxtail seeds in the soil become dormant again (secondary dormancy; summer dormancy) during the warmest part of the year (summer), but low numbers of seeds continue to germinate until the soil temperature becomes cool once again.

Alternating temperatures

Increased germination of foxtail seeds occurs when they are exposed to periods of alternating temperatures, compared to static conditions, during the day (Sells, 1965; Anderson, 1968; James, 1968).

Germination depth in the soil

Foxtail seed germination is greatest at shallow soil depths (Gregg, 1973), e.g. at 1.5 cm and declines with depth to 10 cm. The maximum depth that foxtail seed can emerge from is approximately 14 cm.

Seed longevity in the soil

Yellow foxtail seeds can survive in the soil for up to 30 years (Toole and Brown, 1946; Kivilaan and Bandurski, 1973), although most only survive 13 years (Dawson and Bruns, 1975). Burial of foxtail seed increases both their level of dormancy and their longevity. Seed decay and germination are less when foxtail seed are encased in soil particle aggregates (Pareja *et al.*, 1985).

Physical Environment of the Soil Seed Bank Habitat and Seed Metabolism

The physical conditions of the seed bank environment affect both the immediate behaviour of foxtail seeds, and provide the setting within which foxtail seeds have adapted and evolved. The physical conditions of the soil that appear most important to seed behaviour are interactions among temperature, moisture and gases, especially oxygen.

Water

Foxtail seeds require moisture to germinate, but can tolerate a wide range of moisture conditions prior to germination. Moisture stress affects the different foxtail species differently (Manthey and Nalewaja, 1987). The geographic range of foxtails is, in part, a function of adequate moisture for germination.

Temperature

Foxtail seeds germinate optimally between 20 and 30°C, but some seeds will germinate over a much wider range of temperatures (e.g. 8 to 40°C; data not reported). These soil temperatures are present in seed banks throughout the geographic range of foxtail distribution. The metabolism and oxygen consumption of foxtail seeds and plants increases with increasing temperatures.

Oxygen

The partial pressure of oxygen in air above ground level is 0.21. The oxygen content of agricultural soils was found to vary from about 21% to 19% under a wide range of conditions, including differences due to time of year, tillage, soil depth (Sells, 1965). These soil atmospheric conditions indicate that oxygen is not limiting in the region adjacent to foxtail seeds. An exception to this may be when the seeds are encased in soil particle aggregates sealing them from gas exchange with the soil atmosphere (Pareja *et al.*, 1985).

Oxygen solubility in water

Oxygen solubility increases with decreasing temperature. At 5°C oxygen solubility in water is 0.002 mol l^{-1}, at 20°C it is 0.0015 mol l^{-1}, and at 40°C it is 0.001 mol l^{-1}. Twice as much oxygen is therefore available to foxtail seed embryos in imbibed water at 5°C (early spring) than at 40°C (summer). The role oxygen may play in seed dormancy has been discussed, but the literature is sometimes conflicting and inconclusive (Thornton, 1945; Wareing, 1965; Simpson, 1990; Bewley and Black, 1994; Baskin and Baskin, 1998).

Transduction of Environmental Signals from the Soil Seed Bank to the Enclosed Embryo

The mature foxtail seed enters the soil seed bank habitat as an individual, capable of imbibing water easily, but entry of gases is restricted and regulated by the structure and function of the placental pore. Regardless of the mechanism that regulates foxtail seed dormancy and germinability, water and gas entry into the living part of the seed (the symplast) is restricted to the placental pore region at the base of the seed (Rost, 1971b, 1973, 1975). Water entry via the placental pore is never impeded when unfrozen, but gas entry occurs only when dissolved in imbibition water. The amount of dissolved gas that enters the seed is a function of temperature, solubility and availability from the soil atmosphere. The morphology of the placental pore is identical in dormant and non-dormant caryopsis structures (Rost, 1971b). In this section the morphology of the placental pore at the basal end of the seed will be examined in terms of the structures that water and gases encounter as they enter the seed from the outside.

Seed envelopes: the apoplast

The non-living portion of the placental pore includes residual vascular elements and tracheary elements left from the pedicel connection of the seed to the parental panicle, and the seed envelope structures: glumes, hull (ridges on the lemma, palea; placental pore), caryopsis placental pad and coat (Rost, 1971a,b, 1972, 1973, 1975; Rost and Lersten, 1973).

Glumes

The outermost envelopes of the foxtail seed are the papery glumes that partially surround the seed hull. These absorbant structures protect the seed as well as acting as wick and funnel to pass water to the placental pore region, their point of attachment to the seed. Ridges on the glumes are often at right angles to hull (lemma, palea) surface ridges.

Hull ridges

The seed hull consists of the concave lemma and the palea. Both these structures have ridges on their surface; in some species they are transverse, in others they follow the longitudinal axis. These ridges appear like the drain-board of a sink, and may provide drainage channels for liquids, gases or solid particles in the soil adjacent to the seed. Glume and hull surface ridges may function together to both mix water and air at, and funnel gas-laden water into, the placental pore.

Hull placental pore

At the basal end of the foxtail seed is the site where water and air enter, the placental pore. This hull structure is rounded and hard with a soft centre consisting of remnants of the parental vascular connection into the seed interior – degraded strands of former xylem and phloem tissues (Rost, 1971b). The flaired tissue beneath the placental pad is the region where the caryopsis is connected to the palea. This region is the only water-gas entrance to the seed (unless physically damaged), and may serve as the first spongy filter of debris, fungal spores, and bacterial entry into the seed.

Caryopsis placental pad

At the base of the caryopsis is a thickened region called the placental pad (Rost, 1973). The dark necrotic contents of the placental pad layers make the structure appear as a dark oval-shaped area (about 0.2 mm) when seen from the outside (Dekker *et al.*, 1996). The placental pad may serve as a second spongy filter, yet allow free entry of gas-saturated water.

Caryopsis coat

Immediately beneath the tracheary elements is a thick layer of dark, dense, suberized cells, the caryopsis coat (Rost, 1971). The mature caryopsis coat appears as a filmy cuticle layer, oily to the touch (Rost, 1973). Seen in section it is a gossamer-like film, analogous to the cuticle. The coat's speckled appearance derives from the degradation contents in epidermis pericarp cells (Rost, 1971). The structure of the coat is a complex of many layers formed from crushed cells in various stages of degradation (Rost, 1971, 1973). The expansion of the developing caryopsis causes these cells to become crushed, thereby forming the complex caryopsis coat (Rost, 1971). The many inner layers of coat are continuous around the caryopsis, except around the placental pad (Rost, 1973). The caryopsis coat in the placental pad and placental pore region is different from that around the rest of the caryopsis (see description below). The transfer aleurone cells rest immediately adjacent to the last layer of pad cells. The loose arrangement of the pad cells and the presence of a large number of pits allow for a relatively unimpeded flow of water and solutes into the transfer aleurone cells. These cells transport material from the placental bundle into the immature embryo and endosperm during caryopsis development (Rost and Lersten, 1970; Rost, 1973).

Embryo: the symplast

The living portion of the placental pore includes the endosperm (including the aleurone layer and the aleurone transfer cell layer) and the embryo.

Caryopsis endosperm

The first zygotic tissue that gas-saturated water contacts is the endosperm, which is entirely sealed from the outside by the caryopsis coat, except in the placental pore region. The outermost layer of the endosperm is the aleurone layer. The aleurone layer is structurally and functionally different in the placental pore, the aleurone transfer cell layer.

ALEURONE LAYER. This outermost portion of the endosperm is continuous around the caryopsis. It consists of thick tabular cells, each flattened, somewhat rectangular in surface view, and 25–50 μm in length (Rost, 1971). A thick primary cellulose wall encloses each cell (Rost, 1971b). The matrix of the aleurone cells appears as a grey network intermeshed between storage materials (Rost, 1971). This cell layer has an abundance of protein bodies of various types, lipid bodies (oil droplets or spherosomes), as well as plastids, mitochondria, and other membrane structures, but little starch (Rost and Lersten, 1973). The aleurone layer of endosperm is known to produce enzymes, and is the site where certain processes associated with germination are first initiatated (Rost, 1971a).

TRANSFER ALEURONE CELL LAYER. The aleurone layer is continuous around the entire caryopsis, but adjacent to the placental pad the aleurone cells are strikingly different in appearance – the transfer aleurone cells (Rost, 1971). These cells occur near the base of the caryopsis, adjacent to the end of the coleorhiza where the grain was attached to the parent plant. The cells are enlarged, approximately columnar, and somewhat elongated perpendicular to the fruit coat (Rost and Lersten, 1970, 1973). The thickened portion of the cell wall appears to be heterogeneous, with that part closest to the middle lamella having a fibrous or porous appearance. These specialized aleurone cells have thick walls bearing ingrowths on the outer radial and outer tangential walls which extend into the cell protoplasm (Rost and Lersten, 1970). The ingrowths form a labyrinth, the plasmalemma following the contours of the ingrowths, thereby significantly increasing the membrane surface area of each cell. Internally, the wall has a porous, sponge-like appearance. The wall ingrowths sometimes are very deeply lobed and convoluted (Rost, 1971b). The inner radial and inner tangential parts of the wall lack these ingrowths, and have a middle lamella and typical appearing primary wall (Rost, 1971b), indicating they may not be involved in solute transport (Rost and Lersten, 1970). The aleurone transfer cell wall ingrowths appear similar to those of certain of the transfer cells described by Pate and Gunning (Gunning and Pate, 1969; Pate and Gunning, 1972; O'Brien, 1976; Gunning, 1977). These specialized cells have already been described as playing a role in mature seed of other species (Zee and O'Brien, 1971; Zee, 1972).

Embryo

Within the caryopsis is the embryo (scutellum, coleoptile, coleorhiza). The placental pore, and the transfer aleurone cells are located close to the scutellum and coleorhizal tissues of the embryo at the basal end of the seed. A cementing substance causes the outer epidermis of the coleorhiza and other embryo parts to adhere to the aleurone layer (Rost and Lersten, 1970, 1973). This intimate contact may provide a continuous route for the uptake of gas-saturated water.

Discussion: a Proposed Mechanism of *Setaria* spp. Seed Dormancy and Biogeographical Distribution

Weedy foxtail seed behaviour is an emergent property arising from contributions within the plant organizational hierarchy (i.e. species-group through to individual seed). Foxtail seed behaviour is also an emergent property of the interaction amongst the three different parts of different genomes (i.e. embryo, endosperm, caryopsis coat, placental pore, glumes and hull).

The behavioural, habitat and morphological information presented herein indicates the placental pore region of *Setaria* spp. may play an important regulatory role in the behaviour of seeds. These observations may also provide a theoretical explanation of the global biogeography of the weedy foxtails based on their adaptation to these habitats. Dormancy induction, afterripening, and germination of the seed may all be unified by the regulation of oxygen entry into the embryo by the transfer aleurone cells of the placental pore. The behaviour and biogeography of weedy foxtail species may both be driven by this mechanism.

Therefore, I propose four hypotheses to test the validity of this theory: (i) the biogeographical distribution of weedy foxtails (*Setaria* spp.) is an evolutionary adaptation to changing soil oxygen–water availability; (ii) seed germination is regulated by the functional qualities of the transfer aleurone layer of the placental pore; (iii) seed dormancy is reversible, and is induced when oxygen–water availability and uptake is restricted; and (iv) seed germination and seedling emergence occurs when adequate amounts of oxygen–water reach the embryo. These hypotheses provide the basis for future experimental investigations to elucidate the mechanism(s) of dormancy in *Setaria* spp. seeds in the soil seed bank.

References

Andersen, R.N. (1968) *Germination and Establishment of Weeds for Experimental Purposes.* Weed Science Society of America Handbook. W.F. Humphrey Press, Inc., Geneva, New York.

Baskin, C.C. and Baskin, J.M. (1998) *Seeds – Ecology, Biogeography, and Evolution of Dormancy and Germination.* Academic Press, San Diego, California.

Bewley, J.D. and Black, M. (1994) *Seeds – Physiology of Development and Germination*, 2nd edn. Plenum Press, New York.

Cheng, K. (1973) Radio carbon dates from China: some initial interpretations. *Current Anthropology* 14, 525–528.

Christianson, M.L. and Warnick, D.A. (1984) Phenocritical times in the process of in vitro shoot organogenesis. *Developmental Biology* 101, 382–390.

Dawson, J. and Bruns, V. (1975) Longevity of barnyardgrass, green foxtail, and yellow foxtail seeds in soil. *Weed Science* 23, 437–440.

Dekker, J. (1997) Weed diversity and weed management. *Weed Science* 45, 357–363.

Dekker, J. (1998) Soil seed seed banks and management. *Journal of Crop Production* (in press).

Dekker, J., Dekker, B.I., Hilhorst, H. and Karssen, C. (1996) Weedy adaptation in *Setaria* spp.: IV. Changes in the germinative capacity of *S. faberii* embryos with development from anthesis to after abscission. *American Journal of Botany* 83, 979–991.

Dembinska, M. (1976) Wild corn plants gathered in the 9th to 13th centuries in light of paleobotanical materials. *Folia Quaternaria* 47, 97–103.

Dewet, J. and Harlan, J. (1975) Weeds and domesticates: evolution in the man-made habitat. *Economic Botany* 29, 99–107.

Forcella, F., Wilson, R.G., Renner, K.A., Dekker, J., Harvey, R.G., Alm, D.A., Buhler, D.D. and Cardina, J.A. (1992) Weed seedbanks of the U.S. corn belt: magnitude, variation, emergence, and application. *Weed Science* 40, 636–644.

Forcella, F., Wilson, R.G., Dekker, J., Kremer, R.J., Cardina, J., Anderson, R.L., Alm, D., Renner, K.A., Harvey, R.G., Clay, S. and Buhler, D.D. (1997) Weed seed bank emergence across the corn belt. *Weed Science* 45, 67–76.

Gao, M.J. and Chen, J.J. (1988) Isozymic studies on the origin of cultivated foxtail millet. *Acta Agronomica Sinica* 14, 131–136.

Gregg, W. (1973) Ecology of the annual grass *Setaria lutescens* on old fields of the Pennsylvania Piedmont. *Proceedings of the National Academy of Natural Sciences, Philadelphia* 124, 135–196.

Gunning, B.E.S. (1977) Transfer cells and their roles in transport of solutes in plants. *Science Progress*, 64, 539–568.

Gunning, B.E.S. and Pate, J.S. (1969) 'Transfer cells' plant cells with wall ingrowths, specialized in relation to short distance transport of solutes – their occurrence, structure, and development. *Protoplasma* 68, 107–133.

Hafliger, E. and Scholz, H. (1980) *Grass Weeds 1 – Weeds of the Subfamily Panicoideae.* Ciba-Geigy, Basle, Switzerland, pp. 123–134.

Holm, L., Plucknett, D., Pancho, J. and Herberger, J. (1977) *The World's Worst Weeds: Distribution and Biology.* University of Hawaii Press, Honolulu.

Holm, L., Doll, J., Holm, E., Pancho, J. and Herberger, J. (1997) *World Weeds: Natural Histories and Distribution.* John Wiley & Sons, Inc., New York.

James, A.L. (1968) Some influences of soil atmosphere on germination of annual weeds. PhD thesis, Iowa State University, Ames.

Kawase, K. and Sakamoto, S. (1984) Variation, geographical distribution and genetical analysis of esterase isozymes in foxtail millet, *Setaria italica* (L.) P. Beauv. *Theoretical and Applied Genetics* 67, 529–533.

Kivilaan, A. and Bandurski, R. (1973) The ninety year period for Dr. Beal's seed viability experiment. *American Journal of Botany* 60, 140–145.

Li, C.H., Pao, W.K. and Li, H.W. (1942) Interspecific crosses in *Setaria*. *Journal of Heredity* 33, 351–355.

Li, H.W., Li, C.H. and Pao, W.K. (1945) Cytological and genetical studies of the interspecific cross of the cultivated foxtail millet, *Setaria italica* (L.) Beauv., and the green foxtail millet, *S. viridis* L. *Journal of the American Society of Agronomy* 37, 32–54.

Lorenzi, H.J. and Jeffery, L.S. (1987) *Weeds of the United States and their Control.* Van Nostrand Reinhold Co., New York, pp. 78–80.

Martin, A.C., Zim, H.S. and Nelson, A.L. (1961) *American Wildlife and Plants: a Guide to Wildlife Food Habits.* Dover Publications, New York, 500 pp.

Manthey, D. and Nalawaja, J. (1987) Germination of two foxtail (*Setaria*) species. *Weed Technology* 1, 302–304.

O'Brien, T.P. (1976) Transfer cells. In: Wardlaw, I.F. and Passioura, J.G. (eds) *Transport and Transfer Processes in Plants.* Academic Press, Inc. New York, pp. 59–63.

Pareja, M.R., Staniforth, D.W. and Pareja, G.P. (1985) Distribution of weed seed among soil structural units. *Weed Science* 33, 182–189.

Pate, J.S. and Gunning, B.E.S. (1972) Transfer cells. *Annual Review of Plant Physiology* 23, 173–196.

Prasada Rao, K.E., DeWet, J.M.J., Brink, D.E. and Mengesha, M.H. (1987) Intraspecific variation and systematics of cultivated *Setaria italica*, foxtail millet (Poaceae). *Economic Botany* 41, 108–116.

Rominger, J.M. (1962) Taxonomy of Setaria (Gramineae) in North America. *Illinois Biological Monographs* No. 29, University of Illinois Press.

Rost, T.L. (1971a) Fine structure of endosperm protein bodies in *Setaria lutescens* (*Gramineae*). *Protoplasma* 73, 475–479.

Rost, T.L. (1971b) Structural and histochemical investigations of dormant and non-dormant caryopses of *Setaria lutescens* (*Gramineae*). PhD thesis, Iowa State University, Ames.

Rost, T.L. (1972) The ultrastructure and physiology of protein bodies and lipids from hydrated dormant and nondormant embryos of *Setaria lutescens* (*Gramineae*). *American Journal of Botany* 59, 607–616.

Rost, T.L. (1973) The anatomy of the caryopsis coat in mature caryopses of the yellow foxtail grass (*Setaria lutescens*). *Botanical Gazette* 134, 32–39.

Rost, T.L. (1975) The morphology of germination in *Setaria lutescens* (Gramineae): the effects of covering structures and chemical inhibitors on dormant and non-dormant florets. *Annals of Botany* 39, 21–30.

Rost, T.L. and Lersten, N.R. (1970) Transfer aleurone cells in *Setaria lutescens* (*Gramineae*). *Protoplasma* 71, 403–408.

Rost, T.L. and Lersten, N.R. (1973) A synopsis and selected bibliography of grass caryopsis anatomy and fine structure. *Iowa State Journal of Research* 48, 47–87.

Sells, G.D. (1965) CO_2-O_2 ratios in relation to weed seed germination. PhD thesis, Iowa State University, Ames.

Simpson, G.M. (1990) *Seed Dormancy in Grasses.* Cambridge University Press, Cambridge, UK.

Stanway, V. (1971) Laboratory germination of giant foxtail, *Setaria faberii* Herrm., at different stages of germination. *Proceedings of the Official Seed Analysts* 61, 85–90.

Stapf, O. and Hubbard, C.K. (1930) *Setaria.* In: *Flora of Tropical Africa*, Vol. 9, Prain (eds). London, pp. 768–866.

Silvertown, J.W. (1984) Phenotypic variety in seed germination behavior: the ontogeny and evolution of somatic polymorphism in seeds. *American Naturalist* 124, 1–16.

Taylorson, R.B. and Brown, M.M. (1977) Accelerated after-ripening for overcoming seed dormancy in grass seeds. *Weed Science* 25, 473–476.

Thornton, N.C. (1945) Importance of oxygen supply in secondary dormancy and its relation to the inhibiting mechanism regulating dormancy. *Contributions of the Boyce Thompson Institute* 13, 487–500.

Toole, E. and Brown, E. (1946) Final results of the Duvel buried seed experiment. *Journal of Agricultural Research* 72, 201–210.

Wang, R.L., Wendell, J. and Dekker, J. (1995a) Weedy adaptation in *Setaria* spp.: I. Isozyme analysis of the genetic diversity and population genetic structure in *S. viridis. American Journal of Botany* 82, 308–317.

Wang, R.L, Wendell, J. and Dekker, J. (1995b) Weedy adaptation in *Setaria* spp.: II. Genetic diversity and population genetic structure in *S. glauca, S. geniculata* and *S. faberii. American Journal of Botany* 82, 1031–1039.

Wareing, P.F. (1965) Dormancy in plants. *Science Progress* 53, 529–537.

Williams, R.D. and Schreiber, M.M. (1976) Numerical and chemotaxonomy of the green foxtail complex. *Weed Science* 24, 331–335.

Willweber-Kishimoto, E. (1962) Interspecific relationships in the genus *Setaria. Contributions in Biology*. Kyoto University, Kyoto, Japan, pp. 1–41.

Zee, S.-Y. (1972) Vascular tissue and transfer cell distribution in the rice spikelet. *Australian Journal of Biological Sciences* 25, 411–414.

Zee, S.-Y. and O'Brien, T.P. (1971) Aleurone transfer cells and other structureal features of the spikelet of millet. *Australian Journal of Biological Sciences* 24, 391–395.

VII Applications of Seed Biology

40 Biotechnological Applications of Seed Biology

D.J. MURPHY

Department of Brassica and Oilseeds Research, John Innes Centre, Norwich Research Park, Norwich NR4 7UH, UK

Seed crops are amongst the most important sources of nutrition for human societies. Many seeds also provide actual or potential raw materials for a wide range of non-food products. There is great interest in using the latest techniques of modern biotechnology to improve the range of both edible and industrial products from seed crops in order to benefit the ever increasing global population and to provide substitutes for raw materials, such as petrochemicals, that are currently derived from non-renewable sources. The manipulation of seed oil content has been the subject of much effort over the past 15 years. Most approaches have used transgene technology in order to create new oil profiles in existing major crops such as rapeseed and soybean. These studies have met with mixed success as outlined in the case study of petroselinic acid in transgenic rapeseed. An alternative approach is to use the latest methods of molecular marked assisted selection for the domestication of plants with novel and useful oil compositions. Recent progress in genomics and the application of molecular marker technology now make it feasible to envisage the relatively rapid domestication of entirely new crops such as cuphea, meadowfoam and coriander as important sources of oil-based renewable raw materials, particularly for non-food use.

Introduction

Seeds crops provide the vast majority of edible calories consumed by human societies around the world. The most important nutritional components from seeds are carbohydrates, e.g. from cereals, oils, e.g. from oilseeds, and proteins, e.g. from legumes. Many seeds are also important sources of vitamins such as the antioxidant, vitamin E, which is particularly abundant in oil-rich seeds. Finally, seeds also have the potential to serve as raw materials for a vast range of industrial and pharmaceutical products including plastics, lubricants, paints, cosmetics and therapeutic agents. During the past decade there have been dramatic advances in the understanding of the biosynthetic processes involved in the formation of these useful products in seeds. We have also seen the development of new biotechnological methods which can be used to improve the yield and quality of seed products both for edible and industrial purposes. This review will focus mainly on the application of such techniques to oilseeds but similar methods can and are being used for the improvement of other crop types.

Trends in Oil Crop Production

Oil crops are one of the most valuable traded agricultural commodities and are probably worth well over $100 billion per year. The current volume of traded vegetable oils is over 70 Mt (million tonnes) per year and is predicted to increase to over 100 Mt per year by the year 2010 (Murphy, 1996). This is probably a considerable underestimate of the total volume of oil production since, particularly in developing countries, most vegetable oils are consumed locally rather than entering into international trading networks. Like many other crop types, oil crops are dominated by a very small number of major species. About 75% of globally traded vegetable oil comes from the four major crops, soybean, oil palm, rapeseed and sunflower. Market projections for the next 10–15 years predict that this domination of the global oils market by the 'big four' oil crops will continue and indeed, be reinforced (Murphy *et al.*, 1999). For example, the establishment of new plantations in South East Asia is predicted to double the oil palm production over the next 15 years making it by then the most important source of vegetable oil in the world (Basiron and Thiagarajan, 1998).

The vast majority of vegetable oil production is currently used for edible commodities such as margarines, cooking oils and processed foods. Only about 15% of production goes towards the manufacture of oleochemicals, i.e. industrial products derived from oil crops (Murphy, 1994). Over the next few decades, two important factors will contribute to an expansion in the markets for both edible and industrial oil crops. Firstly, predictions based upon current demographic trends indicate that the global human population will at least double before stabilizing at some time in the mid-21st century. This population increase will be coupled with rising levels of affluence, particularly in many developing countries. These trends will result in increased demands for edible oil products which will be most pronounced in countries such as China and India where recent increases in living standards have already resulted in significantly higher per capita consumption of vegetable oil products (Anon., 1998). Secondly, it is well known that fossil hydrocarbon reserves are a non-renewable resource. The latest estimates from six different national and international agencies predict that world oil production from petroleum reserves will peak at some time between the years 2000 and 2020 and will decline thereafter (Kerr, 1998). In addition to being an important source of energy, e.g. for electricity generation and vehicle fuel, petroleum is also the source of a huge range of petrochemicals. These petrochemicals are the raw materials for the manufacture of products such as plastics, textiles, lubricants, paints, varnishes, etc. Once non-renewable hydrocarbon resources, such as petroleum and oil, are exhausted there will be no other source of such products other than oleochemicals derived from oil crops.

In view of the above trends, the challenge for researchers in the coming years will be to produce oil crops with higher yields in order to satisfy increased demands and also to increase the spectrum of useful products, whether for industrial or edible use, that can be obtained from these oil crops. It will be important to foster and exploit research in three major overlapping areas in order to achieve these important goals. Firstly, it will be necessary to improve our understanding of the biochemical and physiological processes involved in the biosynthesis of plant oils and related compounds such as sterols and vitamin E. Secondly, the use of recombinant DNA technology and related disciplines such as genomics should be extended beyond the 'first generation' oil crops, such as rapeseed and soybean, to other species such as sunflower, linseed and oil palm which may be technically more challenging with regard to transformation (Murphy, 1996). Thirdly, a much higher priority should be given to the development of strategies for the domestication of new oil-producing crop species. The introduction of new species will lessen the dependence of farmers on the very restricted number of major crops grown at present, particularly in the more developed countries. Until now, relatively little effort has gone into domesticating new species for the production of different types of seed oil. Instead, interest has been focused on transferring genes from species which already produce useful novel fatty acids into the major oil crops. However, as discussed later, this strategy may not always be successful. In addition, recent advances in plant genetics, such as map-based cloning and the use of genomics, now make it a realistic option to consider domestication as a viable medium-term alternative to transgenic technology for the production of novel oils for future generations (Murphy, 1998).

Potential and Limitations of Transgene Technology

The manipulation of seed oil content via transgene insertion has been one of the earliest successful applications of modern biotechnology in agriculture. Indeed, in 1995 the first transgenic crop with a modified seed composition was approved for unrestricted commercial cultivation in the USA. This crop was a new rapeseed variety containing an oil enriched in lauric acid which has use in the detergent industry (Murphy, 1996). Many of the larger agricultural companies have made considerable investments in transgene technology particularly as applied to oil crops. One of the major drivers for such investment was a perception that oil quality is a relatively plastic phenotype that could be substantially altered with no effect on other aspects of plant development. It was also expected that significant changes in oil quality, i.e. fatty acid composition, could be effected in crops such as rapeseed by the insertion of one or two key genes. Fortuitously, rapeseed, which is one of the major oil crops, proved to be particularly amenable to tissue culture and regeneration and could therefore be transformed relatively easily. More recently, soybean transformation has also become relatively routine (at least within companies), albeit much more labour intensive than rapeseed transformation (Krebbers *et al.*, 1997). These expectations were to some extent borne out by some significant early successes in producing transgenic rapeseed varieties. The most notable achievement to date is probably the development of rapeseed lines with 40–60% lauric acid in their seed oil. Rapeseed normally accumulates little or no lauric acid (Murphy, 1996). Nevertheless, attempts to achieve higher levels of other novel fatty acids in seed oils have met with much less success and it is apparent that single-gene insertions will often not give the required or even the expected results (Kinney, 1997). There have also been several reports that the presence of novel fatty acids in transgenic plants can sometimes lead to the induction of catabolic pathways which result in the breakdown of the desired novel fatty acid, i.e. the plant recognizes the 'strange' fatty acid and far from tolerating it, may even actively eliminate it from the seed oil (Eccleston and Ohlrogge, 1998; Murphy *et al.*, 1999).

Petroselinic Acid in Transgenic Rapeseed – a Case Study

In our laboratory, we have examined the consequences of producing a novel fatty acid in transgenic rapeseed. Most conventional rapeseed varieties produce an oil enriched in oleic acid, which is an 18 carbon fatty acid with one double bond in the Δ9 position. In contrast, many plants in the *Apiaceae* accumulate high levels of an isomer of oleic acid called petroselinic acid. This fatty acid also has 18 carbons and one double bond but in this case in the Δ6 position. Petroselinic acid is a potential raw material for the manufacture of detergents and polymers and there is a great interest from several large chemical companies in the large scale production of this fatty acid via oilseed crops.

Coriander, which is a member of the *Apiaceae*, accumulates as much as 80% of its seed oil as petroselinic acid. This is due to the activity of a novel Δ4 palmitoyl-ACP desaturase which is not present in rapeseed. We isolated the gene encoding this enzyme from coriander and inserted it under the control of appropriate tissue specific promoters into rapeseed explants. The resulting transgenic plants were then examined for their seed fatty acid composition and related patterns of gene expression.

Expression of the coriander Δ4 palmitoyl-ACP desaturase gene in transgenic rapeseed resulted in the accumulation of over 60% of petroselinic acid in developing embryos during the initial stages of storage oil accumulation as shown in Fig. 40.1. The C16 desaturase produces 16:1 Δ4 but this is then efficiently elongated by the endogenous C16 elongase activity to produce 18:1 Δ6, i.e. petroselenic acid. The embryos continued to accumulate petroselinic acid until the mid-cotyledonary stage after which the amount of petroselinic acid per seed decreased sharply until it only formed 1–3% of the total fatty acids in mature seeds. These results indicated that, although the transgene was expressed at high levels and was able to direct the biosynthesis of petroselinic

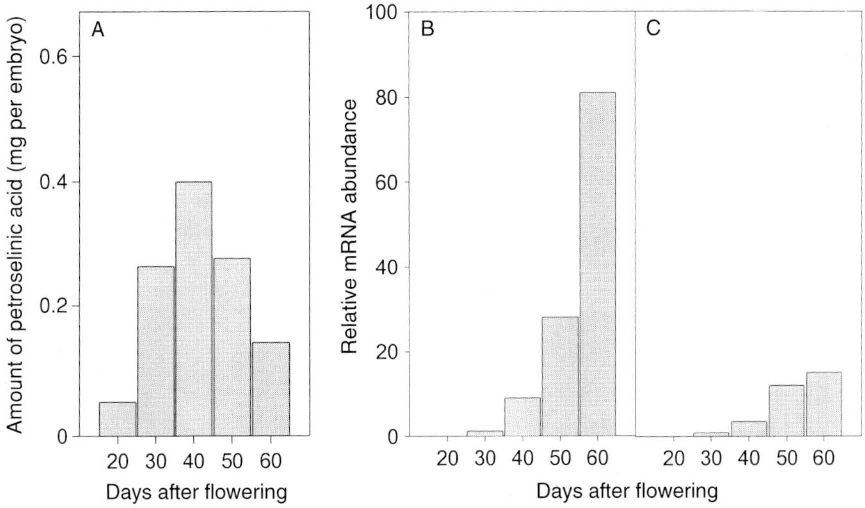

Fig. 40.1. Induction of catabolic gene expression in transgenic rapeseed producing petroselinic acid. A, Transient formation and breakdown of petroselinic acid during seed development. B, Strong induction of mRNAs encoding both isocitrate lyase (shown) and malate synthase (not shown) was found in all five independent transformants expressing the Δ4 desaturase gene. C, Control wild type or transgenic plants transformed with different constructs (e.g. 35S Gus) showed only small rises in levels of ICL and MS transcripts.

acid, additional factors in the transgenic plants resulted in the breakdown of the novel fatty acid and largely negated the activities of the transgene.

In order to elucidate the possible factors responsible for petroselinic acid breakdown in the transgenic plants, the expression profiles of a number of glyoxylate cycle genes were examined. The glyoxylate cycle, coupled with β-oxidation, is the mechanism by which storage fatty acids are broken down to provide energy for seedling development, following seed germination. Such catabolic genes are normally only expressed at very low levels during the storage product accumulation phase of seed development. However, in the transgenic rapeseed plants expressing the coriander desaturase, we observed an early and very strong induction of two glyoxylate cycle genes, isocitrate lyase (ICL) and malate synthase (MS), during embryo development. The induction of 10- to 20-fold higher than normal levels of ICL and MS gene expression (as determined by RT-PCR) was associated with the breakdown of petroselinic acid in the developing embryos of the transgenic rapeseed plants. This phenotype, i.e. the induction of the glyoxylate cycle genes, was strongly correlated with the presence of petroselinic acid in transgenic plants. Non-transgenic plants and transgenic plants harbouring different transgenes did not exhibit this induction of glyoxylate cycle genes. We conclude that there is an association between the production of a novel fatty acid, such as petroselinic acid, and the induction of catabolic pathways, such as β-oxidation and glyoxylate cycle, which result in the breakdown of the novel fatty acid during embryo development.

Why do the transgenic rapeseed plants break down this novel fatty acid? One reason may be that rapeseed is not as efficient as coriander in channelling petroselinic acid away from its cell membranes and towards accumulation in storage lipids. It has recently been shown that transgenic rapeseed plants accumulating another novel fatty acid, lauric acid, are less efficient at segregating this fatty acid from accumulation in membrane lipids than are the species such as *Cuphea* that normally accumulate lauric acid in their seed oils (Wiberg *et al.*, 1997). Accumulation of many novel fatty acids, including both lauric and petroselinic acids, can lead to membrane instability and may trigger protective mechanisms, leading to the removal of these fatty acids. Indeed, there has been another very recent report of the induction of β-oxidation and glyoxylate cycle genes in transgenic rapeseed producing lauric acid (Eccleston and Ohlrogge, 1998). It is also possible that storage lipids themselves may be available for remodelling via acyl exchange reactions as recently reported from several groups (Mancha and Stymne, 1997; Stobart *et al.*, 1997). This may be another mechanism by which novel fatty acids could be removed from a seed oil, for remodelling or breakdown, even after they have been deposited as storage oil bodies. In the future, it will be important to elucidate the mechanisms involved in channelling unusual fatty acids away from membrane lipids and ensuring that such protective catabolic pathways are not induced in transgenic plants. This will be an important objective if we are to realize the biotechnological goal of producing transgenic oil crops with high yields of novel, valuable fatty acids.

Other Challenges with Transgenic Crops

In addition to the technical problems associated with producing novel oils in transgenic crops, it has recently been pointed out that there are considerable challenges involved in the management of such crops (Murphy, 1996). For example, at present there are at least 10–20 different transgenic varieties of rapeseed at various stages of development, either in the laboratory or in field trials (see Table 40.1). All of these varieties look identical and the only differences are in their seed oil compositions. It is these different seed oil profiles that determine that one variety may be used for detergent manufacture, while other varieties may be used for margarines, cosmetics, therapeutic agents or lubricants. There is considerable scope for mixing of different batches of seeds, plants or oil, all of which can result in an adulterated product with little or no commercial value. Clearly, the segregation and identity preservation of such mutually incompatible commodity streams raises formidable challenges at all levels of production, ranging from sowing, harvesting, storage, crushing and processing. The expensive and sophisticated analytical equipment required to differentiate between the different transgenic varieties will not be available to all growers, crushers or processors and therefore there is serious potential for mixing and cross-contamination of different seed lots. Segregation of transgenic seeds and their downstream products may be feasible for low volume, high value niche applications, such as pharmaceuticals, but its

Table 40.1. Transgenic rapeseed and soybean varieties currently in development around the world.

Seed product	Industrial product	First field trials	Development status[a]
40% Stearic (18:0)	Margarine, cocoa butter	1994	***
40% Lauric (12:0)	Detergents	1994	***
60% Lauric (12:0)	Detergents	1996	***
85% Oleic (18:1 Δ_9)[b]	Food, lubricants, inks	1995	***
Petroselinic (18:1 Δ_6)	Polymers, detergents	(?)	**
'Jojoba' wax (C_{20}, C_{22})	Cosmetics, lubricants	1999(?)	**
40% Myristate (14:0)	Detergents, soaps, personal care	1996	**
90% Erucic (22:1)	Polymers, cosmetics, inks, pharmaceuticals	1999(?)	**
Ricinoleic (18:1-OH)	Lubricants, plasticizers, cosmetics, pharmaceuticals	1999(?)	**
Epoxy and acetylinic acids	Paints, varnishes and other coatings	1999(?)	**
Polyhydroxybutyrate	Biodegradable plastics	1999(?)	**
Phytase	Animal feed	1994	***
Industrial enzymes	Fermentation, paper manufacture, food processing	1998(?)	**
Novel peptides	Pharmaceuticals	1995	***

[a]Status: *key genes already cloned; **transgenic plants produced; ***field trials underway.
[b]This product is present in both transgenic rapeseed and soybean. The other products are present only in transgenic rapeseed.

commercial viability for medium to high volume cropping systems remains uncertain.

In many countries, particularly in Europe, there is also considerable consumer resistance to the cultivation of genetically modified crops, particularly for food use. The magnitude of these concerns was probably exacerbated in the past two years by the release into the general commodity stream of transgenic herbicide resistant soybeans developed by the agribusiness company, Monsanto. Soybean products are in some 60% of all processed foods in supermarkets. Since the transgenic soybeans were mixed with non-transgenic soybeans, it was not possible to label such products as had been done previously, e.g. with a transgenic tomato paste developed by Zeneca which was clearly segregated and sold separately from tomato paste made from non-transgenic plants. The impact of consumer concerns has recently been admitted by a senior Monsanto executive who stated:

> 'some European consumer organizations clearly had serious concerns about the beans and felt unprepared for their arrival. Not enough public discussion and debate had occurred about the nature of the soyabeans, the environmental benefits associated with them, and the safety review process for them. I wish

> Monsanto had been more perceptive about this and that we had started sooner and devoted more resources, working with others in the food system, to prompt more public dialogue about the soyabeans and agricultural biotechnology generally. Opinions might still have differed, but less anxiety and confusion would have existed in the discussion.'
>
> (Auxenfans, 1998)

Partially as a result of these consumer concerns France has already imposed a three year moratorium on the release of genetically modified crops and the UK is considering a similar moratorium at present. Although this situation may well change with better public education about genetic research and with more thorough risk-assessment programmes, consumer resistance may well continue to be an important factor which limits the application of transgene technology, at least in the near future.

The final argument against an ever increasing reliance on a very small number of major crops is the concern that large scale monocultures may be more prone to opportunistic infection by pests and diseases as well as reducing biodiversity of both plants and animals at the farm level. It is official policy of the European Union to encourage greater crop diversity and therefore to favour the introduction of new crops rather than the continued increase in the cultivation of existing major crop species.

Impact of Genomics

During the past 10 years a great deal of plant research has concentrated on a single model species namely the cruciferous weed, *Arabidopsis thaliana*. *Arabidopsis* has the virtue of containing a relatively small genome of only about 120 Mb (Meinke *et al.*, 1998). This genome is arranged on five chromosomes with relatively little repetitive DNA. In contrast, the major crop species, maize and barley, have genome sizes of 2500 and 5000 Mb respectively while the hexaploid species, wheat, has a genome of 16,000 Mb. The relatively small genome size of *Arabidopsis* has made it the first plant target for a multinational DNA sequencing project which should cover the entire genome by the end of 2000. Already, physical maps of the *Arabidopsis* genome are available and the vast majority of the estimated 15,000 *Arabidopsis* expressed genes have already been identified as expressed sequence tags (ESTs) (Somerville, 1996).

The coming challenge will be to utilize this formidable genetic resource based on a single relatively simple model plant to effect improvements in the major dicot and monocot crop species. During the past few years, progress has been accelerating rapidly in transferring technologies and knowledge developed in *Arabidopsis* to some of the major crop species. Another encouraging development has been the recent description of extensive synteny between all of the cereal genomes which has revealed that they are composed of very similar chromosome segments (Moore, 1995; Moore *et al.*, 1995). By rearranging these segments slightly and disregarding the repetitive DNA, it is possible to reconstitute the 56 different chromosomes found in wheat, rice, maize, sorghum, millet and sugarcane into a single genomic arrangement. This means that a genetic locus which is mapped into a major cereal crop can also be localized by comparative genome analysis in all of the other major cereal crops including rice and maize. Even more remarkable is the growing appreciation that the order of genes in the genomes of monocots may in some cases be very similar to that of some dicots including *Arabidopsis*. This means that information from the *Arabidopsis* genome sequencing project may be directly applicable to the improvement of agronomically relevant traits in a wide range of plant species, not only in closely related crops such as rapeseed, but also in very distantly related species such as maize and oil palm.

Domestication of New Crop Species

All of our existing major crop species have been through a continual process of domestication and improvement since the beginnings of agriculture more than 10,000 years ago. However, research aimed at the domestication of new species has rarely found favour with the agribusiness industry or with Government funding agencies. The process of domestication is perceived to be extremely slow. It is also obviously limited by the climatic range of the candidate crop. For example, some potential novel oil crops are found in tropical regions and therefore are unlikely ever to be domesticated for cultivation in temperate regions. Nevertheless, given the extraordinary diversity of seed oil contents found in the natural world (Murphy, 1996), it is likely that a species producing useful quantities of a particular seed oil can be found in several different climatic zones. Another potential problem is that the novel crops may have a different growth habit from existing crops and may therefore not be suitable for harvesting using existing equipment. These problems occur in addition to the often serious agronomic difficulties exhibited by many candidate oilseed plants. Nevertheless, we should not be discouraged by such challenges. The application of modern biotechnological methods, such as genome mapping and molecular marker assisted selection, now make it feasible to consider domestication of such species for commercial agriculture within one or two decades (Martin, 1998; Murphy, 1998). This time horizon is well within the normal development lifetime of many other commercial/public ventures, such as the production of most new pharmaceuticals or the completion of large civil engineering projects.

Among the most common agronomic problems faced by new oil crops are asynchronous flowering, premature pod shattering, allogamy, low seed oil content, poor germination rates and low seed yield. Until recently, the solution to such problems was to look for natural variation, e.g. reduced pod shattering or increased oil yield, and to attempt to produce reasonably defined genetic lines in which these characters were well expressed. The production of more uniform inbred lines has now been made much faster and easier by techniques such as double haploids and by the development of molecular maps for at least some of the major species. In the future, new kinds of molecular markers, such as microsatellites, promise to make the process of producing a detailed genetic map for any species (e.g. candidates for domestication) much more rapid and much less expensive than in the past.

The recent advances in genomics and in gene function studies have allowed us to understand the detailed genetic basis of many complex traits, such as flowering time, heights and disease resistance (Murphy, 1998). Many of these complex traits have previously been regarded as being controlled by large numbers of genes, which made them difficult to manipulate by simple Mendelian genetics. However, there are now several striking examples where the vast majority of the variation underlying such complex characters has been mapped on to only a very small number of quantitative trait loci (QTL) (Doebley, 1993; Doebley *et al.*, 1997; Martin, 1998). Once such genes have been mapped in a model species such as *Arabidopsis*, techniques such as positional cloning can be used to isolate the gene of interest and to verify its function in the laboratory. Within the next few years, more and more important agronomic traits will be explicable in terms of a relatively small number of key genes which account for most of the observed variation. Such information can then be used for the selection of plants expressing such genes. For example, in our own laboratories we are currently studying genes regulating characters such as pod shattering, oil yield, oil quality, flowering and canopy architecture. Such research has the potential to provide tools for the much more rapid domestication of new oil crops within the next few years.

Conclusions

During the past few years nearly all the genes encoding enzymes of seed oil biosynthesis have been cloned. Nevertheless there have been many surprising results when these genes are expressed in transgenic plants. This highlights our continuing relative ignorance of the interactions between components of storage lipid biosynthesis and other metabolic pathways *in vivo*. We also know very little about the mechanisms regulating the partitioning of carbon to storage products in sink tissues such as oilseeds. A very promising recent approach currently underway at the John Innes Centre is to identify and map QTL that contribute to characters such as oil yield or fatty acid composition. This can be combined with map-based cloning of the major genes involved and hence the elucidation of their function (Martin, 1998). Such a 'top down' genetic approach may allow for the isolation of higher level regulatory components, e.g. transcription factors, that have already been shown to be important in the control of entire metabolic pathways such as anthocyanin biosynthesis (Murphy, 1998). It is important that this is combined with the 'bottom up' approaches via biochemistry and analysis of individual genes and enzymes, in order to understand fully and hence to be able to modify the complex processes of oil accumulation in seeds and other plant tissues.

It is likely that, in the future, both transgenic oil crops and newly domesticated oil crops will be developed in order to provide the increased amount and diversity of oils which will be required for edible and industrial uses. It is important that we recognize that each of these approaches has both positive and negative aspects. It will be a combination of the two strategies that is most likely to supply the increasing demands for plant oils in the 21st century and beyond.

References

Anon. (1998) China oil, seed imports: up, and up, and up. *Inform* 9, 698–700.

Auxenfans, B. (1998) Monsanto Report on Sustainable Development including Environmental Health and Safety Performance, p. 21.

Basiron, Y. and Thiagarajan, T. (1998) Ten year outlook for: South East Asia. *Inform* 9, 1049–1052.

Doebley, J. (1993) Genetics, development and plant evolution. *Current Opinion in Genetics and Development* 3, 865–872.

Doebley, J., Stec, A. and Hubbard, L. (1997) The evolution of apical dominance in maize. *Nature* 386, 485–488.

Eccleston, V.S. and Ohlrogge, J.B. (1998) Expression of lauroyl-acyl carrier protein thioesterase in *Brassica napus* seeds induces pathways for both fatty acid oxidation and biosynthesis and implies a set point for triacylglycerol accumulation. *The Plant Cell* 10, 613–621.

Kerr, R.A. (1998) The next oil crisis looms large – and perhaps close. *Science* 281, 1128–1131.

Kinney, A.J. (1997) Manipulating flux through plant metabolic pathways. *Current Opinion in Plant Biology* 1, 173–178.

Krebbers, E., Broglie, R., Hitz, B., Jones, T. and Hubbard, N. (1997) Biotechnological approaches to altering seed composition. In: B.A. Larkins and I.K. Vasil (eds) *Cellular and Molecular Biology of Plant Seed Development.* Kluwer Academic Publishers, The Netherlands, pp. 595–633.

Mancha, M. and Stymne, S. (1997) Remodelling of triacylglycerols in microsomal preparations from developing castor bean (*Ricinus communis* L.) endosperm. *Planta* 203, 51–57.

Martin, G.B. (1998) Gene discovery for crop improvement. *Current Opinion in Biotechnology* 9, 220–226.

Meinke, D.W., Cherry, J.M., Dean, C., Rounsley, S.D. and Kournneef, M. (1998) *Arabidopsis thaliana*: a model plant for genome analysis. *Science* 282, 662–682.

Moore, G. (1995) Cereal genome evolution: pastoral pursuits with 'Lego' genomes. *Current Opinion in Genetics and Development* 5, 717–724.

Moore, G., Devos, K.M., Wang, Z. and Gale, M.G. (1995) Grasses, line up and form a circle. *Current Biology* 5, 737–739.

Murphy, D.J. (1994) (ed.) *Designer Oil Crops.* VHC Press, Weinheim, Germany.

Murphy, D.J. (1996) Engineering oil production in rapeseed and other oil crops. *Trends in Biotechnology* 14, 206–213.

Murphy, D.J. (1998) Impact of genomics on improving the quality of agricultural products. In: Dixon, G.K., Copping, L.G. and Livingstone, D. (eds) *Genomics: Commercial Opportunities from a Scientific Revolution.* Society of Chemical Industry, pp. 199–210.

Murphy, D.J., Fairbairn, D.J. and Bowra, S. (1999) Expression of unusual fatty acids in transgenic rapeseed causes induction of glyoxylate cycles genes. *John Innes Centre Annual Report*, pp. 44–46.

Somerville, C. (1996) The physical map of *Arabidopsis* chromosomes. *Trends in Plant Science* 1, 2.

Stobart, K., Mancha, M., Lenman, M., Dahlqvist, A. and Stymne, S. (1997) Triacylglycerols are synthesised and utilised by transacylation reactions in microsomal preparations of developing safflower (*Carthamus tinctorius* L.) seeds. *Planta* 203, 58–66.

Wiberg, E., Banas, A. and Stymne, S. (1997) Fatty acid distribution and lipid metabolism in developing seeds of laurate producing rape (*Brassica napus* L.). *Planta* 203, 341–348.

41 Manipulating Starch Quality in Seeds: a Genetic Approach

T.L. WANG, T.Y. BOGRACHEVA AND C.L. HEDLEY

John Innes Centre, Norwich Research Park, Colney, Norwich NR4 7UH, UK

The main carbohydrate reserve in plants is starch. It forms an important part of nutrition, but also provides a useful raw material for industry. Industrial uses normally involve the chemical modification of the starch to alter its properties. There is an advantage, however, to producing a wider range of natural starches to circumvent the need for chemical modification. If starch structure is to be manipulated to meet future demands, an understanding is needed of the way in which plants produce starch, the physical properties of that starch and its behaviour when processed. The pea is used as a model to understand how plants can be genetically modified to obtain starches with novel properties. Peas contain a unique type of starch and six loci have been identified that influence starch content and its physical properties. Here we describe mutants at each of the loci, the properties of pea starch and the starches of the mutants in comparison with those of conventional starch crops. The potential for the generation of additional variation and novel starches by the creation of double mutants is also discussed.

Introduction

The main carbohydrate reserve in plants is starch. It forms an important part of our nutrition, but also provides a useful raw material for industry. Cereal seeds provide the most widely used starches. The genetic variation available and the ability to chemically modify the polymer make the starch from these crops suitable for a wide range of food ingredients and industrial products (Lillford and Morrison, 1997). Relatively pure cereal starches, however, can be difficult to extract. Furthermore, chemical modification, by forming esters, ethers or cross-links, is frequently required to alleviate shear damage and 'set-back' of the components following processing (Lillford and Morrison, 1997). Such modifications render the product less 'natural' and are becoming increasingly undesirable for environmental reasons. Hence it is worthwhile to investigate

(eds M. Black, K.J. Bradford and J. Vázquez-Ramos)

native starches with inherently useful properties from other crops. One such group of crops is the legumes (Lillford and Morrison, 1997) and peas are a prime candidate within this group to provide novel material. Although not currently viewed as a starch crop, peas have 50% starch by dry weight in their seeds. This starch *per se* is underutilized, since peas, like most legumes, are grown and used primarily as a source of protein, the basis of their commercial value.

In addition to the direct use of the pea seed as a source of starch, this plant is, and has been for many years, a useful research tool. If we are to manipulate starch structure to meet future demands, we need to understand the way in which plants produce starch, the physical structure of the starch granules, and the physico-chemical properties of starches which underpin their functionality and hence industrial uses. Pea is ideal for such work. Indeed, much has been learned already about starch biosynthesis and structure by using pea and the mutants available (Smith *et al.*, 1997; Wang *et al.*, 1998), since its large seed and flower make it ideal for biochemistry, chemistry and genetics. Here we review the genetic variation that is available for such studies, the properties of the starches derived therefrom, and indicate the potential for generating novel starches from peas. Coverage will be brief since detailed reviews have been produced recently (Smith *et al.*, 1997; Wang *et al.*, 1998) and the reader is referred to these for more information.

Pea Mutants

Gregor Mendel made use of a character which affected the shape of the seed in his seminal work defining the laws of heredity (Mendel, 1865). This character determined whether the seed was smooth and round or wrinkled. To date, at least ten loci have been identified that encode genes affecting the wrinkledness of the seed (Hedley and Wang, 1987; Wang *et al.*, 1990; Wang *et al.*, 1998), and five of these, the *rugosus* loci, are known to affect the starch content of the seed. Two of the *rugosus* loci, *r* and *rb*, are represented by naturally occurring (Kooistra, 1962) and chemically-induced mutants (Wang *et al.*, 1990); the other *rugosus* loci are represented by chemically-induced mutants only (Wang *et al.*, 1990). These new loci have been termed, *rug3*, *rug4* and *rug5*. An additional locus (*low amylose* or *lam*) affecting starch composition was identified from the same population of mutagenized seed by using the iodine-staining properties of starch granules as a screen, rather than the shape of the seed (Denyer *et al.*, 1995). The remaining loci either affect the seed coat or have not yet been defined.

The enzymes affected by each of the loci influencing starch composition have been characterized (Table 41.1) and the genes encoding those enzymes have all been cloned and sequenced. Furthermore, one or more alleles at most of the loci have been sequenced and the mutations identified. Each of the enzymes, except for ADPglucose (ADPG) pyrophosphorylase, is encoded by a single locus. For ADPG pyrophosphorylase, there are two loci, one encoding the large and the other the small subunit; the *rb* locus encodes the former.

Table 41.1. Enzyme activities affected by mutant loci.

Mutant	Enzyme activity	Reference
r	Starch-branching enzyme A (SBEA)	Bhattacharyya *et al*., 1990
rb	ADPG pyrophosphorylase	Hylton and Smith, 1992
rug3	Plastidial phosphoglucomutase	Harrison *et al*., 1998
rug4	Sucrose synthase (susI)	Craig *et al*., 1999
rug5	Starch synthase (SSII; 77 kDa)	Craig *et al*., 1998
lam	Granule-bound starch synthase (GBSSI; 59 kDa)	Denyer *et al*., 1995

The starch and amylose contents of the *rugosus* and *lam* mutants are given in Fig. 41.1. The mutants at each locus can be identified readily from their starch composition as they fall into distinct groups (encircled). Furthermore, differences can be identified within each group. For example, within the *r* group the mutant lying away from the remainder has a less severe effect than the rest. Its mutation in the SBE gene is outside the main active site and is in a less well-conserved region (MacLeod, 1994). It is also interesting to note that, if a straight line is drawn on this graph from the origin to the WT, the mutants lying on or close to this line (*rug3*, *rb* and *rug4*) all affect enzymes supplying substrates to the polymerases and branching enzymes. Those falling away from the line (*r*, *rug5* and *lam*) build the polymers themselves. With the substrate suppliers, therefore, as the amount of starch increases, the amount of amylose increases.

These mutants define the pathway to starch in peas. The biosynthesis of starch has been dealt with in detail over the last few years, the latest review being by Smith *et al.* (1997). Briefly, glucose-6-phosphate is imported into the pea plastid, converted to glucose-1-phosphate and thence to ADP-glucose. This is then taken either by a 'waxy'-type granule-bound starch synthase for the production of amylose or another synthase for the production of amylopectin in concert with starch-branching enzymes. The lack of starch in *rug3* mutants (Fig. 41.1) indicates that only glucose-6-phosphate can be imported into pea plastids.

Starch Structure

Starch consists of two polymers: the relatively unbranched amylose and the highly branched amylopectin (Fig. 41.2A). The amylopectin chains are highly complex, form 'clusters' (Fig. 41.2B; Manners, 1989), their short chains can interact with each other to form double helices (French, 1984) and these helices form little crystals (crystallites). According to one model, these crystals can be constructed into lamellae to form the granule (Oostergetel and van Bruggen, 1993). The double helices can been arranged in two different ways with different amounts of bound water to create different polymorph types (Sarko and Wu, 1978), either closely packed (A-type) or more open with more bound water (B-type). These types of arrangement endow a crystalline

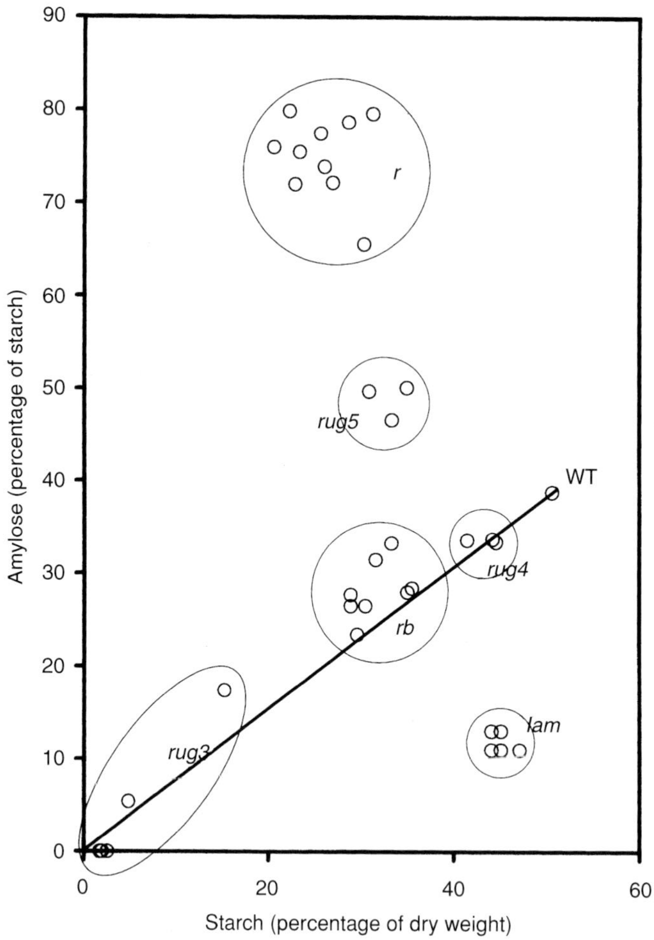

Fig. 41.1. Composition of pea starches. The starch and amylose contents of seeds from different pea mutants are indicated, together with the wild type (WT). Starch and amylose were measured following DMSO extraction of the starch and either enzymatic degradation to glucose for starch content, or iodine staining for amylose content according to Wang *et al.* (1990). This has a tendency to overestimate amylose content in some circumstances. For example, when starch was analysed chromatographically, the amylose in the *lam* mutants was not detectable. Data on *rugosus* lines were from sixth backcross material and from *lam*, second backcross.

structure on starch. Since the granules contain unorganized amylose and not all amylopectin is ordered, there are regions that are disorganized, termed amorphous regions. Hence the granule should be regarded as a semi-crystalline structure (Wang *et al.*, 1998). The semi-crystalline nature of starch granules renders them amenable to studies such as X-ray diffraction to determine crystal structures, differential scanning calorimetry (DSC) for heat flow analysis during melting and gelatinization, NMR for quantification of mobility

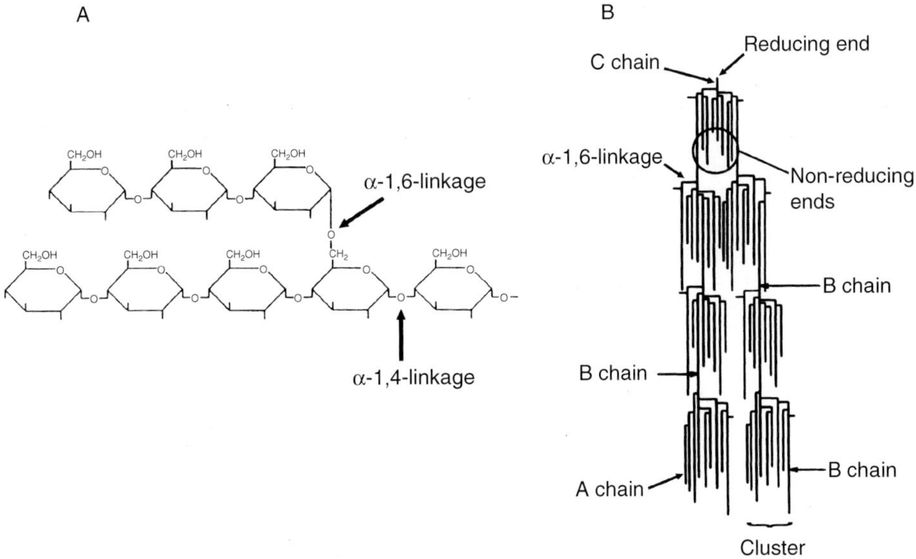

Fig. 41.2. Chemical structure of starch. (A) Polymers of starch indicating branching points. Amylose is essentially an unbranched chain of glucose units with α-1,4-linkages, whereas amylopectin has additional α-1,6-linkages creating a branched polymer. (B) Amylopectin, which consists of a number of different types of polymer chains creating a 'cluster' structure (see text for further details).

and order and polarized light microscopy for visual changes in crystallinity and swelling behaviour (Bogracheva *et al.*, 1998).

Cereal starches are mainly of the A-type of polymorph and hence are called A-type starches, whereas potato starch is of the B-type. Pea starch, however, is a mixture of A and B polymorphs and represents a third type of starch (C-type). In pea, the A-type of polymorph is arranged around the B-type as shown in Fig. 41.3B. This arrangement can be clearly demonstrated when granules are viewed under cross polarizers. Since the crystallites are arranged symmetrically around a central point in the granule, they show birefringence, which appears as white light with a characteristic black cross in the middle (Fig. 41.3A). If pea starch granules are melted slowly in a salt solution to the threshold temperature for the B-type polymorphs and then cooled, only the B-type crystalline structure is lost, indicated by the black centre (Fig. 41.3C). X-ray and DSC studies have shown that the B-type structure is completely lost under these conditions. Heating the granule a second time brings about a single transition in endothermic heat flow, which accounts for A-type polymorphs only (Bogracheva *et al.*, 1998).

Basic differences in starch granules can be identified by microscopy from their shape. Such differences have been observed since the earliest days of the microscope and were first reported for pea starches in 1903 (Gregory, 1903). Round-seeded peas (*RR*) have simple (or smooth) starch granules, whereas wrinkled-seeded (*rr*) peas have compound (or, more correctly, complex) granules. All pea mutants so far examined fall into these two classes: WT, *rb*, *rug4*

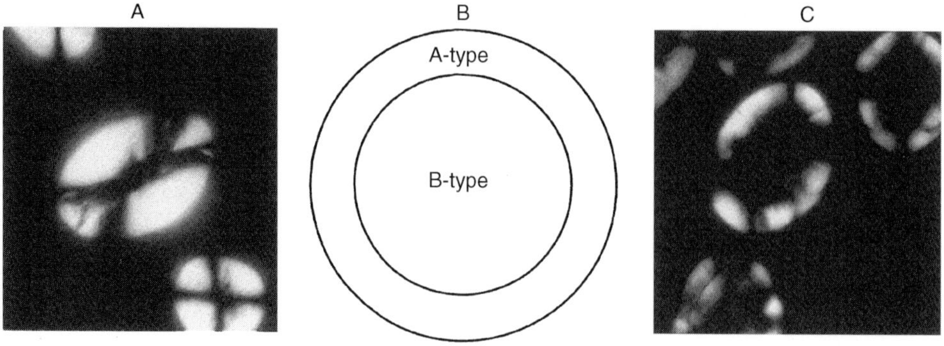

Fig. 41.3. Pea starch structure. The crystalline nature of pea starch is revealed when granules are viewed under polarized light; white areas are crystalline, black are non-crystalline (A). Pea starch is C-type and consists of A and B polymorphs as shown in (B). This arrangement can be seen clearly under polarized light when starch granules are melted slowly as in (C). The central B polymorphs melt first and consequently lose their crystalline structure.

and *lam* all have smooth granules; *r* and *rug5* have complex granules. These differences can be observed readily under the scanning electron microscope (Hedley *et al.*, 1996).

To look in more detail at their structure, starches are extracted slowly in water or weak alkali (Bogracheva *et al.*, 1995) and then analysed in several different ways. As mentioned above, when heated, starches melt, lose their crystalline structure and then swell. This can be viewed directly under polarized light, or can be measured by calculating the heat required during the process using DSC. Furthermore, since starch is semi-crystalline, its structure can be analysed directly using X-ray diffraction methods. Starches from different crops show different characteristics when measured by these techniques. Analysis of starches from maize and potato by wide-angle X-ray diffraction indicates the A- and B-type nature of the starches, each type having different maxima (peaks) missing from a total of 15 in the diffraction angle spectrum (Table 41.2). These spectra can be distinguished from that of pea, since wild-type pea has all 15 peaks present (Bogracheva *et al.*, 1999).

Nevertheless, within a single species there are also major differences. Although most pea starches can be differentiated from those of other species as C-type, there is much variation. For example, starch from the *r* mutant totally lacks two of the 15 peaks in the spectrum (Bogracheva *et al.*, 1999; Table 41.2) which indicates that starch from this mutant is B-type and not C-type. A difference could also be seen when the amounts of the polymorph types were calculated from X-ray data. Three mutants stand out from such an analysis, *r*, *rug5* and *lam* (Table 41.3), all having more B polymorphs. The differences also become very apparent when heat uptake during melting and gelatinization is measured by DSC. Figure 41.4 shows the extremes of behaviour of mutant starches. Even within the smooth granule types there are major differences; *lam* represents one extreme in its shift to a lower temperature for the maximum in heat flow, whereas *rug3* shows the other with a shift to a much higher temperature. The *rug3* effect may be due to the fact that it has the

Table 41.2. Characteristic peaks (in °2θ) from X-ray diffraction analyses of starches*.

Peak no.	Maize	Potato	Pea (*r* mutant)	Pea (WT)
1	–	5.4	5.3	5.5
2	9.8	9.8	9.9	9.8
3	11.1	11.1	11.1	11.1
4	–	13.9	14.0	14.3
5	14.8	14.8	14.8	15.0
6	16.9	16.8	16.9	17.0
7	17.9	–	–	18.0
8	19.7	19.4	19.5	19.6
9	–	22.0	22.0	22.0
10	22.8	–	–	22.7
11	23.8	23.8	23.8	23.7
12	26.3	26.1	26.1	26.3
13	30.3	30.2	30.3	30.3
14	33.4	34.2	34.0	33.6
15	38.3	38.4	38.4	38.5

*X-ray diffraction peaks determined from spectra obtained using a Philips diffractometer with a fixed divergence slit. Data from Bogracheva *et al.* (1999).

Table 41.3. Characteristics of crystallinity of starches from pea mutants*.

Genotype	% A polymorphs	% B polymorphs	B/A
WT	59	45	0.8
rug4	57	39	0.7
rb	58	43	0.7
rug3	63	37	0.6
lam	29	69	2.4
rug5	45	52	1.2
r	nd	73	∞

*X-ray diffraction calculations obtained from measurements taken using a Philips Scientific diffractometer with an automatic divergence slit. Data from Bogracheva *et al.* (1999).

lowest B-type polymorph content of all the mutants. Complex granule types, represented here by the *r* mutant, show no maximum at all, but a gradual drift upwards in value. In general, mutants affecting substrate supply have less effect on polymorph composition than those affecting polymer synthesis, but they still can have major effects on the behaviour of the starch (Bogracheva *et al.*, 1999).

Potential of Pea Starches

It is clear that the physico-chemical properties of pea starches show much variation and differ from those of other species. Moreover, the variation detected

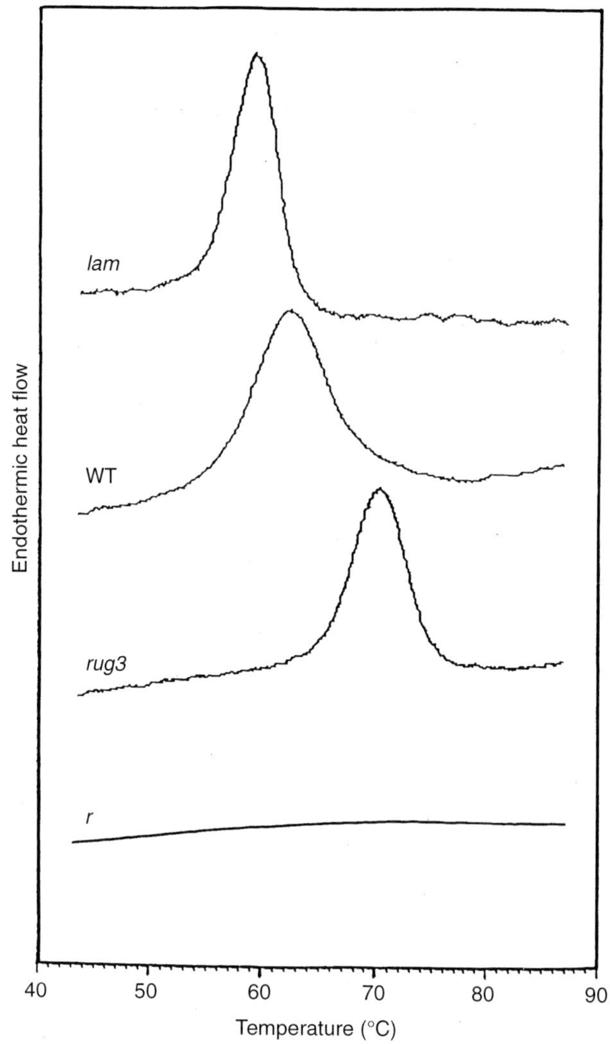

Fig. 41.4. DSC thermograms of pea starch. Water-extracted starch was analysed in suspension at concentrations of between 1.7 and 4%, depending on starch type, using a Setaram Micro-DSC. Data on *rugosus* lines were from sixth backcross material and from *lam,* first backcross. They were taken from Bogracheva *et al.* (1999).

has been due to the single mutants examined so far at the six loci affecting starch. There are, however, a number of different mutants (alleles) bearing different mutations at each locus, as has been mentioned earlier, and some of these show significant differences in their starch composition from the mutants examined to date. It is anticipated, therefore, that starch from such alleles will behave differently in the experimental analyses used. The mutants to date have all been backcrossed six times, apart from those at the *lam* locus which have only reached the fifth generation. In this way, background genetic variation is removed and hence each mutant can be directly compared with other alleles and with lines with mutations at different loci. Six backcrosses are required for the material to be considered near-isogenic. Once this is achieved, the material will represent a unique resource, as similar material is not available in any other crop plant.

To date, we have examined only material with single mutations at single loci. The potential exists to combine the mutations to produce double mutant lines and to study starches from such lines in the absence of background genetic variation. The double mutant, *r/rb*, has been available for some time and is known to have a different starch composition from either of the single mutants (Lloyd *et al.*, 1996). Moreover, combinations with *lam* should have very low amylose contents, if the *Lam* gene is responsible for amylose production (see earlier), but in preliminary analyses of *lam/rug* double mutants, significant quantities of amylose-like material were produced (L. Barber and T.L. Wang, unpublished data) and the amylose content of double mutants more closely resembled those of the corresponding *rugosus* line. Hence, the potential for generating novel variation in native starches must be high. The study of peas (and, most likely, other legumes), therefore, will not only reveal novel starch behaviours, but also will undoubtedly teach us more about starch biosynthesis, structure and functionality.

Acknowledgements

We should thank the BBSRC and MAFF for supporting our work on peas and our colleagues at JIC and IFR for providing information for this article.

References

Bhattacharyya, M.K., Smith, A.M., Ellis, T.H.N., Hedley, C. and Martin, C. (1990) The wrinkled-seed character of peas described by Mendel is caused by a transposon-like insertion in a gene encoding starch branching enzyme. *Cell* 60, 115–122.

Bogracheva, T.Y., Davydova, N.I., Genin, Y.V. and Hedley, C.L. (1995) Mutants at the r and rb loci affect the structure and physico-chemical properties of pea seed starches. *Journal of Experimental Botany* 46, 1905–1913.

Bogracheva, T.Y., Morris, V.J., Ring, S.G. and Hedley, C.L. (1998) The granular structure of C-type pea starch and its role in gelatinization. *Biopolymers* 45, 323–332.

Bogracheva, T.Y., Cairns, P., Noel, T.R., Hulleman, S., Wang, T.L., Morris, V.J., Ring, S.G. and Hedley, C.L. (1999) The effect of mutant genes at the *r*, *rb*, *rug3*, *rug4*, *rug5* and *lam* loci on the granular structure and physico-chemical properties of pea seed starch. *Carbohydrate Polymers* 39, 303–314.

Craig, J., Lloyd, J.R., Tomlinson, K., Barber, L., Edwards, A., Wang, T.L., Martin, C., Hedley, C.L. and Smith, A.M. (1998) Mutations in the gene encoding starch synthase II profoundly alter amylopectin structure in pea embryos. *Plant Cell* 10, 413–426.

Craig, J., Barratt, P., Tatge, H., Déjardin, A., Handley, L., Gardner, C.D., Barber, L., Wang, T.L., Hedley, C.L., Martin, C. and Smith, A.M. (1999) Mutations at the *rug4* locus alter the carbon and nitrogen metabolism of pea plants through an effect on sucrose synthase. *The Plant Journal* 17, 353–362.

Denyer, K., Barber, L.M., Burton, R., Hedley, C.L., Hylton, C.M., Johnson, S., Jones D.A., Marshall, J., Tatge, H., Tomlinson, K. and Wang, T.L. (1995) The isolation and characterisation of novel low-amylose mutants of *Pisum sativum* L. *Plant, Cell and Environment* 18, 1019–1026.

French, D. (1984) Organisation of starch granules. In: Whistler, R.L., BeMiller, J.N. and Paschall, E.F. (eds) *Starch: Chemistry and Technology*. Academic Press, San Diego, pp. 183–247.

Gregory, R.P. (1903) The seed characteristics of *Pisum sativum*. *New Phytologist* 2, 226–228.

Harrison, C.J., Hedley, C.L. and Wang, T.L. (1998) Evidence that the *rug3* locus of pea *Pisum sativum* L. encodes plastidial phosphoglucomutase confirms that the imported substrate for starch synthesis in pea amyloplasts is glucose-6-phosphate. *The Plant Journal* 13, 753–762.

Hedley, C.L. and Wang, T.L. (1987) Seed and foliar mutations in *Pisum*. In: Thomas, H. and Grierson, D. (eds) *Developmental Mutants in Higher Plants*. Cambridge University Press, Cambridge, pp. 219–244.

Hedley, C.L., Bogracheva, T.Y., Lloyd, J.R. and Wang, T.L. (1996) Manipulation of starch composition and quality in peas. In: Fenwick, G.R., Hedley, C.L., Richardson, R.L. and Khokhar, S. (eds) *Agri-Food Quality 95. An Interdisciplinary Approach*. Royal Society of Chemistry, Cambridge, pp. 138–148.

Hylton, C. and Smith, A.M. (1992) The *rb* mutation of peas causes structural and regulatory changes in ADP glucose pyrophosphorylase from developing embryos. *Plant Physiology* 99, 1626–1634.

Kooistra, E. (1962) On the differences between smooth and three types of wrinkled peas. *Euphytica* 11, 357–373.

Lillford, P.J. and Morrison, A. (1997) Structure/function relationship of starches in food. In: Richmond, P., Donald, A.M. and Frazier, P.J. (eds) *Starch: Structure and Function*. Royal Society of Chemistry, Cambridge, pp. 1–8.

Lloyd, J.R., Wang, T.L. and Hedley, C.L. (1996) An analysis of seed development in *Pisum sativum*. XIX. Effect of mutant alleles at the *r* and *rb* loci on starch grain size and on the content and composition of starch in developing pea seeds. *Journal of Experimental Botany* 47, 171–180.

MacLeod, M.R. (1994) Analysis of an allelic series of mutants at the *r* locus of pea. PhD thesis, University of East Anglia, Norwich, UK.

Manners, D. (1989) Recent developments in our understanding of amylopectin structure. *Carbohydrate Polymers* 11, 87–112.

Mendel, G. (1865) Versuche über pflanzen-hybriden. *Verhandlungen des naturforschenden Vereins in Brünn* 4, 3–47.

Oostergetel, G.T. and van Bruggen, E.F.J. (1993) The crystalline domains in potato starch granules are arranged in a helical fashion. *Carbohydrate Polymers* 21, 7–12.

Sarko, A. and Wu, H-C. (1978) The crystal structures of A-, B- and C-polymorphs of amylose and starch. *Starch* 30, 73–78.

Smith, A.M., Denyer, K. and Martin, C. (1997) The synthesis of the starch granule. *Annual Reviews of Plant Physiology and Plant Molecular Biology* 48, 67–87.

Wang, T.L., Hadavizadeh, A., Harwood, A., Welham, T.J., Harwood, W.A., Faulks, R. and Hedley, C.L. (1990) An analysis of seed development in *Pisum sativum*. XIII. The chemical induction of storage product mutants. *Plant Breeding* 105, 311–320.

Wang, T.L., Bogracheva, T.Y. and Hedley, C.L. (1998) Starch: as simple as A, B, C? *Journal of Experimental Botany* 49, 481–502.

42 Identification of Germination-specific Protein Markers and their Use in Seed Priming Technology

D. JOB[1], I. CAPRON[1], C. JOB[1], F. DACHER[2], F. CORBINEAU[2] AND D. CÔME[2]

[1]*Laboratoire mixte CNRS/Rhône-Poulenc (UMR041), Rhône-Poulenc Agro, 14–20 rue Pierre Baizet, 69263 Lyon cédex 9, France;* [2]*Physiologie Végétale Appliquée, Université Pierre et Marie Curie, Tour 53, 1er étage, 4 place Jussieu, 75252 Paris cédex 05, France*

Although interest in seed priming for improving crop establishment is well recognized, a strong limitation in the development of this technology relies upon rather limited knowledge of the biochemical and molecular mechanisms involved in the earliest steps of seed germination. Of particular interest for the design of priming protocols and the development of certification analyses for primed seeds is the characterization of early germination markers that are easy to detect. Using sugarbeet seeds as a model system, two such protein markers have been identified and biochemical assays have been designed to: (i) distinguish between priming protocols (e.g. hydro- versus osmopriming); (ii) continuously monitor the priming treatments and the advancement of germination prior to radicle emergence; and (iii) analyse the influence of priming conditions (temperature, duration, water supply) on priming efficiency.

Introduction

Seed priming (pre-sowing hydration treatments of seeds) is a widely used technique to enhance seed performance, notably with respect to rate and uniformity of germination, thereby enabling better crop establishment (Heydecker *et al.*, 1973; Hegarty, 1978; Heydecker, 1978; for reviews see Bradford, 1986; Parera and Cantliffe, 1994; Taylor *et al.*, 1998). The basis of this technique is that seed water uptake during germination follows a triphasic pattern with an initial rapid imbibition phase (phase 1), followed by a lag period (phase 2, also referred to as germination *sensu stricto*) and finally by a second uptake phase associated with start of seedling growth (phase 3) (Côme, 1980; Côme and

Thévenot, 1982). Since all preliminary processes for germination are presumed to take place during priming (Heydecker, 1978), the objective of seed priming is to perform a controlled water uptake by the seeds up to the end of phase 2, before the radicles protrude from the seed coats. Furthermore, since most seeds are desiccation tolerant up to this developmental stage, the germination process can be arrested by drying. Numerous studies have demonstrated that priming is associated with an increase in protein synthesis (Bray *et al.*, 1989; Dell'Aquila and Bewley, 1989; Davison and Bray, 1991; Dell'Aquila and Spada, 1992) as well as with nucleic acid synthesis and repair (Bray *et al.*, 1989; Clarke and James, 1991; Bray, 1995).

Several methods are used to control seed water uptake during treatments: priming with an osmotically active agent, usually a salt or polyethylene glycol (PEG) (see Bradford, 1986, and references therein), with a water-absorbing carrier (solid matrix) (Taylor *et al.*, 1988), with pure water (prehydration in water) (Tarquis and Bradford, 1992; C. Job *et al.*, 1997) or with only water vapour (drum priming) (Rowse, 1996). The major problem encountered in seed priming is to control seed imbibition to a level permitting pre-germinative processes to proceed but that block radicle emergence. Otherwise, the consequence of drying back the seeds for storage purposes can be a total loss of the treated batch. There is, therefore, strong interest in the characterization of molecular markers for use by the seed industry in the design of priming protocols because optimization of these treatments rests solely on carrying out germination assays, which can only yield *a posteriori* indications on the priming conditions (e.g. duration, water potential and availability, temperature, oxygen availability).

Compared to the extensive literature dealing with methodological improvements and potential application of these techniques to a large number of seed species, only very few markers of priming have been identified. Furthermore, in several cases the associated biochemical and molecular processes have been shown to be restricted to a few cells from the root tips. Hence, detection of these markers requires a careful dissection of the seeds prior to testing, which may complicate their application at the industrial level. This is for example the case for resumption of nuclear replication activity (Bino *et al.*, 1992), β-tubulin synthesis (De Castro *et al.*, 1995), and endo-β-mannanase synthesis (Still *et al.*, 1997) during priming.

11S Globulin Mobilization during Sugarbeet Seed Priming

Using sugarbeet (*Beta vulgaris* L.) seeds as a model system, we recently reported the occurrence during priming of these seeds of an initial mobilization of the seed storage protein 11S globulin (C. Job *et al.*, 1997; D. Job *et al.*, 1997; Chareyre *et al.*, 1998). The 11S globulins, which are stored during seed maturation in protein bodies and are known to be insoluble in low-ionic strength extraction buffers, are composed of two subunits, an acidic A-chain ($M_r \approx 30{,}000$) and a basic B-chain ($M_r \approx 20{,}000$) that are linked covalently by a disulphide bond (Lawrence *et al.*, 1990; Shewry *et al.*, 1995). We found that:

(i) the basic B-subunit of this sugarbeet storage protein becomes the most abundant polypeptide in soluble protein extracts from germinated and primed seeds (see Fig. 42.3), (ii) this priming-induced solubilization of the B-chain results from an endoproteolytic attack on the A-chain (Fig. 42.1), and (iii) there exists a linear relationship between the extent of B-subunit solubilization and the advancement of germination by priming (C. Job *et al.*, 1997; D. Job *et al.*, 1997). A major advantage of this assay relies upon the fact that the B-subunit of 11S globulin is an extremely abundant protein in the seeds, thus making it very easy to detect from whole seed extracts (C. Job *et al.*, 1997; Chareyre *et al.*, 1998).

To investigate further the robustness of this assay for optimization of priming protocols, we have primed sugarbeet seeds according to two different priming treatments and under various incubation conditions. In the first, referred to as prehydration, controlled hydration of the seeds was achieved by incubating them in the presence of known amounts of water. The complete method consists of three steps as follows (C. Job *et al.*, 1997): (i) a washing step of the seeds (4 h in water at 20°C) to remove germination inhibitors from the seed coats, followed by redrying in the air at ambient temperature to initial

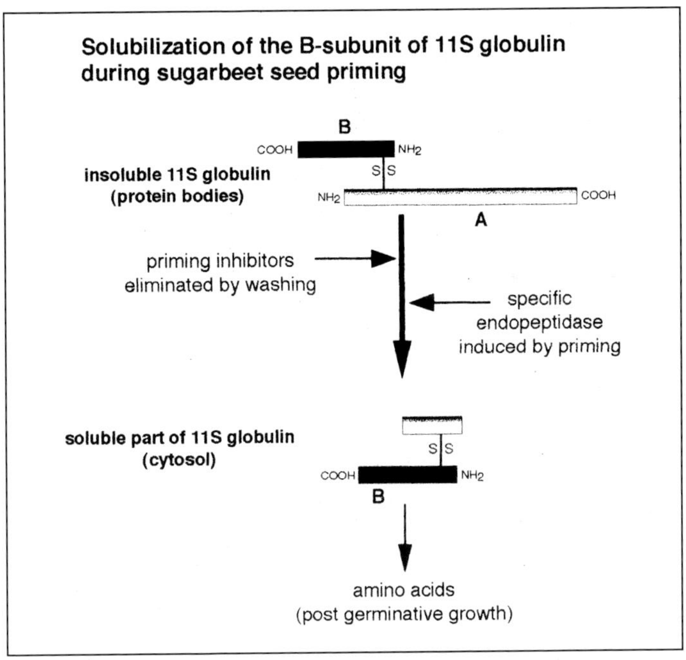

Fig. 42.1. Proposed mechanism for the solubilization of the B-subunit of sugarbeet 11S globulin during seed priming (adapted from C. Job *et al.*, 1997; Chareyre *et al.*, 1998). A and B stand for the acidic and basic subunits of 11S globulin, respectively. This solubilization can be detected from soluble protein samples by SDS-PAGE analysis in the presence of DTT as a disulphide reductant (C. Job *et al.*, 1997) and/or by ELISA using specific antibodies raised against the B-subunit of 11S globulin from sugarbeet (C. Job *et al.*, 1997; D. Job *et al.*, 1997; Chareyre *et al.*, 1998).

moisture content; (ii) an incubation with water at 25°C for 2 days of the washed seeds in plastic tubes sealed with air-tight closures (under these standard conditions the moisture content of the seeds is increased by 30 ± 1%); and (iii) a dehydration of the treated seeds to initial moisture content as described above. In the second protocol, referred to as osmopriming, seeds were placed in a solution of PEG 8000 at −2.0 MPa (Michel and Kaufmann, 1973) for 2 days at 25°C and in air, followed by rinsing and drying to the original moisture content as described by Özbingöl *et al.* (1998) for tomato seeds. Then, hydroprimed or osmoprimed seeds were transferred to water in Petri dishes and germination was recorded at either 5°C or 10°C.

As Fig. 42.2 shows, sugarbeet seeds treated according to either one of the two priming protocols germinated much faster than the untreated control seeds, illustrating the tremendous potential of priming on seed performance, especially when germination is conducted at low temperatures (see Heydecker, 1978). However, under the standard conditions described above (i.e. 2 days incubation at 25°C in air), the prehydration treatment proved superior to the osmopriming treatment, as evidenced from measurements of germination rates and maximum germination percentages (Fig. 42.2).

To characterize more precisely the priming conditions with both techniques, we then performed kinetic analyses by varying the duration of the incubation phase at 25°C. These experiments revealed that the optimal incubation time for both priming techniques was 2 days at 25°C. With the prehydration treatment, none of the seeds germinated by 4 days of incubation. Yet, increasing the prehydration treatment to 5 days may cause some seeds to germinate. Under this latter condition, priming efficiency (as measured by initial rates of germination and final germination percentages) was slightly depressed compared with that observed under optimal priming conditions (data not shown). The influence of incubation time upon priming efficiency

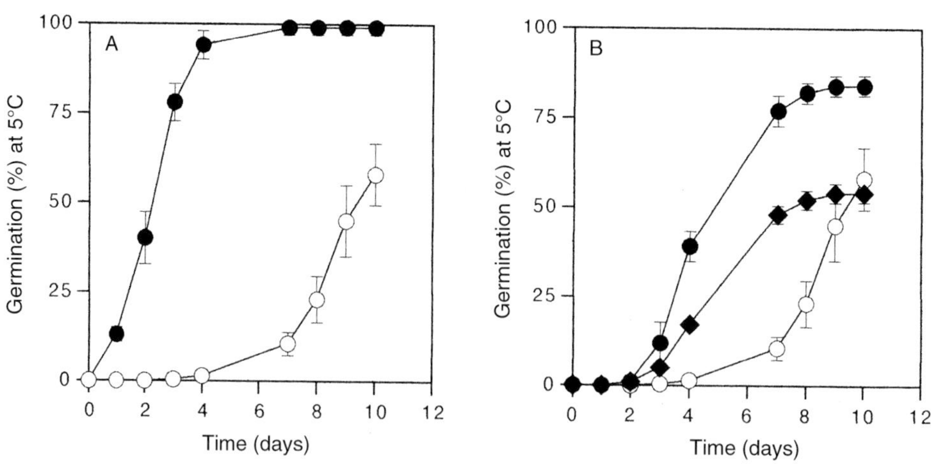

Fig. 42.2. Time courses of germination at 5°C of control unprimed sugarbeet seeds (○) and of seeds primed at 25°C for 2 (●) and 7 days (♦). (A) Hydroprimed seeds; (B) osmoprimed seeds. Means of four replicates ± SD.

could be investigated in more detail with the osmopriming technique since here none of the seeds germinated in up to 14 days of incubation under standard conditions (i.e. incubations in a PEG solution at –2.0 MPa at 25°C in air). Figure 42.2 shows that for incubation times longer than 2 days some seed 'deterioration' was rapidly occurring during osmopriming, leading to a marked decrease in both germination rate and final germination percentage. Thus, following an osmopriming of 7 days at 25°C as much as about 50% of the treated seed population failed to germinate when transferred to water. We verified that this dramatic loss in germination performance was not due to a decreased tolerance of the seeds toward desiccation as seeds approached the time for radicle protrusion during the osmopriming treatment. Indeed, seeds pretreated in the same conditions and transferred to water without redrying germinated similarly to the seeds primed according to the full priming treatment.

Soluble protein extracts from untreated and primed seeds were analysed by SDS-PAGE in the presence of the disulphide reductant DTT and proteins were visualized by Coomassie-blue staining (Fig. 42.3). In agreement with previous results (C. Job *et al.*, 1997; Chareyre *et al.*, 1998), these analyses revealed an accumulation of the B-subunit of 11S globulin during priming (Fig. 42.3). Furthermore, these data were consistent with the conclusion that solubilization during priming of part of the stored B-subunit of 11S globulin occurs independently of the method of priming (C. Job *et al.*, 1997). A comparison of these results with the germination data shown in Fig. 42.2B indicates, however, that although this phenomenon is a characteristic feature of priming,

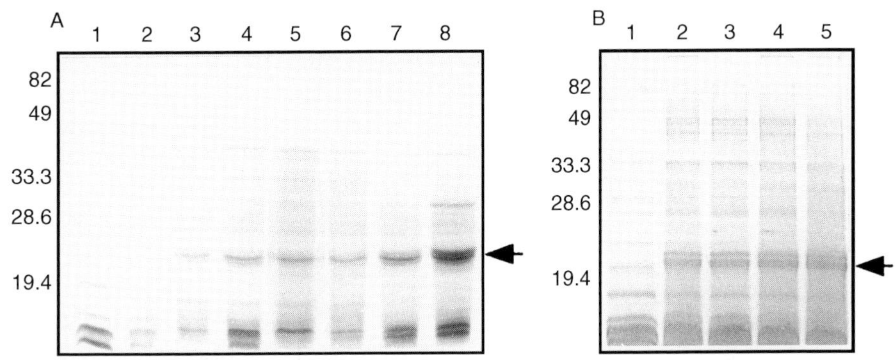

Fig. 42.3. SDS-PAGE profiles after Coomassie-blue staining of soluble protein extracts from untreated and primed sugarbeet seeds. Prior to SDS-PAGE soluble protein samples were incubated at 100°C for 5 min in loading buffer containing both SDS and DTT (SDS-PAGE + DTT conditions). An equal amount of extracts corresponding to 0.1 seed was applied to each lane. The size of molecular mass markers is indicated in kDa. The arrows mark the migration of the B-subunit of 11S globulin. (A) Osmopriming treatment. Analysis of soluble proteins from control untreated seeds (lane 1) and seeds submitted to the osmopriming treatment for 1, 2, 4, 5, 9, 11 and 14 days (lane 2 to 8, respectively). (B) Prehydration treatment. Analysis of soluble proteins from control untreated seeds (lane 1) and seeds submitted to the prehydration treatment for 2, 3, 4 and 5 days (lanes 2 to 5, respectively).

it does not allow us to predict the loss of germination performance observed for osmopriming times longer than 2 days.

Degradation of Seed Biotinylated Protein during Sugarbeet Seed Priming

From the above results, it was of importance to characterize another protein marker for sugarbeet seed priming that would allow discrimination between the positive effects induced by priming itself and the negative effects induced by overpriming on seed performance. To this end, we analysed the temporal accumulation patterns of LEA proteins, that is, the ubiquitous heat-stable proteins that accumulate in late embryogenesis. We chose to investigate this class of proteins because of their rapid degradation in early germination (Dure, 1993). Among these proteins is a unique seed-specific biotinylated protein that we have documented in pea and called SBP65 (for Seed Biotinylated Protein of 65 kDa) (Duval *et al.*, 1994b, 1996). We showed that although in all organisms biotin serves as an essential cofactor for a small number of housekeeping biotin-dependent carboxylases (e.g. acetyl-CoA carboxylase, see Dehaye *et al.*, 1994), SBP65, which is the major biotinylated protein in mature pea seeds, does not exhibit any biotin-dependent carboxylase activity. Instead, this protein behaves as a sink for free biotin during late stages of embryo development and is rapidly degraded during germination (Duval *et al.*, 1994b). A specific biochemical feature of SBP65 is that the protein lacks the consensus sequence evidenced for biotin-dependent carboxylases, thus accounting for the absence of enzyme activity associated with this protein (Duval *et al.*, 1994a). Putative roles for SBP65 are that this protein constitutes a storage form of biotin necessary for germination and/or that, by depleting the free biotin pool during late maturation, it may allow the embryonic cells to enter into a quiescent state.

Figure 42.4 shows that an equivalent of the pea SBP65 exists in sugarbeet. As for LEA proteins, this protein is resistant to heat denaturation (Fig. 42.4), and we verified that it exhibits the same spatial and temporal accumulation patterns as those described for the pea protein. In particular, the protein specifically accumulates in mature seeds and disappears at a high rate during germination (I. Capron *et al.*, unpublished results). Furthermore, as for pea SBP65 (Duval *et al.*, 1994a), this protein does not exhibit the consensus sequence for protein biotinylation evidenced for biotin-dependent carboxylases (C. Job and D. Job, unpublished results). We therefore refer to this sugarbeet protein as SBP (for Seed Biotinylated Protein).

The SBP content per seed, which can be easily evaluated from whole seed extracts by western blotting using the avidin-biotin technology (Duval *et al.*, 1994b), was maximum in untreated seeds and decreased during both seed treatments (Fig. 42.4). It is worth noting, however, that this decrease was much slower during the prehydration treatment (Fig. 42.4B) than during osmopriming (Fig. 42.4A). For example, under our best priming conditions which correspond to a 2-day prehydration treatment (see Fig. 42.2), the SBP content per seed remained nearly at the same level for the hydroprimed seeds

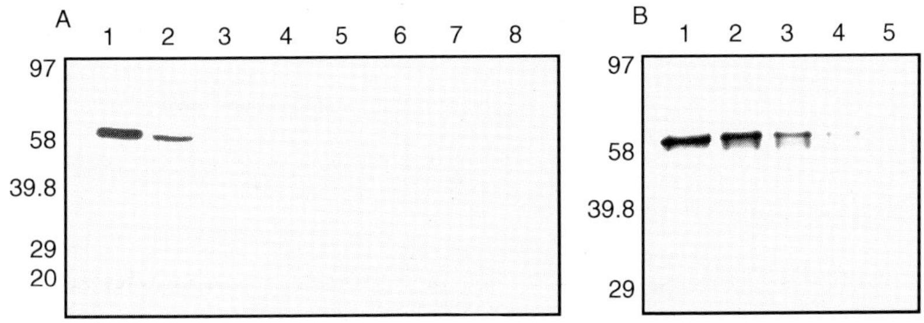

Fig. 42.4. Analysis of biotinylated proteins by SDS-PAGE and Western blotting in control, hydroprimed and osmoprimed sugarbeet seeds. Same soluble protein extracts as shown in Fig. 42.3 were analysed by SDS-PAGE, except that prior to electrophoresis the soluble protein extracts were incubated at 100°C for 5 min in a low ionic strength buffer. Following centrifugation (20,000 *g*, 20 min) the heat-stable protein fractions were recovered in the supernatants. For SDS-PAGE analysis, an equal amount of the heat-stable protein extracts corresponding to 0.1 seed was applied to each lane. Biotinylated proteins were then analysed by Western blotting using peroxidase-labelled streptavidin as a specific reagent for biotin (Duval *et al.*, 1994b). The seed-specific biotin-containing protein of about 60 kDa, which is referred to as SBP (Seed Biotinylated Protein) in the text, is the equivalent of pea SBP65 (Duval *et al.*, 1994a,b, 1996; Dehaye *et al.*, 1994, 1997). The size of molecular mass markers is indicated in kDa. (A) Osmopriming treatment. Analysis of soluble proteins from control untreated seeds (lane 1) and seeds submitted to the osmopriming treatment at 25°C for 1, 2, 4, 5, 9, 11 and 14 days (lanes 2 to 8, respectively). (B) Prehydration treatment. Analysis of soluble proteins from control untreated seeds (lane 1) and seeds submitted to the prehydration treatment at 25°C for 2, 3, 4 and 5 days (lanes 2 to 5, respectively).

compared to the control seeds (Fig. 42.4B). In marked contrast, this protein was rapidly degraded during osmopriming, being at an almost undetectable level after a 2-day treatment (Fig. 42.4A). A comparison of these results with the germination data in Fig. 42.2 suggests, therefore, that the disappearance of SBP correlates with loss of germination performance during osmopriming.

Additional experiments showed that priming efficiency as well as the rate of SBP disappearance were dependent on the temperature at which the osmopriming treatment was carried out. In contrast to recent results on osmopriming of tomato seeds (Özbingöl *et al.*, 1998), the temperature requirements for osmopriming sugarbeet seeds were different to those which are necessary for the germination of untreated seeds, particularly in the 25–35°C temperature range. Thus, although in this range control sugarbeet seeds exhibited their maximum germination performance, the germination rates of osmoprimed seeds were negatively affected (Fig. 42.5A). This behaviour is in agreement with the finding that osmoprimed sugarbeet seeds are very sensitive to deterioration. For the osmoprimed seeds, the SBP content per seed strongly decreased upon increasing the temperature of the treatment, reaching an almost undetectable level at 20°C (Fig. 42.5B). In view of the rapid disappearance of SBP during early germination (I. Capron *et al.*, in

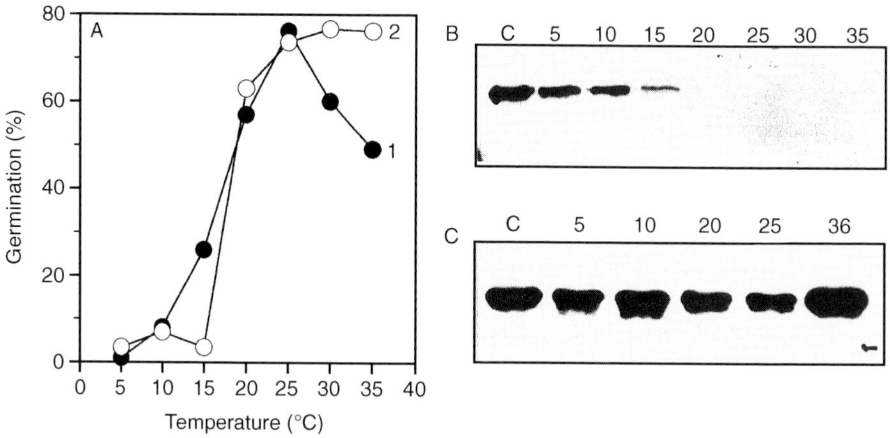

Fig. 42.5. Effects of temperature on germination and SBP contents of control and primed sugarbeet seeds. (A) Germination experiments. Effect of temperature during osmopriming (●). Seeds were osmoprimed for 2 days (complete treatment including redrying) in the indicated temperature range. Then, treated seeds were transferred to water in Petri dishes and germination percentages were scored after 4 days at 5°C. For comparison, the effects of temperature on the germination percentages obtained after 2 days with control unprimed seeds (○) is also shown. (B, C) Analyses of SBP content (see Fig. 42.4) of osmoprimed seeds (B) and hydroprimed seeds (C). The numbers above the lanes correspond to the temperature (°C) at which the treatments were carried out. C, control untreated seeds.

preparation), one possibility to account for the behaviour of osmoprimed seeds could be that during osmopriming sugarbeet seeds are going too far in phase 2 of the germination process and reach a developmental stage where ageing processes are activated if radicle emergence is prevented. Under all conditions analysed, the prehydration treatment was superior to the osmotic treatment. Interestingly, the SBP content of the hydroprimed seeds remained at a high level, whatever the temperature used to conduct the prehydration treatment in the 5–36°C temperature range (Fig. 42.5C).

Conclusions

The present study conducted with sugarbeet seeds has documented both the beneficial effects of priming treatments to improve germination vigour and the difficulties to control such treatments. Kinetic experiments revealed that once these seeds have reached their maximal germination performance they may lose their germination vigour upon prolonged osmopriming treatment and then perform worse than untreated seeds. The mechanism of such a loss in the beneficial effect of priming is at present unknown. Experiments are in progress to investigate whether this osmopriming-induced reduced vigour was due to an acceleration of general ageing processes leading to an irreversible deterioration of the seeds (see Taylor *et al.*, 1998) or to commitment of the osmoprimed seeds to a dormancy state. We found, however, that this loss in germination vigour was not due to a decreased tolerance of the seeds toward desiccation as seeds approached the time for radicle protrusion during priming treatments.

The present work also illustrates the possibility to predict seedlot performance from measurements of selected germination markers. Together, the current results indicate that for a priming treatment of sugarbeet seeds to be the most efficient, there should be an increase in the level of soluble 11S globulin B-subunit per seed but no decrease in that of SBP, the seed-specific biotinylated protein. Both these contents can be easily measured from whole seed extracts by ELISA (Duval *et al.*, 1994b; C. Job *et al.*, 1997; Chareyre *et al.*, 1998).

Acknowledgements

We are grateful to Dr J.F. Seitzer and Dr U. Fisher from KWS (Enbeick, Germany) for many helpful discussions and for a kind gift of sugarbeet seed samples. This work has been supported in part by a grant from the Région Rhône-Alpes (Programme 'Biotechnologies').

References

Bino, R.J., De Vries, J.N., Kraak, H.L. and Van Pijlen, J.G. (1992) Flow cytometric determination of nuclear replication stages in tomato seeds during priming and germination. *Annals of Botany* 69, 231–236.

Bradford, K.J. (1986) Manipulation of seed water relations via osmotic priming to improve germination under stress conditions. *HortScience* 21, 1105–1112.

Bray, C.M. (1995) Biochemical processes during the osmopriming of seeds. In: Kigel, J. and Galili, G. (eds) *Seed Development and Germination.* Marcel Dekker, Inc., New York, pp. 767–789.

Bray, C.M., Davison, P.A., Ashraf, M. and Taylor, R.M. (1989) Biochemical changes during priming of leek seeds. *Annals of Botany* 63, 185–193.

Chareyre, S., Kersulec, A., Job, D. and Job, C. (1998) The use of an ELISA to quantitate the extent of 11S globulin mobilization in untreated and primed sugar beet seed lots. *Comptes Rendus de l'Académie des Sciences Paris* 321, 705–711.

Clarke, N.A. and James, P.E. (1991) The effects of priming and accelerated ageing upon the nucleic acid content of leek seeds and their embryos. *Journal of Experimental Botany* 42, 261–268.

Côme, D. (1980) Problems of embryonal dormancy as exemplified by apple embryo. *Israel Journal of Botany* 29, 145–156.

Côme, D. and Thévenot, C. (1982) Environmental control of embryo dormancy and germination. In: Khan, A.A. (ed.) *The Physiology and Biochemistry of Seed Development, Dormancy and Germination.* Elsevier Biomedical Press, Amsterdam, New York, Oxford, pp. 271–298.

Davison, P.A. and Bray, C.M. (1991) Protein synthesis during osmopriming of leek (*Allium porrum* L.) seeds. *Seed Science Research* 1, 29–35.

De Castro, R., Zheng, X., Bergervoet, J.H.W., De Vos, C.H.R. and Bino, R.J. (1995) β-Tubulin accumulation and DNA replication in imbibing tomato seeds. *Plant Physiology* 109, 499–504.

Dehaye, L., Alban, C., Job, C., Douce, R. and Job, D. (1994) Kinetics of the two forms of acetyl-CoA carboxylase from *Pisum sativum*. Correlation of the substrate specificity of the enzyme and sensitivity towards aryloxyphenoxypropionates herbicides. *European Journal of Biochemistry* 225, 1113–1123.

Dehaye, L., Duval, M., Viguier, D., Yaxley, J. and Job, D. (1997) Cloning and expression of the pea gene encoding SBP65, a seed-specific biotinylated protein. *Plant Molecular Biology* 35, 605–621.

Dell'Aquila, A. and Bewley, J.D. (1989) Protein synthesis in the axes of polyethylene glycol-treated pea seeds and during subsequent germination. *Journal of Experimental Botany* 40, 1001–1007.

Dell'Aquila, A. and Spada, P. (1992) Regulation of protein synthesis in germinating wheat embryos under polyethylene glycol and salt stress. *Seed Science Research* 2, 75–80.

Dure, L. III (1993) The LEA proteins of higher plants. In: Verma, D.P.S. (ed.) *Control of Plant Gene Expression*. CRC Press, Inc., Boca Raton, Florida, pp. 325–335.

Duval, M., DeRose, R.T., Job, C., Faucher, D., Douce, R. and Job, D. (1994a) The major biotinyl protein from *Pisum sativum* seeds covalently binds biotin at a novel site. *Plant Molecular Biology* 26, 265–273.

Duval, M., Job, C., Alban, C., Douce, R. and Job, D. (1994b) Developmental patterns of free and protein-bound biotin during maturation and germination of seeds of *Pisum sativum*. Characterization of a novel seed-specific biotinylated protein. *Biochemical Journal* 299, 141–150.

Duval, M., Loiseau, J., Dehaye, L., Pépin, R., Le Deunff, Y., Wang, T. and Job, D. (1996) SBP65, a seed-specific biotinylated protein behaves as a LEA protein in developing pea embryos. *Comptes Rendus de l'Académie des Sciences Paris* 319, 585–594.

Hegarty, T.W. (1978) The physiology of seed hydration and dehydration, and the relation between water stress and the control of germination: a review. *Plant, Cell and Environment* 1, 101–119.

Heydecker, W. (1978) 'Primed' seeds for better crop establishment? *Span* 21, 12–14.

Heydecker, W., Higgins, J. and Gulliver, R.L. (1973) Accelerated germination by osmotic seed treatment. *Nature* 246, 42–44.

Job, C., Kersulec, A., Ravasio, L., Chareyre, S., Pépin, R. and Job, D. (1997) The solubilization of the basic subunit of sugarbeet seed 11-S globulin during priming and early germination. *Seed Science Research* 7, 225–243.

Job, D., Kersulec, A. and Job, C. (1997) Protéine globuline 11S, utilisable comme marqueur d'imbibition d'une semence au cours de la germination. *International Patent* WO9743418.

Lawrence, D.M., Halmer, P. and Bowles, D.J. (1990) Mobilisation of storage reserves during germination and early seedling growth of sugar beet. *Physiologia Plantarum* 78, 421–429.

Michel, B.E. and Kaufmann, M.R. (1973) The osmotic potential of polyethylene glycol 6000. *Plant Physiology* 51, 914–916.

Özbingöl, N., Corbineau, F. and Côme, D. (1998) Responses of tomato seeds to osmoconditioning as related to temperature and oxygen. *Seed Science Research* 8, 337–384.

Parera, C.A. and Cantliffe, D.J. (1994) Presowing seed priming. *Horticultural Reviews* 16, 109–141.

Rowse, H.R. (1996) Drum priming – a non-osmotic method of priming seeds. *Seed Science and Technology* 24, 281–294.

Shewry, P.R., Napier, J.A. and Tatham, A.S. (1995) Seed storage proteins: structures and biosynthesis. *Plant Cell* 7, 945–956.

Still, D.W., Dahal, P. and Bradford, K.J. (1997) A single-seed assay for endo-β-mannanase activity from tomato endosperm and radicle tissues. *Plant Physiology* 113, 13–20.

Tarquis, A.M. and Bradford, K.J. (1992) Prehydration and priming treatments that advance germination also increase the rate of deterioration of lettuce seeds. *Journal of Experimental Botany* 43, 307–317.

Taylor, A.G., Klein, D.E. and Whitlow, T.H. (1988) SMP: solid matrix priming. *Scientia Horticulturae* 37, 1–11.

Taylor, A.G., Allen, P.S., Bennett, M.A., Bradford, K.J., Burris, J.S. and Misra, M.K. (1998) Seed enhancements. *Seed Science Research* 8, 245–256.

43 A Critical Assessment of the Role of Oligosaccharides in Intracellular Glass Stability

J. BUITINK[1,2], F.A. HOEKSTRA[1] AND M.A. HEMMINGA[2]

[1]*Laboratory of Plant Physiology, Department of Plant Sciences,* [2]*Laboratory of Molecular Physics, Department of Biomolecular Sciences, Wageningen University, Wageningen, The Netherlands*

The role of oligosaccharides in enhancing the stability of intracellular glasses was investigated. Priming *Pisum sativum* seeds for 6 days in a polyethylene glycol solution of −1.0 MPa at 20°C resulted in a change in the sugar composition. In pea axes, total oligosaccharide content decreased from 168 to 27 mg g^{-1} dry weight (DW), while sucrose content increased from 41 to 62 mg g^{-1} DW. Despite the change in oligosaccharide:sucrose ratio, no difference in the glass transition temperature was measured in dry pea axes before and after priming as determined by DSC. Saturation transfer electron spin resonance spectroscopy was used to measure the rotational mobility of a polar spin probe, 3-carboxy-proxyl, which was inserted into the cytoplasm of axes. No difference was found between the rotational mobility in dry untreated axes and that of dry primed axes. There is no significant contribution of oligosaccharides to intracellular glass stability in terms of increasing the glass transition temperature or decreasing the mobility of molecules in the intracellular glass.

Introduction

Oligosaccharides often occur in considerable quantities in dry seeds of many plant species (Amuti and Pollard, 1977). The presence of these sugars appears to correlate with the longevity of seeds (Horbowicz and Obendorf, 1994; Bernal-Lugo and Leopold, 1995; Steadman *et al.*, 1996). There are several hypotheses regarding the function of oligosaccharides in seeds. They may play a role in the protection of membranes and proteins, in a similar way as described for disaccharides (Crowe *et al.*, 1992), or prevent crystallization of sucrose (Caffrey *et al.*, 1988). Moreover, they are thought to be involved in the formation of stable glasses by increasing the glass transition temperature (T_g) and thereby increasing the viscosity of the glassy cytoplasm (Leopold *et al.*, 1994; Bernal-Lugo and Leopold, 1995). The underlying concept for the role of

 Seed Biology: Advances and Applications (eds M. Black, K.J. Bradford and J. Vázquez-Ramos)

glasses in storage stability is that the high viscosity of intracellular glasses will slow down ageing reactions (Leopold *et al.*, 1994; Sun, 1997; Buitink *et al.*, 1998a,b). Indeed, using saturation transfer electron spin resonance spectroscopy (ST-ESR), a relation between mobility of molecules in the glassy cytoplasm and longevity was found for *Typha latifolia* pollen and pea seeds (Buitink *et al.*, 1998b).

Seed priming, i.e. the pre-imbibition of seeds in osmotic solution, is known to considerably improve seed quality by enhancing rates of germination and seedling uniformity. However, a drawback of this treatment is the often reduced longevity of primed seeds (Tarquis and Bradford, 1992; Saracco *et al.*, 1995). The causes for this reduced longevity are not well understood. Hoekstra *et al.* (1994) found that priming of cauliflower seeds resulted in a decrease in oligosaccharide content. Considering the proposed role of oligosaccharides in increasing glass stability, it is possible that the reduced longevity of primed seeds may be attributed to the decrease in oligosaccharide content resulting in a decrease in cytoplasmic viscosity.

In this study, we investigated whether changes in sugar composition after priming lead to changes in the properties of intracellular glasses, such as a decrease in T_g (DSC) and an increase in the mobility of molecules in the cytoplasm (ST-ESR).

Material and Methods

Seeds of pea (*Pisum sativum* cv. Karina) were primed for 6 days in a polyethylene glycol solution of −1.0 MPa at 20°C. After priming, the seeds were rinsed with demineralized water and dried at 3% RH for 2 days at room temperature. Subsequently, seeds were used for sugar content determination, differential scanning calorimetry (DSC) measurements and ST-ESR experiments.

Soluble sugars of isolated axes were extracted in a solution of 80% methanol (v:v) containing lactose as internal standard according to Hoekstra *et al.* (1994). Sugar concentrations were determined by HPLC using pulsed amperometric detection on a Carbopac PA-1 column (Dionex Corp., Sunnyvale, California, USA). Data are averages of duplicated extractions.

Glass transition temperatures (T_g) of pea axes were determined using a Perkin-Elmer (Norwalk, Connecticut, USA) Pyris-1 DSC. The T_g values were determined as the onset of the temperature range over which the change in specific heat occurred in heating scans recorded at a scanning rate of 10°C min^{-1}. Different sample water contents were achieved by equilibrating the tissues over different saturated salt solutions for at least 72 h.

Labelling of isolated axes with the spin probe 3-carboxy-proxyl (CP) was performed according to Buitink *et al.* (1998b). After labelling, axes were dried in an airflow of 3% RH for 24 h, then stored over saturated salt solutions for 6 days to obtain various water contents. For each ST-ESR measurement, 20 mg of tissue was sealed in a glass capillary. Spectra were recorded on a Bruker X-band ESR spectrometer (Bruker Analytik, Rheinstetten, Germany, model 300E). Instrument settings were according to Buitink *et al.* (1998b). After the measurements, tissues were removed from the capillary and the water contents were determined. Water contents were determined gravimetrically by weighing the samples before and after heating at 96°C for 36–48 h.

Results and Discussion

The longevity of seeds often has been related to their oligosaccharide contents (Horbowicz and Obendorf, 1994; Bernal-Lugo and Leopold, 1995; Steadman *et al.*, 1996). To establish whether priming of pea seeds results in changes in sugar composition, sucrose and oligosaccharide contents were determined in axes before and after priming (Table 43.1). Priming for 6 days at −1.0 MPa resulted in an increase in sucrose and a decrease in oligosaccharide content. The ratio between the oligosaccharide and the sucrose content in pea axes decreased from 4.2 in the untreated axes to 0.4 in the primed axes. Similar changes in soluble sugar contents after priming were found previously in cauliflower seeds (Hoekstra *et al.*, 1994).

In model systems made of sugar mixtures, changes in the composition result in changes in glass properties (Levine and Slade, 1988). It has been shown that the T_g is dependent on the molecular weight of the components forming the glass. For example, the T_g of dry sucrose glasses is approximately 63°C, whereas the T_g of dry raffinose glasses is around 100°C (Levine and Slade, 1988; Wolkers *et al.*, 1998). In untreated and redried primed pea axes, the T_g was measured as a function of water content using DSC (Fig. 43.1). No difference could be found between the T_g values of untreated and primed axes, despite the differences in sugar composition (Table 43.1). Furthermore, no change was observed in the apparent heat capacity that is associated with the glass transition (data not shown).

The viscosity of a raffinose glass is higher than that of a sucrose glass, under the same conditions of water content and temperature (Levine and Slade, 1988). Although no difference was found in T_g between untreated and primed pea axes, the viscosity or molecular mobility in the cytoplasm may still be different. Using ST-ESR, changes in molecular mobility can be estimated by determining the rotational correlation time (τ_R) of a nitroxide spin probe that is inserted into the cytoplasm of seed tissues (Buitink *et al.*, 1998b). The τ_R corresponds to the lifetime of the probe in a given orientation. ST-ESR spectroscopy

Table 43.1. Soluble sugar contents in untreated and redried primed pea axes. Seeds were incubated for 6 days in a solution of polyethylene glycol (−1.0 MPa) at 20°C then dried. Data are expressed as mg sugar g^{-1} DW and are averages of duplicate extractions.

Treatment	Sucrose	Raffinose	Stachyose	Verbascose
Untreated	40.5	14.5	73.6	80.0
Primed	62.5	15.1	6.5	5.7

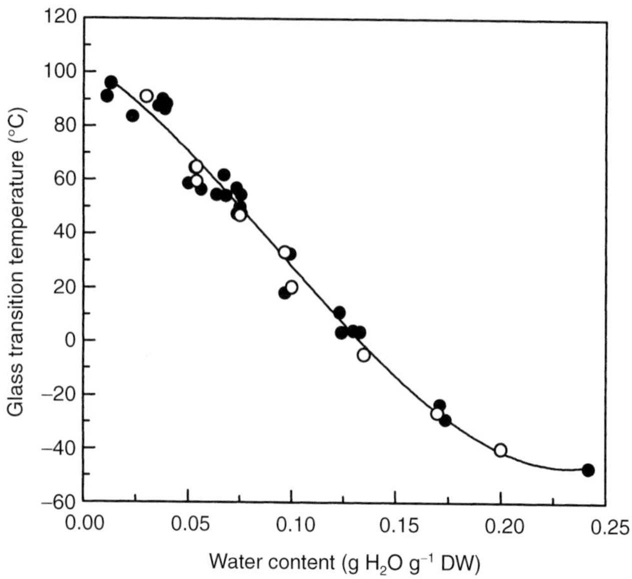

Fig. 43.1. State diagram of untreated (●) or primed pea (○) axes. Priming was achieved by incubating seeds for 6 days at −1.0 MPa at 20°C. The T_g values were determined as the onset of the temperature range over which the change in specific heat occurred in heating scans recorded at a scanning rate of 10°C min^{-1}.

Table 43.2. Rotational correlation time (τ_R) of the spin probe 3-carboxy-proxyl in the cytoplasm of untreated and primed pea axes at different water contents at 30°C. The water contents were chosen so that τ_R is determined within and out of the glassy state (see Fig. 43.1). Priming was performed as explained in Table 43.1. Data (± SD) are averages of three experiments. Water contents are expressed as g H_2O g^{-1} DW and τ_R in µs.

Treatment	Water content	Rotational correlation time
Untreated	0.07 ± 0.00	88 ± 1.8
	0.12 ± 0.01	24 ± 2.1
Primed	0.07 ± 0.01	91 ± 3.0
	0.12 ± 0.01	22 ± 2.9

has been applied previously to determine rotational motions of nitroxide spin probes in glassy sugar systems (Hemminga and Van den Dries, 1998). The τ_R of a polar spin probe (CP) in untreated or primed pea axes was determined at 30°C at two water contents (Table 43.2). These water contents represent conditions under which the cytoplasm is in or out of the glassy state (Table 43.2, compare with Fig. 43.1). At both water contents, no significant difference was measured in τ_R between untreated and primed seeds, implying that the priming and redrying treatment does not affect the molecular mobility in the cytoplasm.

The above observations do not support the hypothesis that oligosaccharides stabilize the intracellular glass by decreasing the molecular mobility. Apparently, other molecules besides soluble sugars play an important role in intracellular glass formation. Another role proposed for oligosaccharides in seeds is prevention of crystallization of sucrose *in vivo* (Caffrey *et al.*, 1988). However, attempts to detect crystallization *in vivo* within the cytoplasm have failed (Sun and Leopold, 1993). Most likely, the complex mixture of cytoplasmic solutes will prevent crystallization of sucrose, regardless of the presence of oligosaccharides. Furthermore, oligosaccharides might contribute to the protection of membranes according to the water replacement hypothesis (Crowe *et al.*, 1992). However, this hypothesis was put forward to explain the role of soluble sugars in desiccation tolerance. Yet, after priming, pea seeds are still desiccation tolerant. Further investigations will reveal whether the membrane structure and stability is affected by priming. However, it might be that there is no specific role for oligosaccharides in longevity. Oligosaccharides could simply be an indicator of seed maturity and could serve as a storage reserve (Kuo *et al.*, 1988; Hoekstra *et al.*, 1994).

Conclusions

We investigated the role of oligosaccharides in the stability of intracellular glasses using primed pea seeds as a model system. During priming, the oligosaccharide content decreased, but this change did not affect the intracellular glass properties in terms of glass transition temperature and mobility of a spin probe in the cytoplasm. A role for oligosaccharides in longevity, if any, does not appear to be mediated by changing the characteristics of the intracellular glass.

Acknowledgements

We thank Dr O. Leprince for critically reading the manuscript. This work was financially supported by the Netherlands Technology Foundation (STW), and was co-ordinated by the Life Sciences Foundation.

References

Amuti, K.S. and Pollard, C.J. (1977) Soluble carbohydrates of dry and developing seeds. *Phytochemistry* 16, 529–532.

Bernal-Lugo, I. and Leopold, A.C. (1995) Seed stability during storage: raffinose content and seed glassy state. *Seed Science Research* 5, 75–80.

Buitink, J., Walters, C., Hoekstra, F.A. and Crane, J. (1998a) Storage behavior of *Typha latifolia* pollen at low water contents: interpretation on the basis of water activity and glass concepts. *Physiologia Plantarum* 103, 145–153.

Buitink, J., Claessens, M.M.A.E., Hemminga, M.A. and Hoekstra, F.A. (1998b) Influence of water content and temperature on molecular mobility and intracellular glasses in seeds and pollen. *Plant Physiology* 118, 531–541.

Caffrey, M., Fonseca, V. and Leopold, A.C. (1988) Lipid-sugar interaction. *Plant Physiology* 86, 754–758.

Crowe, J.H., Hoekstra, F.A. and Crowe, L.M. (1992) Anhydrobiosis. *Annual Review of Physiology* 54, 579–599.

Hemminga, M.A. and Van den Dries, I.J. (1998) Spin label applications to food science. In: Berliner, L.J. (ed.) *Biological Magnetic Resonance, Vol. 14: Spin Labeling: The Next Millennium.* Plenum Publishing Corporation, New York, pp. 339–366.

Hoekstra, F.A., Haigh, A.M., Tetteroo, F.A.A. and van Roekel, T. (1994) Changes in soluble sugars in relation to desiccation tolerance in cauliflower seeds. *Seed Science Research* 4, 143–147.

Horbowicz, M. and Obendorf, R.L. (1994) Seed desiccation tolerance and storability: dependence on flatulence-producing oligosaccharides and cyclitols – a review and survey. *Seed Science Research* 4, 385–405.

Kuo, T.M., VanMiddlesworth, J.F. and Wolf, W.J. (1988) Content of raffinose oligosaccharides and sucrose in various plant seeds. *Journal of Agricultural and Food Chemistry* 36, 32–36.

Leopold, A.C., Sun, W.Q. and Bernal-Lugo, I. (1994) The glassy state in seeds: analysis and function. *Seed Science Research* 4, 267–274.

Levine, H. and Slade, L. (1988) Water as a plasticizer: physico-chemical aspects of low-moisture polymeric systems. In: Franks, F. (ed.) *Water Science Reviews.* Cambridge University Press, Cambridge, Vol. 3, pp. 79–185.

Saracco, F., Bino, R.J., Bergervoet, J.H.W. and Lanteri, S. (1995) Influence of priming-induced nuclear replication activity on storability of pepper (*Capsicum annuum* L.) seed. *Seed Science Research* 5, 25–29.

Steadman, K.J., Pritchard, H.W. and Dey, P.M. (1996) Tissue-specific soluble sugars in seeds as indicators of storage category. *Annals of Botany* 77, 667–674.

Sun, W.Q. (1997) Glassy state and seed storage stability: the WLF kinetics of seed viability loss at T-T_g and the plasticization effect of water on storage stability. *Annals of Botany* 79, 291–297.

Sun, W.Q. and Leopold, A.C. (1993) The glassy state and accelerated aging of soybeans. *Physiologia Plantarum* 89, 767–774.

Tarquis, A.M. and Bradford, K.J. (1992) Prehydration and priming treatments that advance germination also increase the rate of deterioration of lettuce seeds. *Journal of Experimental Botany* 43, 307–317.

Wolkers, W.F., Oldenhof, H., Alberda, M. and Hoekstra, F.A. (1998) A Fourier transform infrared study of sugar glasses: application to anhydrobiotic higher plant cells. *Biochimica et Biophysica Acta* 1379, 83–96.

44 Improvement of Tomato Seed Germination by Osmopriming as Related to Energy Metabolism

F. CORBINEAU, N. ÖZBINGÖL, D. VINEL AND D. CÔME

Physiologie Végétale Appliquée, Université Pierre et Marie Curie, Tour 53, 1er étage, 4 place Jussieu, 75252 Paris cédex 05, France

In dry, unprimed tomato seeds, ATP represents only 2.1% of the adenylic nucleotide pool, and the energy charge (EC) and the ATP/ADP ratio are very low (0.11 and 0.12, respectively). Imbibition of seeds in a polyethylene glycol-8000 solution at −1 MPa results in sharp increases in ATP (60% of the adenylic nucleotide pool), EC (0.77–0.78) and the ATP/ADP ratio (1.75–2.32). The level of ATP is strongly reduced after drying the primed seeds, but the EC (0.28–0.33) and the ATP/ADP ratio (0.32–0.48) remain higher than in dry unprimed seeds. The energy metabolism of dried primed seeds is also much more intense during the first 4 h of subsequent imbibition in water than that of unprimed ones. To improve markedly the subsequent germination of tomato seeds, osmopriming treatment requires at least 5% oxygen in the atmosphere, and its maximal improving effect is obtained in atmospheres containing more than 10% oxygen. The treatment is also totally ineffective in the presence of a respiratory inhibitor (NaN_3) at high concentration (0.5 or 1 mM). Results obtained with seeds primed in reduced oxygen tensions or in the presence of NaN_3 at various concentrations show that the beneficial effect of priming increases with increasing EC and ATP/ADP ratio, and is optimal for values higher than 0.75 for EC and 1.7 for the ATP/ADP ratio.

Introduction

Osmopriming consists of the presoaking of seeds in an osmotic solution, usually a salt or polyethylene glycol (PEG) solution, in order to control their water uptake and prevent radicle protrusion (Bray, 1995). This treatment, followed by dehydration of the seeds, has been demonstrated to improve subsequent germination in water of seeds of numerous species (Brocklehurst and Dearman, 1983; Bradford, 1986, 1995; Karssen *et al.*, 1989). After priming, seeds germinate in a wider range of temperatures (Brocklehurst and Dearman, 1983; Bradford, 1986; Corbineau *et al.*, 1994; Özbingöl *et al.*, 1998) and are less

(eds M. Black, K.J. Bradford and J. Vázquez-Ramos)

sensitive to oxygen deprivation (Smok *et al.*, 1993; Corbineau *et al.*, 1994; Özbingöl *et al.*, 1998). The efficiency of priming depends on the conditions (water potential, temperature, oxygen availability) and the duration of the treatment (Karssen *et al.*, 1989; Bujalski *et al.*, 1993; Bradford, 1995; Bray, 1995). In tomato seeds, for example, there exists a linear relationship between the median germination rate after priming and the accumulated hydrothermal priming time (Bradford and Haigh, 1994; Bradford, 1995), and the osmotic pretreatment requires at least 3–5% oxygen in the atmosphere to improve subsequent germination at 15°C (Özbingöl *et al.*, 1998).

The beneficial effect of osmopriming is associated with various biochemical, cellular and molecular events (Karssen *et al.*, 1989; Bray, 1995). Most studies have dealt with changes in nucleic acid and protein synthesis (Bray, 1995), and the induction of the cell cycle (Lanteri *et al.*, 1994; Özbingöl *et al.*, 1997) during priming. The respiratory activity of the seeds is also enhanced by priming during the first hours of subsequent imbibition in water (Halpin-Ingham and Sundstrom, 1992; Smok *et al.*, 1993; Chojnowski *et al.*, 1997), and the ATP level is increased (Mazor *et al.*, 1984). Various studies have shown that the respiratory activity of seeds is high during priming (Fu *et al.*, 1988; Smok *et al.*, 1993; Dahal *et al.*, 1996). However, it decreases with decreasing water potential (Ibrahim *et al.*, 1992; Dahal *et al.*, 1996). In tomato seeds, respiration rate measured during phase II of the germination process is reduced exponentially with decreasing water potential, and there exists an exponential relationship between the respiration activity and the median germination rate (Dahal *et al.*, 1996). Oxygen uptake measured before radicle protrusion is highly correlated with subsequent germination rate in pea (Carver and Matthews, 1975), maize (Woodstock and Grabe, 1967) and pepper (Halpin-Ingham and Sundstrom, 1992) but is not a good indicator of the germination ability in other species (Côme and Corbineau, 1989).

The aims of the present work were: (i) to investigate the changes in energy metabolism in tomato seeds during osmopriming; (ii) to determine whether reduction of respiratory activity during priming because of a decrease in oxygen tension or the supply of a respiratory inhibitor (NaN_3) resulted in a decrease in the efficiency of the treatment; and (iii) to study the effects of drying and reimbibition in water of primed seeds on their energy metabolism.

Materials and Methods

Plant material and germination assays

Experiments were carried out with seeds of tomato (*Lycopersicon esculentum* Mill., cv Elko) supplied by Clause Semences (France).

Germination ability was tested in darkness at 15°C, in samples of 100 seeds placed in 9 cm Petri dishes (25 seeds per dish, four replicates) on a layer of cotton wool moistened with distilled water. A suboptimal temperature of 15°C, instead of 25–30°C (the thermal optimum), was chosen because of better expression of the effects of priming. A seed was regarded as germinated when the radicle had pierced the seed coat. Germination counts were made daily up to 12 days. The results presented are the means of the germination percentages obtained in four replicates ± standard deviation (SD). Germination rate is expressed as the time to reach 50% germination (T_{50}).

Priming treatment

For osmopriming treatment, seeds were placed for various durations on a PEG-8000 solution at –1 MPa at 15°C, in atmospheres containing 0 (pure nitrogen) to 21% oxygen (air) according to the procedure developed by Côme and Tissaoui (1968). The concentration of the PEG solution giving a water potential of –1 MPa at 15°C was calculated according to Michel and Kaufmann (1973). In order to determine whether the respiratory activity of the seeds was involved in the efficiency of priming, the osmotic treatment was also carried out in the air in the presence of the respiratory inhibitor NaN_3 at the concentrations of 0.5 and 1 mM.

After priming, seeds were rinsed with distilled water for about 30 s and dried at 20°C and 55% relative humidity for 7 days. The moisture content of redried primed seeds was close (9% dry weight basis) to that of the control unprimed seeds (8% dry weight basis).

Adenosine phosphate assays

Adenosine phosphates were extracted from two seeds according to Olempska-Beer and Bautz-Freeze (1984). ATP, ADP and AMP contents of the extracts were measured using the bioluminescence method with a pico-ATP biophotometer as previously described by Pradet (1967). The results obtained are expressed in nmol per g dry matter and are the means of 10 to 16 measurements ± SD.

The energy charge (EC) was calculated by the ratio (ATP + 0.5 ADP)/(ATP + ADP + AMP) defined by Atkinson (1968).

Results

Effects of osmopriming on the germination rate

Figure 44.1A shows the effects of 1, 3 and 7 days of osmopriming on the subsequent germination of seeds on water at 15°C. Unprimed seeds germinated at 95% within 12 days and the time to obtain 50% germination (T_{50}) was 129 h. Priming enhanced the germination rate; the T_{50} was only 54 h and 36 h after 3 and 7 days of priming, respectively.

To improve subsequent germination at 15°C, the osmotic treatment required at least 3–5% oxygen in the atmosphere, and the maximal improving effect was obtained in atmospheres containing more than 10% of this gas (Fig. 44.1B).

Energy metabolism during the osmotic treatment

In dry, unprimed seeds, AMP, ADP and ATP levels were respectively 337, 71 and 9 nmol per g of dry matter, ATP representing only 2.1% of the total adenylate pool (Fig. 44.2). The energy charge and the ATP/ADP ratio were then very low (0.11 and 0.12, respectively) (Table 44.1).

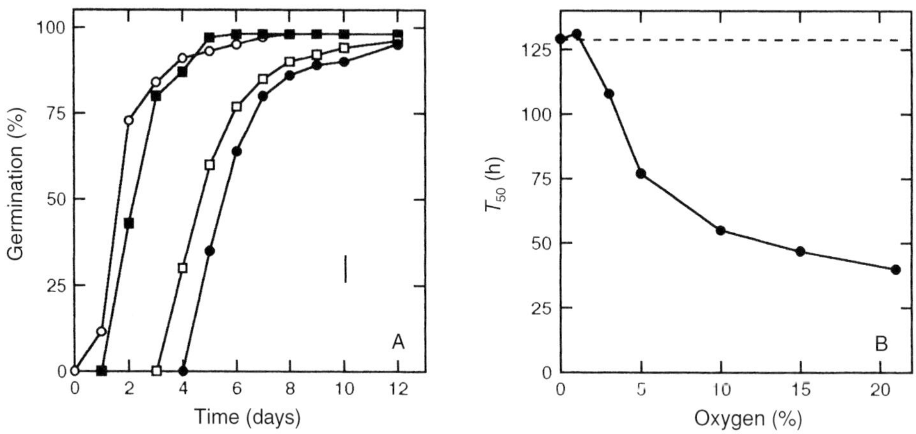

Fig. 44.1. Effects of osmopriming on the germination of seeds at 15°C. A, time courses of germination of control unprimed seeds (●) and seeds primed for 1 (□), 3 (■) and 7 days (○). Vertical bar denotes the largest SD. B, effects of oxygen concentration during priming on the time to obtain 50% germination (T_{50}) with seeds primed for 7 days. Dotted line corresponds to T_{50} for control unprimed seeds.

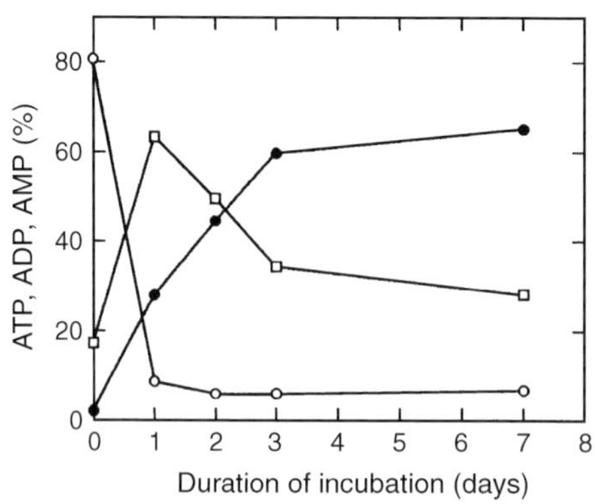

Fig. 44.2. Effects of the duration of incubation of seeds on the PEG solution on their ATP (●), ADP (□) and AMP (○) contents expressed as % of the total adenylic nucleotides. Means of 10 to 16 measurements.

Incubation of seeds on a PEG solution at –1 MPa did not result in an increase in the adenylate pool (data not shown), but induced a sharp decrease in AMP and an increase in ATP (Fig. 44.2), and then an increase in EC and in the ATP/ADP ratio (Table 44.1). After 7 days of incubation, for example, ATP level corresponded to 65.1% of the adenylate pool, EC reached 0.78 and the value of the ATP/ADP ratio was the highest (2.32).

Table 44.2 shows that ATP content of seeds increased with oxygen tension during the osmotic treatment and reached its highest values from 10% of this gas. Energy charge value increased from 0.66 in 3% oxygen to 0.74–0.79 in atmospheres containing at least 5% oxygen. The ATP/ADP ratio, which did not exceed 1.00 in 3–5% oxygen, doubled in the highest concentrations of this gas. Comparison with the effect of osmotic treatment on subsequent germination (cf. Fig. 44.1B) shows that the efficiency of priming was maximal when EC and the ATP/ADP ratio were close to 0.78–0.79 and 2.1–2.3, respectively.

Effects of NaN_3 during osmopriming

In this experiment, seeds were placed for 7 days at 15°C on a PEG-solution at –1 MPa with or without 0.5 or 1 mM NaN_3. They were then dried and transferred to water at the same temperature to follow their germination. Table 44.3 shows that the application of NaN_3 during priming resulted in a decrease in ATP content, EC and the ATP/ADP ratio, and in an increase in ADP level. ATP corresponded to only 38 and 18% of the total adenylic nucleotides in seeds

Table 44.1. Effects of the duration of incubation of seeds on the PEG solution on their EC and ATP/ADP ratio. Means of 10 to 16 measurements.

Duration of incubation (days)	EC	ATP/ADP
0	0.11	0.12
1	0.58	0.43
2	0.70	0.91
3	0.77	1.75
7	0.78	2.32

Table 44.2. Effects of oxygen concentration during 7 days of incubation of seeds on the PEG solution on their ATP, ADP and AMP contents, their EC and the ATP/ADP ratio. Means of 10 to 16 measurements ± SD when indicated.

	Nucleotides (nmol g^{-1} DW)				
Oxygen (%)	ATP	ADP	AMP	EC	ATP/ADP
3	142 ± 20	150 ± 19	35 ± 10	0.66	0.94
5	182 ± 24	184 ± 17	6 ± 5	0.74	1.00
10	218 ± 26	103 ± 24	22 ± 6	0.79	2.11
21	234 ± 17	101 ± 8	24 ± 7	0.78	2.32

primed in the presence of NaN_3 at the concentrations of 0.5 and 1 mM, respectively, whereas ADP reached 52% and 74% of the adenylate pool under the same conditions.

Osmopriming with NaN_3 had no stimulatory effect on subsequent germination. It even resulted in a delayed germination since T_{50} reached 187 and 237 h (Table 44.3) as against 129 h for control, untreated seeds (cf. Fig. 44.1B).

Energy metabolism of primed seeds after drying and reimbibition

Results presented in Table 44.4 show that drying of seeds after 3 or 7 days of priming resulted in a large decrease in ATP (11–16% of the adenylate pool) and an increase in AMP (50–54% of the adenylate pool), but did not change markedly the ADP content (33–34% of the adenylate pool) (compare with Fig. 44.2 for non-redried primed seeds). However, EC and the ATP/ADP ratio remained higher in dry primed seeds (0.28–0.33 and 0.32–0.48, respectively) than in unprimed ones (0.11 and 0.12). Moreover, the energy metabolism of dried primed seeds was much more intense after 4 h of subsequent imbibition in water than that of unprimed ones (Table 44.4). This remaining stimulatory

Table 44.3. Effects of NaN_3 during osmopriming of seeds for 7 days on their ATP, ADP and AMP contents, their EC and the ATP/ADP ratio, and the time to obtain 50% germination (T_{50}) after transfer to water. Means of 10 to 16 measurements (energy metabolism) or four measurements (T_{50}) ± SD when indicated.

	Nucleotides (nmol g^{-1} DW)					
NaN_3 (mM)	ATP	ADP	AMP	EC	ATP/ADP	T_{50} (h)
0	234 ± 17	101 ± 8	24 ± 7	0.78	2.32	40
0.5	138 ± 25	204 ± 32	52 ± 10	0.61	0.68	187
1.0	77 ± 10	326 ± 30	21 ± 8	0.57	0.24	237

Table 44.4. Adenylic nucleotide contents (% of the total adenylic nucleotides), EC and ATP/ADP ratio of seeds primed for 0, 3 or 7 days and then dried and reimbibed for 4 h in water. Means of 10 to 16 measurements ± SD.

		Nucleotides (%)				
Seeds	Duration of priming (days)	ATP	ADP	AMP	EC	ATP/ADP
Primed then dried	0 (control, dry)	2.1 ± 0.4	17.4 ± 1.7	80.7 ± 2.0	0.11 ± 0.01	0.12 ± 0.05
	3	11.1 ± 3.0	34.6 ± 7.7	54.3 ± 10.1	0.28 ± 0.06	0.32 ± 0.09
	7	16.4 ± 4.1	33.6 ± 7.0	50.0 ± 10.2	0.33 ± 0.07	0.48 ± 0.08
Primed, dried then imbibed	0 (control, imbibed)	7.7 ± 1.5	17.6 ± 3.1	74.7 ± 4.3	0.17 ± 0.03	0.44 ± 0.07
	3	26.7 ± 4.9	21.6 ± 2.4	51.7 ± 6.5	0.38 ± 0.08	1.24 ± 0.10
	7	35.4 ± 5.2	29.7 ± 3.6	34.9 ± 5.8	0.50 ± 0.10	1.19 ± 0.09

effect of priming on energy metabolism after drying was higher in seeds primed for 7 days than in those treated for only 3 days.

Discussion and Conclusions

As for various other species, such as carrot, celery and onion (Brocklehurst and Dearman, 1983), leek (Corbineau *et al.*, 1994) and sunflower (Smok *et al.*, 1993; Chojnowski *et al.*, 1997), osmopriming stimulates germination of tomato seeds at suboptimal temperatures (Fig. 44.1A). It has been shown also by Özbingöl *et al.* (1998) that after priming, tomato seeds germinate better in a wide range of temperatures and in hypoxia. The beneficial effect of priming increases with increasing duration of the treatment (Fig. 44.1A) and is maximal after 5–7 days (Özbingöl *et al.*, 1998). To be efficient, osmopriming requires oxygen, and its maximal effect is observed in atmospheres containing more than 10% of this gas (Fig. 44.1B, Özbingöl *et al.*, 1998). It is also totally ineffective in the presence of NaN_3 at high concentrations (Table 44.3).

In dry, unprimed tomato seeds, 80% of the adenylic nucleotides are in the monophosphate form (AMP), and ATP represents only 2.1% of the adenylate pool (Fig. 44.2). Imbibition of seeds during osmotic treatment is associated with a fast increase in ATP and a sharp decrease in AMP (Fig. 44.2) which result in an increase in EC and ATP/ADP ratio (Table 44.1). As in imbibing unprimed seeds, this activation of the energy metabolism is probably linked to an oxygen-dependent, cyanide-sensitive pathway (Hourmant and Pradet, 1981; Raymond *et al.*, 1982). Data presented by Leopold and Vertucci (1989) demonstrate that mitochondria become active when the moisture content of seeds reaches 14% in apple, 20% in maize, 24% in soybean and 26% in pea.

Results obtained with seeds primed in reduced oxygen tensions (Table 44.2) and in the presence of a respiratory inhibitor at high concentrations (Table 44.3) demonstrate that the beneficial effect of priming on germination of tomato seeds increases with increasing energy metabolism; it is optimal when EC and ATP/ADP ratio are higher than about 0.75 and 1.7, respectively. ATP might participate in protein synthesis as well as in nucleic acid synthesis and repair (Bray, 1995). In tomato seeds, induction of the cell cycle requires at least 5% oxygen (Özbingöl *et al.*, 1997), which corresponds to an energy charge higher than 0.74 (Table 44.2), and most of the cells are arrested at the G1 phase of the cell cycle in the presence of NaN_3 (Özbingöl *et al.*, 1997).

Drying back tomato seeds after priming results in a large reduction in ATP level and EC (Table 44.4). Similar observations were made during dehydration of recalcitrant *Araucaria* embryos (Corbineau *et al.*, 1997) and primed seeds of kohlrabi, spinach, eggplant and pepper (Mazor *et al.*, 1984). However, ATP level, EC and ATP/ADP ratio remain higher in dry, primed tomato seeds (Table 44.4) than in dry unprimed ones (Fig. 44.2 and Table 44.1) when the moisture content of both types of seeds is very similar. It must be noted that the higher ATP content of redried primed seeds is maintained for at least 4 or 6 months of storage at 20°C (Özbingöl, 1998).

As in sunflower (Smok *et al.*, 1993; Chojnowski *et al.*, 1997) and pepper (Halpin-Ingham and Sundstrom, 1992), priming enhances the respiratory activity of tomato seeds transferred onto water after drying (Özbingöl, 1998). It also enhances ATP synthesis, EC and ATP/ADP ratio during imbibition (Table 44.4). This stimulatory effect of priming on the energy metabolism increases with increasing duration of the osmotic treatment. It might result from an increase in number of mitochondria during priming as observed in leek embryos (Bray, 1995). In the latter, the higher level of ATP is associated with an increase in UTP and UDP-glucose contents (Bray, 1995). However, the results obtained do not allow us to establish a strict correlation between the respiratory activity or the energy metabolism during the first hours of imbibition and the subsequent germination rate.

References

Atkinson, D.E. (1968) The energy charge of adenylate pool as a regulatory parameter. Interaction with feed back modifiers. *Biochemistry* 7, 4030–4034.

Bradford, K.J. (1986) Manipulation of seed water relations via osmotic priming to improve germination under stress conditions. *HortScience* 21, 1105–1112.

Bradford, K.J. (1995) Water relations in seed germination. In: Kigel, J. and Galili, G. (eds) *Seed Development and Germination.* Marcel Dekker, New York, Basel, Hong Kong, pp. 351–396.

Bradford, K.J. and Haigh, A.M. (1994) Relationship between accumulated hydrothermal time during seed priming and subsequent seed germination rates. *Seed Science Research* 4, 63–69

Bray, C.M. (1995) Biochemical processes during the osmopriming of seeds. In: Kigel, J. and Galili, G. (eds) *Seed Development and Germination.* Marcel Dekker, New York, Basel, Hong Kong, pp. 767–789.

Brocklehurst, P.A. and Dearman, J. (1983) Interactions between seed priming treatments and nine seed lots of carrot, celery and onion. I. Laboratory germination. *Annals of Applied Biology* 102, 577–584.

Bujalski, W., Nienow, A.W., Maude, R.B. and Gray, D. (1993) Priming responses of leek (*Allium porrum* L.) seeds to different dissolved oxygen levels in the osmoticum. *Annals of Applied Biology* 122, 569–577.

Carver, M.F.F. and Matthews, S. (1975) Respiratory measurements as indicators of field emergence ability in peas. *Seed Science and Technology* 3, 871–879.

Chojnowski, M., Corbineau, F. and Côme, D. (1997) Physiological and biochemical changes induced in sunflower seeds by osmopriming and subsequent drying, storage and aging. *Seed Science Research* 7, 323–331.

Côme, D. and Corbineau, F. (1989) Some aspects of metabolic regulation of seed germination and dormancy. In: Taylorson, R.B. (ed.) *Recent Advances in the Development and Germination of Seeds.* Plenum Press, New York, Oxford, pp. 165–179.

Côme, D. and Tissaoui, T. (1968) Induction d'une dormance embryonnaire secondaire chez le pommier (*Pirus malus* L.) par des atmosphères très appauvries en oxygène. *Comptes Rendus de l'Académie des Sciences, Paris* 266, Série D, 477–479.

Corbineau, F., Picard, M.A. and Côme, D. (1994) Germinability of leek seeds and its improvement by osmopriming. *Acta Horticulturae* 371, 45–52.

Corbineau, F., Salmen Espindola, L., Vinel, D. and Côme, D. (1997) Cellular and metabolic events associated with dehydration of recalcitrant *Araucaria angustifolia* embryos. In: Ellis, R.H., Black, M., Murdoch, A.J. and Hong, T.D. (eds) *Basic and Applied Aspects of Seed Biology*. Kluwer Academic Publishers, Dordrech, Boston, London, pp. 715–721.

Dahal, P., Kim, N.-S. and Bradford, K.J. (1996) Respiration and germination rates of tomato seeds at suboptimal temperatures and reduced water potentials. *Journal of Experimental Botany* 47, 941–947.

Fu, J.R., Lu, X.H., Chen, R.Z., Zhang, B.Z., Liu, Z.S., Li, Z.S. and Cai, D.Y. (1988) Osmoconditioning of peanut (*Arachis hypogea* L.) seeds with PEG to improve vigour and some biochemical activities. *Seed Science and Technology* 16, 197–212.

Halpin-Ingham, B. and Sundstrom, F.J. (1992) Pepper seed water content, germination response and respiration following priming treatment. *Seed Science and Technology* 20, 589–596.

Hourmant, A. and Pradet, A. (1981) Oxidative phosphorylation in germinating lettuce seeds (*Lactuca sativa*) during the first hours of imbibition. *Plant Physiology* 68, 631–635.

Ibrahim, A.E., Roberts, E.H. and Murdoch, A.J. (1992) Viability of lettuce seeds. II. Survival and oxygen uptake in osmotically controlled storage. *Journal of Experimental Botany* 34, 631–635.

Karssen, C.M., Haigh, A., van der Toorn, P. and Weges, R. (1989) Physiological mechanisms involved in seed priming. In: Taylorson, R.B. (ed.) *Recent Advances in the Development and Germination of Seeds*. Plenum Press, New York, London, pp. 269–280.

Lanteri, S., Saracco, F., Kraak, H.L. and Bino, R.J. (1994) The effects of priming on nuclear replication activity and germination of pepper (*Capsicum annuum*) and tomato (*Lycopersicon esculentum*) seeds. *Seed Science Research* 4, 81–87.

Leopold, A.C. and Vertucci, C.V. (1989) Moisture as a regulator of physiological reaction in seeds. In: Stanwood, P.C. and McDonald, M.B. (eds) *Seed Moisture*. Crop Science Society of America, Madison, pp. 51–67.

Mazor, L., Perl, M. and Negbi, M. (1984) Change in some ATP-dependent activities in seeds during treatment with polyethylene glycol and during the redrying process. *Journal of Experimental Botany* 35, 1119–1127.

Michel, B.E. and Kaufmann, M.R. (1973) The osmotic potential of polyethylene glycol 6000. *Plant Physiology* 51, 914–916.

Olempska-Beer, Z. and Bautz-Freeze, E. (1984) Optimal extraction conditions for high performance liquid chromatography determination of nucleotides in yeast. *Annals of Biochemistry* 140, 236–245.

Özbingöl, N. (1998) Evénements cellulaires et métaboliques associés à la stimulation de la germination des graines de tomate (*Lycopersicon esculentum* Mill.) par un traitement de prégermination. Thesis, University Pierre et Marie Curie, Paris, France.

Özbingöl, N., Corbineau, F., Groot, S. and Côme, D. (1997) Improvement of tomato seed quality by priming as related to cell cycle. In: Pech, J.C., Latché, A. and Bouzayen, M. (eds) *Plant Sciences 1997*. SFPV, Paris, pp. 120–121.

Özbingöl, N., Corbineau, F. and Côme, D. (1998) Response of tomato seeds to osmoconditioning as related to temperature and oxygen. *Seed Science Research* 8, 377–384.

Pradet, A. (1967) Etude des adénosines-5' mono-, di- et tri-phosphates dans les tissus végétaux. I. Dosage enzymatique. *Physiologie Végétale* 5, 209–221.

Raymond, P., Hourmant, A., Leblanc, J.M., Al-Ani, A. and Pradet, A. (1982) Mécanismes régulateurs d'ATP au cours des premières phases de la germination. *Bulletin de la Société Botanique de France* 129, 91–97.

Smok, M.A., Chojnowski, M., Corbineau, F. and Côme, D. (1993) Effects of osmotic treatment on sunflower seed germination in relation with temperature and oxygen. In: Côme, D. and Corbineau, F. (eds) *Fourth International Workshop on Seeds. Basic and Applied Aspects of Seed Biology*, Vol. 3. ASFIS, Paris, pp. 1033–1038.

Woodstock, L.W. and Grabe, D.F. (1967) Relationships between seed respiration during imbibition and subsequent growth in *Zea mays* L. *Plant Physiology* 42, 1071–1076.

45 Bio-osmopriming Tomato (*Lycopersicon esculentum* Mill.) Seeds for Improved Seedling Establishment

J.E. Warren and M.A. Bennett

Department of Horticulture and Crop Science, The Ohio State University, Columbus, OH 43210, USA

Bio-osmopriming (BOP) is a combination of osmoconditioning and biopriming procedures that simultaneously hydrates seeds and applies a bacterial coating. The bacterium applied is *Pseudomonas aureofaciens* AB254. Processing tomato (cv. 'OH 8245') seedlots of 5 g each received one of four treatments; untreated, osmoprimed, AB254 coated, and bio-osmoprimed. Osmoprimed treatments soaked the seeds in aerated –0.8 MPa $NaNO_3$ for 7 days and then dried them back to their original moisture content (approx. 14%). Seeds were bio-osmoprimed by soaking in aerated –0.8 MPa $NaNO_3$ for 4 days at which time a mixture of nutrient broth, polyalkylene glycol, and bacterial stock were added. Seeds were then hydrated for an additional 3 days. In the absence of pathogen pressure, no differences in germination were observed between BOP and conventional osmopriming. Bio-osmopriming treatments consistently contained 10^5 bacteria colony forming units (cfu) per seed compared to 10^8 cfu for AB254 coated treatments. When seeds were planted in soilless media infected with *Pythium ultimum*, AB254 coatings protected tomato seeds from infection equally as well as the fungicide metalaxyl. Bio-osmopriming also provided protection, but at a slightly lower rate. This technique improves the chances of the seed lot establishing a healthy stand.

Introduction

Improved germination under unfavourable soil conditions is an important safeguard against yield losses in direct-seeded crops. Osmoprimed seed provides earlier and more uniform germination as well as improved performance under environmental stresses such as salinity (Wiebe and Muhyaddin, 1987), excessively high or low temperatures (Valdes *et al.*, 1985; Bradford, 1986; Pill and Finch-Savage, 1988; Osburn and Schroth, 1989) and reduced water availability (Frett and Pill, 1989). Work has been reported on osmoprimed tomato seeds using osmotic potentials from –0.25 to –1.75 MPa with the species showing a variety of responses to different osmotica. Seeds performed better if they were primed in the lowest strength osmoticum that prevented germination (Haigh and Barlow, 1987). In several studies, osmoconditioning tomato seed improved uniformity and low temperature germination (Alvarado *et al.*, 1987; Pill *et al.*, 1991). These attributes, combined with the reduced rates of damping-off associated with *Pseudomonas aureofaciens* AB254, suggest that a bio-osmopriming treatment could promote rapid and more uniform germination under a wider range of soil temperatures while providing disease resistance and improved growth associated with bacterial coatings (Bennett, 1998). Sweet corn seeds coated with *P. aureofaciens* exhibited control of *Pythium* equal to seeds treated with the fungicide metalaxyl (Callan *et al.*, 1991).

This study had three objectives. The first was to determine the effectiveness of *P. aureofaciens* AB254 in controlling damping-off of tomato seedlings caused by *Pythium ultimum* and to compare differences in effectiveness between various application techniques such as coating and bio-osmopriming. This bacterial strain has been used successfully to control *Pythium* on a variety of vegetables but, as yet, has not been tested on tomato. The second objective was to combine osmopriming and biopriming (coating) into a single procedure accomplishing the essential elements of both enhancement techniques. The third objective was to determine how these two forms of application affect the storage life of the coating. Seeds of each treatment were removed from storage at 4 and 8 months to assess which application technique retained the highest percentage of the original bacterial population. Seeds were also examined using scanning electron microscopy (SEM) to look for physical differences in colony morphology between these two application techniques.

Materials and Methods

The processing tomato (cultivar OH 8245) seed lot used in this research was produced in 1995 and supplied by Dr David Francis, tomato breeder/geneticist at Ohio State University. The osmopriming apparatus consists of a laboratory air source, fibreglass filter, moisturizing flask, valve bank, and priming flask (Fig. 45.1). Air flowed from the air source through a fiberglass filter that

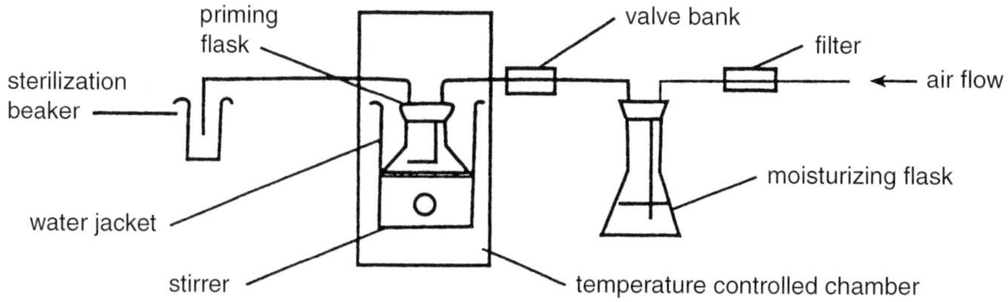

Fig. 45.1. Diagram of priming equipment.

removed any debris and continued on to the moisturizing flask. The moisturizing flask was a 1 l Erlenmeyer flask fitted with a two hole stopper. Air flowed in through one hose down to the bottom of the flask and was bubbled through approximately 500 ml distilled water. The moist air was then exhausted through the other hose. This moisturizing flask increased the humidity of the air flowing into the priming flask to reduce evaporation which changes the osmotic potential of the solution (Akers and Holley, 1986). The priming flask consisted of a 500 ml Erlenmeyer flask containing 325 ml priming solution and fitted with a two hole stopper. This solution was aerated by an air hose bent into a circle around the inside perimeter of the flask. It is important to have the air hose touch the priming flask at as few points as possible to reduce the number of seeds trapped between the hose and the glass. Air flow into the priming flask was regulated by two valves in tandem. Air enters the valve bank from the moisturizing flask and then either flows into the priming flask or is vented. Air flow is regulated by adjusting the amount of air exhausted out of the system. Two valves were needed so that pressure was released and does not build up in the system and cause air line failure. The priming flask rests inside a 1 l beaker filled with water to the level of the priming solution. This created a water jacket which provided constant temperatures during the treatment period. The priming flask was kept in a chamber at 20°C to control temperature. A slow mixing of the solution was provided by a magnetic stirrer beneath the beaker.

The bio-osmopriming apparatus was similar to the osmopriming equipment except that it contained a sterilization beaker that received waste air from the priming flask. This small beaker contained 250 ml of a mixture of approximately 60% distilled water, 30% ethyl alcohol, and 10% vanilla extract, which was refreshed on a regular basis. This prevented development of airborne bacteria and counteracted odours created by bacterial digestion of the nutrient broth.

This study consisted of four treatments with 5 g of seed per treatment: (i) untreated; (ii) osmoprimed; (iii) AB254 coated; and (iv) bio-osmoprimed. Prior to use, all apparatus from the moisturizing flask on was sterilized in 10% ethyl alcohol for 24 h. Seeds in all treatments (T) were sterilized using 10% ethyl alcohol for 5 min and then dried between sheets of germination paper under ambient conditions for 24 h. Sterilized seeds were essentially free of microbes at this point. In T2, seeds were osmoprimed in a dark, aerated, –0.8 MPa $NaNO_3$ solution for 1 week at 20°C. The seeds were then rinsed in distilled water to remove any remaining salts and dried by placing them on a mesh screen and left under ambient conditions for 24 h. Once the wet seeds were placed on the screens, the screens were tipped so that any excess moisture ran off one corner.

Seeds in T3 were coated by inoculation with *Pseudomonas aureofaciens* AB254. Original stock cultures were supplied by Dr Nancy Callan at the Montana State University Western Agricultural Research Center (WARC). The bacteria were cultured on trypticase soy agar and then harvested using a bent glass rod and approx. 5 ml of distilled water per Petri dish. One plate was used per 5 g seed lot. From the harvested bacteria, a 1.5% methyl cellulose suspension of AB254 was made, poured onto the seeds and stirred. The pubescence on the surface of tomato seeds absorbed a great deal of water and made even coverage difficult without this additional water. After all the seeds are thoroughly coated, they were placed between sheets of germination paper (Anchor Paper Co., St Paul, Minnesota) and left to dry for 24 h under ambient conditions. Placing the seeds between sheets of paper reduced the possibility of airborne contamination.

In T4 (bio-osmopriming), the seeds were placed in the priming flask containing 325 ml of aerated –0.8 MPa $NaNO_3$ solution. The seeds were left in this solution at 20°C for 4 days. At that time, 42 ml of 800% nutrient broth (64 g l^{-1}, Difco), 0.1 ml polyalkylene glycol, and 0.2 ml bacterial stock were added to the flask and left for an additional 3 days to allow for bacterial proliferation. The mixture of all of these added compounds was of approximately the same osmotic potential as the existing $NaNO_3$ solution to maintain the integrity of the osmotic solution. On day 7, the seeds were removed from the priming flask and placed on mesh screens under ambient conditions and dried for 24 h. When the bio-osmoprimed seeds were removed from the system, they contained more moisture than inoculated treatments so screens, rather than germination paper, were used for better air circulation. For treatments 2, 3 and 4, seeds were stirred several times during drying to prevent clumping and uneven drying.

The bacterial stock used in the coating and bio-osmopriming treatments contained King's B broth and 80% glycerol at a ratio of 3.5:2. Using the techniques of N.W. Callan, D.E. Mathre and J.B. Miller (personal communication), storage vials were prepared by pipetting 2 ml King's B broth into 1 dram vials. These vials were then autoclaved with the caps loosened. After cooling, the bacteria were added using sterile technique and allowed to multiply for 2 days at room temperature. Each day, the caps were loosened to allow air exchange and then tightened and shaken. On the second day 1.15 ml 80% glycerol was added. The vials were then shaken a final time to homogenize the mixture and were stored at –20°C. When retrieving the slurry from storage, the time the vials spent out of the freezer was minimized to minimize loss of bacteria.

Bacterial colony forming units (cfu) per seed were counted for T3 and T4 to evaluate the quality of the coating using the techniques described by Callan *et al.* (1990). Samples were diluted in a phosphate buffer solution consisting of 8.5 g l^{-1} NaCl, 11.4 g l^{-1} $K_2HPO_4 3H_2O$, and 6.8 g l^{-1} KH_2PO_4. Under sterile conditions, three samples of five seeds each were placed in 5 ml phosphate buffer solution and vortexed for 3 s every 7 min for 30 min. The resulting solution was diluted from 10^{-3} to 10^{-6} of the original solution and plated on trypticase soy agar (45 g l^{-1}, Difco). After 1.5 days, colonies resembling AB254 were counted.

This research had several experimental projects outlined below.

Experiment 1. Germination of bio-osmoprimed and conventionally osmoconditioned seeds

Seeds were either osmoconditioned or bio-osmoprimed using the techniques described above. The germination characteristics of bio-osmoprimed seeds were compared against those of osmoprimed and untreated seeds to determine if this new procedure achieved osmoconditioning. Seeds (40) were placed into 9 cm plastic Petri dishes containing one disk of blue blotter paper (Anchor Paper Co., St Paul, Minnesota) moistened with distilled water. The seeds were placed on a thermogradient table with ten settings from 10 to 30°C in 2°C increments. There were three replications with 40 seeds per plate per temperature. Seeds exhibiting radicle protrusion were counted after 3, 5 and 7 days.

Experiment 2. Colony forming units's/seed applied using bio-osmopriming vs. inoculation

Four tomato seed lots were bio-osmoprimed, osmoprimed and then AB254 coated, or only AB254 coated using the techniques described above. Samples were taken from each seedlot and cfu per seed counted. Colony forming units reported were the average of three 5-seed samples. *Cucurbitaceae* species were also bio-osmoprimed using this technique.

Experiment 3. Evaluations of AB254 ability to control Pythium ultimum *compared to the fungicide metalaxyl when applied using various techniques*

Four treatments were used: control, AB254 coated, bio-osmoprimed, and metalaxyl treated. Control, AB254 coated, and metalaxyl treated seeds were osmoprimed prior to further treatment to eliminate priming effects so that the biological effects could be evaluated. AB254 inoculated and bio-osmoprimed treatments were performed as described above. Metalaxyl (Apron 25W, Ciba-Geigy, Greensboro, North Carolina) treated seeds were treated with 0.3 mg a.i. g^{-1} of seed (Anon., 1985). The fungicide was combined with 8 ml distilled water to facilitate coating. The seeds were then dried between sheets of germination paper for 24 h. Cell trays (200 cells) were filled with sterilized soilless mix inoculated with corn meal agar containing *P. ultimum*. Seeds of each of these four treatments were sown in randomized blocks, watered and placed in a germinator at 5°C for 24 h to permit infection. At the end of this period, they were removed and placed in growth chambers at 25°C for germination. Seedlings possessing emerged radicles were counted after 7 days. The soilless mix/agar blend at the time of sowing contained 1.9×10^4 *Pythium* propagules per gram on average. Propagules per gram were determined by plating three samples of 1.0 mg inoculated soilless mix on PDA (39 g l^{-1}, Difco). *Pythium* colonies were counted after 24 h and confirmed by locating the aplersporia characteristic of *P. ultimum*. Koch's postulates were followed to determine that *P. ultimum* was the pathogen responsible for the death of the seeds.

Experiment 4. Storage characteristics of bacteria on bio-osmoprimed vs. inoculated seeds

Seeds of each treatment were placed in storage at 5°C and 50% RH. After 6 months, due to a mechanical failure, the seeds were placed in a sealed container along with a pouch of lithium chloride which provided very low humidity. Colony forming units per seed were counted after 4 and 8 months.

Results

Experiment 1. Germination of bio-osmoprimed and conventionally osmoprimed seeds

Bio-osmoprimed seeds performed similar to osmoprimed seeds, dramatically improving germination after 3 days when compared to untreated seeds (Table 45.1). Both treatments exhibited a slight reduction in overall germination due to the death of marginal seeds within the seed lot and a trend toward improved germination under cold temperatures (10–14°C) (Table 45.2). Total percent germination was not significantly different between treatments (Table 45.1). Since these treatments stressed the seeds by slowing imbibition, low vigour seeds within the seed lot were killed resulting in slightly lower emergence in growth chamber studies. This may not be of practical importance since low vigour seeds are unlikely to establish viable seedlings under field conditions. Another possible cause of death for these seeds could be that they were damaged by contact with the stirring rod.

Experiment 2. Colony forming units per seed applied using bio-osmopriming and inoculation

The tomato seed is ideal for applying beneficials, having a high surface area to volume ratio and a pubescence which provides an abundance of binding sites as well as protection for the bacteria. While both of these factors contribute to the high number of colony forming units (cfu) on a seed this size, bio-osmopriming can also be used to apply inoculum to smoother coated seeds such as in the cucurbit family (Table 45.3). Little difference was noted among bio-osmoprimed tomato seed lots with cfu per seed being between 3 and 8×10^5. Coating tomato seeds with AB254 and methyl-cellulose yielded $4–8 \times 10^8$ cfu per seed (Table 45.4). These values were similar whether the seed was untreated or osmoprimed prior to inoculation. The reason for the difference in bacterial concentration between bio-osmopriming and coatings was that the bio-osmopriming solution is too dilute to contain higher populations

Table 45.1. Percentage germination for bio-osmoprimed and osmoprimed 'OH 8245' tomato seeds after 3, 5 and 7 days on a thermogradient table, averaged over ten temperature ranges from 10 to 30°C.

	Seedling emergence (days)			
Seed treatment	3	5	7	Ungerminated
Untreated	24.0	42.5	11.0	22.3
Osmoprimed	65.0	4.5	4.0	26.3
Bio-osmoprimed	62.3	6.3	6.0	25.5
LSD (0.05)	2.50	3.09	2.83	NS

Table 45.2. Percentage germination for bio-osmoprimed and osmoprimed 'OH 8245' tomato seeds placed on a thermogradient table at various temperatures for 7 days.

	Temperature settings (average temperature (°C) per plate)									
Treatment	1 (30.0)	2 (27.8)	3 (25.5)	4 (23.3)	5 (21.1)	6 (18.9)	7 (16.7)	8 (14.4)	9 (12.2)	10 (10.0)
Untreated	95.5	96.8	96.3	95.5	96.0	94.5	94.3	78.8	15.0	5.3
Osmoprimed	93.8	94.3	93.3	90.8	93.0	91.3	91.8	73.8	24.3	11.0
Bio-osmoprimed	92.8	91.5	90.8	88.5	89.0	89.3	87.3	73.5	33.8	17.3
LSD (0.05)	NS	5.17	NS	NS	NS	NS	6.94	NS	10.07	NS

Table 45.3. Colony forming units (cfu) per seed achieved by bio-osmopriming seed of various *Cucurbitaceae* species for 4 days in aerated –0.8 MPa $NaNO_3$ and then adding a mixture of nutrient broth, polyalkylene glycol, and bacterial stock and hydrating for an additional 3 days.

Species	Cfu per seed
Cucumber (*Cucumis sativus* L.)	6.03×10^5
Muskmelon (*Cucumis melo* L.)	4.00×10^7
Zucchini (*Cucurbita pepo* L.)	8.20×10^7

Table 45.4. Colony forming units (cfu) per seed achieved by biopriming (coating) and bio-osmopriming 'OH 8245' tomato seed.

Treatment	Cfu per seed
Bioprimed (with methylcellulose)	6.60×10^8
Bioprimed (without methylcellulose)	5.00×10^6
Bio-osmoprimed no. 1	7.70×10^5
no. 2	6.05×10^5
no. 3	3.08×10^5

of bacteria. Solutions with higher bacterial concentration would become too viscous and begin to bind the seeds to the apparatus and to themselves. While bio-osmopriming yields far less cfu per seed, the following experiment demonstrates that the bacteria applied in that manner can be effective.

Experiment 3. Evaluation of AB254's ability to control **Pythium ultimum** *compared to the fungicide metalaxyl when applied using various techniques*

In this experiment, AB254 was as effective as the fungicide metalaxyl in controlling damping-off caused by *P. ultimum*. Germination tests were performed to determine how seeds treated in these two ways performed under pathogen pressure. Sterile soilless media were inoculated with *Pythium* and sown with seeds that were either bio-osmoprimed or osmoprimed and then treated with one of the following three treatments: AB254 coated, metalaxyl treated, and untreated.

Emergence of untreated osmoprimed seeds sown into sterile soilless medium inoculated with *Pythium* was only 56%, while the emergence of treated seeds was signficantly higher, ranging from 73 to 80%. Emergence of bio-osmoprimed seeds was slightly lower than that of the metalaxyl and bioprimed (AB254 coated) seeds (Table 45.5), but this was a small difference in the level of control compared to the huge difference in bacterial numbers (Table 45.4). Possible explanations of this may be: (i) 10^8 bacteria per seed is excessive and the number of bacteria required for adequate control on tomato is much smaller; or (ii) that the placement of the bacteria applied during bio-osmopriming is significantly different making the smaller populations more effective, either by the bacteria entering the seed or through a more strategic colonization of the surface. Most likely, both of these explanations are possible. There are limits to the amount of microbial life that a seed can support (Osburn *et al.*, 1989). Inoculum beyond that amount dies and can become a nutrient source for beneficials and pathogens (Dandurand and Knudsen, 1993). Bioprimed and bio-osmoprimed seeds were examined under an electron microscope looking for differences in colony morphology and location on the tomato seed. Bacterial colonies of AB254 were consistently found to be uniformly distributed over the seed surface and hairs of bio-osmoprimed tomato seeds, while less uniformity was observed on bioprimed seeds (Fig. 45.2).

Table 45.5. Percentage final seedling emergence of 'OH 8245' tomato seeds following various seed treatments when sown in soilless mix inoculated with *Pythium ultimum*.

Seed treatment	Seedling emergence (%)
Osmoprimed only	56
Metalaxyl treated	80
Bioprimed (AB254 coated)	77
Bio-osmoprimed	73

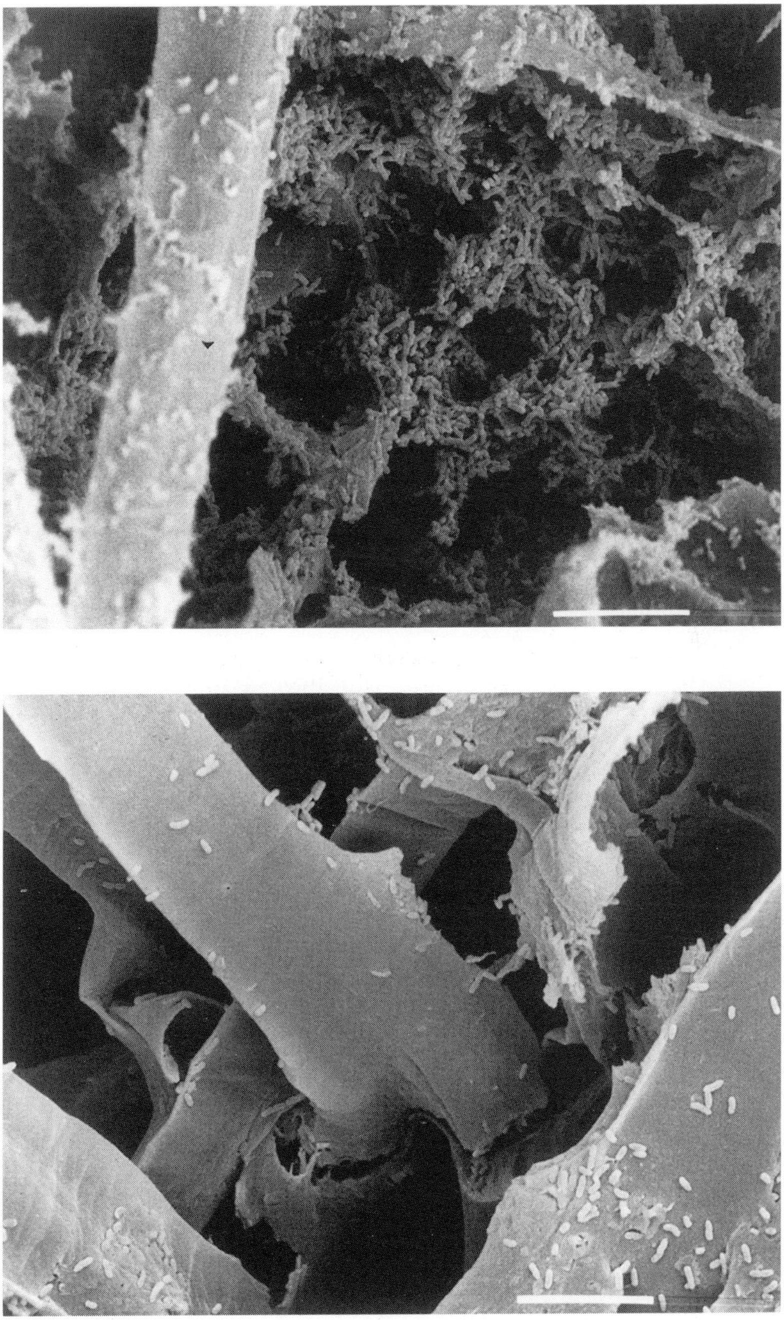

Fig. 45.2. Scanning electron micrographs of a bio-osmoprimed tomato seed at 1700× magnification. The upper micrograph shows the uniform population of *Pseudomonas aureofaciens* AB254 on seed hairs and surface, while excessive (not shown) or inadequate levels of beneficial bacteria (lower micrograph) were commonly seen on bioprimed tomato seeds.

Table 45.6. Bacterial populations after 4 and 8 months storage.

	Colony forming units (cfu) per seed		
	Initial	4 months	8 months
Bio-osmoprimed lots			
1	7.7×10^5	6.6×10^5	3.3×10^4
2	6.1×10^5	4.7×10^5	1.4×10^4
Bioprimed lots			
1	7.5×10^8	1.9×10^8	8.7×10^6
2	5.6×10^8	1.2×10^8	8.4×10^6

Availability of nutrients on the spermosphere, temperature, and interactions between antagonists (e.g. pseudomonads) and seed pathogens may all influence spatial colonization patterns of bacteria on seed (Fukui *et al.*, 1994).

Experiment 4. Storage characteristics of bacteria on bio-osmoprimed vs. bioprimed seeds

Bacterial numbers remained high over 4 months storage, then fell after 8 months for both treatments. This fall may be attributed to the low humidity in which seeds were stored from 6 months onward. However, bio-osmoprimed seeds retained a higher percentage after 8 months than bioprimed seeds (Table 45.6). Again, this may be due to more strategic colonization of the seed. Optimal storage of bio-osmoprimed seed is more challenging because of the greater humidity during storage than would be desired. Cold and dry conditions associated with good seed storage are not beneficial for bacteria.

In summary, biological control organisms present a unique opportunity for preventing soil-borne diseases by providing organic control of pathogens and potentially creating naturally suppressive soils. Unfortunately at this time, formulations and delivery systems have not developed to the degree where they can compete with chemicals under all situations. Bio-osmopriming is a step toward improving the effectiveness of biologicals by incorporating physiological improvements which improve germination and contribute to disease prevention. Bio-osmopriming has been found to both successfully prime the seed and inoculate it with beneficial bacteria with an improved storage life compared to inoculated treatments. While applying only a fraction of the cfu of coatings, bio-osmopriming still controls tomato seedling damping-off caused by *P. ultimum*.

Acknowledgements

Salaries and research support provided by state and federal funds appropriated to the Ohio Agricultural Research and Development Center, The Ohio State

University. Additional support was provided by a grant from the Ohio Vegetable and Small Fruit Research and Development Program.

References

Akers, S.W. and Holley, K.E. (1986) SPS: a system for priming seeds using aerated polyethylene glycol or salt solutions. *HortScience* 21, 529–531.

Alvarado, A.D., Bradford, K.J. and Hewitt, J.D. (1987) Osmotic priming of tomato seeds: effects on germination, field emergence, seedling growth, and fruit yield. *Journal of the American Society for Horticultural Science* 112, 427–432.

Anon. (1985) *Label for Ridomil 2E Fungicide.* Ciba-Geigy Corporation, Greensboro, NC 27419.

Bennett, M.A. (1998) The use of biologicals to enhance vegetable seed quality. *Seed Technology* 20, 198–208.

Bradford, K.J. (1986) Manipulation of seed water relations via osmotic priming to improve germination under stress conditions. *HortScience* 21, 1105–1112.

Callan, N.W., Mathre, D.E. and Miller, J.B. (1990) Biopriming seed treatment for biological control of *Pythium ultimum* pre-emergence damping-off in sh2 sweet corn. *Plant Disease* 74, 368–372.

Callan, N.W., Mathre, D.E. and Miller, J.B. (1991) Yield performance of sweet corn seed bio-primed and coated with *Pseudomonas fluorescens* AB254. *HortScience* 26, 1163–1165.

Dandurand, L.M. and Knudsen, G.R. (1993) Influence of *Pseudomonas fluorescens* on hyphal growth and biocontrol activity of *Trichoderma harzianum* in the spermosphere and rhizosphere of pea. *Phytopathology* 83, 265–270.

Frett, J.J. and Pill, W.G. (1989) Germination characteristics of osmotically primed and stored impatiens seeds. *Scientia Horticulturae* 40, 171–179.

Fukui, R., Poinar, E.I., Bauer, P.H., Schroth, M.N., Hendson, M., Wang, X.-L. and Hancock, J.G. (1994) Spatial colonization patterns and interaction of bacteria on inoculated sugar beet seed. *Phytopathology* 84, 1338–1345.

Haigh, A.M. and Barlow, E.W.R. (1987) Germination and priming of tomato, carrot, onion, and sorghum seeds in a range of osmotica. *Journal of the American Society for Horticultural Science* 112, 202–208.

Osburn, R.M. and Schroth, M.N. (1989) Effect of osmopriming sugar beet seed on germination rate and incidence of *Pythium ultimum* damping-off. *Plant Disease* 73, 21-24.

Osburn, R.M., Schroth, M.N., Hancock, J.G. and Hendson, M. (1989) Dynamics of sugar beet seed colonization by *Pythium ultimum* and pseudomonas species: effects on seed rot and damping-off. *Phytopathology* 79, 709–716.

Pill, W.G. and Finch-Savage, W.E. (1988) Effects of combining priming and plant growth regulator treatments on the synchronization of carrot seed germination. *Annals of Applied Biology* 113, 383–389.

Pill, W.G., Frett, J.J. and Morneau, D.C. (1991) Germination and seedling emergence of primed tomato and asparagus seeds under adverse conditions. *HortScience* 26, 1160–1162.

Valdes, V.M., Bradford, K.J. and Mayberry, K.S. (1985) Alleviation of thermodormancy in coated lettuce seeds by seed priming. *HortScience* 20, 1112–1114.

Wiebe, H.J. and Muhyaddin, T. (1987) Improvement of emergence by osmotic seed treatments in soils of high salinity. *Acta Horticulturae* 198, 91–100.

46 Use of Threshold Germination Models under Variable Field Conditions

W.E. FINCH-SAVAGE, K. PHELPS, L. PEACH AND J.R.A. STECKEL

Horticulture Research International, Wellesbourne, Warwick CV35 9EF, UK

The potential of four threshold models to predict germination patterns recorded in 15 different field environments was investigated. Thermal time tended to underestimate the time to germination and hydrothermal time tended to overestimate the time to germination. The prediction was improved under some conditions by incorporating hydrothermal priming time to take account of advancement towards germination below the base water potential. However, a threshold model which was modified to allow progress toward germination to be unaffected by water potentials above the base water potential provided the most accurate prediction. These results suggest that seeds progressed towards germination faster in variable field conditions than would be expected from laboratory experiments in constant conditions of temperature and water potential.

Introduction

The timing and uniformity of seedling emergence of field vegetable crops directly influences both their yield and monetary value (Finch-Savage, 1995). Much of the variation in seedling emergence can be accounted for by the timing of germination (Finch-Savage and Phelps, 1993), which is largely determined by the patterns of temperature and water potential following sowing. Threshold models such as thermal time (Garcia-Huidobro *et al.*, 1982) and hydrothermal time (Gummerson, 1986; Bradford *et al.*, 1993; Bradford, 1995) have been used to explain and describe the germination response of seeds to temperature and water potential and have added much to our understanding of how seeds behave. However, these models have been derived and tested against data collected under constant conditions.

Finch-Savage *et al.* (1998) investigated the potential of threshold models, to predict germination patterns that occur under the variable conditions

(eds M. Black, K.J. Bradford and J. Vázquez-Ramos)

experienced in horticultural seed beds. In the present work we speculate on questions that were raised in that study and test an alternative approach.

Materials and Methods

The data used in the present work was collected in experiments described in detail by Finch-Savage *et al.* (1998) and summarized below.

Field experiment

Germination was recorded on samples of 50 seeds removed from the seed bed at intervals after sowing from each of the three replicate sub-plots within a split-plot field experiment design. Main plots were five sowing occasions (30 March; 8 April; 6 May; 3 June; 3 August) and sub-plots were three irrigation treatments. Irrigation treatments were no irrigation, 12.5 mm applied more than 90°Cd (> 1°C) after sowing (pre-emergence), or 12.5 mm applied 48 h before sowing (pre-sowing) plus pre-emergence. Each sub-plot had rows of carrot (*Daucus carota* L. cv. Nantura) seeds sown 15 mm deep.

Using recorded soil moistures as initial values, a model developed for these soils (Walker and Barnes, 1981) was used to estimate soil moisture and temperature at sowing depth at 6-hourly intervals following sowing. The model utilized daily air temperature and rainfall measurements made at an agrometeorological station within 0.5 km of the plots. A water release curve was determined, using pressure membrane apparatus, and used to convert estimates of soil moisture produced by the model to soil water potential.

Models and methods of application

The ability of different threshold models to describe seed germination recorded under variable field conditions was tested. The models described were applied to variable field conditions of water potential and temperature in the following way. Germination for any fraction of the population G was assumed to occur after $n(G)$ 6 h (0.25 day) periods; $T(i),\Psi(i)$ are the temperature and water potential for each period; T_b is the base temperature and $\Psi_b(G)$ is the base water potential for the fraction G of the population. Germination does not occur below these bases.

Model 1

Thermal time ($\theta(G)$; Garcia-Huidobro *et al.*, 1982):

$$\theta(G) \geq 0.25 \sum_{i=0}^{n(G)} (T(i) - T_b) \quad T(i) > T_b$$

Model 2

Hydrothermal time (θ_{HT}; Gummerson, 1986):

$$\theta_{HT} \geq 0.25 \sum_{i=0}^{n(G)} (T(i) - T_b)(\Psi(i) - \Psi_b(G)) \quad T(i) > T_b,\ \Psi(i) > \Psi_b(G)$$

Model 3

Modified threshold model (θ_{HT}; Finch-Savage *et al.*, 1998):

$$\theta_{HT} \geq 0.25 \sum_{i=0}^{n(G)} (T(i) - T_b)(0 - \Psi_b(G)) \quad T(i) > T_b,\ \Psi(i) > \Psi_b(G)$$

A further threshold model, hydrothermal-priming time (θ_{HTP}; Tarquis and Bradford, 1992; Bradford and Haigh, 1994) takes account of the advancement towards germination that can occur below T_b and $\Psi_b(G)$, for example during priming, by accumulating: $(T - T_{min})(\Psi - \Psi_{min}(G))$ where T_{min} and Ψ_{min} are the minimum T and Ψ for metabolic advancement. Bradford (1995) suggested that progress towards germination in conditions where Ψ falls below $\Psi_b(G)$ may be described by the sum of progress in both hydrothermal and hydrothermal priming time. We interpreted this suggestion and applied it to variable conditions of water potential and temperature in the following way.

Model 4

The seed is considered in three states depending on the conditions (Fig. 46.1). In state R ($T > T_b$ and $\Psi > \Psi_b(G)$) the seeds will progress in hydrothermal time to radicle emergence. In state P (where ($T_{min} < T < T_b$) or ($\Psi_{min}(G) < \Psi < \Psi_b$)) the seeds can progress towards germination in hydrothermal priming time, but radicle emergence cannot occur. In state Q ($T < T_{min}$ or $\Psi < \Psi_{min}(G)$) the seeds become quiescent and there is no progress towards germination.

Rates of progress towards germination $\frac{dy_1}{dt}, \frac{dy_2}{dt}$ are defined below for states P and R.

$$\frac{dy_1(G)}{dt} = \frac{(T - T_{min})(\psi - \psi_{min}(G))}{\theta_{HTP}} \quad \text{in state } P$$

$$= 0 \text{ otherwise}$$

$$\frac{dy_2(G)}{dt} = \frac{(T - T_b)(\psi - \psi_b(G))}{\theta_{HT}} \quad \text{in state } R$$

$$= 0 \text{ otherwise}$$

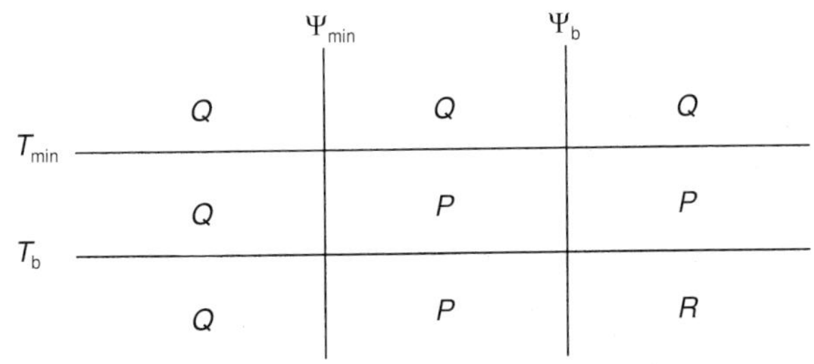

Fig. 46.1. Schematic diagram to illustrate different seed states.

Germination occurs at time x provided that the seed is in state R.

$$R\left(\int_{o}^{x}\frac{\mathrm{d}y_1(G)}{\mathrm{d}t}.\,\mathrm{d}t+\int_{o}^{x}\frac{\mathrm{d}t_2(G)}{\mathrm{d}t}.\,\mathrm{d}t\right)=1 \qquad (46.1)$$

Germination parameters under constant conditions

Data set 1

Germination of carrot seeds (cv. Nandor) was recorded on moist absorbent paper (Whatman, grade 181) at approximately 5°C intervals between 5 and 35°C and at a range of six water potentials (Ψ) between 0.0 and −1.2 MPa at 20°C. Temperature was monitored continuously by thermistors placed, like seeds, on moist paper. Water potentials were established using polyethylene glycol (PEG 8000, BDH Ltd, Poole, UK) solutions made up according to Michel (1983). The same solution volume to filter paper ratio was used for germination as that in the vapour pressure osmometer (model 5100C; Wescor Inc., Logan, Utah, USA) used to measure water potential. Thus the problem of filter paper exclusion of PEG pointed out by Hardegree and Emmerich (1990) was avoided. The germination data collected were used to calculate thermal time $\theta(50)$ = 56°Cd), hydrothermal time (θ_{HT}) = 47.71 MPa °Cd, T_b = 2.15°C and $\Psi_b(50)$ = −0.92 MPa.

Data set 2

Using seeds from the same lot the effect of osmotic priming in an aerated PEG 8000 solution at −1.5 MPa for 10 days at 15°C on germination at approximately 5°C intervals between 5 and 20°C was determined. For simplicity G was set to 50 and θ_{HTP} was estimated. If seeds of data set 1 progressed according to hydrothermal time and data set 2 according to hydrothermal-priming time followed by hydrothermal time, the following equation can be derived from equation 46.1:

$$\theta_{HTP}=\frac{\theta_{HT}(T_P-T_{min})(\psi_P-\psi_{min})t_{pi}}{\theta_{HT}-(T_w-T_b)(0-\psi_b)(t_{gi}-t_{pi})} \qquad (46.2)$$

where Ψ_p is the water potential used for priming at temperature T_p for t_{pi} days and T_w is the temperature during subsequent imbibition. Germination occurs after a total of t_{gi} days of priming and subsequent imbibition. The minimum water potential and temperature were taken to be −2.4 MPa and 0°C respectively (see Results and Discussion) to give a hydrothermal priming time of 355 MPa °Cd.

Results and Discussion

Finch-Savage *et al.* (1998) have shown that both thermal and hydrothermal time adequately describe carrot germination data collected under constant conditions. However, when applied to data collected under variable conditions in the field these models did not accurately describe the data (Fig. 46.2). Thermal time consistently underestimated the time to germination (Fig. 46.2a), whereas hydrothermal time overestimated the time to germination except under very moist conditions (Fig. 46.2b). We discuss below how this could happen, but it is important to remember that the soil water potential data used here to predict germination is itself a simulation and it is notoriously difficult to model soil moisture accurately near the soil surface.

The surface layers of the soil can dry rapidly so that the seeds experience sub-optimal moisture conditions before germination. It is therefore not surprising that thermal time did not accurately describe germination patterns of seeds sown 12.5 mm deep in the field. There are at least three reasons why a poor fit could be obtained using the hydrothermal time model. Firstly it is an assumption of hydrothermal time that T_b, $\Psi_b(G)$ and θ_{HT} are intrinsic properties of the seed, but there is no reason to believe this is always true. For example, Ni and Bradford (1992) have shown that seeds appear to adapt physiologically when exposed for prolonged periods to low Ψ, by increasing osmotic potential and lowering Ψ_b. However, under most sets of variable field conditions in the present work prolonged dry periods did not occur.

A second implicit assumption is that seeds wet up and dry as rapidly and to the same extent as the surrounding soil and that the impact of this on the rate of progress towards germination, including radicle emergence, is equal throughout the germination process. A third implicit assumption is that there is no progress towards germination when T and Ψ go below the bases for radicle growth. However, seed priming is known to advance seed metabolism both below T_b (e.g. Coolbear *et al.*, 1987) and below Ψ_b (e.g. Khan, 1992). Using different threshold models, it is possible to investigate the importance of these assumptions.

It can be argued that under good horticultural practice seeds are sown into moist soil so that initial imbibition can be rapid and complete. This contrasts with laboratory studies where seeds are exposed during imbibition and subsequent germination to constant sub-optimal water potentials. Finch-Savage and

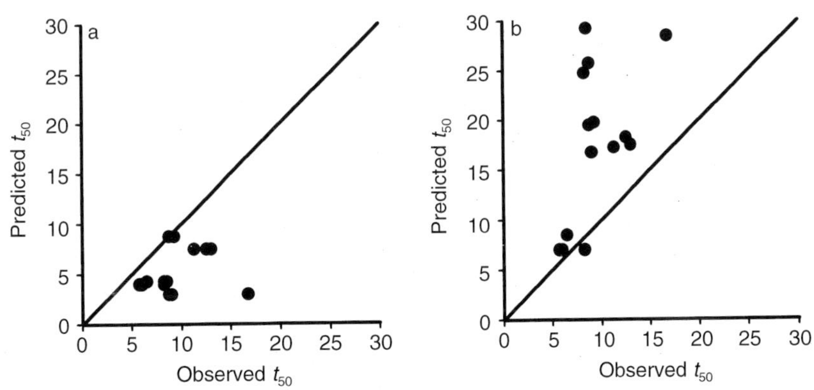

Fig. 46.2. Observed time to 50% germination (t_{50}) and predictions using (a) thermal time (model 1) and (b) hydrothermal time (model 2).

Phelps (1993) suggest that under the above horticultural conditions the seeds, once imbibed, may progress towards germination little affected by Ψ provided it remained above $\Psi_b(G)$. This situation is described by model 3, where time accumulates more quickly than in hydrothermal time, but it is subjected to similar constraints. The prediction of time to germination in variable field conditions using this model is much more accurate (Fig. 46.3a). However, this model does not usefully describe the seeds' physiological responses because it takes no account of changes that are known to occur below Ψ_b.

Only the end point of the germination process is observed in experiments to determine germination rate (reciprocal of time to germination) and therefore the effect of Ψ below $\Psi_b(G)$ cannot be directly observed. However, the effect of such a barrier has been shown in seed priming studies where the progress of germination to radicle emergence is prevented (e.g. Khan, 1992). This situation is also likely to occur in a seed bed where the surface layers of the soil repeatedly wet up and then dry below (Ψ_b) before germination (Fig. 46.4). Model 4 was developed in an attempt to take account of this by accumulating time above a minimum temperature and water potential for metabolic advancement as in hydrothermal priming time. A very similar Ψ_{min} of *c.* −2.4 MPa was found for the two small-seeded vegetables so far studied, lettuce (Tarquis and Bradford, 1992) and tomato (Bradford and Haigh, 1994; Cheng and Bradford, 1999). The results of Gray *et al.* (1990) provide evidence of a priming effect in carrot at −2.0 MPa, but not at −3.0 MPa or lower, suggesting a similar Ψ_{min} exists in carrot. Previous work has shown that in carrot seeds a priming effect of 0°C is possible (W.E. Finch-Savage, unpublished). These values of −2.4 MPa and 0°C were initially adopted as minima for calculating of θ_{HTP} in equation 46.1.

Although model 4, using these values, improves the prediction of germination time compared with hydrothermal time in some situations, the prediction under drier conditions is still poor (Fig. 46.3b). Altering the values of minima in equation 46.1 had relatively little impact on the prediction. Indeed, it was necessary to reduce θ_{HTP} by a factor of 10 to improve the prediction. There are other potential ways of modelling and combining hydrothermal-priming time

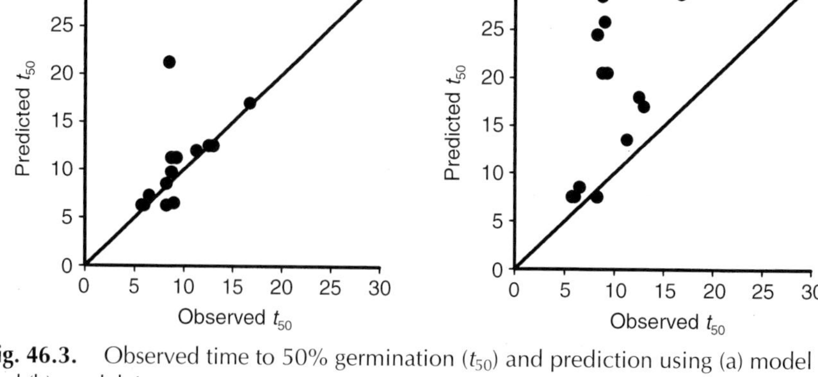

Fig. 46.3. Observed time to 50% germination (t_{50}) and prediction using (a) model 3 and (b) model 4.

and hydrothermal time that may improve the prediction. However, further inspection of Fig. 46.4 shows, as with data sets at other sowings, that the time spent between Ψ_b and Ψ_{min} is limited and much of that time Ψ tends towards Ψ_{min} and thus in practice little θ_{HTP} accumulates.

Model 3 was more successful in predicting germination time compared to the other threshold models compared here. This suggests that the seeds progressed towards germination faster in variable seed-bed conditions than

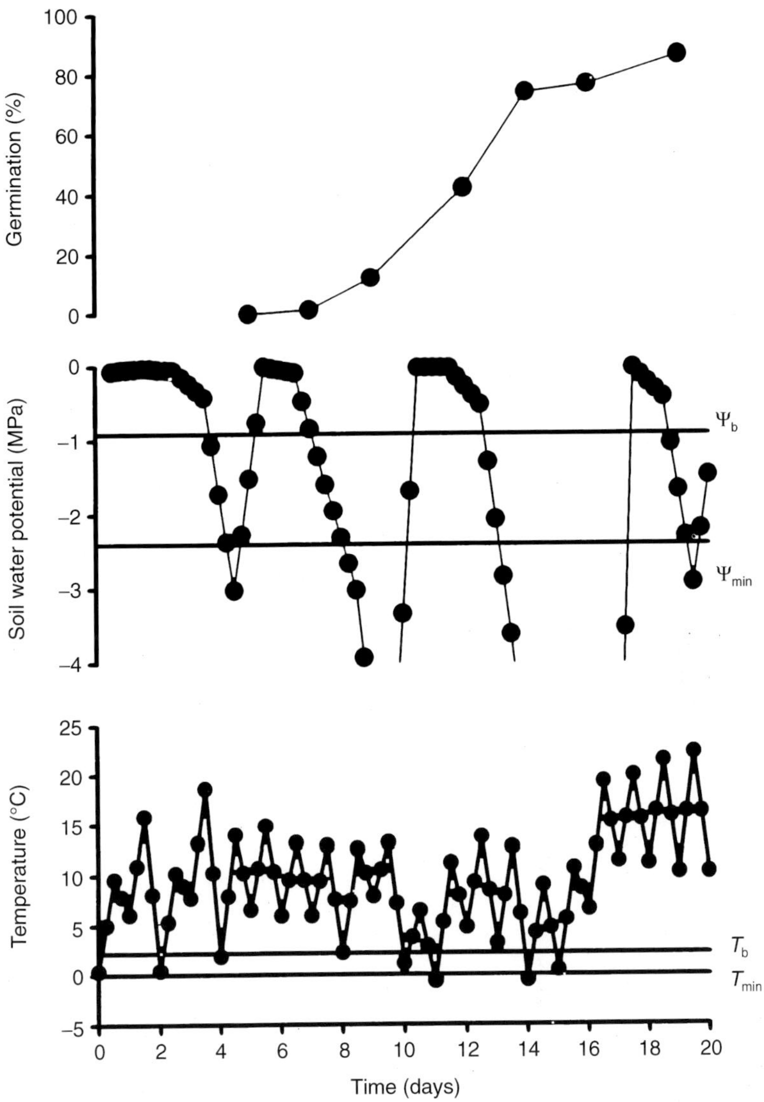

Fig. 46.4. An example of soil water potential and temperature patterns following sowing and the resulting germination curve. Temperature and water potential bases and minima are shown as horizontal lines.

would be expected from laboratory experiments in constant conditions of temperature and water potential. This result could be explained in several ways. For example, seeds may resist water loss so that they do not dry as quickly as the surrounding soil or they may acquire water that condenses on them in a cycling environment. The seed response to water potential may not remain the same throughout all stages of germination or once the seeds have become fully moist they subsequently germinate faster at sub-optimal water potentials than if they were imbibed and remained at that water potential. These possibilities highlight the need for more experimentation under variable conditions. In addition, more definitive field testing of these models requires accurate measurement, rather than stimulation, of water potential at sowing depth next to the seed. Current developments in soil moisture sensors should make this possible in the near future.

Acknowledgements

We thank the Ministry of Agriculture Fisheries and Food for funding this work and A. Walker for use of his model.

References

Bradford, K.J. (1995) Water relations in seed germination. In: Kigel, J. and Galili, G. (eds) *Seed Development and Germination.* Marcel Dekker Inc., New York, pp. 351–396.

Bradford, K.J. and Haigh, A.M. (1994) Relationship between accumulated hydrothermal time during seed priming and subsequent seed germination rates. *Seed Science Research* 4, 63–69.

Bradford, K.J., Dahal, P. and Ni, B.R. (1993) Quantitative models describing germination responses to temperature, water potential and growth regulators. In: Côme, D. and Corbineau, F. (eds) *Proceedings of the Fourth International Workshop on Seeds: Basic and Applied Aspects of Seed Biology.* ASFIS, Paris, Angers, France, pp. 239–248.

Cheng, Z. and Bradford, K.J. (1999) Hydrothermal time analysis of tomato seed germination responses to priming treatments. *Journal of Experimental Botany* 50, 89–99.

Coolbear, P., Newell, A.J. and Bryant, J.A. (1987) An evaluation of the potential of low temperature pre-sowing treatments of tomato seeds as a means of improving germination performance. *Annals of Applied Biology* 110, 185–194.

Finch-Savage, W.E. (1995) Influence of seed quality on crop establishment, growth and yield. In: Basra, S. (ed.) *Seed Quality. Basic Mechanisms and Agricultural Implications.* Haworth Press, New York, pp. 361–384.

Finch-Savage, W.E. and Phelps, K. (1993) Onion (*Alium cepa* L.) seedling emergence patterns can be explained by the influence of soil temperature and water potential on seed germination. *Journal of Experimental Botany* 44, 407–414.

Finch-Savage, W.E., Steckel, J.R.A. and Phelps, K. (1998) Germination and post-germination growth to carrot seedling emergence: predictive threshold models and sources of variation between sowing occasions. *New Phytologist* 139, 505–516.

Garcia-Huidobro, J., Monteith, J.L. and Squire, G.R. (1982) Time temperature and germination of pearl millet. I. Constant temperature. *Journal of Experimental Botany* 33, 287–295.

Gray, D., Steckel, J.A. and Hands, L.J. (1990) Responses of vegetable seeds to controlled hydration. *Annals of Botany* 66, 227–235.

Gummerson, R.J. (1986) The effect of constant temperatures and osmotic potentials on the germination of sugar beet. *Journal of Experimental Botany* 37, 729–741.

Hardegree, S.P. and Emmerich, W.E. (1990) Effect of polyethylene glycol exclusion on the water potential of solution-saturated filter paper. *Plant Physiology* 92, 462–466.

Khan, A.A. (1992) Preplant physiological seed conditions. In: Janick, J. (ed.) *Horticultural Reviews* 14. John Wiley and Sons, New York, pp. 131–181.

Michel, B.E. (1983) Evaluation of the water potentials of solutions of polyethylene glycol 8000 both in the absence and presence of other solutes. *Plant Physiology* 72, 66–70.

Ni, B.R. and Bradford, K.J. (1992) Quantitative models characterizing seed germination responses to abscisic acid and osmoticum. *Plant Physiology* 98, 1057–1068.

Tarquis, A. and Bradford, K.J. (1992) Prehydration and seed priming treatments that advance germination also increase the rate of deterioration of lettuce seed. *Journal of Experimental Botany* 43, 307–317.

Walker, A. and Barnes, A. (1981) Simulation of herbicide persistence in soil; a revised computer model. *Pesticide Science* 12, 123–132.

Index

Note: Page numbers in *italic* text refer to figures and tables